# VIRUS AS POPULATIONS

# VIRUS AS POPULATIONS

---

## Composition, Complexity, Quasispecies, Dynamics, and Biological Implications

### SECOND EDITION

ESTEBAN DOMINGO

*Centro de Biología Molecular "Severo Ochoa" (CSIC-UAM)*
*Cantoblanco, Madrid, Spain*

ELSEVIER

**ACADEMIC PRESS**
An imprint of Elsevier

Academic Press is an imprint of Elsevier
125 London Wall, London EC2Y 5AS, United Kingdom
525 B Street, Suite 1650, San Diego, CA 92101, United States
50 Hampshire Street, 5th Floor, Cambridge, MA 02139, United States
The Boulevard, Langford Lane, Kidlington, Oxford OX5 1GB, United Kingdom

**Notices**

Knowledge and best practice in this field are constantly changing. As new research and experience
broaden our understanding, changes in research methods, professional practices, or medical
treatment may become necessary.

Practitioners and researchers must always rely on their own experience and knowledge in evaluating
and using any information, methods, compounds, or experiments described herein. In using such
information or methods they should be mindful of their own safety and the safety of others, including
parties for whom they have a professional responsibility.

To the fullest extent of the law, neither the Publisher nor the authors, contributors, or editors, assume
any liability for any injury and/or damage to persons or property as a matter of products liability,
negligence or otherwise, or from any use or operation of any methods, products, instructions, or ideas
contained in the material herein.

**Library of Congress Cataloging-in-Publication Data**
A catalog record for this book is available from the Library of Congress

**British Library Cataloguing-in-Publication Data**
A catalogue record for this book is available from the British Library

ISBN: 978-0-12-816331-3

For information on all Academic Press publications visit our
website at https://www.elsevier.com/books-and-journals

*Publisher:* Andre Gerhard Wolff
*Acquisition Editor:* Kattie Washington
*Editorial Project Manager:* Pat Gonzalez
*Production Project Manager:* Sreejith Viswanathan
*Cover Designer:* Greg Harris

Typeset by TNQ Technologies

Working together
to grow libraries in
developing countries
www.elsevier.com • www.bookaid.org

# Contents

# Foreword

Viruses are minute organisms that, although mostly invisible, rule the world. Scientists may argue whether these obligatory intracellular organisms belong in the tree of life or whether they were the first replicating entities on earth, but all agree that they have decisively influenced evolution. The vast majority of the planet's 1031 viruses live in the oceans, whereas the terrestrial viruses make up only a very small fraction. They were first discovered, however, because their "ghosts" were exposed through diseases in plants and mammals (researchers were unable to "see" these ghosts until electron microscopy revealed them in the late 1930s). Following the discovery of tobacco mosaic virus and foot-and-mouth disease virus in 1898, a bewildering number of viruses have been identified many of which cause terrible human diseases, triggering immense fear and admiration. How can these minute creatures live so efficiently and, during pandemics, how can they spread so rapidly across an entire globe?

In principle, all viruses have the same, simple architecture: a relatively small piece of nucleic acid (the DNA or RNA genome), containing all information for proliferation, that is surrounded by a proteinaceous shell for protection and cell attachment, often in combination with a membranous layer. Yet the differences in sizes and shapes of virions for numerous different virus families are astounding. Even more mind-boggling is the wealth of genetic information stored in their small genomes, which directs very different strategies of viral proliferation leading to cell transformation, to cell death, or the death of the entire host. Throughout his professional life, Esteban Domingo has studied the highly complex issues of viral genetics, which qualifies him second to none to summarize the entire field.

Domingo begins his narrative with a short course in general virology, worthy of study for anyone involved in teaching or carrying out research in virology. This is followed by an introduction to the overwhelming, complex secrets of genetic information hidden in even the smallest viral genomes. Domingo's text is not dogmatic. Superbly written, it offers facts and hypotheses and encourages the readers to arrive themselves at the "truth," however short-lived that truth may be. The reader is pushed to ask questions like what exactly is a virus and where does it come from? How can we understand infection and viral diseases? How can we prevent their terror?

Beyond experimental breakthroughs that have informed our growing understanding of the molecular nature of viruses and their replication in cells and tissues of their respective hosts, the more recent decades have brought forth new concepts in how we perceive viruses. A shocking surprise in virus research was the landmark discovery by Hans Eggers and Igor Tamm and by Esteban Domingo and Charles Weissmann of the high mutation rate in viral replication, particularly of RNA viruses. The realization of error-prone replication of viruses heralded a paradigm shift in understanding molecular evolution. It changed the landscape of all studies regarding viral replication and pathogenesis. Rather than fixed entities operating with a single, unique nucleic acid sequence, individual species and serotypes of viruses have come to be viewed as genetically heterogeneous, complex populations comprised of a consensus sequence genotype, called by Manfred Eigen and John Holland "quasispecies." As a result of the error-prone replication and subsequent selection of mutations that confer virus fitness in response to cell-specific host factors, immune modulation, or environmental factors, viruses have evolved to exist as quasispecies populations. With these observations as a backdrop, Esteban Domingo offers a comprehensive, up-to-date treatise on the mutable nature of viruses and how this property affects virus–host interactions, viral fitness and adaptation, viral pathogenesis, and the ever-changing prospects for antiviral therapies, including the intriguing possibility of making use of what is called an error catastrophe.

The narrative on the error-prone replication of viruses is amplified by bringing Darwinian principles to bear on the generation of variant genomes (and, as a result, viral quasispecies) and how these virus populations behave in terms of random drift, competition, and adaptation/selection for fitness in different environments. This latter topic is addressed in depth via a discussion of the interactions of virus populations with their hosts and how biological parameters such as receptor usage, codon usage/codon pair frequencies, and the host immune response can provide selective pressure and resultant costs to viral fitness. In a subsequent section of this volume, viral fitness is addressed head-on, with a discussion of how genomic sequences of

viruses impact the fitness landscape during viral replication and how subpopulations of virus quasispecies may impact the molecular memory of an evolving quasispecies.

In the second half of this volume, Domingo describes different experimental systems and approaches used to analyze changes in virus population composition resulting from perturbations and selective forces applied during infection and replication. For example, persistent viral infections in cell culture allow the study of experimental evolution, while plaque-to-plaque transfers of viruses experimentally recapitulate infectivity bottlenecks that produce low-fitness viruses which may have extreme phenotypes. In addition, these experimental systems provide tractable platforms to test more general principles of genetics, for example, Muller's ratchet or the Red Queen hypothesis, during viral replication. This book then moves to describe the study of virus transmission and evolution in nature and the many parameters that impact evolution rates. Emergence and reemergence of pathogens is discussed in the context of the complexity of behavior of viral populations intertwined with nonlinear events derived from environmental, sociologic, and ecological factors.

The final sections of this book are devoted to a discussion of some of the more tangible implications of viral quasispecies/sequence evolution in considering its impact on disease prevention and strategies for vaccine design and antiviral therapeutics. This latter topic is analyzed at both the theoretical and pragmatic levels as part of what Domingo calls "virus as moving targets," due to the dynamic nature of viral genome sequences and the proteins they encode. Domingo then concludes this illuminating monograph by bringing the discussion of quasispecies and population dynamics full circle to a focus on nonviral systems (cells, cancer, and other infectious agents) that leave the reader with a number of "big-picture" concepts related to genetic variation, random versus selected replication events, and information theory. It is the universal nature of these concepts that makes this volume essential reading for virologists, evolutionary biologists, population geneticists, and any others who wish to be exposed to an intense level of scholarship on a fascinating topic. *Fiat lux!*

**Bert Semler**
Irvine, California
**Eckard Wimmer**
Stony Brook, New York

# Preface for the second edition

A second edition of a science book is an indication that it deals with a timely subject that attracts the interest of the scientific community. Virtually each of the topics (ranging from origin of life, role of viruses in biological evolution, quasispecies dynamics, to its implications for viral disease emergence and control, or how cancer cell adaptation resembles the rapidly evolving RNA viruses) maintain their interest. Despite promise of eradication of several viral diseases, the problems of selection of escape mutants, and the inaccessibility of treatments in many areas of the world persist. Furthermore, recognition of differences among individual components of biological collectivities (be them viruses or cells) is becoming increasingly apparent. "Complexity" is the key term, and the one that provides a mystery dose to the understanding of populations as wholes. This is what the second edition of *Virus as Populations* is about.

My main task for the second edition has been to update observations, emphasize some points, correct mistakes, and add literature references. Each chapter should be understandable in its own, although I have aimed at connecting concepts through references to other chapters. The new edition is faithful to the conviction that the main interest of this book lies in the concepts, as the supporting experiments and models change month after month. I was very pleased that Elsevier was positive about the updating of *Virus as Populations*; book preparation was facilitated by the generous and efficient involvement of Pat Gonzalez, Kattie Washington, and Linda Versteeg-Buschman.

The Acknowledgment section of the first edition remains entirely valid. Some scientists who contributed to the foundations of the topics covered in this book are no longer with us. John J. Holland (1929−2013) and Christof K. Biebricher (1941−2009) passed away before the first edition was written. Unfortunately, the great Manfred Eigen (1927−2019) has to be added. One of his last (probably the very last) pieces of scientific writing was his Foreword for volume 293 of *Current Topics in Microbiology and Immunology* that I edited with Peter Schuster. In that volume, as well as in pages 317 to 498 of volume 47 of *European Biophysics Journal* (dedicated to him on the occasion of his 90th birthday), the width of Eigen's contributions to molecular evolution becomes evident. Without Eigen, Holland, and Biebricher, this book would not exist. Equally sad was the passing away of José Antonio Melero (1949−2018), an outstanding virologist and colleague from Instituto de Salud Carlos III who maintained a vivid interest in science and virology, including the subjects of this book.

*Virus as Populations* could not have been written without the involvement and generous help of Celia Perales (now with her own group), Ana Isabel de Avila, and Isabel Gallego, with the participation María Eugenia Soria of Celia's team. My deepest appreciation goes to them and to the students who have been or are involved in our research.

It is my hope that this book continues being a useful introduction to students of many scientific areas, as well as to professionals whose experience tells them that it is time to understand biological entities (mainly, but not only, viruses) at the population level.

**Esteban Domingo**
Cantoblanco, Madrid, September 2019

# Acknowledgments

This book is a long story, with a lot to be acknowledged. I begin with the most immediate and then go back in time, with some deviations from linearity. The present core team in my laboratory at Centro de Biología Molecular "Severo Ochoa" (CBMSO) in Madrid, composed of Celia Perales, Ana Isabel de Avila, and Isabel Gallego, have done an immense job that has permitted the timely completion of this book (photography at the end of the Acknowledgments). In addition to her scientific leadership, Celia has unique skills to convert scientific concepts into images. Hers are all the original figures in this book. Ana and Isabel have diligently complemented my computer age inabilities by keeping track of end-note and professional typing, always alert to mistakes and inconsistencies in my drafts. Celia, Ana, and Isabel have worked on this book while continuing with experiments and acting as a survival brigade in very difficult times for Spanish science due to drastic budget restrictions. I am deeply thankful to the three of them.

I am also indebted to Eckard Wimmer and Bert Semler for taking time to read this book and to write a Foreword. I deeply appreciate their help, not only now but also at different stages of my activity as a picornavirologist. We shared friendship with John Holland, a decisive name in the scientific contents of this book.

Many thanks go also to Elsevier staff for their support and involvement. In particular, to Elizabeth Gibson in the early stages of book proposal, Jill Leonard for her continuous encouragement, Halima Williams for pushing me the right dose (never beyond a catastrophe threshold), and Julia Haynes for an intelligent and positive attitude during the last stages of book production. I am indebted to Elsevier for publishing this book and also for long-time support reflected in the publication of the two editions of "Origin and Evolution of Viruses" and my involvement in the editorial board of Virus Research initially with Brian Mahy, and recently with Luis Enjuanes, Alina Helsloot, and distinguished colleagues.

Our laboratory has had the privilege to engage in multiple collaborations that have broadened the scope of our research while maintaining the focus on implications of virus complexity. We belong to an active network of experts in liver disease (Spanish acronym CIBERehd), thanks to the interest of Jordi Gomez, Jaume Bosch, Juan Ignacio Esteban, and Josep Quer in our work on viral quasispecies and model studies with hepatitis C virus (HCV). Our current cooperation involves liver disease experts of different institutions, particularly Josep Quer, Josep Gregori, Javier Garcia-Samaniego, Aurora Sanchez-Pacheco, Antonio Madejón, Manuel Leal, Antonio Mas, Pablo Gastaminza, Xavier Forns, and their teams. Our connection with HCV goes back to the early 1990s when Jordi Gomez and Juan Ignacio Esteban from Hospital Vall d'Hebrón in Barcelona came to our laboratory to discuss the results that became the first evidence of HCV quasispecies in infected patients (the now classic Martell et al., 1992, J.Virol. 66, 3225–3229). The cell culture system for HCV was implemented in Madrid with the help of Charles Rice and the efforts of Celia Perales and Julie Sheldon. These ongoing studies are a source of knowledge and inspiration for us.

Before HCV, there were 35 years of research on foot-and-mouth disease virus (FMDV). About 100 students, postdocs, and visitors contributed to the research whose major aim was to explore if quasispecies dynamics applied to animal viruses. I cannot mention all people involved, but most of them are coauthors of publications quoted in the book. Juan Ortín, the now famous influenza virus (IV) expert, and I organized and shared our first independent laboratory ("down at the corner" of C–V block) at CBMSO, an important institution for Spanish biomedical sciences created in the 1970s under the auspices of Eladio Viñuela, Margarita Salas, David Vázquez, Carlos Asensio, Federico Mayor Zaragoza, Severo Ochoa, and others. Eladio suggested to Juan and I to work on IV and FMDV because of the unknowns underlying antigenic variation of these viruses and the problem of producing effective vaccines. And so we did. Juan on IV and me on FMDV. Mercedes Dávila joined as laboratory assistant and remained in the FMDV laboratory for 31 years (!), only to leave as a result of an appointment to manage a central facility at CBMSO. The first graduate students were Francisco Sobrino (Pachi) and Juan Carlos de la Torre. They initiated the FMDV quasispecies work at the pre-PCR age. Mauricio G. Mateu was determinant to expand the scope of our research because at the suggestion and with help of Luis Enjuanes, we extended the studies of genetic variation to antigenic variation by

producing and using monoclonal antibodies against FMDV. In addition to our own antibodies, we received some from Panaftosa, Brasil, and Emiliana Brocchi from Brescia, Italy. Collaborations were established with Panaftosa and Centro de Virología Animal and INTA-Castelar from Argentina. We also began productive projects with Ernest Giralt and David Andreu on peptide antigens, with Ignasi Fita, Nuria Verdaguer on the structure of antigen–antibody complexes, and David Stuart on the structure of the FMDV serotype used in our research. The collaboration with Nuria Verdaguer and her team continues to date on viral polymerases, particularly with the work of Cristina Ferrer-Orta, in a line of work initiated by Armando Arias at the biochemical level.

The first nucleotide sequences of our FMDV reference virus were obtained by Nieves Villanueva and Encarnación Martínez-Salas. In the late 1980s, Cristina Escarmís (my wife and the person who introduced nucleotide sequencing in Spain and in our laboratory) joined our group to expand our sequencing know-how and to work on the molecular basis of Muller's ratchet. Using routinely nucleotide sequences rather than T1-fingerprints (a transition facilitated by John Skehel, with a summer visit to Mill Hill) was as exciting a change as we experience nowadays seeing deep sequencing data. Additional help to the FMDV research was obtained from Joan Plana, Eduardo L. Palma, María Teresa Franze-Fernández, Elisa C. Carrillo, Alex Donaldson, Joaquín Dopazo, Andrés Moya, Pedro Lowenstein, and María Elisa Piccone, among others.

In the 1980s, there were two events that led to the connection of our laboratory with those of John Holland and Manfred Eigen. One was the publication by John Holland and colleagues in 1982 of the *Science* (1982) article on *"Rapid evolution of*

*RNA genomes.*" The paper was brought to me by my colleague from Instituto de Salud Carlos III José Antonio Melero. What I thought was the forgotten Qβ work of Zürich suddenly revived in a *Science* article that proposed a number of biological implications of high mutation rates and rapid RNA evolution. Not without hesitation, some years later, I wrote a letter to John. As he was extremely receptive, we talked on the phone and met for the first time at the International Congress of Virology in Edmonton, Canada, in 1987. Juan Carlos de la Torre went to work with John as a postdoctoral student, and then I spent a sabbatical stay in 1988–89, followed by several other visits to UC San Diego. Our friendship and shared view of how RNA viruses work lasted until his death in 2013. John and his team (which at the time of my visits included David Steinhauer, David Clarke, Elizabeth Duarte, Juan Carlos de la Torre, Isabel Novella, Scott Weaver, and several students, and visits of Santiago Elena and Josep Quer) were pioneers in linking basic concepts of population genetics with viral evolution. Some of the experiments marked the beginning of fitness assays and lethal mutagenesis, so important in current virology. John was a great support to our work in Madrid in those times of incredulity of high mutation rates and quasi-species. I miss his timely and encouraging comments enormously. The visits to San Diego opened also links with the Scripps Research Institute at La Jolla, when Juan Carlos de la Torre joined Michael B. A. Oldstone department. Each visit to Scripps is an important scientific stimulus, and the friendship and support of Juan Carlos and Michael remain unforgettable.

The contact with Manfred Eigen began in a coincidental manner. The Colombian physicist Antonio M. Rodriguez Vargas invited several European and US scientists to participate in the first Latin American School of Biophysics held in Bogotá in 1984. Invited speakers included the members of Sol Spiegelman's team, Fred Kramer and Donald Mills, and Christof Biebricher from Göttingen. This was the first time I met Christof, and our friendship lasted until his death in 2009. He introduced me to Manfred Eigen and I participated in several Max-Planck Winter Seminars that Manfred organizes at Klosters, Switzerland. I will never forget the discussions with Christof on the theory—experiment interphase of quasispecies, at freezing temperatures combined with a hot soup at a mountain restaurant hut. One of the years, John Drake, the pioneer of mutation rates, joined us in extremely lively discussions. At Klosters, I met Peter Schuster with whom I have kept contact since. The discussions with Peter on quasispecies and error threshold have been extremely helpful and clarifying. The Winter Seminars were a key in convincing me that our Madrid research was on the right track, and that I could survive giving a talk in front of Nobel Prize awardees which at the time I perceived as an achievement.

The transdisciplinary flavor of the seminars at Klosters revived years later in Madrid when in the 1990s the physicist Juan Perez Mercader, at the suggestion of Federico Morán, invited me to join in the organization of a new center termed Centro de Astrobiología (CAB), in Torrejón de Ardoz, near Madrid. Exciting discussions with many scientists conformed a new interest in the nascent science of Astrobiology with participation of noted scientists such as Andrés Moya, Ricard Solé, Federico Morán, Ricardo Amils, David Hochberg, Alvaro Giménez, Luis Vázquez, Ricardo García-Pelayo, Ramón Capote, and Francisco Anguita, among others. The science at CAB opened several collaborations with Susanna Manrubia, Ester Lázaro, and Carlos Briones that continue today, with a monthly seminar with Francisco Montero,

Cecilio López-Galíndez, and other colleagues as participants.

The view on viral populations that this book conveys had its origin in work with bacteriophage Qβ in the laboratory of Charles Weissmann in Zürich. I firmly believe that despite the great recognition that Charles has had as a scientist due to many achievements in different fields, the early Qβ work that contributed the first site-directed mutagenesis protocols, the birth of reverse genetics, and the first evidence of high mutation rates and quasispecies dynamics is still underappreciated. Perhaps, as the noted molecular biologist and virologist Richard Jackson once put it, the work came 20 years too early. Fundamental groundwork on the early genetics of bacteriophage Qβ was performed in Zürich by Martin Billeter, Hans Weber, Eric Bandle, Donna Sabo, Tadatsugu Taniguchi, and Richard Flavell. Years later, when Weissmann read Holland's 1982 paper in *Science*, he told me that a new field of research had begun.

I come back to the present. The reason to name so many people in previous paragraphs is not to publicize a biosketch that will interest only my family (and minimally). The reason is that each of the persons, institutions, events, and encounters mentioned has had some influence in the contents of this book. The idea of writing a book occurred to me years back when I produced a chapter for Fields Virology that I had to reduce by a factor of five to fit the required length. Other incentives came from

From left to right Isabel Gallego, Ana Isabel de Avila, Celia Perales, and Esteban Domingo during a discussion of a figure for this book at Centro de Biología Molecular Severo Ochoa *Photography by José Antonio Pérez Gracia.*

what would be usually considered inconsequential episodes. For example, about 10 years ago, before a talk in France, one of the host scientists told me that he was eager to hear my "new" ideas about virus evolution, when I had been working on quasispecies for almost 30 years! In addition, at a meeting on antiviral agents, somebody told me that he was unable to select a resistant viral mutant, and he found a good idea my suggestion of increasing the viral population size in the serial passages. These and a few other events reinforced in me the thought that perhaps a book could be of some use. This book is intended to be an introduction to fundamental concepts related to the role that viral population numbers play in several features of virus biology. Related aspects of a given concept are explained in different chapters. This is why many boxes and cross references among chapters are included. In genetics language, I have aimed at complementation among chapters. The specific examples (some of which attain a considerable degree of detail and are understandably biased toward my expertise) are just an excuse to illustrate general points. Hopefully, the reader will be able to apply them to specific cases.

Before ending, I have additional names to thank for their support on different occasions. At the risk of forgetting some names, they are Simon Wain-Hobson, Noel Tordo, Jean Louis Virelizier, Marco Vignuzzi, Carla Saleh, and other colleagues at Pasteur Institute in Paris, Rafael Nájera, Miguel Angel Martínez, Enrique Tabarés, Albert Bosch, Rosa Pintó, Raul Andino, Craig Cameron, Karla Kirkegaard, Olen Kew, Ernst Peterhans, Etienne Thiry, Paul-Pierre Pastoret, Francisco Rodriguez-Frias, Xavier Forns, Luis Menéndez-Arias, Vincent Soriano, Noemi Sevilla, Steven Tracy, David Rowlands, Louis M. Mansky, Roberto Cattaneo, Stefan G. Sarafianos, Kamalendra Singh, Ulrich Desselberger, Alexander Plyusnin, Margarita Salas, Jesús Avila, Antonio García Bellido, Carlos López Otín, and Pedro García Barreno. In addition, my thanks go to board members of the Spanish Society of Virology and of the Royal Academy of Sciences of Spain. Last but not least, colleagues and staff at CBMSO, and funding agencies, who have made our work possible. My deepest appreciation goes to all.

**Esteban Domingo**
**Cantoblanco, Madrid, Summer 2015**
**For Iker, Laia, Héctor, Jorge**

# Introduction to virus origins and their role in biological evolution

## Abbreviations

**AIDS** acquired immune deficiency syndrome
**APOBEC** apolipoprotein B mRNA editing complex
**CCMV** cowpea chlorotic mottle virus
**dsRNA** double-stranded RNA
*E. coli* Escherichia coli
**eHBV** endogenous hepatitis B viruses
**HBV** hepatitis B virus
**HCV** hepatitis C virus
**HDV** hepatitis delta virus
**HIV-1** human immunodeficiency virus type 1
**ICTV** International Committee on Taxonomy of Viruses
**Kbp** thousand base pairs
**mRNA** messenger RNA
**PMWS** postweaning multisystemic wasting syndrome
**RT** reverse transcriptase
**RdRp** RNA-dependent RNA polymerase
**ssRNA** single-stranded RNA
**T7** bacteriophage T7
**tRNA** transfer RNA
**UV** ultraviolet

## 1.1 Considerations on biological diversity

To approach the behavior of viruses acting as populations, we must first examine the diversity of the present-day biosphere and the physical and biological context in which primitive viral forms might have arisen. Evolution pervades nature. Thanks to new theories and to the availability of powerful instruments, new experimental procedures, and increasing computing power—which together constitute the very roots of scientific progress—we know that the physical and biological worlds are constantly evolving. Several classes of energy have gradually shaped matter and living entities, basically as the outcome of random events and Darwinian natural selection in its broadest sense. The identification of DNA as the genetic material and the advent of genomics in the second half of the twentieth century unveiled an astonishing degree of diversity within the living world that derives mainly from combinations of four classes of nucleotides. Biodiversity, a term coined by O. Wilson in 1984 and emphasized by T. Lovejoy and others, is a feature of all living beings, be differentiated multicellular organisms, single-cell organisms, or subcellular genetic elements, among them the viruses. Next-generation sequencing methods developed at the beginning of the 21st century allow thousands of sequences from the same biological sample (a microbial community in a soil or ocean sample, a tumor, or an infected host) to be determined. These procedures have documented the presence of myriads of variants in a "single biological entity" or in "communities of biological entities." Differences extend to individuals that belong to the same biological group, be it *Homo sapiens*, *Drosophila melanogaster*, *Escherichia coli*, or human immunodeficiency virus type 1 (HIV-1). No exceptions have been described. Diversity is extensive and not restricted to the genotypic level. It also affects phenotypic traits.

During decades, in the first half of the twentieth century, population genetics had as one of its tenets that genetic variation due to mutation had for the most part been originated in a remote past. It was generally thought that the present-day diversity was essentially brought about by the reassortment of chromosomes during sexual reproduction. This view was weakened by the discovery of extensive genetic polymorphisms, first in *Drosophila* and humans, through secondary analyses of electrophoretic mobility of enzymes, detected by in situ activity assays to yield zymograms that were displayed as electromorphs. These early studies on allozymes were soon extended to other organisms. Assuming that no protein modifications had specifically occurred in some individuals, the results suggested the presence of several different (allelic) forms of a given gene among individuals of the same species, be it humans, insects, or bacteria. In the absence of information on DNA nucleotide sequences, the first estimates of heterogeneity from the numbers of electromorphs were collated with the protein sequence information available. An excellent review of these developments (Selander, 1976) ended with the following premonitory sentence on the role of molecular biology in unveiling evolutionarily relevant information: "Considering the magnitude of this effect, we may not be overfanciful to think that future historians will see molecular biology more as the salvation for than, as it first seemed, the nemesis of evolutionary biology."

The conceptual break was confirmed and accentuated when molecular cloning and nucleotide sequencing techniques produced genomic nucleotide sequences from multiple individuals of the same biological species. Variety has shaken our classification schemes, opening a debate on how to define and delimit biological "species" in the microbial world. From a medical perspective, it has opened the way to "personalized" medicine, so different are the individual contexts in which disease processes (infectious or other) unfold. Diversity is a general feature of the biological world, with multiple implications for interactions in the environment, and also for human health and disease (Bernstein, 2014).

## 1.2 Some questions of current virology and the scope of this book

Viruses (from the Latin "virus," poison) are no exception regarding diversity. The number of different viruses and their dissimilarity in shape and behavior is astounding. Current estimates indicate that the total number of virus particles in our biosphere reaches $10^{32}$, exceeding by one order of magnitude the total number of cells. Viruses are found in surface and deep-sea and lake waters, below the Earth surface, in any type of soil, in deserts, and in most environments designated as extreme regarding ionic conditions (i.e., hypersaline) and temperature (thermophilic) (Breitbart et al., 2004; Villarreal, 2005; Lopez-Bueno et al., 2009; Box 1.1). The viruses that have been studied are probably a minimal and biased representation of those that exist, with at least hundred thousand mammalian viruses awaiting discovery (Anthony et al., 2013; Epstein and Anthony, 2017). This is because high-throughput screening procedures have only recently become available, and also because prevention of disease has provided the main incentive to study viruses. Disease-associated viruses are those most described in the scientific literature.

---

BOX 1.1

## SOME NUMBERS CONCERNING VIRUSES IN THE EARTH BIOSPHERE

- Total number of viral particles: $\sim 10^{32}$. This is 10 times more than cells, and they are equivalent to $2 \times 10^8$ tons of carbon.
- Virus particles in 1 $cm^3$ of seawater: $\sim 10^8$.
- Virus particles in 1 $m^3$ of air: $\sim 2 \times 10^6$ to $40 \times 10^6$.
- Rate of viral infections in the oceans: $\sim 1 \times 10^{23}$/s.

- A string with the viruses on Earth would be about $\sim 2 \times 10^8$ light-years long ($\sim 1.9 \times 10^{24}$ m). This is the distance from Earth of the galaxy clusters Centaurus, Hydra, and Virgo.

*Based on: Suttle (2007), Whon et al. (2012), and Koonin and Dolja (2013).*

---

Current virology poses some general and fascinating questions, which are not easily approachable experimentally. Here are some:

- What is the origin of viruses?
- Did they originate before or after a cellular world was in place?
- What selective forces have maintained multiple viruses as parasites of unicellular and multicellular organisms?
- Do viruses exist essentially as selfish parasites, or have they played constructive roles in the biosphere?
- Why have a few viral forms not outcompeted most other forms? Or have they?
- Have viruses been maintained as modulators of the population numbers of their host species?
- Does virus variation play a role in the unfolding of viral disease processes?
- Is it quasispecies dynamics that characterize RNA and many DNA viruses, a remnant of an adaptive strategy that presided all life forms in the remote past?

- Is the behavior of present-day viruses at the population level only an inheritance of their origins or a present-day necessity?

This book deals with some of these issues, mainly those that are amenable to experimental testing. Topics covered include molecular mechanisms of genetic variation, with emphasis on high mutation rates, Darwinian principles acting on viruses, quasispecies dynamics and its implications, consequences for virus-host interactions, fitness as a relevant parameter, experimental model systems in cell culture, ex-vivo and in vivo, long-term virus evolution, the current situation of antiviral strategies to confront quasispecies swarms, and conceptual extensions of quasispecies to nonviral systems. These subjects have as a common thread that Darwinian natural selection has an immediate imprint on them, observable in the time scale of days or even hours. The capacity for rapid evolution displayed by viruses represents an unprecedented and often underappreciated development in biology: the direct observation of Darwinian principles at play within short times.

Evolution is defined as a change in the genetic composition of a population over time. In this book evolution will be used in its broader sense to mean any change in the genetic composition of a virus over time, irrespective of the time frame involved, and the transience of the change. We treat as evolution both the genome variation that poliovirus has undergone from the middle of the 20th century to present days and the changes that the same virus undergoes within an infected individual. It is remarkable that only a few decades ago virus evolution (or for that matter microbial evolution in general) was not considered a significant factor in viral pathogenesis. Evolution was largely overlooked in the planning of strategies for microbial disease control. A lucid historical account of the different perceptions of virus evolution, including early evidence of phenotypic variation of viruses, with emphasis on the impact of the complexity of RNA virus populations, was written by J.J. Holland (2006). Despite having been largely ignored by virologists, the present book was partly stimulated by the conviction that the concept of complexity is pertinent to the understanding of viruses at the population level, having direct connections with viral disease and disease control.

## 1.3 The staggering ubiquity and diversity of viruses: limited morphotypes

Despite pleomorphism in cells and viruses (presence or not of envelopes, and viruses being spherical or even displaying a lemon-like shape, or being elongated), the size of viral particles and their host cells tends to be commensurate with the amount of genetic material that they contain and transmit to progeny (Fig. 1.1). The 2017 report from International Committee on Taxonomy of Viruses (ICTV) divides viruses into 10

Orders, and each of them is subdivided into several families, subfamilies, genera, species, and isolates (https://talk.ictvonline.org/), and each isolate includes a multitude of variants. The task of classifying viruses meets with considerable hurdles and requires periodic revisions by the ICTV, an organization whose role has been essential to provide conceptual order in the vast viral world. One of its objectives is the assignment of newly discovered viruses to the adequate group. A remarkable number of isolates remain unclassified, an echo of the

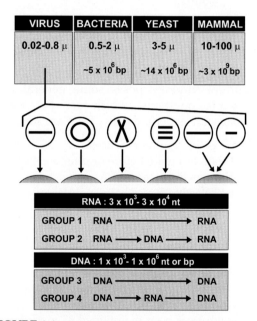

| VIRUS | BACTERIA | YEAST | MAMMAL |
|-------|----------|-------|--------|
| 0.02–0.8 $\mu$ | 0.5–2 $\mu$ | 3–5 $\mu$ | 10–100 $\mu$ |
|  | ~5 x $10^6$ bp | ~14 x $10^6$ bp | ~3 x $10^9$ bp |

| RNA : 3 x $10^3$ – 3 x $10^4$ nt | | |
|---|---|---|
| GROUP 1   RNA $\longrightarrow$ RNA | | |
| GROUP 2   RNA $\longrightarrow$ DNA $\longrightarrow$ RNA | | |

| DNA : 1 x $10^3$ – 1 x $10^6$ nt or bp | | |
|---|---|---|
| GROUP 3   DNA $\longrightarrow$ DNA | | |
| GROUP 4   DNA $\longrightarrow$ RNA $\longrightarrow$ DNA | | |

FIGURE 1.1    Representative average diameter values and genome complexity of viruses and some cell types. Diameters are expressed in microns ($\mu$), length of DNA in base pairs (bp), and of RNA in nucleotides (nt). Viral genomes can be linear, circular, diploid, segmented, or bipartite (multipartite in general; genome segments encapsidated in separate particles); in the latter case at least two particles, each with one kind of genomic segment, must infect the same cell for progeny production. The bottom boxes describe four groups of viruses according to the type of nucleic acid that acts as replicative intermediate.

natural diversity of viruses, even among the limited subset that has been isolated and characterized.

Viruses can be divided into two broad groups: those that have RNA as genetic material, termed the *RNA viruses*, and those that have DNA as genetic material termed the *DNA viruses*. They can have linear, circular, or segmented genomes of single-stranded or double-stranded nucleic acid (Fig. 1.1). All evidence suggests that the polynucleotide chain (or chains) that constitute the viral genome has all the information to generate infectious progeny in a cell, as evidenced by the production of infectious poliovirus from synthetic DNA copies assembled to represent the genomic nucleotide sequence (Cello et al., 2002).

With regard to the concepts of genome stability versus variation addressed in this book, it is helpful to divide viruses into four groups, depending on whether it is DNA or RNA the type of genetic material, which acts as a replicative intermediate in the infected cell (bottom gray shaded boxes in Fig. 1.1). The nucleic acids written in the four schemes are those involved in the flow of genetic information (indicated by arrows), not in gene expression since all of them use messenger RNA (mRNA) for virus-specific protein synthesis. Mistakes in the form of misincorporation of nucleotides during the replication steps indicated by arrows are transmitted to progeny genomes. RNAs produced by transcription to serve solely as mRNAs are essential for gene expression and virus multiplication, but misincorporations in such transcripts are not transmitted to progeny. It could be considered that some mRNA molecules when synthesized from the corresponding RNA or DNA template may acquire mutations and that these mutated molecules (e.g., an mRNA encoding a viral polymerase), when translated, may produce a polymerase with lower copying fidelity that will evoke additional mutations; we will ignore this

possibility since a single mRNA molecule should have a rather limited contribution to the overall genetic variation of a replicating virus population with thousands of polymerase and template molecules in the replication complexes (or replication factories).

Group 1 (with a replicative scheme abbreviated as RNA→RNA) includes RNA viruses whose genomic replication cycle involves only RNA. They are sometimes called *riboviruses*. Examples are the influenza viruses, hepatitis A and C viruses, poliovirus, coronaviruses, Ebola virus, foot-and-mouth disease virus, or tobacco mosaic virus, among many other important human, animal, and plant pathogens. Their replication is catalyzed by an RNA-dependent RNA polymerase (RdRp) encoded in the viral genome, often organized as a replication complex with viral and host proteins in cellular membrane structures.

Group 2 (RNA→DNA→RNA) comprises the retroviruses [such as HIV-1, the acquired immune deficiency syndrome (AIDS) virus, and several tumor viruses] that retrotranscribe their RNA into DNA. Retrotranscription is catalyzed by reverse transcriptase (RT), an RNA-dependent DNA polymerase encoded in the retroviral genome. It reverses the first step in the normal flow of expression of genetic information from DNA to RNA to protein, once known as the *dogma* of molecular biology. This enzyme was instrumental in the understanding of cancer, and for genetic engineering and the origin of modern biotechnology. As a historical account of the impact of H. Temin's work (codiscoverer of RT with D. Baltimore), the reader is referred to Cooper et al. (1995). Retroviruses include a provirus stage in which the viral DNA is integrated into host DNA. When silently installed in cellular DNA, the viral genome behaves mostly as a cellular gene.

Group 3 (DNA→DNA) contains most DNA viruses, such as herpesviruses, poxviruses, iridoviruses, and papillomaviruses, and the extremely large viruses of amebae (i.e., Mimivirus, Megavirus, or Pandoravirus, generically termed *giant viruses*) (La Scola et al., 2008; Colson et al., 2017). Their replication is catalyzed by a DNA-dependent DNA polymerase either encoded in the viral genome or in the cellular DNA. Cellular DNA polymerases are involved in the replication of DNA viruses that do not encode their own DNA polymerase.

Finally, Group 4 (DNA→RNA→DNA) includes viruses which despite having DNA as genetic material, produce an RNA as a replicative intermediate, the most significant examples being the human and animal hepatitis B viruses (HBVs) and the cauliflower mosaic virus of plants, termed *hepadnaviruses*.

Most viruses, from the more complex DNA viruses [i.e., 1200 Kbp (1000 base pairs) for the ameba Mimivirus, 752 Kbp for some tailed bacteriophages, and up to 370 Kbp for poxviruses, iridoviruses, and herpesviruses], the virophages that are parasites of the giant DNA viruses, the simplest DNA viruses (the circular single-stranded 1760 residue DNA of porcine circovirus), RNA bacteriophages (4220 nucleotides of ssRNA for bacteriophage Qβ), or subviral elements (viroids, virusoids, satellites, and helper-dependent defective replicons) show remarkable genetic diversity. However, RNA viruses that replicate entirely via RNA templates (Group 1 in Fig. 1.1); retroviruses (Group 2); and the hepadnaviruses (Group 4) display salient genetic plasticity, mainly in the way of a high rate of introduction of point mutations (Chapter 2). Their mutability may be an inheritance of universal flexibility that probably characterized primitive RNA or RNA-like molecules, thought to have populated an ancestral RNA world at an early stage of life on Earth (Section 1.4.2). Thus, the presence of RNA at any place in the replicative schemes (Group 1, 2, and 4 in Fig. 1.1) implies error-prone replication and the potential

of very rapid evolution. "Potential" must be underlined because high error rates do not necessarily result in rapid long-term evolution in nature (Chapter 7).

The extent of genetic variation and its biological consequences have been less investigated for DNA viruses than for RNA viruses. The available data suggest that DNA viruses are closer to RNA viruses than suspected only a few years ago, regarding their capacity of variation and adaptation. This is particularly true of the single-stranded DNA viruses of animals and plants. Evolutionary theory predicts that high-fidelity polymerase machinery is necessary to maintain the informational stability of complex genomes (those that carry a large amount of genetic information). This necessity accomplished by proofreading-repair and postreplicative-repair activities that assist the replicative DNA polymerases and their cellular and viral DNA progeny.

Our current capacity to sample many thousands viral genomes in short times (a trend that is continuously expanding) is revealing an astonishing number of slightly different viral genomes within a single infected host, and even within an organ or within individual cells of an organ! Intrahost diversity of viruses can be the result not only of diversification within the host but also of coinfection with different viruses (or variants of one virus), or an infection that triggers reactivation of a related or unrelated virus from a latent reservoir, or combined effects of these mechanisms. In turn, interhost (long-term) virus diversification can result from selection acting on variants generated by mutation, recombination or reassortment, and random sampling events (independent of selection) within hosts and in host-to-host transmission, or their combined effects (Chapters 2 and 3).

A different picture of diversity is obtained by comparing the morphological characteristics of viral particles (termed *virions*). The hundreds of thousands of bacterial and archaeal viruses that have been recognized can be assigned to as few as 20 morphotypes. The capsids of nonenveloped (naked) viruses display helical or icosahedral symmetry that determines the architecture of the virion. Variation in size and surface protein distributions can be attained from limited protein folds and the same symmetry principles (Mateu, 2013, Fig. 1.2). Divergent primary amino acid sequences in proteins can fold in closely related structures. The "structural space" available to viruses particles is much more restricted than the "sequence space" available to viral genomes (Abrescia et al., 2012). Sequence space and its mapping into a phenotypic space are key concepts for the understanding of evolutionary mechanisms (Chapter 3).

Three-dimensional structures of entire virions or their constituent proteins can provide an overview of phylogenetic lineages and evolutionary steps in cases in which the information cannot be attained by viral genomics (Ravantti et al.,

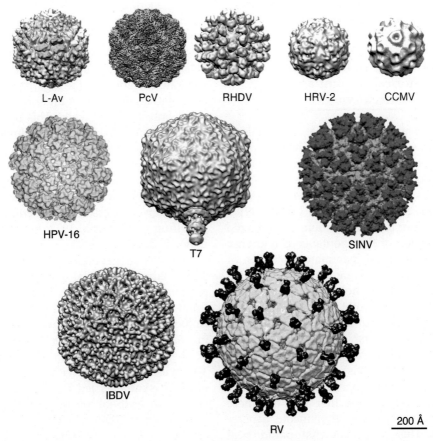

**FIGURE 1.2** Examples of spherical bacterial, fungal, plant, and animal virus particles reconstructed from cryo-electron microscopy images. L-Av, *Saccharomyces cerevisae* virus; PcV, *Penicillum Chrysogenum* virus; RHDV, rabbit hemorrhagic disease virus; HRV-2 human rhinovirus type 2; CCMV, cowpea chlorotic mottle virus; HPV-16, human papillomavirus type 16; T7, head of bacteriophage T7; SINV, Sindbis virus; IBVD, Infectious bursal disease virus; RV, rotavirus. *Picture modified from one kindly supplied by J.R. Castón, Carrascosa, J.L., 2013. The basic architecture of viruses. In: Mateu, M.G. (Ed.), Structure and Physics of Viruses. Springer, Dordrecht, Heidelberg, New York, London, pp. 53—75 with permission.*

2013). Yet, minor genetic modifications that do not affect the phylogenetic position of a virus or the structure of the encoded viral proteins in any substantial manner can nevertheless have major consequences for traits as important as host range or pathogenicity. How such minor changes in viruses can have major biological consequences may relate to the historical role of viruses in an evolving, ancestral, pre-cellular biosphere. To further address this issue, we need to examine how viruses may have originated. This, in turn, begs the question of the origin of life and the possible involvement of viruses in early life development.

## 1.4 Origin of life: a brief historical account and current views

An understanding of the mechanisms involved in the origin of life may help in penetrating into the origin of previral entities, defined as the precursors of the viruses we isolate in modern times. Different notions on the origin of life have been held in human history, often linked to religious debate. Opinions have ranged from a conviction of the spontaneous and easy generation of life from inanimate materials, or its beginning from a unique and rare combination of small prebiotic molecules, or being the result of a lengthy prebiotic process, or its inevitability as the outcome of the evolution of matter in our universe (or "sets" of universes with the adequate physical parameters, according to some cosmological models).

As little as 150 years ago (a time not distant from the discovery of the first viruses), there was a general belief in the spontaneous generation of life. This was somewhat paradoxical because chemists of the seventeenth century divided chemistry into mineral chemistry, vegetal chemistry, and animal chemistry. J.J. Berzelius put together animal and plant chemistry and named the resulting discipline "organic" chemistry, which he distinguished from "inorganic" chemistry (Berzelius, 1806). He formulated what was known as the *central dogma of chemistry*: "The generation of organic compounds from inorganic compounds, outside a living organism, is impossible." The classical experiments of L. Pasteur provided definitive proof that, at least under the prevailing conditions on present-day Earth, "life comes from life" (Pasteur, 1861). He established what was considered the "central dogma of biology": "The generation of a whole living organism from chemical compounds, outside a living organism is impossible." The requirement of life to generate life was, however, extended to the belief that "living" and "nonliving" were two separate categories in the organization of matter, and that organic compounds could be synthesized only by living cells. This doctrine, called *vitalism* dominated biology for almost a century, and in a modified manner, it continues today regarding the interpretation of mental activity in humans (matter and spirit as "substance dualism"). Historical views on the origin of life have been addressed in several publications (Rohlfing and Oparin, 1972; Bengtson, 1994; de Duve, 2002; Eigen, 2002, 2013; Lazcano, 2010).

Dogmas are generally not to stay. "Vitalism" was shattered by the chemical synthesis of organic compounds from inorganic precursors (urea by F. Wöhler in 1828, acetic acid by H. Kolbe in 1845, hydrocarbons by D. Mendeleev in 1877, and several other compounds by M. Berthelot in the second half of the 19th century). The evidence that no "vital force" was needed for such syntheses led F.A. Kekulé to write in his classic textbook on organic chemistry published during 1859—60: "We have come to the conviction that ... no difference exists between organic and inorganic compounds." From then on, organic chemistry became the chemistry of "carbon compounds." We know now that living entities are made of the same chemical elements found in the mineral world. Of the 118 elements in the periodic table (in 2019), 59 are found in the human body.

A key experiment carried out in 1953 by S. Miller, working with H.C. Urey showed that components of biological molecules could be obtained from inorganic precursors. He mimicked the conditions thought to be prevalent in the primitive Earth, and mixed hydrogen ($H_2$), ammonium ($NH_3$), and methane ($CH_4$) in a sealed reactor with an influx of water vapor. Synthesis of a number of organic compounds occurred under the influence of electrical discharges. The de novo synthesized chemicals included amino acids (glycine, alanine, aspartic acid, and glutamic acid), formic, acetic, propionic and fatty acids, cyanide, and formaldehyde (Miller, 1953, 1987). Several researchers followed the Miller's approach using other starting chemical mixes and confirmed that key components of the macromolecules that are associated with living materials (notably purines, pyrimidines, and amino acids) could be made from precursors, which were abundant in the primitive Earth or its atmosphere. Today, variant versions of Miller's protocol (including additional starting chemicals, aerosol spread of chemicals, freeze-thaw cycles, different sources of energy, electron beams, etc.) produce interesting information on the synthesis of organic molecules (Dobson et al., 2000; Miyakawa et al., 2002; Bada and Lazcano, 2003; Ruiz-Mirazo et al., 2014).

Intense ultraviolet (UV) irradiation may have contributed to the synthesis of compounds relevant to life: ammonia, methane, ethane, carbon monoxide, formaldehyde, sugars, nitric acid, and cyanide. Complex organic compounds (notably aromatic hydrocarbons and alcohols) are found in interplanetary dust, comets, asteroids, and meteorites, and they can be generated under the effect of cosmic and stellar radiation. Thus, many organic compounds could have been produced within the Earth atmosphere or away from it, and be transported to the Earth

surface by meteorites, comets, or rain, to become the building blocks for additional life-prone organic molecules. Places at which peptide bond formation and prebiotic evolution could have been favored are hydrothermal systems and the interface between the ocean and the atmosphere (Chang, 1994; Horneck and Baumstark-Khan, 2002; Ehrenfreund et al., 2011; Parker et al., 2011; Danger et al., 2012; Griffith and Vaida, 2012; Ritson and Sutherland, 2012; Nakashima et al., 2018).

A key issue from prebiotic syntheses is the degree of oxidation of the primitive Earth atmosphere. Records of an early surface environment dated 3.8 billion years ago were found in meta-sediments of Isua, Greenland. These materials suggest that the surface temperature of the Earth was below 100°C, with the presence of liquid and vapor water, and gases supplied by intense volcanism ($CO_2$, $SO_2$, and $N_2$). The composition of primitive rocks, together with theoretical considerations, suggest a neutral redox composition of the Earth atmosphere (with relative gas abundances: $N_2$, $CO_2 > CO \gg CH_4$, $H_2O \gg H_2$, $SO_2 > H_2S$) around the time of the origin of primeval forms of life. The possible presence of a general or localized reducing atmosphere ($N_2$, $CO > CH_4 > CO_2$, $H_2O$, $\sim H_2$, $H_2S > SO_2$) is still debated, but increasingly viewed as unlikely. In an oxidative atmosphere, yields of amino acids, nucleotides, and sugars would be lower. Either these diminished yields were sufficient to attain critical levels of relevant building blocks, or an earlier reducing atmosphere may have accumulated them, among other possibilities (Trail et al., 2011). In spite of the validity of "life comes from life" in the current Earth environment, the experimental facts suggest that there is no barrier for the generation of life from nonlife, provided suitable environmental conditions are met. In this line, A.I. Oparin proposed that a "primitive soup" could well have been the cradle of life on Earth, as described in his famous treatise on the origin of life (Oparin,

1938), a concept that had already been sketched by C. Darwin.

The "protein first" versus "nucleic acid first" for the origin of life is still a contended issue (Falk and Lazcano, 2012). Although during decades there was a preference for nucleic acids due to their superior capacity for self-organization to perpetuate inheritable messages through base pairings, recently both views have accumulated arguments in their favor. An integrative "metabolism first" view based on mutually catalytic networks of small molecules displaying the capacity to replicate and evolve is gaining adepts (Lancet et al., 2018). The building blocks of nucleic acids have been more difficult to obtain from primeval chemicals than the building blocks of proteins. Peptides of about 20 amino acids in length could have been easily formed under prebiotic conditions (Fox and Dose, 1992), and peptides or their derived multimers had a potential to display catalytic activities at a proto-metabolic stage. Peptide amyloids have been proposed as an alternative to nucleic acids for the origin of life in what has been termed *the amyloid-word hypothesis* (Greenwald et al., 2018). Interestingly, amyloid aggregates display features of quasispecies, not in the way of mutant distributions found in viruses (Chapter 3) but as alternative protein conformations and conformation heterogeneity that confers functional diversity (see Chapter 10 on the collective behavior of prions). The protein-versus nucleic acid-first alternatives may not be incompatible. Short nucleotide and amino acid polymers might have cooperated to cross a complexity threshold for self-sustained replication to arise and evolve (Kauffman, 1993). As soon as peptide- or protein-based catalytic activities developed, they had to be coordinated with oligo- or polynucleotide replication. This integration of genotypic information with its phenotypic expression may be achieved through hypercyclic couplings, as proposed by M. Eigen and P. Schuster (Eigen and Schuster, 1979; Eigen, 2013), to render incipient life forms sustainable.

### 1.4.1 Early synthesis of oligonucleotides: a possible ancestral positive selection

Substantiating the abiotic synthesis of nucleotide- or amino acid-based polymers has been arduous relative to the synthesis of monomeric organic molecules. However, early work by L. Orgel and his colleagues documented that polynucleotides could be synthesized from activated nucleotides in the absence of an enzyme (Miller and Orgel, 1974), and recent work has established that there are multiple chemical pathways for abiotic and nonenzymatic nucleotide synthesis (Adamala and Szostak, 2013; Ruiz-Mirazo et al., 2014). Mineral surfaces (clay minerals, zeolites, manganates, hydroxides, etc.) have been proposed as key participants in the origin of life, by providing scaffolds for the synthesis of nucleotide and amino acid polymers (Bernal, 1951; Gedulin and Arrhenius, 1994; Ruiz-Mirazo et al., 2014; Nakashima et al., 2018; Pedreira-Segade et al., 2018)). The relevant features of clays are their adsorption power, ordered structure, capacity to concentrate organic compounds, and ability to serve as polymerization templates. Adsorption onto mineral surfaces may lower the activation energy of intermolecular reactions. Minerals with an excess positive charge on their surfaces might have played a role in the evolution of RNA-like or RNA-precursor molecules. It has been suggested that mineral-organic complexes (chimeras between materials once thought to belong to unmixable categories) could have been the first living organisms endowed with a genetic program (Cairns-Smith, 1965).

Bond formation and chain extension, which are needed to synthesize nucleic acids and proteins prebiotically, are inhibited in liquid water, thus favoring mineral surfaces as potential biogenic sites. Polymer formation could have also been facilitated by heating and drying, with key involvement of the dried phases in catalysis (Towe, 1994; Horneck and Baumstark-Khan, 2002). On different clay types, nucleotides and amino acid polymers of dozens of residues have been synthesized, providing a realistic scenario for a transition toward primitive self-replicating entities (Ferris et al., 1996). One of the first episodes of Darwinian positive selection could have operated prebiologically through differential surface binding, followed by the selection of autocatalytic heteropolymers (Tkachenko and Maslov, 2018). We arrive at the paradox that while primitive polymerization reactions might have required dry surfaces, later forms of life evolved in a water-rich environment, with water being the major component of cells. "Origin" and "establishment and evolution" of life have been considered either as linked processes that obey similar rules, or distinct events whose investigation requires dissimilar approaches. A distinction between "origin" and "establishment" will become pertinent also when addressing the origin versus the evolution

of viruses later in this chapter, and the mechanisms of viral disease emergence in Chapter 7.

### 1.4.2 A primitive RNA world

A simplified overview of a course of events that led to the origin of biological systems is depicted in Fig. 1.3. The prebiotic synthesis of potential building blocks—which might have been initiated earlier than 5000 million years ago—renders plausible the existence of a pre-RNA era that was then replaced by an RNA world in the late Hadean early Archean periods on Earth. This stage should have been followed by one in which RNA was complemented by DNA as a repository of genetic information (Bywater, 2012). Polymers other than DNA and RNA are also capable of encoding evolvable inheritable information (Pinheiro et al., 2012;

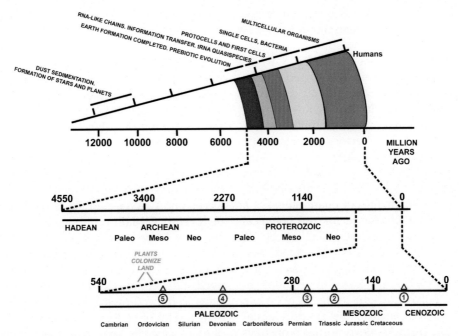

**FIGURE 1.3** Schematic representation of geological eons and eras in Earth history, and major prebiological and biological transitions. Time (horizontal lines) is expressed as millions of years before present, counted from the big bang (estimated at ∼ 13,800 million years ago). The time at which plants colonized land (green cone) and major mass extinctions (red triangles) are numbered: 1, end Cretaceous extinction; 2, late Triassic extinction; 3, end Permian extinction; 4, late Devonian extinction; 5, end Ordovician extinction. A proposed sixth extinction presently underway is not indicated. *Illustration by C. Perales and E. Domingo from information retrieved from references given in the text.*

Robertson and Joyce, 2012). Heterogeneous nucleic acid molecules (including mixtures of ribo- and deoxyribo-polymers) can give rise to functional nucleic acids (Gilbert, 1986; Lazcano, 1994a; Gesteland et al., 2006; Derr et al., 2011; Szostak, 2011; Trevino et al., 2011). RNA enzymes (ribozymes), such as RNA ligases can evolve from random-sequence RNAs (Joyce, 2004; Sczepanski and Joyce, 2014). The critical polymerization reaction involves the formation of a phosphodiester bond and release of pyrophosphate—analogous to the reactions catalyzed by the present-day RdRps—and represent an incipient, primitive anabolism (Eigen, 1992; Lazcano, 1994a; Orgel, 2002, 2004; Dworkin et al., 2003; Joshi et al., 2011). In support of a possible link between catalytic RNA activities and solid mineral surfaces in the origin of life (as summarized in Section 1.4.1), the catalytic activity of the hammerhead ribozyme of the Avocado Sun Blotch viroid was maintained when bound to the clay mineral montmorillonite (Biondi et al., 2007).

Minimum requirements for an RNA world would be the presence of ribozymes and mechanisms for the intake of energy-rich molecules (Orgel, 2002, 2004). The inherently low copying fidelity of the putative ribozyme polymerases, estimated in $10^{-2}$ to $10^{-4}$ errors per nucleotide copied (Wochner et al., 2011) should have ensured genetic variation for selection to act upon variant RNA molecules. Chirality (the existence of two mirror images or enantiomers of a molecule) poses a challenge for the chemical origin of biological molecules (Caglioti et al., 2011; Ruiz-Mirazo et al., 2014; Sczepanski and Joyce, 2014). Present-day biological systems use only D-ribose (D from "dextro" or rotation of the plane of polarized light to the right) while chemical condensation reactions produce equal amounts of the D- and L- (levo) forms. Nonenzymatic template-dependent reactions can be inhibited by the incorrect enantiomer. This led to the proposal that analogs devoid of enantiomeric forms, such as glycerol derivatives, could

have been the basis of the most primitive genetic systems (Schwartz and Orgel, 1985). Theoretical studies support the notion that initial achiral conditions can evolve toward chirality, in what has been defined as an extension of punctuated equilibrium to prebiological evolution (Gleiser et al., 2008).

Probably, very little, if anything, remains in our present-day biosphere of a primitive RNA world, let alone traces of the prolonged process that went from chemistry to the first replicating organizations, so profound have been the changes experienced by the Earth and its surroundings for over four billion years (Cantine and Fournier, 2018). Contemporary catalytic RNAs (found in the ribosome, as part of some protein complexes, and in some viroids), as well as nucleotide-like coenzymes, have been regarded as possible molecular remnants of a primitive RNA world (Lazcano, 1994a). Some authors consider the possibility that cells whose genetic material is made of RNA (Forterre, 2005) may still hide in some remote sites of our planet (Yarus, 2010). For a transition from a purely RNA (or RNA-like) world to an extended scenario with the participation of proteins, the presence of a transfer RNA (tRNA) quasispecies and the generation of a genetic code endowed with evolutionary potential must have been critical (Eigen, 1992; Koonin and Novozhilov, 2017). Initial theories of how the genetic code might have arisen were put forward by F. Crick, L. Orgel, and C. Woese in the middle of the twentieth century. Main proposals included a stereochemical fit between some amino acids and the corresponding bases (or codons to be), progressive evolution from a one nucleotide to a three nucleotides code, gradual incorporation of amino acids in the coding system, and the frozen accident model of codon universality (for a review of early concepts, see Crick, 1968). New insights on the code origin have come from integrating the knowledge of the mechanisms of protein synthesis with likely events in the RNA world. Primitive tRNA quasispecies (existing around 4000 million years ago, late Hadean, early Archean, Fig. 1.3) and tRNA aminoacylating ribozymes should have evolved to fit the genetic code, which was later expanded in coevolution with the translation machinery (Szathmary, 1999; Rodin et al., 2011; Caetano-Anolles et al., 2013). The age of the genetic code has been estimated in $3.8 \pm 0.5$ thousand million years (Eigen, 2013), and the present code structure is remarkably redundant. Effects of redundancy are to minimize the deleterious effects of mutations, and to contribute to flexibility of functional involvement of mRNA secondary structure (Shabalina et al., 2006; Koonin and Novozhilov, 2017; see also Chapter 2).

The advent of DNA as an informational macromolecule that was physically more stable than RNA opened the way to an increase of complexity (in the sense of the amount of genetic information) of the genetic material. DNA allowed integration of modules to form the first "chromosomes" and transcriptional regulation prior to protein expression. As in current evolutionary virology, perhaps the most challenging problem to understand early life is to identify the selective constraints that influenced the course of events. In contrast to the present-day environmental changes confronted by viruses (multiple and complex, but amenable to experimentation; Chapters 4 and 6), the conditions that permitted primitive genetic entities to acquire expanded coding and signaling capacities defy our imagination. Some aspects are considered next.

## 1.4.3 Life from mistakes, information from noninformation: origin of replicons

The most salient attributes of living matter are reproduction, evolvability, energy conversion (metabolism), and compartmentalization. What selective forces might have led to the integration of these features? Concerning reproduction, there must have been a critical transition from the absence of any inheritable instruction (despite the presence of primitive polymers) to the first molecules endowed with "inheritable" information, for example, a macromolecule capable of self-copying. Such a molecule should have had an immense selective advantage over surrounding macromolecular peers devoid of the capacity for reproduction. The transition from "no inheritable information" to "inheritable information" is essential for the origin of life. Current evidence suggests that the process that allowed such critical transformation was slow and inaccurate. Slow because the catalytic RNAs selected in the laboratory are about 10 million-fold slower than most protein enzymes (Jeffares et al., 1998; Yarus, 2010). Inaccurate because preenzymatic nucleotide polymerization would rarely display error rates below $10^{-1}$ to $10^{-2}$ mutations per nucleotide copied (Inoue and Orgel, 1983). No "predators" such as degrading enzymes were present to impede a slow accumulation of variant replicating molecules. The slow sequence exploration might have taken place in microenvironments shielded from damaging radiation until points in sequence space compatible with self-organization, replication, and adaptation were encountered. The adaptive potential of mutant distributions of polynucleotides is a key concept in the quasispecies theory of the origin of life (Eigen and Schuster, 1979), and a signature of present-day viruses (Chapter 3). Replicative inaccuracy and heterogeneity appear as recurrent requirements for the major transitions and adaptability of the forms generated in the course of prebiological and biological evolution. Once a primitive molecular "memory" was implemented, in the words of M. Eigen: " ....information generates itself in feedback loops via replication and selection, the objective being 'to be or not to be'" (Eigen, 1994, 2013). In those times, this was the only simple requirement: to be or not to be.

This singular transition resulted in the first replicating entities that are also termed *replicators* or *replicons*. They were selected for replicability, stability, and evolvability with trade-offs (acquisition of benefits for one of the three traits at some cost for another trait) likely play a role at this stage (see Chapter 4 for trade-offs in virus evolution). Optimization of primitive replicons should have been facilitated by dual genotypic and phenotypic features in the same molecule (Eigen, 2013). That RNA itself is both genotype and phenotype is a feature of present-day RNA viruses. That is, the genomic RNA determines phenotypic traits, independently of its protein-coding activity, for example, through its involvement in RNA-RNA and RNA-protein interactions. "Priming" of polynucleotide synthesis, in the sense as we know it today, should not have been a limitation since circular RNA or RNA-like molecules could fold partially to prime their own copying. The term "replicon" currently refers to any genetic element that encodes sufficient information to be copied (i.e., viruses, plasmids, etc.), even if the copying is carried out by (or in conjunction with) elaborate cell-dependent machinery. Virtual replicons are used in computer simulations to learn about the dynamics of natural living systems (Adami, 1998; Eigen, 2013).

The environment in which primitive replicons had to self-organize about 4000 million years ago was very different from the environment we have today on Earth. The sun was about 25% −30% less luminous than the present-day sun, yet it produced more ultraviolet (UV) light. Due to the absence of oxygen and ozone layer, the UV radiation that reached the Earth was 10- to 100-fold more intense than today, with the difference being accentuated for the radiation in the 200−280 nm range. Studies of the conversion of UV radiation into DNA-damage equivalents suggest a two to three logarithm larger biologically relevant UV radiation during the time of the putative RNA world as compared with today's radiation (Canuto et al., 1982; Chang, 1994; Horneck and Baumstark-Khan, 2002). In such an environment, radiation-related mutational input could have had drastic effects on replicating entities in ways that can be

only roughly anticipated from the present-day chemistry. Even the simplest present-day RNA genetic systems, despite their small target size, would undergo severe radiation damage. Reconstruction of protein enzyme-free nucleic acid synthesis under the radiation conditions prevalent on Earth during the RNA world development, during late Hadean and early Archean eras, offers a fascinating challenge and opportunity of experimental research for the rising field of Astrobiology.

It has been considered that the time elapsed since the Earth attained a life-friendly environment until protocells arose (from about 4500 million to about 3500 million years ago, Fig. 1.3) was insufficient for life development. This led to the panspermia theory, which proposes that life has an extraterrestrial origin. Panspermia in different forms has been defended by noted scientists, such S. Arrhenius in the early twentieth century and later by F. Crick and L. Orgel (discussed by de Duve, 2002). In addition to the time estimates for life generation being arbitrary, our present understanding of how error-prone replication can facilitate evolvability and exploration of novel biological functions (Chapters 2 and 3) converts a 1 million-year time period in a long time for life to originate and initiate multiple branches for its development.

## 1.4.4 Uptake of energy and a second primitive positive selection

Energy conversion is an essential feature of life. The primitive precellular organizations might have obtained energy either from organic molecules captured from the external environment (heterotrophy) or from metabolites they synthesized endogenously using external energy (autotrophy). One line of thought considers that it is more likely that the first cells were heterotrophs, and that only later they evolved toward autotrophy, in the form of photosynthesis, which represented a major transition in the repertoire of biosynthetic pathways (Nakashima et al., 2018). Fermentation reactions were likely the first ones exploited to break energy-rich bonds, as a source of energy for primordial biochemical reactions.

An alternative view is that the first cellular organism was an autotroph, in particular, a chemoautotroph (also termed *lithotroph*) that used inorganic compounds to obtain energy. One of these proposals is that the formation of pyrite from hydrogen sulfide was used by primitive cells as an energy source, resulting from reactions such as $FeS + H_2S \rightarrow FeS_2 + 2H^+ + 2e^-$ (Wächtershäuser, 1994). Positive charges on pyrite crystals could accumulate negatively charged molecules (e.g., the products of $CO_2$ fixation) and undergo reductive reactions. The system might have selected surface-bound polymers rather than monomers, and given rise to biochirality (selection of one enantiomeric form over another) because of the chiral structure of pyrite (Wächtershäuser, 1988). Then, an evolution toward an Archean carbon-fixation cycle would occur, a precursor of the metabolic cycles found in present-day archaeal and bacterial organisms (Ruiz-Mirazo et al., 2014). The picture may have been more complex, as judged by the success of mixotrophic organisms which are capable of switching from phototrophy (light-mediated break-down of $CO_2$ for their metabolism) to heterotrophy in some extreme environments of the present-day Earth (Laybourn-Parry and Pearce, 2007).

The integration of early replication systems and metabolism was probably favored by some compartmentalization of replicative-metabolic units through lipid bilayers (Carrara et al., 2012; Stano et al., 2013; Ruiz-Mirazo et al., 2014; Hanczyc and Monnard, 2017). Here a second decisive positive selection event might have entered the scene at the stage of formation of proto-cells and the first individual cells (3600−3200 million years ago, Fig. 1.3; Eigen, 1992). Lipid bilayers endowed with a splitting capacity should have been strongly selected at this stage of life development because of their power to spread. This underlines the concept that Darwinian selection need not be associated exclusively with template-copying processes. Membrane traffic and reorganizations are essential for the life cycle of many present-day viruses (Huotari and Helenius, 2011). Virus particle stability and capacity to spread, associated with membrane capture and interactions, might have derived from the early genomes that exploited membrane-based organelles to achieve functional diversification. Compartmentalization is considered one of the key developments to initiate a cell-based life (Morowitz, 1992; Hanczyc and Monnard, 2017).

## 1.4.5 Definitions of life

Despite obvious difficulties in reproducing physical and chemical processes that were likely involved in the origin of life, there is sufficient evidence to render probable that the most primitive organizations that we would now consider as "living" resulted from the assembly of simple organic compounds that attained a required level of complexity (Kauffman, 1993). The various facets that distinguish living from inanimate matter are recapitulated in the definitions of life that scientists from different backgrounds in physics, chemistry, or biology have proposed (Box 1.2). Some definitions underline entropy requirements, while others emphasize self-organization, evolvability, metabolism, or regulated complexity. Their disparity reflects the intricacies of a rather enigmatic unfolding of matter.

A. Lazcano summarized the current situation in the arduous search of the origin of life: "There are still unsolved problems but they are not completely shrouded in mystery, and this is no minor scientific achievement. Why should we feel disappointed by our inability to even foresee the possible answers to these luring questions? As the Greek poet Konstantinos Kevafis once

---

**BOX 1.2**

## Some definitions of life and living organisms

- Life is the property of a system that continuously draws negative entropy (maintains orderliness), and delays decay into thermodynamic equilibrium (Schrödinger, 1944).
- Life is an expected, collectively self-organized property of catalytic polymers (Kauffman, 1993).
- Life is a property of any population of entities possessing those properties that are needed if the population is to evolve by natural selection (Maynard Smith and Szathmáry, 1999).
- Life is what is common to all living beings. This answer is not a tautology, as it allows many attributes to be excluded from the definition of life (de Duve, 2002).
- Life is descent with modification. Replication is life's ultimate chemical and physical survival strategy (Yarus, 2010).

- Life is a self-sustained chemical system capable of Darwinian evolution (NASA Astrobiology Institute).
- Living organisms are metabolic-replicating systems composed of molecules and cells which are subject to spontaneous changes in structure and function due to mutations (Demetrius, 2013).
- Life in three statements: (1) Life is not represented by any fundamental physical structure. (2) Life is an overall organization that is governed by functional rather than by structural principles. (3) In order for life to come about, there must exist some physical principle that controls complexity (Eigen, 2013).
- Life, whatever else it may be, is certainly a regularity among material processes (Eigen, 2013).

---

wrote, *Odysseus should be grateful not because he was able to return home, but on what he learned on his way back to Ithaka*. It is the journey that matters" (Lazcano, 1994b; for a general overview of the history of life on Earth, centered in paleontology, see Cowen, 2005).

After this brief survey of the origin of life, we can now examine theories on how, when, and why viruses arose and became active actors in our biosphere.

## 1.5 Theories of the origins of viruses

Although not in a linear fashion, the number of nucleotides or base pairs in the genetic material—that presumably reflects the amount of genetic information relevant to confer phenotypic traits—increased as evolution led to differentiated organisms. The major theories of the origin of viruses are divided into two opposite categories: those that attribute virus origins to the early development of life, and those that propose that viruses arose when a cellular life was already in place (Fig. 1.4). These two broad views that we may term "viruses *without* cells" and "viruses *from* cells" are not irreconcilable, although reconstruction of ancestral developments is challenging. They can be divided into five main theories—not all independent or mutually exclusive— which are summarized next.

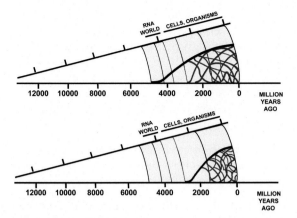

**FIGURE 1.4** Two possible courses of events regarding when viruses first appeared and participated in the evolution of the biosphere. The scheme of time frames and major biological events (RNA world, first cells, and organisms) are those displayed in Fig. 1.3. According to the upper diagram, viruses (or previrus-like entities) arose together with the first (precellular) replicating entities. According to the second diagram, viruses (or previrus-like entities) arose when a cellular life had already been established. Presence of virus is generically represented by the external, thick, black curves. The internal red, wavy lines represent generation, dominance, and extinction of multiple viral lineages whose numbers and true dynamics will remain unknown. *Illustration by C. Perales and E. Domingo from information retrieved from the different models of virus origin and references included in the text.*

## 1.5.1 Viruses are remnants of primeval genetic elements

- Viruses were involved in the origin of life. Some viruses are the descendants of primitive RNA or RNA-like replicons that preceded cellular forms.

Because of their limited genetic complexity, RNA viruses and subviral RNA elements have been considered possible descendants of the primitive replicating entities that predated cell-based life forms (upper diagram in Fig. 1.4). As early as the beginning of the twentieth century, H.J. Muller, L.T. Troland, and J.B.S. Haldane suggested that viruses represented primordial life forms. Influenced by the discovery of bacteriophages by F. d'Herelle, J.B.S. Haldane proposed viruses as intermediates between the prebiotic soup and primitive cells (reviewed in Lazcano, 2010). In those times, knowledge of viruses was superficial from today's perspective, and lent to daring proposals coherent with viruses being perceived as simple. Despite simplicity and replication being features that could be also be attributed to primitive life forms, we have to distinguish the role that virus-like entities might have played in the establishment of early life from the possibility that present-day viruses reflect how early life might have been.

An origin independent of cells is suggested by the presence of a number of "virus hallmark proteins" in viruses, which do not have counterparts in present-day cells, at least in the sequences represented in data banks. A number of protein folds (such as the double jelly-roll fold) found in the nucleoplasmic large DNA viruses (and related structures) are also present in viruses that infect the archaeon *Sulfolobus*, and in some single-stranded and double-stranded RNA viruses that infect eukaryotic cells. These unique structures and other features of viruses, such as genome size and its composition, constitute a "virus innate self" (Krupovic and Bamford, 2007), that has been taken as an indication that viruses were constructed from modules that were different from those captured to build cells.

The fact that some viral proteins do not have cellular homologs (or are only distantly related to cellular homologs) has favored the view that a virus world preceded the origin of cells (Forterre, 2005; Villarreal, 2005; Koonin et al., 2006; Koonin, 2009). In this line of thought, the present-day RNA viruses are considered remnants of an RNA world; no other biological group that has been identified uses RNA as genetic material. Likewise, transposable elements, repeat sequences of mobile elements (LTR-retrotransposons, LINEs, SINEs, ALUs) and of telomeres and centromeres are considered to be likely of a viral origin [reviewed in Witzany, 2012)].

Other present-day RNA genetic elements that include ribozymes as constituents [i.e., plant viroids, self-replicating RNAs of about 165—430 nucleotides in length (Hadidi et al., 2017) or the defective delta agent, also termed *hepatitis delta virus* (HDV)] might be a vestige of primitive genetic elements (Robertson et al., 1992). HDV is dependent on hepatitis B virus (HBV) for the completion of its infectious cycle (Taylor and Pelchat, 2010). The HDV genome is a mosaic RNA consisting of a viroid-like RNA and an RNA region whose complementary RNA (antigenomic strand) encodes two forms of a structural protein termed *the delta antigen*. Both the genomic and antigenomic RNAs possess a strong secondary structure with about 70% paired nucleotides. The delta antigen is encapsidated by the HBV surface antigen as a component of HDV particles. Thus, HDV appears to be the result of an RNA conjunction between a viroid-like RNA and an mRNA-coding region. Such conjoined RNAs might have been the precursors of the modern eukaryotic genome partition into coding sequences (exons) and intervening sequences (introns) (Sharp, 1985; Robertson, 1992, 1996; Chao, 2007; Taylor and Pelchat, 2010). The structure of the HDV genome seems to echo processes that originated with primitive RNAs selected for their ability to replicate that incorporated a protein-coding moiety through recombination with other RNAs. Replication-competent and protein-coding chimeras might have signaled a relevant intermediate step toward more complex RNA and DNA genomes. The advent of an enzyme that could copy RNA into DNA to carry out reverse transcription should have been instrumental in generating primitive DNA viruses and other DNA-based genomes of increased complexity (Lazcano et al., 1992).

The discovery of the chimeric structure of the HDV genome illustrates how insights into the origin and early evolution of life can be gained from current genomics, despite lacking experimental approaches to recreate episodes that led to virus origins. Even if experiments could be designed, the time frame involved would occupy several generations of scientists, which is not feasible given the current research grant system.

## 1.5.2 Viruses are the result of regressive microbial evolution

• Viruses originated from regressive evolution of microbes with a cellular organization and became parasites of cells.

This theory is quite the opposite of the previous one because it presupposes that a cellular world was the source of viruses (lower diagram in Fig. 1.4). One of its conceptual pillars is that the intricacies of virus-host relationships unveiled by molecular virology render unlikely that present-day RNA viruses are remnants of an ancestral RNA viral world. Also, the conditions prevailing in the RNA world did not necessitate that a primitive replicon displays rapid replication—a trait of most present-day viruses—for it to become established, because of the scarcity of predators.

The theory of a cellular origin of viruses was already put forward in the twentieth century when it was evidenced that complex DNA viruses encoded enzymes and immunomodulatory proteins that had cellular counterparts. The virus-generating cells could be autonomously functional from the onset (as autonomous as individual cells can be) or belong to a class of simple cells that parasitized functionally more advanced cells.

One of the mechanisms by which viruses may have originated from cells is by regression or reduction. It implies that the number of genes of the virus-precursor cells diminished gradually or in a step-wise fashion to a minimum number compatible with replication at the expense of other cells that supplied basic components for gene expression and energy capture. The giant DNA viruses (Nasir et al., 2012; Colson et al., 2017) have been regarded as a possible missing link between a primitive DNA cell and the DNA viruses, and some of them include part of the translation apparatus (Abrahao et al., 2018).

The capacity to spread, so inherent to the concept of the virus, might have been first selected as a positive trait for cells, which then regressed toward a subcellular transmissible form. Prokaryotic cells can spread effectively among differentiated eukaryotic hosts, and tumor cells have been regarded as transmissible parasites (Banfield et al., 1965; Murgia et al., 2006; Pearse and Swift, 2006). The observation that some "infectious cells" can be disseminated by insects provides a model for an early origin of arthropod-borne viruses. It has been recognized that transmission of the virus from an infected cell into a recipient cell need not involve a prolonged stay of the virus in the extracellular environment. Infection "synapses" allow intimate cell-to-cell contacts through which virus transmission takes place. It is estimated that synapse-mediated transmission maybe 100-fold more efficient than the transmission of viral particles released into the extracellular environment. This is one of several mechanisms of bloc (multiple particles) transmission, which has several implications for viral quasispecies dynamics (Chapter 3). Acquisition of capacity for long-range transmission in space and time should have provided a sufficient selective advantage for a "cell-to-external environment-to-cell" transmission to coexist with "cell-to-cell" transmission in our biosphere.

The possibility that RNA viruses derive from some "organism" that used RNA as genetic material was suggested initially by D. Baltimore (1980). The evidence that RNA-dependent RNA synthesis is rare in cells suggests that either RNA viruses derived their replicates from a now probably extinct "RNA organism" or that the viral replicates evolved from cellular DNA polymerases (Baltimore, 1980; Forterre, 2005, 2006a,b; Yarus, 2010).

### 1.5.3 Viruses are liberated autonomous entities

- Viruses originated from cellular DNA or RNA that evolved to embody autonomous replication, and an extracellular step in their replication cycle.

An alternative to the regressive or reduction of virus origin from cells is the escape theory. It proposes that part of the genome complement of cells found a way toward autonomous replication in the sense that they could replicate on their own as parasites of cells. The capacity to replicate could come from the parental cell or be acquired externally. The kind of autonomy attained implied an intracellular phase of their multiplication cycle and an extracellular phase in the form of transmissible particles (virions). This new way of life should have been positively selected if an increased capacity of cell-to-cell transfer conferred an advantage to the cells regarding the acquisition of new traits for functional diversification while maintaining a capacity to sustain virus multiplication.

Decades ago, the view that viruses originated from subcellular organelles was a favored one (see e.g., Joklik, 1974). Despite discernible sequence identity between some viral and mitochondrial DNA sequences, no evidence of viruses having functions encoded in cellular organelles and not in chromosomal DNA has been obtained, perhaps reflecting an earlier relationship between viruses and primitive free-living cells rather than between viruses and modern (eukaryotic) cells (see also Section 1.7)

### 1.5.4 Viruses are elements for long-term coevolution

- Viruses are as ancient as cells, and coevolved with cells or even with precellular genomic organizations, with which they shared functional modules.

Current genomics of viruses and their host organisms (Bushman, 2002; Mount, 2004; Hacker and Dobrindt, 2006) tends to favor a long history of coevolution between viruses and cells. The structure and functions of many viral proteins do not deviate in any salient way from cellular counterparts. This applies to proteins involved in genome replication, and in proteolytic processing of proteins and polyprotein precursors. In contrast, the "virus innate self" that groups viral proteins and protein folds not shared with cellular counterparts, and that provides a powerful argument for a virus origin independent from cells (Section 1.5.1), does not exclude a coevolutionary mechanism even if all modules were not common to cellular and viral entities over a several million years period (see section 1.7). Comparative genomics suggests that the exchange of functional and structural modules through lateral gene transfers, together with fine adjustments mediated by mutation, have contributed to the coadaptation of cells and autonomous replicons over ancient evolutionary periods (Gorbalenya, 1995; Holland and Domingo, 1998; Jalasvuori and Bamford, 2008; Villarreal, 2008). Even mechanisms that prompt viral variation in cell tropism (Chapter 4) have parallels in differentiated organism As an example, a two amino acids insertion into ectodysplasin—a member of the tumor necrosis-binding family—alters its receptor specificity, and the differential expression of the two protein versions plays a role in epidermal morphogenesis (Yan et al., 2000). Furthermore, as the number of three-dimensional structures for viral cellular enzymes has increased to reach thousands, structural similarities between key cellular and viral enzymes (polymerases, proteases) have become apparent.

Baltimore (1980) proposed that a limited number of "archetypal" proteins could be responsible for RNA virus function. He named as "archetypal" a "positive virus polymerase," a "negative virus polymerase," a set of "surface" proteins, and several "proteases," among other proteins and regulatory elements. The argument

that "archetypal" modules could spread among positive- and negative-strand RNA viruses was based on features and mechanisms now recognized as much more profuse than in 1980: the multifunctionality of viral proteins, their capacity to diversify by mutation, and the existence of RNA recombination (Chapter 2). Regulatory strategies were also likely shared by cells and viruses. Small micro-RNAs that now populate the cellular world can act as molecular switches for RNA viral genomes to modulate their replication and gene expression (van Rij and Andino, 2008; Diaz-Toledano et al., 2009). A likely course of events is that the long coevolution of protocells and primitive virus-like elements gradually shaped both cells at the individual and organismal level and the precursors of present-day viruses.

### 1.5.5 Viruses from vesicles

- "Protoviruses" might have originated in primitive vesicles.

Early protocellular communities probably lacked a cell wall or other compartmentalization barriers, an absence that allowed fluid transfers of metabolites and genetic material (Woese, 2002). The most primitive vesicles based on lipid structures inaugurated the distinction between "internal" and "external" milieu and might have evolved to contain self-replicating macromolecules (Jalasvuori and Bamford, 2008; Adamala and Szostak, 2013; Hanczyc and Monnard, 2017). Vesicles located in favorable microenvironments of a primitive Earth had the potential to exchange small molecules between the "inside" and "outside," and export materials toward other vesicles. Exchanges could assist in the coupling between genome replication and an incipient metabolism.

Depending on their composition, lipid vesicles could form and remain stable at temperatures of about 100°C. Their transfer to lower temperatures might have modulated their permeability prior to the stage at which peptide or protein transporters were inserted as membrane components. Here again, "heterogeneities" became important: membranes made of mixtures of amphiphiles display increased thermostability, permeability, and tolerance to divalent cations. Budding vesicles endowed with traits such as growth, division, and permeability should have been positively selected for their ability to spread favorable replicating molecules and mediators of protocellular functions. Selected protoviral vesicles became gradually dispensable for the spreading of beneficial genes, and viruses evolved from being solely beneficial entities into also displaying a parasitic behavior that exploited cell resources (compare with Sections 1.7 and 1.8). According to this model, when cells became independent units surrounded by lipid-based membranes, viruses promoted the selection of cells that expressed peptidoglycan molecules on their surface to decrease or prevent virus infection. This transition could mark the onset of an arms race behavior that has been associated with a survival strategy of many present-day viruses. A cell wall allowed vertical transmission of genetic information and rendered the cell metabolically independent. In addition, it provided an osmotic environment suitable for energy production. Genetic information and molecular devices for energy capture, storage, and use were equally important for a sustainable cellular organization. Thus, according to this theory, viruses originated from protoviral elements whose main function was to spread useful genes horizontally. The evolved viruses acted as selective agents to promote microbial evolution and then became established in an increasingly differentiated cellular world (Villarreal, 2005, 2008; Hendrix, 2008; Jalasvuori and Bamford, 2008).

Geological studies indicate that mass extinctions of a brief duration (less than 100,000 years, a short geological time!) occurred at several points when multicellular organisms populated the Earth (some of them indicated in Fig. 1.3).

The end-Permian, end-Triassic, and end-Cretaceous extinctions rank among the most drastic, resulting in profound environmental perturbation (Burgess et al., 2014 and references therein). As depicted in the form of red upward and downward curved lines in Fig. 1.4, perturbations associated with viral extinctions (due to massive host extinctions) and severe bottleneck events have periodically blurred traces of many viruses unknown to us.

Mechanisms implied by each of the five theories summarized here might have participated in the origin of viruses as we know them today, and any model will remain tentative for several obvious reasons. Viruses have not left a fossil record amenable to analysis with current technology. Moreover, viral genomes can evolve at very high rates in response to environmental necessities (Chapter 7), and reconstructions of ancient Earth environments are necessarily imprecise. For these reasons, viruses, independently on when they became active actors in the biosphere (Fig. 1.4), are unlikely to have maintained molecular signatures that could shed light on their remote past.

## 1.6 Teachings from mycoviruses

The viruses that infect fungi, the mycoviruses, recapitulate key issues on virus origins and diversity (Son et al., 2015; Roossinck, 2019). Two main theories of their origin are that they evolved from plant viruses that invaded fungi or that they have coexisted and coevolved with their host fungi for a long time, in line with the long coevolution model of virus origins (Section 1.5.4). They are diverse regarding the genetic material that can be single-stranded or double-stranded RNA or DNA. They can be either asymptomatic, produce disease, or modify critical host traits, such as virulence, again reflecting a broad range of host interactions observed with other viral groups. The presence of a dsRNA mycovirus in a fungus is essential for the latter to confer heat tolerance to some plants; these findings have defined a three-way symbiosis required for an important phenotypic trait (Marquez et al., 2007; Roossinck, 2013) (see also Section 1.7.3).

## 1.7 Being alive versus being part of life

An often asked question is whether viruses are alive or not. The answer is debated, as reflected in the various definitions of virus (Box 1.3). Two opposite views coexist. One of them considers viruses as macromolecular aggregates, that is, viruses are regarded as "chemicals." A second proposal is championed by the "ribovirocell" concept of P. Forterre, which implies that a viable, virus-infected cell contains two different organisms that coexist symbiotically: the cell that produces virus and the replicating virus itself. Another facet of the same duality is manifested when viruses are considered merely as perturbing chemicals

with no participation in the tree of life versus viruses being actually important players in the tree of life. The dual character of viruses as "alive" during intracellular replication and "not alive" outside the cell was stressed in some early literature (e.g., Davis et al., 1968).

As discussed in Section 1.4, if life is best defined as a conglomerate of complementary features, viruses display two of these features: the capacity to replicate and to evolve. Thus, although it is debatable whether viruses qualify as "alive," they are (and most likely have historically been) an integral part of life and the construction of life. Viruses have influenced the tree of life, as we know it, with lateral gene transfers—some mediated by viruses—being a key element in its architecture and differentiation of the different domains of life (Ciccarelli et al., 2006; Weiss et al., 2018). Because of the difficulties of defining "life" unambiguously, the reader might have noticed that in previous sections, the question addressed has been "how is life" rather than "what is life." The question of "why" have viruses persisted in the biosphere is addressed next.

---

### BOX 1.3

#### Some definitions of virus

- Viruses are strictly intracellular and potentially pathogenic entities with an infectious phase and (1) possessing one type of nucleic acid, (2) multiplying in the form of their genetic material, (3) unable to grow and to undergo binary fission, and (4) devoid of a Lipmann system (Lwoff, 1957).
- Viruses are entities whose genomes are elements of nucleic acid that replicate inside living cells using the cellular synthetic

machinery and causing the synthesis of specialized elements that can transfer the viral genome to other cells (Luria et al., 1978).
- Viruses are replicating microorganisms that are among the smallest of all life forms (first two editions of *Fields Virology*).
- Viruses are transmissible deoxyribonucleic acid (DNA) or ribonucleic acid (RNA) genetic elements that require a cell for multiplication (Domingo and Perales, 2014).

## 1.7.1 Role of viruses in the evolution of the biosphere

Two general models have been proposed to explain the maintenance of viruses in the biological world:

- Viruses have persisted because they parasitized opportunistically any cellular niche which was compatible with their replication apparatus, and which provided adequate resources.

In this view, viruses are "selfish" replicating elements that became successful when increasingly efficient polymerization activities became part of their life cycles. However, several observations indicate that what was once considered purely "selfish" (also referred to as "junk") cellular DNA may not be that useless after all. Given the intimate connections between cells and viruses, the view that viruses primarily exist because they are mere "selfish" entities seems unlikely. A different issue is that under some particular environments, a virus displays a "selfish" element-like behavior, as one of the outcomes of their having been positively selected. An example, is the blind replication at the expense of elimination of hosts in disease processes, although this is necessarily a transient behavior.

- The presence of viruses was positively selected because they promoted cellular variation and functional diversification.

This proposal relates to some of the theories of virus origins (Section 1.5) and the early evolution of life that implied a role for virus-like entities. According to this view, viruses, together with other subcellular genetic elements (plasmids, retroelements, etc.), penetrated the genetic material of ancestral cell forms, acted as agents of lateral gene transfers, and modified the expression profiles of the recipient cells. Probably, there has been (and there still is) a constant flow of genes between cells and viruses and other mobile genetic elements. The abundance of endogenous retroviruses in the mammalian genomes (García-Montojo et al., 2018) is a clear symptom of such a genetic flow. A nonfunctional viral infectivity factor (Vif) [the HIV-1 protein that can counteract the mutagenic activity of the apolipoprotein B mRNA editing complex (APOBEC) proteins; see Section 2.7 in Chapter 2] was found in the remnant of a rabbit endogenous retrovirus termed *rabbit endogenous lentivirus type K* (Katzourakis et al., 2007). About 8% of the human genome is made of retroviral-like elements. Present-day human endogenous retroviruses probably contribute to pluripotency of human cells (Santoni et al., 2012), and genome regulation (Chuong, 2018). The fusion of cells from the placenta is mediated by syncytin, a protein in HERV-W endogenous retroviruses (Chuong, 2018; Villarreal, 2005).

In addition to the promotion of gene transfers to construct key cellular components, viruses probably acted as selective agents for cells to evolve defense mechanisms against viruses, and this may have originated new cellular functions. Also, viruses could favor the survival of some cell types over others, based on differential cell susceptibility to virus infection, thus contributing to cellular diversification. The need to escape viral infection may have furnished novel cell surface receptor proteins through a selection of cellular escape mutants (Buckling and Rainey, 2002; Saren et al., 2005). Some experimental systems consisting of persistently infected cells in which the cells and the resident virus coevolve (Chapter 6) illustrate how viruses could act as selective agents to promote cellular variation. Such variation would not necessarily involve exchanges of genetic material between the virus and the cells, provided sufficient genetic variation of cells took place.

Multicellular organisms devoid of viral entities should have endured a long-term disadvantage over an alternative scenario with the coexistence of cells and viruses. Selection by viruses need not be restricted to cells, and it can be extended to entire host organisms and their populations. An abundance of hosts may promote viral epidemiological fitness, which is a factor in viral disease emergence (Chapters 5 and 7). Host subpopulations may be selected by their resistance to epidemic outbreaks by highly pathogenic viruses, such as in the 1918 influenza pandemics or currently with the AIDS or Zika pandemics, and Ebola outbreaks in some parts of Africa. Traditionally, plagues decimated the human population and acted as selective agents for differential survival of individuals.

Selfish-opportunistic and selected-functional are not incompatible models of virus maintenance. Once the instruction to replicate had been positively selected, selfish elements could ensue. As we learn about viral and cellular genomics, the current promiscuity and diversity of viruses (Section 1.3) appear as complementary agencies to promote general biological evolution following Darwinian mechanisms. Viruses might have contributed to the DNA replication machinery of cells, to the formation of the eukaryotic cell nucleus, and to a number of developmental processes (Baranowski et al., 2001; Bushman, 2002; Bacarese-Hamilton et al., 2004; Mallet et al., 2004; Villarreal, 2005, 2008; Forterre, 2006a). Cells are a necessity for viruses and viruses are promoters of cell diversity and as a consequence, of cellular differentiation (compartmentalization and functional specialization).

## 1.7.2 Current exchanges of genetic material

Present-day viruses reveal several mechanisms of exchange of genetic material that might have roots in early cellular evolution. Temperate bacteriophages (the prototypic example being E. coli phage λ) integrate their genomic DNA in the DNA of their host bacteria. The uptake of cellular genes by viruses has been amply documented in transducing bacteriophages (those that can transfer DNA from one bacterium to

another), as well as in RNA and DNA tumor viruses. Even RNA viruses that are not known to include a reverse transcription step in their replication cycle can incorporate host RNA sequences. Replication-competent, cytopathic variants of bovine viral diarrhea virus (a type species of the genus Pestivirus of the important family of pathogens Flaviviridae) can acquire cellular mRNA sequences in their genome, via nonhomologous recombination (Meyers et al., 1989). Insertion of 28S ribosomal RNA sequences into the hemagglutinin gene of influenza virus increased its pathogenicity (Khatchikian et al., 1989). Some defective-interfering particles of Sindbis virus included cellular tRNA sequences at their 5′-ends (Monroe and Schlesinger, 1983). Sequences related to some flaviviruses can persist in an integrated form into the DNA of the insect vectors *Aedes albopictus* and *Aedes aegypti* (Crochu et al., 2004). Endogenous hepatitis B viruses (eHBVs) have been identified in the genomes of birds and land vertebrates (amniotes), crocodilians, snakes, and turtles. The evidence is that eHBVs are more than 207 million years old and that ancient HBV-like viruses infected animals during the Mesozoic Era (Suh et al., 2014, Fig. 1.4). The existence of alternative mechanisms for the integration of viral genetic material into cellular DNA suggests an ancient origin and a selective advantage of exchanges of genetic information in shaping a diverse and adaptable cellular world (Eigen, 1992, 2013; Gibbs et al., 1995; Villarreal, 2005, 2008).

## 1.7.3 Symbiotic relationships

Symbiosis is an important determinant of coevolutionary interactions between hosts and their parasites (Vorburger and Perlman, 2018). Frequent symbiotic and mutualistic interactions have been established with viruses. Human endogenous retroviruses can protect human tissues and the developing fetus against infection by some exogenous retroviruses (Ryan, 2004).

Some bacteria require bacteriophage to express virulence determinants (Tinsley et al., 2006). Symbiosis can be established between bacteriophages and animals (Barr et al., 2013). Several plant RNA viruses delay the symptoms of abiotic stress, such as those produced by drought and frost (dehydration, osmotic stress, and oxidative stress). Protection is mediated by increased levels of osmoprotectants and antioxidants in the infected plants (Xu et al., 2008).

Symbiotic relationships represent a state of local equilibrium between viruses and hosts, triggered by compatibility and occasionally by mutual benefits. An arms race implied by the virus-host interactions described in previous sections might have been the *modus vivendi* for viruses, only interrupted by occasional armistices. Alternatively, symbiotic and mutualistic interactions might have been the norm, only interrupted by occasional defections by killer personalities that have become the key actors of hospital wards and virology textbooks (Roossinck, 2011; Li and Delwart, 2011). As discussed in connection with natural counterparts of the transition toward error catastrophe in viruses (Chapter 9), cellular editing activities such as those displayed by some of the adenosine deaminase acting on double-stranded RNA and APOBEC proteins are also part of the innate immune response against some viruses. In turn, viruses have evolved multiple functions to counteract the host immune response (Chapter 4). The recruiting of cellular functions to confront viruses attests of transient losses of an equilibrated coexistence. Some middle- and long-term equilibrium between virus and host population numbers must be continuously restored by selective events; otherwise, this book would not have been written.

The evolutionary origin of defense mechanisms against viruses can be regarded as a response to an excessive number of virus-cell interactions. Superinfection exclusion is one of the mechanisms used by present-day cells to limit replication of a virus when another one is actively replicating (or incorporated) into the same cell. Exclusion has a biochemical interpretation in the competition of two viral entities for cellular resources, but it might have been boosted by early cellular adaptation to limit viral invasions (Chapter 4). Likewise, components of the intrinsic and innate immune response that prevent infection and disease might have also been endowed with activities that promoted cell variation for adaptability.

## 1.8 Virus and disease

The two opposite views of the activity of viruses in the biosphere (i.e., opportunistic occupation of any suitable cellular niche or intimate cooperative coevolution with host cells) would be expected to produce a different proportion of pathogenic viruses. Opportunistic invasions should lead mainly to disease-prone viruses, while long coevolutionary periods should lead to a dominance of nonpathogenic viruses. Not all viruses that have been characterized are pathogenic, and in fact, only a minority of those that exist might be. However, since only a limited number of the viral genomes predicted by metagenomic surveys have been characterized, it is not possible to adventure a proportion of beneficial or neutral versus harmful viruses. In the course of investigations on poliomyelitis, a search for related viruses was undertaken, and a number of new viruses later to be known as *echoviruses* were discovered. They were isolated because they caused cytopathology to cells in culture. The virus-containing samples were from individuals that did not show symptoms of a viral infection. The new isolates were designated as "orphans," meaning viruses without the disease. The term echovirus derives from the enteric cytopathogenic human orphan virus. Some of them, or their close relatives, were later associated with disease syndromes, but others were not. Viruses, as diverse as circoviruses, polyomaviruses, or herpesviruses colonize a considerable proportion of animals, and only some of the virus types are the direct cause of disease [e.g., postweaning multisystemic wasting syndrome (PMWS) by porcine circovirus type 2, or cancer by some polyomaviruses, among many other examples]. Disease potential is unrelated to viral genome size. PMWS is associated with the smallest mammalian DNA virus genome of only 1.7 Kb, while the almost 100-fold larger herpes virus genomes can coexist with immunocompetent humans without the noticeable disease.

The pathogenic character of a virus depends on the intricacies of virus-host interactions that are poorly understood. The more the knowledge of virus-host interactions, the border between the virus being pathogenic or nonpathogenic becomes fuzzier. Viruses may not damage essential cell functions but may affect dispensable cellular functions. Studies by M.B.A. Oldstone and his colleagues on persistent infections of lymphocytic choriomeningitis virus in neuroblastoma cells demonstrated that the resident virus altered the expression of differentiated cell trait while preserving vital functions (Oldstone et al., 1977). This is one of many examples of virus-induced modifications of the so-called *luxury* functions of cells (Oldstone, 1984). Disease manifestations depend on multiple host and viral factors (allelic forms of several genes, microbiome composition, immune status, coinfections with viruses or other agents, and so on). The dual infections with HIV-1 and hepatitis C virus (HCV) during the end of the 20th century accelerated the evolution of HCV-associated liver disease.

## 1.9 Viral and cellular dynamics and the tree of life

In the following chapters, molecular mechanisms of virus variation are reviewed in connection with virus survival and host interactions.

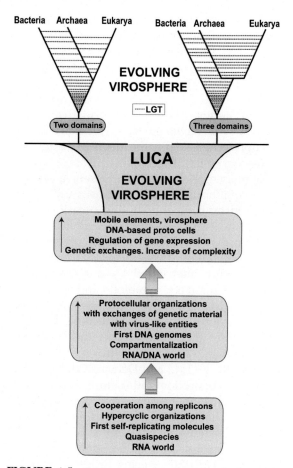

**FIGURE 1.5** Scheme of likely occurrences in the origin and early evolution of life and virus involvement, according to current genomics and mechanisms of biological evolution. The time frame involved goes from about 4200 million years ago (bottom) until about 2000 million years ago (top) (compare with geological eras in Fig. 1.3). In each box, the arrow indicates a succession of major developments, probably with overlapping phases. An evolving virosphere is portrayed as having influenced the establishment of the major cellular domains of life. Two of several proposed cell differentiation routes into life domains are drawn. LGT (lateral gene transfer) are depicted as horizontal discontinuous lines, assumed to be more abundant during pre-cellular and early cellular evolution. *Illustration taken from Domingo, E., Perales, C., 2019. Virus evolution. In: Microbial Ecology and Evolution. Encyclopedia of Microbiology. fourth ed. Elsevier, Amsterdam), with permission.*

A reasonable hypothesis is that mechanisms similar to those observed in the present-day biosphere and virosphere were also in operation during the multiple transitions from a primitive replicon-based biosphere to a cellular world differentiated into different domains of life (Fig. 1.5). Although this point will again be examined as part of an overview in the closing chapter of the book (Chapter 10), here, some of the observations on the current biosphere dynamics that bear on the difficulties encountered in reconstructing the sequential transitions toward cellular differentiation are outlined. The following circumstances could have blurred the relationships among evolving previral and precellular entities and among viral genomes under construction, from the perspective of current genomics: (i) episodes of evolution at rates higher than average might historically have accelerated evolution and extinction of viral and cellular genetic elements. (ii) Experimental evolution has documented that viruses often have available multiple mutational pathways to gain fitness in changing environments. This could well apply to other genome assemblies at different stages of complexity. The genome assemblies and linkages we observe are a set among many others possible that could have existed for prolonged time periods but that are now extinct. A related aspect is that the acquisition of genomic modules from other cells or viral elements could render prior acquisitions obsolete or even detrimental. In consequence, an observed absence of shared genomic regions among different viral or cellular lineages from different life domains does not exclude that the shared domains had been transiently (transiently meaning million years!), with no trace in present-day viruses or cells. (iii) Bottleneck events could have erased biological organizations that had a similar degree of adaptation than others that survived. (iv) Fitness variation associated with genomic changes (due to mutation, recombination or lateral gene transfers) is environment-dependent, and the changes in the

environment over millions of years might have been dramatic. Such changes might have affected a limited number of local environments or most environments globally. Furthermore, most of these potentially perturbing events might have taken place many times in unpredictable successions, affecting local and few or general and many environments, thus rendering reconstruction of past events from current genomics problematic (Domingo and Perales, 2019). Again, these contingencies will be more apparent when different facets of viral dynamics are dissected in the following chapters.

## 1.10 Overview and concluding remarks

Contrary to the usual practice in experimental virology, the contents of this chapter have forced a considerable degree of uncertainty, and at times, speculative argumentation. It was, however, a necessary exercise, since at least part of what we see today in viruses must have roots in the origin of life and the role of viruses in life development during epochs that humans have not witnessed. We are left with trying to reconstruct. This chapter is excitingly as close to physics and chemistry as it is to biology. Many of the questions addressed are open to debate, and they will probably remain open for a long time. Interestingly, the terms "heterogeneity" and "complexity" have appeared several times in this chapter, anticipating features of present-day viruses discussed in the upcoming chapters. Complex populations, be they of viruses, cells, protocells, lipid vesicles, primitive replicons, or peptide soups, are the raw materials on which natural selection can act. Primitiveness does not imply simplicity. Complexity was likely a constant trait for early and modern life, both for the extinguished unknown viruses and for those we strive to understand and control (see Summary Box).

## Summary box

- Life is characterized by four integrated features: replication, evolvability, metabolism, and compartmentalization. It is debated whether replication or metabolism was the dominant triggering factor.
- At least two ancestral positive selection events might have contributed to the development of life: the selection of replicating over nonreplicating polymers, and selection of splitting-prone versus nonsplitting-prone membrane vesicles.
- Present-day viruses may be descendants from the primitive replicons that participated in early life prior to cellular organizations, or from structured cells by escape or reduction.
- Key features of many present-day viruses may be an inheritance of primitive replicons, notably error-prone replication, spread through membrane structures, and tendency to engage in genome integration and transfers.
- The geological record indicates multiple, brief, and drastic mass extinction events in Earth's history. Such events anticipate mass extinction of the resident viruses. A dynamics of emergences, reemergences, and extinctions might have operated historically at a grand-scale, with mechanisms similar to those we observe with present-day viruses. Present virosphere dynamics suggests difficulties to reconstruct events that spanned millions of years.

# References

Abrahao, J., Silva, L., Silva, L.S., Bou Khalil, J.Y., Rodrigues, R., et al., 2018. Tailed giant Tupanvirus possesses the most complete translational apparatus of the known virosphere. Nat. Commun. 9, 749.

Abrescia, N.G., Bamford, D.H., Grimes, J.M., Stuart, D.I., 2012. Structure unifies the viral universe. Annu. Rev. Biochem. 81, 795–822.

Adamala, K., Szostak, J.W., 2013. Nonenzymatic template-directed RNA synthesis inside model protocells. Science 342, 1098–1100.

Adami, C., 1998. Introduction to Artificial Life. Springer-Verlag Inc., New York.

Anthony, S.J., Epstein, J.H., Murray, K.A., Navarrete-Macias, I., Zambrana-Torrelio, C.M., et al., 2013. A strategy to estimate unknown viral diversity in mammals. mBio 4, e00598, 13.

Bacarese-Hamilton, T., Mezzasoma, L., Ardizzoni, A., Bistoni, F., Crisanti, A., 2004. Serodiagnosis of infectious diseases with antigen microarrays. J. Appl. Microbiol. 96, 10–17.

Bada, J.L., Lazcano, A., 2003. Perceptions of science. Prebiotic soup—revisiting the Miller experiment. Science 300, 745–746.

Baltimore, D., 1980. Evolution of RNA viruses. Ann. N. Y. Acad. Sci. 354, 492–497.

Banfield, W.G., Woke, P.A., Mackay, C.M., Cooper, H.L., 1965. Mosquito transmission of a reticulum cell sarcoma of hamsters. Science 148, 1239–1240.

Baranowski, E., Ruiz-Jarabo, C.M., Domingo, E., 2001. Evolution of cell recognition by viruses. Science 292, 1102–1105.

Barr, J.J., Auro, R., Furlan, M., Whiteson, K.L., Erb, M.L., et al., 2013. Bacteriophage adhering to mucus provide a non-host-derived immunity. Proc. Natl. Acad. Sci. U.S.A. 110, 10771–10776.

Bengtson, S., 1994. In: Early Life on Earth. Nobel Symposium No. 84. Columbia University Press, New York.

Bernal, J.D., 1951. The Physical Basis of Life. Routledge and Kegan Paul, London.

Bernstein, A.S., 2014. Biological diversity and public health. Annu. Rev. Public Health 35, 153–167.

Berzelius, J.J., 1806. Föreläsningar i Djurkemien, vol. 1–2. Tryckte hos Carl Delén, Stockholm.

Biondi, E., Branciamore, S., Fusi, L., Gago, S., Gallori, E., 2007. Catalytic activity of hammerhead ribozymes in a clay mineral environment: implications for the RNA world. Gene 389, 10–18.

Breitbart, M., Miyake, J.H., Rohwer, F., 2004. Global distribution of nearly identical phage-encoded DNA sequences. FEMS Microbiol. Lett. 236, 249–256.

Buckling, A., Rainey, P.B., 2002. Antagonistic coevolution between a bacterium and a bacteriophage. Proc. Biol. Sci. 269, 931–936.

Burgess, S.D., Bowring, S., Shen, S.Z., 2014. High-precision timeline for Earth's most severe extinction. Proc. Natl. Acad. Sci. U.S.A. 111, 3316–3321.

Bushman, F., 2002. Lateral DNA Transfer. Mechanisms and Consequences. Cold Spring Harbor Laboratory Press, Cold Spring Harbor, New York.

Bywater, R.P., 2012. On dating stages in prebiotic chemical evolution. Naturwissenschaften 99, 167–176.

Caetano-Anolles, G., Wang, M., Caetano-Anolles, D., 2013. Structural phylogenomics retrodicts the origin of the genetic code and uncovers the evolutionary impact of protein flexibility. PLoS One 8, e72225.

Caglioti, L., Micskei, K., Palyi, G., 2011. First molecules, biological chirality, origin(s) of life. Chirality 23, 65–68.

Cairns-Smith, A.G., 1965. The origin of life and the nature of the primitive gene. J. Theor. Biol. 10, 53–88.

Cantine, M.D., Fournier, G.P., 2018. Environmental adaptation from the origin of life to the last universal common ancestor. Orig. Life Evol. Biosph. 48, 35–54.

Canuto, V.M., Levine, J.S., Augustsson, T.R., Imhoff, C.L., 1982. UV radiation from the young Sun and oxygen and ozone levels in the prebiological palaeoatmosphere. Nature 296, 816–820.

Carrara, P., Stano, P., Luisi, P.L., 2012. Giant vesicles "colonies": a model for primitive cell communities. Chembiochem 13, 1497–1502.

Castón, J.R., Carrascosa, J.L., 2013. The basic architecture of viruses. In: Mateu, M.G. (Ed.), Structure and Physics of Viruses. Springer, Dordrecht, Heidelberg, New York, London, pp. 53–75.

Cello, J., Paul, A.V., Wimmer, E., 2002. Chemical synthesis of poliovirus cDNA: generation of infectious virus in the absence of natural template. Science 297, 1016–1018.

Chang, S., 1994. The planetary setting of prebiotic evolution. In: Bengtson, S. (Ed.), Early Life on Earth. Nobel Symposium No. 84. Columbia University Press, New York, pp. 10–23.

Chao, M., 2007. RNA recombination in hepatitis delta virus: implications regarding the abilities of mammalian RNA polymerases. Virus Res. 127, 208–215.

Chuong, E.B., 2018. The placenta goes viral: retroviruses control gene expression in pregnancy. PLoS Biol. 16, e3000028.

Ciccarelli, F.D., Doerks, T., von Mering, C., Creevey, C.J., Snel, B., et al., 2006. Toward automatic reconstruction of a highly resolved tree of life. Science 311, 1283–1287.

Colson, P., La Scola, B., Levasseur, A., Caetano-Anollés, G., Raoult, D., 2017. Mimivirus: leading the way in the Discovery of giant viruses of amoebae. Nat. Rev. Microbiol. 15, 243–254.

Cooper, G.M., Temin, R.G., Sugden, B.V., 1995. The DNA Provirus. Howard Temin's Scientific Legacy. ASM Press, Washington, DC.

Cowen, R., 2005. History of Life, fourth ed. Blackwell Publishing, Malden, USA.

Crick, F.H., 1968. The origin of the genetic code. J. Mol. Biol. 38, 367–379.

Crochu, S., Cook, S., Attoui, H., Charrel, R.N., De Chesse, R., et al., 2004. Sequences of flavivirus-related RNA viruses persist in DNA form integrated in the genome of Aedes spp. mosquitoes. J. Gen. Virol. 85, 1971–1980.

Danger, G., Plasson, R., Pascal, R., 2012. Pathways for the formation and evolution of peptides in prebiotic environments. Chem. Soc. Rev. 41, 5416–5429.

Davis, B.D., Dulbecco, R., Eisen, H.N., Ginsberg, H.S., Wood, W.B., 1968. Microbiology. Hoeber Medical Division, Harper and Row, New York.

de Duve, C., 2002. Life Evolving. Molecules, Mind and Meaning. Oxford University Press, Oxford.

Demetrius, L.A., 2013. Boltzmann, Darwin and directionality theory. Phys. Rep. 530, 1–85.

Derr, J., Manapat, M.L., Rajamani, S., Leu, K., Xulvi-Brunet, R., et al., 2011. Prebiotically plausible mechanisms increase compositional diversity of nucleic acid sequences. Nucleic Acids Res. 40, 4711–4722.

Diaz-Toledano, R., Ariza-Mateos, A., Birk, A., Martinez-Garcia, B., Gomez, J., 2009. In vitro characterization of a miR-122-sensitive double-helical switch element in the 5′ region of hepatitis C virus RNA. Nucleic Acids Res. 37, 5498–5510.

Dobson, C.M., Ellison, G.B., Tuck, A.F., Vaida, V., 2000. Atmospheric aerosols as prebiotic chemical reactors. Proc. Natl. Acad. Sci. U.S.A. 97, 11864–11868.

Domingo, E., Perales, C., 2014. Virus Evolution. Encyclopedia of Life Sciences. John Wiley & Sons, Ltd, Chichester, UK. https://doi.org/10.1002/9780470015902.a0000436.pub3.

Domingo, E., Perales, C., 2019. Virus evolution. In: Microbial Ecology and Evolution. Encyclopedia of Microbiology, fourth ed. Elsevier, Amsterdam.

Dworkin, J.P., Lazcano, A., Miller, S.L., 2003. The roads to and from the RNA world. J. Theor. Biol. 222, 127–134.

Ehrenfreund, P., Spaans, M., Holm, N.G., 2011. The evolution of organic matter in space. Philos. Transact. A Math. Phys. Eng. Sci. 369, 538–554.

Eigen, M., 1992. Steps Towards Life. Oxford University Press, Oxford.

Eigen, M., 1994. On the origin of biological information. Biophys. Chem. 50, 1.

Eigen, M., 2002. Error catastrophe and antiviral strategy. Proc. Natl. Acad. Sci. U.S.A. 99, 13374–13376.

Eigen, M., 2013. From Strange Simplicity to Complex Familiarity. Oxford University Press, Oxford.

Eigen, M., Schuster, P., 1979. The Hypercycle. A Principle of Natural Self-Organization. Springer, Berlin.

Epstein, J.H., Anthony, S.J., 2017. Viral discovery as a tool for pandemic preparedness. Rev. Sci. Tech. 36, 499–512.

Falk, R., Lazcano, A., 2012. The forgotten dispute: AI Oparin and HJ Muller on the origin of life. Hist. Philos. Life Sci. 34, 373–390.

Ferris, J.P., Hill Jr., A.R., Liu, R., Orgel, L.E., 1996. Synthesis of long prebiotic oligomers on mineral surfaces. Nature 381, 59–61.

Forterre, P., 2005. The two ages of the RNA world, and the transition to the DNA world: a story of viruses and cells. Biochimie 87, 793–803.

Forterre, P., 2006a. The origin of viruses and their possible roles in major evolutionary transitions. Virus Res. 117, 5–16.

Forterre, P., 2006b. Three RNA cells for ribosomal lineages and three DNA viruses to replicate their genomes: a hypothesis for the origin of cellular domain. Proc. Natl. Acad. Sci. U.S.A. 103, 3669–3674.

Fox, S.W., Dose, K., 1992. Molecular Evolution and the Origins of Life. W.H. Freeman, San Francisco.

García-Montojo, M., Doucet-O'Hare, T., Henderson, L., Nath, A., 2018. Human endogenous retroviruses–K (HML-2): a comprehensive review. Crit. Rev. Microbiol. https://doi.org/10.1080/1040841X.2018.1501345.

Gedulin, B., Arrhenius, G., 1994. Sources of geochemical evolution of RNA precursor molecules: the role of phosphate. In: Bengtson, S. (Ed.), Early Life on Earth. Nobel Symposium No. 84. Columbia University Press, New York, pp. 91–106.

Gesteland, R.F., Cech, T.R., Atkins, J.F., 2006. The RNA World. Cold Spring Harbor Laboratory Press, Cold Spring Harbor, New York.

Gibbs, A., Calisher, C., García-Arenal, F., 1995. Molecular Basis of Virus Evolution. Cambridge University Press, Cambridge.

Gilbert, W., 1986. The RNA world. Nature 319, 618.

Gleiser, M., Thorarinson, J., Walker, S.I., 2008. Punctuated chirality. Orig. Life Evol. Biosph. 38, 499–508.

Gorbalenya, A.E., 1995. Origin of RNA viral genomes; approaching the problem by comparative sequence analysis. In: Gibbs, A., Calisher, C.H., Garcia-Arenal, F. (Eds.), Molecular Basis of Virus Evolution. Cambridge University Press, Cambridge, pp. 49–66.

Greenwald, J., Kwiatkowski, W., Riek, R., 2018. J. Mol. Biol. 430, 3735–3750.

Griffith, E.C., Vaida, V., 2012. In situ observation of peptide bond formation at the water-air interface. Proc. Natl. Acad. Sci. U.S.A. 109, 15697–15701.

Hacker, J., Dobrindt, U., 2006. Pathogenomics: Genome Analysis of Pathogenic Microbes. Wiley-VCH Verlag GmbH&Co KGaA, Weinheim.

Hadidi, A., Flores, R., Randles, J., Palukaitis, P. (Eds.), 2017. Viroids and Satellites. Academic Press, Boston.

Hanczyc, M.M., Monnard, P.A., 2017. Primordial membranes: more than simple container boundaries. Curr. Opin. Chem. Biol. 40, 78–86.

Hendrix, R.W., 2008. Evolution of dsDNA tailed phages. In: Domingo, E., Parrish, C., Holland, J.J. (Eds.), Origin and Evolution of Viruses, second ed. Elsevier, Oxford, pp. 219–228.

Holland, J.J., 2006. Transitions in understanding of RNA viruses: an historical perspective. Curr. Top. Microbiol. Immunol. 299, 371–401.

Holland, J., Domingo, E., 1998. Origin and evolution of viruses. Virus Genes 16, 13–21.

Horneck, G., Baumstark-Khan, C., 2002. Astrobiology, the Quest for the Conditions of Life. Springer, Berlin.

Huotari, J., Helenius, A., 2011. Endosome maturation. EMBO J. 30, 3481–3500.

Inoue, T., Orgel, L.E., 1983. A nonenzymatic RNA polymerase model. Science 219, 859–862.

Jalasvuori, M., Bamford, J.K., 2008. Structural co-evolution of viruses and cells in the primordial world. Orig. Life Evol. Biosph. 38, 165–181.

Jeffares, D.C., Poole, A.M., Penny, D., 1998. Relics from the RNA world. J. Mol. Evol. 46, 18–36.

Joklik, W.K., 1974. Evolution in viruses. In: Carlile, M.M., Skehel, J.J. (Eds.), Evolution in the Microbial World. Cambridge University Press, Cambridge, pp. 293–320.

Joshi, P.C., Aldersley, M.F., Price, J.D., Zagorevski, D.V., Ferris, J.P., 2011. Progress in studies on the RNA world. Orig. Life Evol. Biosph. 41, 575–579.

Joyce, G.F., 2004. Directed evolution of nucleic acid enzymes. Annu. Rev. Biochem. 73, 791–836.

Katzourakis, A., Tristem, M., Pybus, O.G., Gifford, R.J., 2007. Discovery and analysis of the first endogenous lentivirus. Proc. Natl. Acad. Sci. U.S.A. 104, 6261–6265.

Kauffman, S.A., 1993. The Origins of Order. Self-Organization and Selection in Evolution. Oxford University Press, New York, Oxford.

Khatchikian, D., Orlich, M., Rott, R., 1989. Increased viral pathogenicity after insertion of a 28S ribosomal RNA sequence into the haemagglutinin gene of an influenza virus. Nature 340, 156–157.

Koonin, E.V., Senkevich, T.G., Dolja, V.V., 2006. The ancient Virus World and evolution of cells. Biol. Direct 1, 29.

Koonin, E.V., 2009. On the origin of cells and viruses: primordial virus world scenario. Ann. N. Y. Acad. Sci. 1178, 47–64.

Koonin, E.V., Dolja, V.V., 2013. A virocentric perspective on the evolution of life. Curr. Opin. Virol. 3, 546–557.

Koonin, E.V., Novozhilov, A.S., 2017. Origin and evolution of the universal genetic code. Annu. Rev. Genetics 51, 45–62.

Krupovic, M., Bamford, D.H., 2007. Putative prophages related to lytic tailless marine dsDNA phage PM2 are widespread in the genomes of aquatic bacteria. BMC Genomics 8, 236.

La Scola, B., Desnues, C., Pagnier, I., Robert, C., Barrassi, L., et al., 2008. The virophage as a unique parasite of the giant mimivirus. Nature 455, 100–104.

Lancet, D., Zidovetzki, R., Markovitch, O., 2018. Systems protobiology: origin of life in lipid catalytic networks. J. R. Soc. Interface 15, 20180159.

Laybourn-Parry, J., Pearce, D.A., 2007. The biodiversity and ecology of Antarctic lakes: models for evolution. Philos. Trans. R. Soc. Lond. B Biol. Sci. 362, 2273–2289.

Lazcano, A., 1994a. The RNA world, its predecessors, and its descendants. In: Bengtson, S. (Ed.), Early Life on Earth. Nobel Symposium No. 84. Columbia University Press, New York, pp. 70–80.

Lazcano, A., 1994b. The transition form nonliving to living. In: Bengtson, S. (Ed.), Early Life on Earth. Nobel Symposium No. 84. Columbia University Press, New York, pp. 60–69.

Lazcano, A., 2010. Historical development of origins research. Cold Spring Harb. Perspect. Biol. 2 (11), a002089.

Lazcano, A., Valverde, V., Hernandez, G., Gariglio, P., Fox, G.E., et al., 1992. On the early emergence of reverse transcription: theoretical basis and experimental evidence. J. Mol. Evol. 35, 524–536.

Li, L., Delwart, E., 2011. From orphan virus to pathogen: the path to the clinical lab. Curr. Opin. Virol. 1, 282–288.

Lopez-Bueno, A., Tamames, J., Velazquez, D., Moya, A., Quesada, A., et al., 2009. High diversity of the viral community from an Antarctic lake. Science 326, 858–861.

Luria, S.E., Darnell Jr., J.E., Baltimore, D., Campbell, A., 1978. General Virology, second ed. John Wiley & Sons, New York.

Lwoff, A., 1957. The concept of virus. J. Gen. Microbiol. 17, 239–253.

Mallet, F., Bouton, O., Prudhomme, S., Cheynet, V., Oriol, G., et al., 2004. The endogenous retroviral locus ERVWE1 is a bona fide gene involved in hominoid placental physiology. Proc. Natl. Acad. Sci. U.S.A. 101, 1731–1736.

Marquez, L.M., Redman, R.S., Rodriguez, R.J., Roossinck, M.J., 2007. A virus in a fungus in a plant: three-way symbiosis required for thermal tolerance. Science 315, 513–515.

Mateu, M.G., 2013. Structure and Physics of Viruses. Springer, Dordrecht, Heidelberg, New York, London.

Maynard Smith, J., Szathmáry, E., 1999. The Origins of Life. From the Birth of Life to the Origins of Language. Oxford University Press, Oxford.

Meyers, G., Rumenapf, T., Thiel, H.J., 1989. Ubiquitin in a togavirus. Nature 341, 491.

Miller, S.L., 1953. A production of amino acids under possible primitive Earth conditions. Science 117, 528–529.

Miller, S.L., 1987. Which organic compounds could have occurred on the prebiotic Earth? Cold Spring Harb. Symp. Quant. Biol. 52, 17–27.

Miller, S.L., Orgel, L.E., 1974. The Origins of Life on Earth. Prentice-Hall, Englewood Cliffs, New Jersey.

Miyakawa, S., Yamanashi, H., Kobayashi, K., Cleaves, H.J., Miller, S.L., 2002. Prebiotic synthesis from CO atmospheres: implications for the origins of life. Proc. Natl. Acad. Sci. U.S.A. 99, 14628–14631.

Monroe, S.S., Schlesinger, S., 1983. RNAs from two independently isolated defective interfering particles of Sindbis virus contain a cellular tRNA sequence at their 5' ends. Proc. Natl. Acad. Sci. U.S.A. 80, 3279–3283.

Morowitz, H.J., 1992. Beginnings of Cellular Life: Metabolism Recapitulates Biogenesis. Yale University Press, New Haven and London.

Mount, D.W., 2004. Bioinformatics Sequence and Genome Analysis. Cold Spring Harbor Laboratory Press, Cold Spring Harbor, New York.

Murgia, C., Pritchard, J.K., Kim, S.Y., Fassati, A., Weiss, R.A., 2006. Clonal origin and evolution of a transmissible cancer. Cell 126, 477–487.

Nakashima, S., Kebukawa, Y., Kitadai, N., Igisu, M., Matsuoka, N., 2018. Geochemistry and the origin of life: from extraterrestrial processes, chemical evolution on Earth, fossilized life's records, to natures of the extant life. Life 8, 39.

Nasir, A., Kim, K.M., Caetano-Anolles, G., 2012. Giant viruses coexisted with the cellular ancestors and represent a distinct supergroup along with superkingdoms Archaea. Bacteria and Eukarya. BMC Evol. Biol. 12, 156.

Oldstone, M.B.A., 1984. Virus can alter cell function without causing cell pathology: disordered function leads to imbalance of homeostasis and disease. In: Notkins, A.L., Oldstone, M.B.A. (Eds.), Concepts in Viral Pathogenesis. Springer-Verlag, New York, pp. 269–276.

Oldstone, M.B., Holmstoen, J., Welsh Jr., R.M., 1977. Alterations of acetylcholine enzymes in neuroblastoma cells persistently infected with lymphocytic choriomeningitis virus. J. Cell. Physiol. 91, 459–472.

Oparin, A.I., 1938. Origin of Life. Dover Publications, New York. English version 1953.

Orgel, L.E., 2002. Is cyanoacetylene prebiotic? Orig. Life Evol. Biosph. 32, 279–281.

Orgel, L.E., 2004. Prebiotic chemistry and the origin of the RNA world. Crit. Rev. Biochem. Mol. Biol. 39, 99–123.

Parker, E.T., Cleaves, H.J., Callahan, M.P., Dworkin, J.P., Glavin, D.P., et al., 2011. Prebiotic synthesis of methionine and other sulfur-containing organic compounds on the primitive Earth: a contemporary reassessment based on an unpublished 1958 Stanley Miller experiment. Orig. Life Evol. Biosph. 41, 201–212.

Pasteur, L., 1861. Memoire sur les corpuscules organisés qui existent dans l'atmosphere. Examen de la doctrine des génerations spontanés. Ann. Sci. Nat. 16, 5–98, 4esue series.

Pearse, A.M., Swift, K., 2006. Allograft theory: transmission of devil facial-tumour disease. Nature 439, 549.

Pedreira-Segade, U., Michot, L.J., Daniel, I., 2018. Effects of salinity on the adsorption of nucleotides onto phyllosilicates. Phys. Chem. Chem. Phys. 20, 1938–1952.

Pinheiro, V.B., Taylor, A.I., Cozens, C., Abramov, M., Renders, M., et al., 2012. Synthetic genetic polymers capable of heredity and evolution. Science 336, 341–344.

Ravantti, J., Bamford, D., Stuart, D.I., 2013. Automatic comparison and classification of protein structures. J. Struct. Biol. 183, 47–56.

Ritson, D., Sutherland, J.D., 2012. Prebiotic synthesis of simple sugars by photoredox systems chemistry. Nat. Chem. 4, 895–899.

Robertson, H.D., 1992. Replication and evolution of viroid-like pathogens. Curr. Top. Microbiol. Immunol. 176, 213–219.

Robertson, H.D., 1996. How did replicating and coding RNAs first get together? Science 274, 66–67.

Robertson, M.P., Joyce, G.F., 2012. The origins of the RNA world. Cold Spring Harb. Perspect. Biol. 4 (5), a003608.

Robertson, B.H., Jansen, R.W., Khanna, B., Totsuka, A., Nainan, O.V., et al., 1992. Genetic relatedness of hepatitis A virus strains recovered from different geographical regions. J. Gen. Virol. 73, 1365–1377.

Rodin, A.S., Szathmary, E., Rodin, S.N., 2011. On origin of genetic code and tRNA before translation. Biol. Direct 6, 14.

Rohlfing, D.L., Oparin, A.I., 1972. Molecular Evolution: Prebiological and Biological. Plenum Press, New York and London.

Roossinck, M.J., 2011. The good viruses: viral mutualistic symbioses. Nat. Rev. Microbiol. 9, 99–108.

Roossinck, M.J., 2013. Plant virus ecology. PLoS Pathog. 9, e1003304.

Roossinck, M.J., 2019. Evolutionary and ecological links between plant and fungal viruses. New Phytol. 221, 86–92.

Ruiz-Mirazo, K., Briones, C., de la Escosura, A., 2014. Prebiotic systems chemistry: new perspectives for the origins of life. Chem. Rev. 114, 285–366.

Ryan, F.P., 2004. Human endogenous retroviruses in health and disease: a symbiotic perspective. J. R. Soc. Med. 97, 560–565.

Santoni, F.A., Guerra, J., Luban, J., 2012. HERV-H RNA is abundant in human embryonic stem cells and a precise marker for pluripotency. Retrovirology 9, 111.

Saren, A.M., Ravantti, J.J., Benson, S.D., Burnett, R.M., Paulin, L., et al., 2005. A snapshot of viral evolution from genome analysis of the tectiviridae family. J. Mol. Biol. 350, 427–440.

Schrödinger, E., 1944. What is Life? Cambridge University Press, Cambridge.

Schwartz, A.W., Orgel, L.E., 1985. Template-directed synthesis of novel, nucleic acid-like structures. Science 228, 585–587.

Sczepanski, J.T., Joyce, G.F., 2014. A cross-chiral RNA polymerase ribozyme. Nature 515, 440–442.

Selander, R.K., 1976. Genetic variation in natural populations. In: Ayala, F.J. (Ed.), Molecular Evolution. Sinauer Associates Inc., Sunderland, Massachusetts, pp. 21–45.

Shabalina, S.A., Ogurtsov, A.Y., Spiridonov, N.A., 2006. A periodic pattern of mRNA secondary structure created by the genetic code. Nucleic Acids Res. 34, 2428–2437.

Sharp, P.A., 1985. On the origin of RNA splicing and introns. Cell 42, 397–400.

Son, M., Yu, J., Kim, K.H., 2015. Five questions about mycoviruses. PLoS Pathog. 11, e1005172.

Stano, P., D'Aguanno, E., Bolz, J., Fahr, A., Luisi, P.L., 2013. A remarkable self-organization process as the origin of primitive functional cells. Angew. Chem. Int. Ed. Engl. 52, 13397–13400.

Suh, A., Weber, C.C., Kehlmaier, C., Braun, E.L., Green, R.E., et al., 2014. Early mesozoic coexistence of amniotes and hepadnaviridae. PLoS Genet. 10, e1004559.

Suttle, C.A., 2007. Marine viruses—major players in the global ecosystem. Nat. Rev. Microbiol. 5, 801–812.

Szathmary, E., 1999. The origin of the genetic code: amino acids as cofactors in an RNA world. Trends Genet. 15, 223–229.

Szostak, J.W., 2011. An optimal degree of physical and chemical heterogeneity for the origin of life? Philos. Trans. R. Soc. Lond. B Biol. Sci. 366, 2894–2901.

Taylor, J., Pelchat, M., 2010. Origin of hepatitis delta virus. Future Microbiol. 5, 393–402.

Tinsley, C.R., Bille, E., Nassif, X., 2006. Bacteriophages and pathogenicity: more than just providing a toxin? Microb. Infect. 8, 1365–1371.

Tkachenko, A.V., Maslov, S., 2018. Onset of natural selection in populations of autocatalytic heteropolymers. J. Chem. Phys. 149, 134901.

Towe, K.M., 1994. Earth's early atmosphere: constraints and opportunities for early evolution. In: Bengton, S. (Ed.), Early Life on Earth. Nobel Symposium No. 84. Columbia University press, New York, pp. 36–47.

Trail, D., Watson, E.B., Tailby, N.D., 2011. The oxidation state of Hadean magmas and implications for early Earth's atmosphere. Nature 480, 79–82.

Trevino, S.G., Zhang, N., Elenko, M.P., Luptak, A., Szostak, J.W., 2011. Evolution of functional nucleic acids in the presence of nonheritable backbone heterogeneity. Proc. Natl. Acad. Sci. U.S.A. 108, 13492–13497.

van Rij, R.P., Andino, R., 2008. The complex interactions of viruses and the RNAi machinery: a driving force in viral evolution. In: Domingo, E., Parrish, C.R., Holland, J.J. (Eds.), Origin and Evolution of Viruses, second ed. Elsevier, Oxford, pp. 161–182.

Villarreal, L.P., 2005. Viruses and the Evolution of Life. ASM Press, Washington, DC.

Villarreal, L.P., 2008. The widespread evolutionary significance of viruses. In: Domingo, E., Parrish, C.R., Holland, J.J. (Eds.), Origin and Evolution of Viruses, second ed. Elsevier, Oxford, pp. 477–516.

Vorburger, C., Perlman, S.J., 2018. The role of defensive symbionts in host-parasite coevolution. Biol. Rev. Camb. Philos. Soc. 93, 1747–1764.

Wächtershäuser, G., 1988. Before enzymes and templates: theory of surface metabolism. Microbiol. Rev. 52, 452–484.

Wächtershäuser, G., 1994. Vitalists and virulists: a theory of self-expanding reproduction. In: Bengtson, S. (Ed.), Early Life on Earth. Columbia University Press, New York, pp. 124–132.

Weiss, M.C., Preiner, M., Xavier, J.C., Zimorski, V., Martin, W.F., 2018. The last universal common ancestor between ancient Earth chemistry and the onset of genetics. PLoS Genet. 14, e1007518.

Whon, T.W., Kim, M.S., Roh, S.W., Shin, N.R., Lee, H.W., et al., 2012. Metagenomic characterization of airborne viral DNA diversity in the near-surface atmosphere. J. Virol. 86, 8221–8231.

Witzany, G., 2012. From molecular entities to competent agents: viral-infection-derived consortia act as natural genetic engineers. In: Witzany, G. (Ed.), Viruses: Essential Agents of Life. Springer + Science Business Media, Dordrecht, pp. 407–419.

Wochner, A., Attwater, J., Coulson, A., Holliger, P., 2011. Ribozyme-catalyzed transcription of an active ribozyme. Science 332, 209–212.

Woese, C.R., 2002. On the evolution of cells. Proc. Natl. Acad. Sci. U.S.A. 99, 8742–8747.

Xu, P., Chen, F., Mannas, J.P., Feldman, T., Sumner, L.W., et al., 2008. Virus infection improves drought tolerance. New Phytol. 180, 911–921.

Yan, M., Wang, L.C., Hymowitz, S.G., Schilbach, S., Lee, J., et al., 2000. Two-amino acid molecular switch in an epithelial morphogen that regulates binding to two distinct receptors. Science 290, 523–527.

Yarus, M., 2010. Life from an RNA world. The Ancestor Within. Harvard University Press, Cambridge, Massachusetts and London, England.

# Molecular basis of genetic variation of viruses: error-prone replication

## Abbreviations

**A** adenine
**ADAR** adenosine deaminase acting on double-stranded RNA
**AIDS** acquired immune deficiency syndrome
**AMV-RT** avian myeloblastosis virus reverse transcriptase
**APOBEC** apolipoprotein B mRNA editing complex
**BHK** baby hamster kidney cells
**bp** base pairs
**C** cytosine
**cDNA** complementary DNA
**CVB3** coxsackievirus B3
**DI** defective interfering
**DI RNAs** defective interfering RNAs
*E. coli* Escherichia coli
**FMDV** foot-and-mouth disease virus
**G** guanine
**HBV** hepatitis B virus
**HCV** hepatitis C virus
**HIV-1** human immunodeficiency virus type 1
**HIV-RT** human immunodeficiency virus reverse transcriptase
**HPV** human papillomavirus
**IFN** interferon
**I** inosine
**IRES** internal ribosome entry site
**Kb** kilobase
**miRNA** micro-RNA
**mRNA** messenger RNA
**NTP-dependent** nucleoside triphosphate-dependent
**OPV** oral poliovirus vaccine
**PCR** polymerase chain reaction
**PV** poliovirus
**RdRp** RNA-dependent RNA polymerase
**rRNA** ribosomal RNA
**RT** reverse transcriptase
**siRNA** small interfering RNA
**T** thymine
**U** uracil
**VSV** vesicular stomatitis virus

## 2.1 Universal need for genetic variation

Genetic change was a prerequisite for the early life forms to be generated and maintained (Chapter 1), and it is also a requirement for the evolution of present-day life. We may willingly or inadvertently modify selective pressures, but genetic change is rooted in all replication machineries. The results of genetic modifications, regarding relative dominances of variant forms, are guided by selective pressures and random events. The replicative machinery itself has probably been influenced by natural selection; as an example, polymerases devoid of a capacity to generate variants should have endured a long-term selective disadvantage. However, once the replicative machinery was established, the mechanisms of variation acted independently of the selective pressures applied or to come.

Viruses use the same molecular mechanisms of genetic variation than other forms of life: mutation (that encompasses point mutations and insertions-deletions of different lengths), hypermutation, several types of recombination, and genome segment reassortment. Mutation is observed in all viruses, with no known exceptions. Recombination is also widespread, but its role in the generation of diversity appears to vary among viruses. Its occurrence was soon accepted for DNA viruses, but it was considered uncertain for the RNA viruses. Pioneering studies of poliovirus (PV) by P. D. Cooper, V. I. Agol and colleagues, and of foot-and-mouth disease virus (FMDV) by A.M. King and colleagues provided the first evidence of recombination in RNA. The present perception is that recombination is more widespread than thought only a few decades ago and that its frequency and the types of genomic forms it generates are varied among viruses. For example, it appears that positive-strand RNA viruses recombine more easily than negative-strand RNA viruses to give rise to mosaic genomes of standard length. Several negative-strand RNA viruses, however, can yield defective genomes through recombination, frequently characterized by deletions in their RNA. A connection between the structure of replication complexes—as viewed by X-ray diffraction or high-resolution cryo-electron microscopy—and the propensity to produce defective genomes has not been established. Defective genomes are increasingly perceived not only as unavoidable side-products of blind replicative imperfections but as classes of genome subpopulations that perform relevant biological roles for the standard, infectious viruses. Genome segment reassortment, a type of variation close to chromosomal exchanges in sexual reproduction, is an adaptive asset of segmented viral genomes, as continuously evidenced by the ongoing evolution of the influenza viruses. The three modes of virus genome variation are not incompatible, and reassortant-recombinant-mutant genomes are continuously

arising in present-day viruses. The potential for genetic variation of RNA and DNA viral genomes is remarkable, and it is the ultimate molecular mechanism that lies at the origin of the virus diversity delineated in Chapter 1.

## 2.2 Molecular basis of mutation

Mutation is a localized alteration of a nucleotide residue in a nucleic acid. It generally refers to an inheritable modification of the genetic material. In the case of viral genomes, mutations can result from different mechanisms: (i) template miscopying (direct incorporation of an incorrect nucleotide); (ii) primer-template misalignments that include miscoding followed by realignment, and misalignment of the template relative to the growing chain (polymerase "slippage" or "stuttering"); (iii) activity of cellular enzymes (i.e., deaminases), or (iv) chemical damage to the viral nucleic acids (deamination, depurination, depyrimidination, reactions with oxygen radicals, direct and indirect effects of ionizing radiation, photochemical reactions, etc.) (Naegeli, 1997; Bloomfield et al., 2000; Friedberg et al., 2006).

The basis of nucleotide misincorporation during template copying (defined as the incorporation of a nucleotide different from that expected from the template residue at that position) lies mainly in the electronic structure of the bases that make up DNA (adenine, A; guanine, G; cytosine, C; thymine, T) or RNA (with uracil, U instead of T). Each base includes potential hydrogen-bonding donor sites (amino or amino protons) and hydrogen-bonding acceptor sites (carbonyl oxygens or aromatic nitrogens) that contribute to standard Watson-Crick base pairs (Fig. 2.1), as well as wobble base pairs (nonstandard Watson-Crick, but fundamental for RNA secondary structure and mRNA translation) (Fig. 2.2). The conformation of the purine and pyrimidine bases is highly dynamic. Amino and methyl groups rotate about the bonds that link them to the ring structure. In dilute solution,

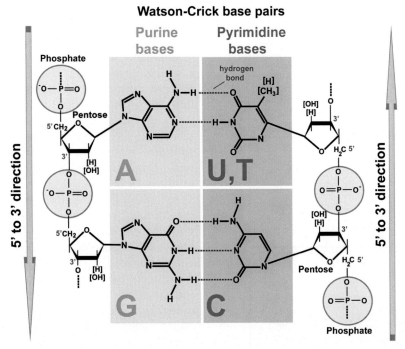

**FIGURE 2.1** The standard Watson-Crick base pairs in DNA (A-T and G-C, with deoxyribose as pentose) and RNA (A-U and G-C, with ribose as pentose). Phosphodiester bonds of two potential polynucleotide chains of different polarity (outer arrows) are indicated.

hydrogen bonds are established with water, and they can be displaced by nucleotide or amino acid residues to give rise to nucleotide-nucleotide or nucleotide-amino acid interactions. The strength difference between hydrogen bonds established in a polynucleotide chain with water, and their strength between two bases in separate polynucleotide chains determines whether a double-stranded polynucleotide will be formed.

Purine and pyrimidine bases can acquire different charge distributions and ionization states. As a consequence, in addition to the standard Watson-Crick and wobble, other base pairs are found in naturally occurring nucleic acids (notably cellular rRNA and tRNA) and in synthetic oligonucleotides (A-U or A-T Hoogsteen, and A-G, C-U, G-G, and U-U pairs, as well as interactions involving ionized bases). One of the types of electronic redistribution leads to

tautomeric changes, such as the keto-enol and amino-imino transitions, which modify the hydrogen-bonding properties of the base; tautomeric imino and enol forms of the standard bases can produce non-Watson-Crick pairs. The proportion of the alternative tautomeric forms can be influenced by modifications in the purine and pyrimidine rings, which, in turn, can favor either the *syn* or *anti* conformation of a nucleoside, which is defined by the torsion angle of the bond between the 1' carbon of the ribose and either N1 in pyrimidines or N9 in purines (Fig. 2.1). The *anti*-conformation is usually the most stable in standard nucleotides and polynucleotides. The transition from the *anti* to the *syn* conformation may alter the hydrogen-bonding properties of the base, thereby inducing mutagenesis (Bloomfield et al., 2000; Suzuki et al., 2006). The understanding of conformational

## Wobble base pairs

FIGURE 2.2    Examples of a class of non-Watson-Crick base pairs termed wobble base pairs. The drawing is similar to that of Fig. 2.1, except that the sugar residues and phosphodiester bonds have been omitted. Hydrogen bonds (*discontinuous lines in red*) are shown between I (inosine) and C, U, and A, and between G and U. Wobble base pairs are important for codon-anticodon interactions, as described in the text.

Base-base interactions are not only responsible for part of the mutations that occur during genome replication, but also for the formation of double-stranded nucleic acids, either within the same polynucleotide chain or between two different chains. Transitions from a coil-like into an organized double-stranded (or other) structure are functionally relevant for both RNA and DNA. In the case of RNA, double-stranded regions in the adequate alternation with single-stranded regions, determine key catalytic or macromolecule-attracting abilities, as for example, ribozyme activities (Chapter 1), the internal ribosome entry site (IRES) of several viral and cellular mRNAs, or multitudes of functional RNA-protein interactions (Denny and Greenleaf, 2018).

Adjacent base stacking due to electronic interactions (rather than hydrophobic bonds as once thought), contribute also to the stability of double-helical regions in nucleic acid molecules. Structural transitions due to alternative stacking conformations, particularly within polypurine or polypyrimidine tracts, can affect nucleic acid-protein interactions. In turn, replication machineries (typically including viral and host proteins gathered in membrane structures) may also be affected by nucleic acid conformations; such effects are important in virology regarding consequences for mutant generation in a given template sequence context. These considerations on structural transitions are relevant to the nonneutral character of silent (also termed synonymous) mutations (those in open-reading frames that do not result in an amino acid substitution), a point to be addressed in the next section. Transitions from a single-stranded into a double-stranded nucleic acid structure and the relative stability of the two forms depend on multiple factors that include the nucleotide sequence of the nucleic acid, its being a ribo- or a deoxyribose-polynucleotide, temperature, ionic environment, and ionic strength. Positively charged counterions neutralize negatively charged phosphates, and favor duplex stability [as an overview of

effects on the base-pairing tendencies of nucleoside analogs in the context of the active site of a polymerase is very relevant to the design of specific mutagenic analogs for viral polymerases in lethal mutagenesis-centered antiviral approaches (Chapter 9).

physical and chemical properties of nucleic acids and their nucleotide components, see (Bloomfield et al., 2000)].

## 2.3 Types and effects of mutations

Mutations resulting from any of the mechanism just summarized can be divided into transitions, transversions (both referred to as point mutations), and insertions and deletions (referred to as *indels*) (Fig. 2.3). The latter occurs preferentially at homopolymeric tracts and also at short, repeated, sequences which are prone to misalignment mutagenesis (Fig. 2.4). An example is an editing mechanism for some viral mRNAs, such as the phosphoprotein mRNA of the *Paramyxovirinae* [(Kolakofsky et al., 2005) and references therein]. Other examples in vivo are hot spots for variation in reiterated sequences in complex DNA genomes (Yamaguchi et al., 1998; Barrett and McFadden, 2008; McGeoch et al., 2008), or the insertion of two amino acids (often Ser-Ser, Ser-Gly, or Ser-Ala between residues 69 and 70 of the HIV-1 reverse transcriptase), in concert with HIV-1 resistance to nucleoside inhibitors (Winters and Merigan, 2005) (Chapter 8). Hairpin structures in RNA and DNA may also induce deletions as a result of slippage mutagenesis (Pathak and Temin, 1992; Viguera et al., 2001). Transition mutations occur more frequently than either transversions or *indels* during virus replication. Nucleotide discrimination at the catalytic site of viral polymerases fits this observation because of the more likely replacement of a purine or pyrimidine nucleotide by its structurally more similar nucleotide. In some cases, however, an abundance of *indels* and similar numbers of transitions and transversions have been recorded (Cheynier et al., 2001; Malpica et al., 2002). The molecular bases of such unexpected behavior regarding mutational spectra are not well understood.

The generation of point mutations and *indels* is subject to thermodynamic and quantum-mechanical uncertainties inherent to atomic fluctuations, rendering mutagenesis a highly unpredictable event, thus introducing stochasticity (randomness) in a key motor of evolution: the generation of diversity at the molecular level (Domingo et al., 1995; Eigen, 2013).

**FIGURE 2.3** Major types of mutations in RNA (U) and DNA (T): four transitions and eight transversions. Below, a means to indicate point mutations, insertion or deletions (known as *indels*). A genome is depicted as an elongated rod. Symbols on the rod (*cross, circle, and line*) represent mutations. Hypermutation is generally associated with a high frequency of specific mutation types (*crosses and lines*). A region inserted or deleted from the genome is depicted as an empty rod.

**FIGURE 2.4** Misalignment mutagenesis. (A) Production of one or two nucleotide deletion or insertion at a homopolymeric tract in an RNA template. The basis of these events is a displacement of template or product residues during RNA synthesis. (B) A specific example of an internal oligoadenylated extension in the genome of FMDV subjected to plaque-to-plaque transfers. See text for biological implications and references.

The effect of mutations on the structure and function of proteins is extremely relevant to penetrate into the mechanisms that drive virus evolution since selection acts on phenotypes that are often embodied in protein molecules. Silent or synonymous mutations are those that do not give rise to an amino acid substitution despite being located in a protein-coding region of a genome. Their occurrence is due to the degeneracy of the genetic code: the same amino acid can be coded for by two or more triplets (codons), with the exception of AUG for methionine and UGG for tryptophan. Synonymous mutations are not necessarily selectively neutral, neutral meaning that they have no discernible consequence for any viral function. The assumption that synonymous mutations are selectively neutral, and the fact that the early

comparison of nucleotide sequences of homologous genes showed a dominance of synonymous over nonsynonymous mutations, contributed to the foundations of the neutral theory of molecular evolution. This theory attributes the evolution of organisms at the molecular level mainly to the random drift of genomes carrying neutral or quasi- (or nearly-) neutral mutations (King and Jukes, 1969; Kimura, 1983, 1989). The terms quasi-neutral or nearly-neutral may seem ambiguous to molecular biologists. However, in the formulation of the neutral theory they had a precise meaning of the selection coefficient (a parameter that measures fitness differences) being lower than the inverse of the effective population size, with minor variations in the equations of some formulations (Kimura, 1983).

Despite random drift of genomes playing an important role in molecular evolution, evidence gathered over the last decades renders untenable the assumption that synonymous mutations are neutral. Evidence to the contrary has been obtained with viruses and cells, including mutations in the human genome that may affect enhancer functions (Hirsch and Birnbaum, 2015), mRNA folding (Faure et al., 2017; Mittal et al., 2018), and microRNA targeting (Brest et al., 2011), among other processes (Novella, 2004; Novella et al., 2004; Parmley et al., 2006; Hamano et al., 2007; Resch et al., 2007; Lafforgue et al., 2011; Nevot et al., 2011, 2018; Supek, 2016). There are several mechanisms by which synonymous mutations can affect virus behavior: alteration of cis-acting regulatory elements in viral genomes, decrease of the stability of duplex structures within the RNA genome or between viral sequences and miRNAs or siRNAs, or changes of viral gene expression [splicing precision or translation fidelity through the modification of RNA-RNA or RNA-protein interactions; reviewed in (Martínez et al., 2016)].

Synonymous codons use different tRNAs for protein synthesis, and different tRNAs do not have the same relative abundance in different host cell types. Thus, the rate of protein synthesis, an important phenotypic trait for cells and viruses, can be affected by the frequency of alternative synonymous codons present in mRNAs (Richmond, 1970; Akashi, 2001). Not only codon bias, but also specific codons or codon combinations may affect ribosome speed to regulate the folding of nascent proteins during translation (Makhoul and Trifonov, 2002; Rocha, 2004; Aragones et al., 2010; Brule and Grayhack, 2017). As a consequence, generation of rare codons by mutation of abundant codons (or vice versa) can modify viral fitness (Chapter 5). Rare codons may also limit the fidelity of amino acid incorporation when the frequency of the required aminoacyl-tRNAS is low (Ling et al., 2009; Zaher and Green, 2009; Czech et al., 2010). The frequency of codon pairs in RNA genomes is also a fitness determinant relevant to the preparation of attenuated viral vaccines.

To complicate matters further, a synonymous mutation may be neutral or quasi-neutral in one environment, but it may contribute to selection in a different environment, because of the phenotypic effects of RNA structure and codon usage. Neutrality is relative to the environment.

Regarding the effects of mutations (Box 2.1), the following general statements are applicable to viruses:

- Although difficult to prove due to the limited number of environments used for experimentation, truly neutral mutations (i.e., with no influence on the virus in any environment) are probably very rare. This applies to synonymous, as well as to nonsynonymous mutations.
- Mutations resulting in chemically conservative amino acid substitutions are more likely to be tolerated than those leading to chemically different amino acids. Tolerability (quantified by substitution matrices among amino acids in protein evolution) should be distinguished from neutrality. A tolerated mutation may cause a reduction in fitness, which is nevertheless compatible with virus replication.
- A conservative amino acid substitution may have important biological consequences.
- The effect of any individual mutation is context-dependent in two ways: it may depend on other mutations in the same genome (epistasis, see also Section 2.8 and Chapter 5) or on the mutant cloud that surrounds the genome harboring the mutation (effects of complementation, cooperation, or interference, discussed in Section 3.8 of Chapter 3).
- The previous points do not deny the influence of random drift of genomes on intrahost and interhost evolution. The currently most accepted view is that positive and negative selection and random drift occur continuously during virus evolution (Chapter 3).

---

BOX 2.1
———————

### The effects of mutations on viruses

### In noncoding regions

Mutations may affect stem-loop or other secondary and higher-order structures involved in regulatory processes through nucleic acid-nucleic acid or nucleic acid-protein interactions. The primary sequence in nonstructured, noncoding regions may also be functionally relevant.

### In coding regions

- Synonymous or silent mutations do not affect the amino acid sequence of the encoded protein.
- Nonsynonymous mutations give rise to an amino acid substitution in the encoded protein.
- Some mutations may generate a stop codon, leading to a truncated protein.

### Regarding functional effects

- Neutral mutations are those that have no functional effects.
- Nonneutral mutations can have a broad range of fitness effects: from nearly complete tolerability to lethality.
- Most proteins are multifunctional. A nonsynonymous mutation can affect one but not other functions performed by the same protein.

### Context-dependence

The effect of a mutation may be context-dependent in two manners: it may be affected by other mutations in the same genome (epistasis) or by other genomes of the surrounding mutant spectrum.

---

## 2.4  Inferences on evolution drawn from mutation types
————————————————————

The proportion of transition versus transversion mutations may depend initially on the specific replication machinery of a virus that tends to produce some mutation types preferentially over others. For a given virus, short-term evolution is often reflected in the dominance of transitions, a dominance which is less apparent when distantly related sequences of the same virus are compared. The effect of evolutionary distance on the transition to transversion ratio was observed in the FMDV genome sequence comparisons carried out in our laboratory over several decades, that ranged from analyses of mutant spectra relative to their corresponding consensus sequence to independent viral isolates from disease outbreaks separated by several

decades [review of the work on FMDV evolution in (Domingo et al., 1990, 2003)]. These two levels of sequence comparisons (within mutant spectra vs. independent isolates) can be highly significant, as discussed in Chapters 3 and 7.

The proportion of synonymous and nonsynonymous mutations that have mediated the diversification of viral genomic sequences that belong to the same phylogenetic lineage is often considered informative of the underlying evolutionary forces. Probably because of the rooted (albeit uncertain) notion that biological function is more likely to reside in protein than in DNA or RNA, the ratio of nonsynonymous substitutions (corrected per nonsynonymous site in the sequence under study) ($d_n$), to the number of synonymous substitutions per synonymous site ($d_s$), termed $\omega$ ($\omega = d_n/d_s$) is calculated to infer the dominant mode of evolution (Nei and Gojobori, 1986).

When $\omega = 1$ the evolution is considered neutral, when $\omega < 1$ purifying (or negative) selection is dominant, and when $\omega > 1$ positive (or directional) selection prevails (Yang and Bielawski, 2000). The types of selection undergone by viruses are discussed in Section 3.4 of Chapter 3.

There are several reasons to be cautious about the significance of $\omega$: (i) synonymous mutations need not be neutral, for reasons discussed in Section 2.3. (ii) In the course of evolution, important but transient events of positive selection (termed episodic positive selection) due to one or a few amino acid substitutions may be accompanied by a larger number of synonymous, tolerated mutations. In this situation, $\omega$ computes as $\omega < 1$, thus indicative of purifying selection despite a critical role of positive selection triggered by one or few nonsynonymous mutations in the evolutionary outcome (Crandall et al., 1999). (iii) In a striking proof of the above arguments, statistically significant mutational biases led to a value of $\omega$ indicative of positive selection in an in vitro evolution experiment simulating pseudogene evolution in which positive selection was not possible (Vartanian et al., 2001); this study represents a warning which is rarely mentioned when discussing the limitations of conclusions based on the value of $\omega$. (iv) A synonymous change may permit the mutant codon to acquire a relevant nonsynonymous change through a point mutation. The term *quasisynonymous* has been used to describe codons that encode the same amino acid, but that has a different evolutionary potential regarding the amino acids that they can access through a point mutation. Alternative codons for a given amino acid approximate a replicative system to points of sequence space from which a phenotypically relevant change has a different probability (Chapters 3 and 4). (v) Finally, $\omega$ was initially proposed to compare distantly related rather than closely related genomes, as is often the case in the short-term evolution of viruses (Kryazhimskiy and Plotkin, 2008).

For all these reasons, $\omega$ values as a diagnostic of forces mediating DNA and RNA virus evolution must be regarded only as indirect and suggestive, not as a definitive parameter. Despite these arguments, use of $\omega$ to propose a model of virus evolution continues being surprisingly unchallenged in the literature of virus evolution. We use $\omega$ only in a limited way in subsequent chapters because, in addition to the limitations just listed, it does not help in the interpretation of critical evolutionary events regarding viruses. Related shortcomings apply to other tests of neutrality developed to interpret the origin of DNA polymorphisms in the years following the summit of the neutralist-selectionist controversy (Fu, 1997; Achaz, 2009).

## 2.5 Mutation rates and frequencies for DNA and RNA genomes

Mutation rates quantify the number of misincorporations per nucleotide copied, irrespective of the fate (increase or decrease in frequency) of the mutated genome produced. A mutation rate for a genomic site measures a biochemical event dictated by the replication machinery and environmental parameters that affect the catalytic properties of the polymerase. In contrast, a mutant (or mutation) frequency describes the proportion of a mutant (or a set of mutants) in a genome population. The frequency of a mutant will depend on the rate at which it is generated (given by the mutation rate) and on its replication capacity relative to other genomes in the population (Drake and Holland, 1999) (Fig. 2.5). A specific mutation may be produced at a modest rate, but then be found at high frequency because the mutation is advantageous for genome replication in that environment. The converse situation may also occur. Some mutational hot spots (in the sense of genomic sites where mutations tend to occur with high probability) may never be reflected among the repertoire of mutations found in a

FIGURE 2.5  Scheme that illustrates the difference between mutation rate and mutant frequency. Residue A in a template residue (top) can be misread to incorporate a C, A, or G into the complementary strand (discontinuous lines), at a rate of $10^{-4}$, $10^{-5}$, and $10^{-5}$ substitutions per nucleotide, respectively. The replicative capacity of the newly generated templates (with G, U, and C, *continuous lines*) will determine widely different mutant frequencies with G > C > U.

genome population because of the selective disadvantage they inflict upon the genome harboring them.

A very significant example is the elongation of an internal oligoadenylate tract located between the two functional AUG initiation codons in the FMDV genome. The homopolymeric tract constitutes a hot spot for variation due to polymerase slippage (Fig. 2.4B). The elongation of the internal oligoadenylate was dramatic because it sextuplicated the number of adenylate residues present at that site; it was only observed when FMDV was subjected to repeated plaque-to-plaque (bottleneck) transfers, not large population passages. In fact, this drastic genetic modification has not been recorded among natural isolates of the virus. The molecular instruction to elongate the oligoadenylate was very strong because it was observed in many independent biological clones subjected to bottleneck transfers (Escarmís et al., 1996). Despite qualifying as a hot spot for variation, the first event in

fitness recovery when the clones were subjected to large population passages was the reversion of the elongated tract to its original size (Escarmís et al., 1999). The interpretation of these findings, to be further analyzed in Chapter 6, is that during plaque-to-plaque transfers the negative selection to eliminate unfit genomes is less intense than during large, highly competitive population passages. Again, a clear molecular instruction to elongate a homopolymeric track may not be reflected in a high frequency of the affected genomes. Therefore, although mutation rates and frequencies for viruses bear some relationship, rates cannot be inferred from frequencies and vice versa (Fig. 2.5).

The first calculations of mutation rates for cellular organisms and for some DNA bacteriophages were carried out by J.W. Drake, who pursued comparative measurements that have generally supported a difference between mutation rates for DNA and RNA viruses. The rates estimated for bacteriophages λ and T4 were about 100 times higher than those of their host *E. coli*. An approximately constant rate of 0.003 mutations per genome per replication round was calculated for a number of DNA-based microbes (Drake, 1991), an observation sometimes referred to as "Drake's rule." This rather surprising constancy suggests that different DNA organisms have accommodated the template-copying fidelity of their replication machineries to achieve a narrow window in the mutational load measured as mutations fixed per genome, a remarkable fitting of biochemistry with evolutionary needs. The basal mutation rate in mammalian cells has been estimated at about $10^{-10}$ substitutions per nucleotide and cell generation [reviewed in (Naegeli, 1997; Domingo et al., 2001; Friedberg et al., 2006)] (Table 2.1).

The synonymous mutation rate measured with experimental populations of bacteria has been assumed to reflect the neutral mutation rate (despite limitations explained in Section 2.3). Values for *E. coli* have ranged from $2 \times 10^{-11}$ up

**TABLE 2.1** Mutation rates and frequencies for RNA and DNA genomes.

| Viriods | $2 \times 10^{-3}$ |
| --- | --- |
| RNA viruses | $10^{-5}$ to $10^{-3}$ |
| Retroviruses | $10^{-6}$ to $10^{-4}$ |
| DNA viruses | $10^{-8}$ to $10^{-3}$ |
| Cellular DNA | $10^{-9}$ to $10^{-11}$ |

Values are expressed as substitutions per nucleotide. The range of values is the most likely according to several independent studies. No distinction is made between mutation rates and frequencies. See text for comments and references.

to $5 \times 10^{-9}$ substitutions per synonymous site per generation (Ochman et al., 1999) with $5 \times 10^{-10}$ as the most likely estimate (Lenski et al., 2003). The latter value is in agreement with a rate of $3 \times 10^{-10}$ to $4 \times 10^{-10}$ substitutions per base pair and generation based on whole-genome deep sequencing of an experimentally evolved lineage of *Myxococcus xanthus* (Velicer et al., 2006). There are biological phyla for which no mutation rates have been calculated. From current knowledge, we can assume that mutation rates in cells and viruses depend on the replicative machinery (generally a multiprotein complex that includes the relevant viral polymerase with additional viral and host proteins and membrane structures) and on multiple environmental parameters (template nucleotide sequence context, ionic environment, temperature, metabolites in interaction with components of the replication apparatus, etc.). Whether bacteria are in the exponential or stationary phase of growth can affect intracellular metabolites and proton exchange rates which, in turn, may alter the proportion of tautomeric forms in nucleotides and misincorporation tendencies (Friedberg et al., 2006). The sequence context of the template nucleic acids (presence of repeated sequences that can induce misalignment mutagenesis or G-C vs. A-T rich regions in relation to relative nucleotide substrate abundances, etc.) may impel or attenuate mutability. Insertion

elements may enhance mutation rates at neighboring sites in a bacterial genome (Miller and Day, 2004). Despite these influences, vesicular stomatitis virus (VSV) displayed comparable mutation rates in several host cells (Combe and Sanjuan, 2014) suggesting that there is a limited range of average error rates needed for a virus to maintain fitness (Chapters 5 and 9).

In addition to the general environmental and sequence context consequences for template-copying fidelity that may affect any genome type, mutation rates for DNA viruses will also be influenced by: (i) whether the DNA polymerase that catalyzes viral DNA synthesis includes or lacks a functional proofreading-repair activity. High copying fidelity is typical of DNA polymerases involved in cellular DNA replication (Bebenek and Ziuzia-Graczyk, 2018), and low copying fidelity is generally a feature of DNA polymerases involved in DNA repair (Friedberg et al., 2006; Ganai and Johansson, 2016). Thus, repair of lesions that by themselves might not be mutagenic may lead to the introduction of mutations during the error-prone repair process. (ii) Expression of proteins active in repair encoded in the viral genome, such as uracil-DNA glycosylase, DNA repair endonucleases, etc. (iii) The mechanism of viral DNA replication, particularly the occurrence of double-stranded versus single-stranded DNA in replicative intermediates. (iv) Intracellular site of replication and the availability of postreplicative DNA repair proteins (regarding both intracellular location and concentration) to the viral replication factories. Little is known of the spatial relationships and relative affinities of cellular and viral proteins and structures that may critically affect polymerase fidelity. Comparative measurements of mutation rates at specific genome sites of DNA viruses are needed, as a first step to define the cellular and biochemical influences on the fidelity of DNA virus genome replication.

General genetic variability affecting the entire virus genome should be distinguished from localized variability at hot spots in a genome. Even the extremely complex human genome shows genetic instability at specific loci, some associated with genetic disease (Domingo et al., 2001; Alberts et al., 2002; Bushman, 2002). Genome size is a parameter pertinent to biological behavior, not only because it imposes a commensurate copying fidelity, but also because it affects the impact of genetic heterogeneity within infected organisms and upon the invasion of new hosts (Chapter 3).

Mutation frequencies measured by subjecting virus to a specific selective agent (e.g., mutants that escape the neutralizing activity of a monoclonal antibody or mutants that escape inhibition by a drug) span a broad range of values ($10^{-3}$ to $10^{-8}$) for DNA and RNA viruses (Smith and Inglis, 1987; Sarisky et al., 2000; Domingo et al., 2001) (Table 2.1). The technical details of any procedure used to calculate a mutation frequency should be carefully evaluated to translate its meaning to the genome level. Important variables are the efficacy of the antibody or drug (which will be concentration-dependent) or the possibility of phenotypic hiding-mixing in the escape mutants to be quantified (Holland et al., 1989; Valcarcel and Ortin, 1989). Unexpected low levels of escape mutants (that would imply $< 10^{-6}$ substitutions per nucleotide) for an RNA virus can mean either a general or site-specific high polymerase fidelity, a selective disadvantage of the genome that harbors the mutation or, when a phenotypic alteration is measured, the requirement of two or more mutations to produce the alteration. Conversely, a high mutation frequency for a DNA virus whose replication is catalyzed by a high-fidelity DNA polymerase may mean that either repair activities were not functional or that the mutant displayed a selective advantage and overgrew the wild type prior to the measurement of its frequency. Mathematical treatments that take into account reversion of a low fitness mutant and its competition with wild-type virus have been used to calculate mutation rates (Batschelet et al., 1976; Coffin, 1990).

Despite difficulties and limitations in the calculations, independent genetic and biochemical methods with different viruses support mutation rates for RNA viruses in the range of $10^{-3}$ to $10^{-5}$ substitutions per nucleotide copied [as representative articles and reviews see (Batschelet et al., 1976; Domingo et al., 1978, 2001; Steinhauer and Holland, 1986; Eigen and Biebricher, 1988; Varela-Echavarria et al., 1992; Ward and Flanegan, 1992; Mansky and Temin, 1995; Preston and Dougherty, 1996; Drake and Holland, 1999; Sanjuan et al., 2010; Bradwell et al., 2013)] (Table 2.1). A few early studies indicated unusual low mutation rates or frequencies for some RNA viruses. As discussed in some of the reviews listed above, there are technical reasons to suggest that such values were probably underestimates of the true average mutations rates or frequencies. Obviously, it cannot be excluded that some genomic sites or viruses under a given environment might be unusually refractory to introduce mutations, but most evidence supports the range of values listed in Table 2.1. The near million-fold higher mutation rates for RNA viruses than cellular DNA, whose biological implications were presciently anticipated by J. Holland and colleagues (Holland et al., 1982), have been confirmed. That is, for RNA viruses of genome length between 3 Kb and 32 Kb, an average of 0.1-1 mutation is introduced per template molecule copied in the replicating population. Unless most mutations impeded viral replication, a continuous input of mutant genomes is expected, as indeed found experimentally (Chapter 3).

High mutation rates for RNA genomes are also supported by measurements of template-copying fidelity by RNA polymerases, reverse transcriptases, and DNA polymerases devoid of $3–5'$ proofreading exonuclease (or under conditions in which, such exonuclease is not functional) [(Steinhauer et al., 1992; Varela-Echavarria et al., 1992; Mansky and Temin, 1995; Domingo et al., 2001; Friedberg et al., 2002, 2006; Menéndez-Arias, 2002), and references therein]. In vitro fidelity tests may be based on genetic or biochemical assays using homopolymeric or heteropolymeric template-primers. Measurements include the kinetics of incorporation of an incorrect versus the correct nucleotide directed by a specific position of a template or the capacity of a polymerase to elongate a mismatched template-primer 3' end, [these and other assays have been reviewed (Menéndez-Arias, 2002)] (see also Section 2.6). Differences between related enzymes (i.e., AMV RT is more accurate than HIV-1 RT), and the fact that amino acid substitutions in the polymerases affect nucleotide discrimination, demonstrate that proofreading-repair activities together with the structure of the polymerase and replication complexes are determinants of template-copying fidelity.

## 2.5.1 Undesired consequences of the confusion between mutation rates and mutation frequencies

The term mutation rate if often used in a light manner in the literature of virus evolution, probably driven by nomenclature from classical population genetics. It is used to mean mutation frequency, rate of evolution, and sometimes to mean mutation rate in its real sense (as explained in Section 2.5). A particularly risky habit is to use mutation rate when what is measured is a mutation frequency. Some studies have claimed that they have a replication system devoid of selection, and therefore, the number of mutations counted corresponds to the true mutation rate of the system. This is incorrect. There is no replicative system devoid of selection because at least selection to maintain replication is in continuous operation. Furthermore, as taught by quasispecies dynamics (Stadler, 2016), supported by mutational waves in hepatitis C virus upon prolonged replication in a constant cellular environment (Moreno et al., 2017), the mutant spectrum per se is part of the environment. Since

the mutant spectrum is constantly changing, so is the environment in which replication takes place. Unfortunately, some studies have proposed the existence of mutational cold spots (sites at which mutations as a biochemical event occur at a very low frequency) ignoring that negative selection might have eliminated newly arising mutations. These false conclusions imply the existence of genomic regions particularly suitable as drug or antibody targets because they cannot mutate. These are the types of studies and incorrect conclusions that keep perpetuating the problem of control of viral diseases by encouraging antiviral and vaccine designs doomed to failure (Chapters 8 and 9).

## 2.6 Evolutionary origins, evolvability, and consequences of high mutation rates: fidelity mutants

The amino acid substitutions in the core polymerase that affect fidelity can be located either close to or away from the active site of the enzyme. The change in fidelity can reach almost one order of magnitude, but virus viability is not compromised. Thus, error rates themselves can be subjected to selection, as supported by theoretical studies on evolvability (Earl and Deem, 2004). Early studies documented heterogeneity in the mutation rates among individual plaque isolates of influenza A virus (Suarez et al., 1992). It is not clear whether mutation rates of viruses have evolved to procure a balance between adaptability and genetic stability, or whether other selective constraints have imposed the observed values. It has been suggested that because of the generally deleterious nature of most mutations, the adaptive value of the high mutation rates for RNA viruses is debatable and that there might have been a trade-off between replication rate and copying fidelity (Elena and Sanjuan, 2005). Mutation rates would be a consequence of rapid RNA replication, and an increase in copying fidelity would come at a cost, resulting in a lower replication rate. A connection between elongation and error rate has been suggested by results with some viral and cellular polymerases [review (Kunkel and Erie, 2005)]. In an early study with the poliovirus RdRp in vitro, an increase in the error frequency was observed when the pH and $Mg^{2+}$ ion conditions were modified, and the decreased fidelity correlated with increased RNA elongation rate (Ward et al., 1988). A possible connection between elongation rate and copying fidelity cannot be ruled out, but current evidence points to template-copying fidelity as being the result of multiple factors, not necessarily linked to the rate of genome replication (Vignuzzi and Andino, 2010; Campagnola et al., 2015; Domingo and Perales, 2019).

There is ample support for an adaptive value of high mutation rates for RNA viruses, independently of their biochemical origins. A poliovirus mutant, whose RdRp displayed a three-to five-fold higher fidelity than the wild-type enzyme, replicated at a slightly lower rate than wild-type virus in cell culture but displayed a strong selective disadvantage regarding the invasion of the brain of susceptible mice (Pfeiffer and Kirkegaard, 2005; Vignuzzi et al., 2006). The impediment to cause neuropathology was due at least partly to the limited complexity of the mutant spectrum since its broadening through mutagenesis restored the capacity to produce neuropathology. These and other studies have provided evidence that mutant spectrum complexity, by virtue of its impact on fitness, can be a virulence determinant. The work by M. Vignuzzi, J. Pfeiffer, R. Andino, C. Cameron, K. Kirkegaard, and their colleagues on poliovirus fidelity mutants opened a much-needed branch of research in virus evolution and quasispecies implications. As proof of this statement, the field of fidelity mutants is rapidly expanding (Bordería et al., 2016), and references to the information they provide will be made in several chapters.

Theoretical models and experimental observations suggest that mechanisms for error correction had to evolve to maintain functionality of increasingly complex genomes (Swetina and Schuster, 1982; Eigen and Biebricher, 1988; Domingo et al., 2001; Eigen, 2002, 2013) (here complexity means genome size, provided no redundant information is encoded). The coronaviruses have the largest genomes among the known RNA viruses, with 30–33 kb. This is about 10-fold more genetic information than encoded in the simple RNA bacteriophages, such as MS2 or Qβ. Coronaviruses are replicated by complex RNA-dependent RNA polymerases, which include a domain that corresponds to a $3'−5'$ exonuclease, proofreading-repair activity. The protein displays exonuclease activity in vitro, and its inactivation affects viral RNA synthesis (Minskaia et al., 2006), and results in increases of about 15-fold in the average mutation frequency (Eckerle et al., 2007, 2010). A coronavirus mutant devoid of this repair function is more susceptible to lethal mutagenesis than the corresponding, nonmutated virus (Smith et al., 2013; Smith and Denison, 2013), as expected from a connection between replication accuracy and proximity to an error threshold for the maintenance of genetic information (Chapter 9). Thus, it is likely that a proofreading activity evolved (or was captured from a cellular counterpart) in RNA genomes, whose genomic complexity was in the limit compatible with the fidelity achievable by standard RNA replicases. It would be interesting to discover new RNA viruses with a single RNA molecule longer than 30 Kb as a genome to analyze whether they have evolved more accurate core polymerases or exhibit a proofreading-repair function during replication. Toward the other end of the RNA size scale, viroid RNAs display a mutation rate higher than (or close to the highest) recorded for RNA viruses, consistent with the correlation between genome size and template-copying accuracy (Gago et al., 2009).

Studies with bacteria have identified some of the factors that successively increase copying fidelity. It has been estimated that during *E. coli* DNA replication the error rate would be $10^{-1}$ to $10^{-2}$ mutations per nucleotide copied if accuracy relied only upon the strength of interactions provided by base pairing (Section 2.2). The error rate would decrease to $10^{-5}$ to $10^{-6}$ with base selection and proofreading-repair, to about $10^{-7}$ with the contribution of additional proteins present in the replication complex, and to about $10^{-10}$ misincorporations per nucleotide with the participation of postreplicative mismatch correction mechanisms (Naegeli, 1997; Kunkel and Erie, 2005; Friedberg et al., 2006). Reduction of bacterial genome size results in the increase of mutation frequency (Nishimura et al., 2017). The error rate of the bacteriophage $\phi$29 DNA polymerase is about $10^{-6}$ without the proofreading exonuclease activity, and it decreases to $10^{-8}$ with the correcting activity [(de Vega et al., 2010) and references therein]. Postreplicative repair pathways act on double-stranded DNA, but not (or very inefficiently) on RNA or DNA-RNA hybrids. Therefore, the known postreplicative repair systems that operate in cellular DNA do not make a significant contribution to error correction in RNA viruses (Steinhauer et al., 1992).

The importance of copying fidelity for complex genomes is reflected in the fact that more than 100 proteins are directly or indirectly involved in the repair of the human genome (Liu et al., 2017). Elevated mutation rates in the range of those operating for RNA viruses would be lethal for large DNA genomes. Localized genetic modification occurs physiologically in processes, such as somatic hypermutation and class-switch recombination in B cells of the germinal centers, as mechanisms of diversification of immunoglobulin genes (Upton et al., 2011; Methot et al., 2017). Chromosomal instability has long been associated with cancer (Gatenby and Frieden, 2004; Stratton et al., 2009). Surveys have been (and are currently being) used to identify genes associated with chromosomal instability and their role in aging and disease (Aguilera and Garcia-Muse, 2013; Vijg et al., 2017) (see also Chapter 10). While uncontrolled high mutability is deleterious for differentiated cellular organisms, it constitutes a *modus vivendi* for a great majority of viruses.

Despite its attractiveness, definitive proof of the hypothesis of a direct relationship between error rate and limited genome complexity will require additional functional and biochemical studies. Exceptions to the absence of repair activities in simple genetic elements have been described. A satellite RNA of the plant virus turnip crinkle carmovirus evolved a 3'-end RNA repair mechanism. It implicates the synthesis of short oligoribonucleotides by the viral replicase using the 3'-end of the viral genome as a template. The mechanism consists probably of template-independent priming at the 3'-end of the damaged RNA to generate wild type, negative strand, and satellite RNA (Nagy and Simon, 1997). A reversible, NTP-dependent excision of the 3' residue of the nascent nucleic acid product has been described in some retroviruses and hepatitis C virus (Meyer et al., 1998; Jin et al., 2013). This activity is important for drug resistance, and it may also modulate the overall fidelity of some polymerases. It cannot be excluded that some type of point mutation correction may operate in RNA genetic elements of less than 30 Kb. Such putative mechanisms may even diminish mutation rates that would otherwise be prohibitively deleterious, and they do not overshadow high mutation rates as a feature of RNA and some DNA genomes (Table 2.1).

Limited copying fidelity in the absence of proofreading-correction mechanisms can be regarded as an unavoidable consequence of the molecular mechanisms involved in template copying by viral polymerases. Most nucleic acid polymerases share a structure that resembles a right hand, with fingers, palm, and thumb domains (Fig. 2.6A). Three-dimensional structures of viral RdRps and RTs indicate that interactions between the incoming nucleotide or residues of the template-primer with amino acids of the polymerase must permit displacement of the growing polymerase chain along

**FIGURE 2.6** Polymerization by viral polymerases. (A) The structure of nucleic acid polymerases resembles that of a right hand. This was first evidenced with the structure of the Klenow fragment of the *E. coli* DNA polymerase I by T.A. Steitz and colleagues in 1985. On the left, the Klenow fragment is represented with colored fingers, palm and thumb domains, next to an open right hand. The structure on the right is that of PV RdRp, next to a closed right hand. Courtesy of N. Verdaguer and L. Vives-Adrián (the hand is that of L. Vives-Adrián). (B) The structure of the ternary complex between FMDV 3D, an RNA molecule, and UTP as the substrate (PDB id. 2E9Z). The left panel is a front view of the complex, depicting the polymerase chain as a yellow ribbon, the RNA in dark blue (template) and cyan (primer). The incoming UTP and the pyrophosphate product are shown in atom type, and 2 $Mg^{2+}$ ions as magenta balls. The right panel is the same complex in a top-down orientation. (Figure courtesy of C. Ferrer-Orta and N. Verdaguer). (C) Scheme of the minimum number of steps involved in nucleotide incorporation. The first step consists of the binding of polymerase $^0E$ to the template-primer $R_n$ (elongated up to nucleotide n) to form a complex $^0ER_n$. Formation of the activated complex $ER_n$ is governed by the rate of constant $k_{assembly}$ ($k_a$). The activated $ER_n$ complex binds a nucleotide NTP with an apparent binding affinity given by $K_{d,app}$ to form the $ER_nNTP$ complex. Catalysis to covalently incorporate the NTP to the growing primer chain to yield $ER_{n+1}$ and pyrophosphate ($PP_i$) is governed by the rate constant $k_{pol}$. Other constants depicted in the scheme are the inactivation rate constant ($k_{inact}$) of $^0E$, and dissociation of E from RNA ($k_{off}$, $ER_n$ and $k_{off}$, $ER_{n+1}$). *Based on Arias, A., Arnold, J.J., Sierra, M., Smidansky, E.D., Domingo, E., et al., 2008. Determinants of RNA-dependent RNA polymerase (in)fidelity revealed by kinetic analysis of the polymerase encoded by a foot-and-mouth disease virus mutant with reduced sensitivity to ribavirin. J. Virol 82, 12346–12355, and previous studies with PV polymerase 3D by C. E. Cameron and his colleagues.*

the channel located at the palm domain of the polymerase (Steitz, 1999; Ferrer-Orta et al., 2006; Wu and Gong, 2018) (Fig. 2.6B). The polymerase of classical swine fever virus and other pestiviruses, such as bovine viral diarrhea virus includes an N-terminal extra-domain of about 100 amino acids; its interaction with the palm domain is important for template copying fidelity (Liu et al., 2018). If interactions around the catalytic site to ensure the correct nucleotide incorporation were so strong as to preclude misincorporations, the movement of the growing polynucleotide chain would be hampered. Again, this compromise suggests a match between biochemical and evolutionary needs.

The orientation of the triphosphate moiety of the incoming nucleotide substrate is important for nucleotide incorporation (Menéndez-Arias, 2002; Graci and Cameron, 2004; Ferrer-Orta et al., 2009). One of the several steps involved in nucleotide incorporation is the formation of a ternary complex (polymerase with template-primer and the incoming nucleotide) that undergoes a conformational change (reorientation of the divalent ion-complexed triphosphate moiety of the incoming nucleotide). This conformational change activates the complex for phosphoryl transfer, to link the nucleoside-monophosphate to the $3'$-terminus of the primer (or growing chain). Steps involved in the nucleotide incorporation are represented in Fig. 2.6C. Both the conformational change and the relative rate of phosphoryl transfer for an incorrect nucleotide versus the correct nucleotide influence the error rate at each site of the growing chain. Critical kinetic constants in Fig. 2.6C that are determined experimentally to quantify relative nucleotide incorporations and misincorporations are $K_{d,app}$ (expressed as $\mu M$), $k_{pol}$ (expressed as $s^{-1}$), and the ratio $k_{pol}/K_{d,app}$ ($\mu M^{-1} s^{-1}$) termed the catalytic efficiency. The ratio of $k_{pol}/K_{d,app}$ for the incorporation of an incorrect nucleotide to $k_{pol}/K_{d,app}$ for a correct nucleotide gives the frequency of that particular

misincorporation, and an assessment of polymerase fidelity [(Castro et al., 2005) and references therein]. Modifications of polymerase residues by site-directed mutagenesis, combined with comparisons of the relevant structures, have identified critical amino acid residues involved in template-copying fidelity.

High-fidelity mutants are frequently obtained by selecting viruses resistant to mutagenic nucleotide analogs such as the antiviral agent ribavirin (Beaucourt and Vignuzzi, 2014). Limited incorporation of a deleterious nucleotide can be attained either through specific discrimination against the analog (ribavirin or other) or through a general decrease of all types of misincorporations, that is, a high-fidelity phenotype. Structural modifications of viral polymerases that lead to high fidelity have inspired the design of mutant viral polymerases displaying either an increase or decrease of copying fidelity achieved through a single amino acid substitution (Wainberg et al., 1996; Menéndez-Arias, 2002; Mansky et al., 2003; Pfeiffer and Kirkegaard, 2003; Arnold et al., 2005; Domingo, 2005; Vignuzzi et al., 2006; Coffey et al., 2011; Gnadig et al., 2012; Meng and Kwang, 2014; Rozen-Gagnon et al., 2014; Borderia et al., 2016). The capacity of the virus to evolve at higher or lower rates than their ancestors is achievable through modest numbers of mutations (limited movements in sequence space, Chapter 3), again emphasizing the evolvability of mutation rates.

The rates of mutation and recombination need not be independent. M. Vignuzzi and colleagues have shown that a mutator Sindbis virus displays a higher recombination rate and enhanced production of DI particles than the wild type virus (Poirier et al., 2015). A connection between mutation and recombination rates strengthens the evolutionary consequences of the modifications of template copying fidelity that can be achieved through a single amino acid substitution.

## 2.7 Hypermutagenesis and its application to generating a variation: APOBEC and ADAR activities

Some viral genomes either isolated from biological samples or evolved in cell culture show biased mutation types (e.g., monotonous G → A or C → U substitutions in the same genome), generally at frequencies of around $10^{-2}$ substitutions per nucleotide (10- to 1000-fold higher than standard mutation rates and frequencies) (Table 2.1). Biased hypermutation was first observed in some defective interfering (DI) RNAs of vesicular stomatitis virus (VSV) (Holland et al., 1982), and in variant forms of measles virus, associated with postmeasles neurological disease, such as subacute sclerosing panencephalitis (Cattaneo and Billeter, 1992). Hypermutation is mainly due to the activity of cellular deaminases, such as the apolipoprotein B mRNA and the editing complex (APOBEC), or the adenosine deaminase acting on double-stranded RNA (ADAR) families, that are involved in cellular editing and regulatory functions (Sheehy et al., 2002; Santiago and Greene, 2008; Nishikura, 2010; Stavrou et al., 2014; Pfaller et al., 2018; Venkatesan et al., 2018). In the event of a viral infection, such cellular functions can become part of an innate defense mechanism against the invading virus. Viral proteins (i.e., Vif in HIV-1) bind some APOBEC proteins, thus inhibiting mutagenesis and permitting virus survival (Sheehy et al., 2002). In oncoretroviruses, retroviruses, and hepatitis B virus (HBV), the APOBEC-3 cytidine deaminase acts on single-stranded DNA and results mainly in G → A and C → U hypermutation, that may affect 40% −100% of the G residues. The preferred sequence context for G hypermutation in HIV-1 observed in vivo is GpA > GpG > GpT ≈ GpC. The specific dinucleotide context of the hypermutated sites provides a means to distinguish genomes that have undergone hypermutation by cellular activities from those that are heavily mutated by other mechanisms, such as the action of mutagenic agents (Chapter 9).

APOBEC 3 proteins play a role in cancer through cytidine deaminase mutagenesis and generation of double-strand breaks in chromosomal DNA (Wang et al., 2016). APOBEC 3 levels in the cell may be regulated by cellular and viral proteins, for example, human papillomavirus (HPV) oncoprotein E that stabilizes APOBEC 3A in human keratinocytes that may promote cervical cancer (Westrich et al., 2018).

The ADAR-associated hypermutation was identified in negative-strand RNA riboviruses and results mainly in A → G and U → C hypermutation. It is originated by A → I (inosine) modifications in double-stranded viral RNA catalyzed by ADAR-1 L, one of more than 100 proteins inducible by type I IFN (Maas et al., 2003). Inosine can be recognized as G by the replication machinery (Valente and Nishikura, 2005), although it can form wobble base pairs also with A and U (Fig. 2.2). Hypermutation can contribute to genetic variation of viruses (Hirose et al., 2018).

There are additional mechanisms of hypermutagenesis. Higher than average mutation frequencies can occur as a result of replication in the presence of biased concentrations of the standard nucleotide substrates; this has been applied to the in vitro generation of genes mutated at frequencies of $10^{-1}$ to $10^{-2}$ (mutagenic PCR), as a powerful tool to study sequence-function relationships and functional robustness of nucleic acids and proteins (Meyerhans and Vartanian, 1999). Error-prone PCR has been used in experiments of in vitro evolution of nucleic acid enzymes to generate heterogeneous collections of nucleic acid sequences to select for molecules capable of catalyzing specific reactions (Joyce, 2004) (Chapter 1).

## 2.8 Error-prone replication and maintenance of genetic information: instability of laboratory viral constructs

High mutation rates have practical implications in laboratory studies on the behavior of virus mutants obtained by molecular cloning of a biological sample, or constructed by site-directed mutagenesis. A transition mutation that causes a strong fitness decrease but that still allows residual RNA genome replication will most likely revert following infection or transfection of cells with the mutant construct and subsequent viral replication. Double or triple mutants (preferentially including transversions) should be engineered (when possible according to the genetic code) to study the behavior of a viral mutant with an amino acid replacement of interest that may produce a fitness decrease. As an example, a C → U transition found in an open reading frame of an RNA virus may convert a Pro into a Ser (CCG → UCG). Since Ser will revert to Pro through a U → C transition in the triplet (a common type of misincorporation by most polymerases), Ser should be engineered to be encoded by AGU; in the course of replication, reversion to Pro would require at least two transversions since the codons for Pro are CCU, CCC, CCA, or CCG. Thus, if effects derived from the difference in the primary sequence of the RNA or codon bias do not intervene in the behavior of the viral genome, codons with a high genetic barrier to reversion should be engineered for studies involving viral replication.

In general, deletions revert at a much lower frequency than point mutations, and when appropriate for the question under study, a deletion should be introduced within the gene of interest to probe gene function in reversegenetics studies. High mutation rates also imply that infection or transfection with debilitated mutant viruses may result in progeny with sequences that differ from the input. V. I. Agol and colleagues have coined the term *quasi-infectious* to refer to mutant viruses that are capable of yielding progeny, but the progeny differs from the initial genome (pseudorevertants) (Gmyl et al., 1993; Agol and Gmyl, 2018). The difference between the input mutant and the rescued progeny virus will depend on the type of genetic lesion in the input virus and its consequences

for virus multiplication. A single point mutation that decreases replication is likely to evolve to yield a true revertant (return to the original sequence) upon replication. If the same reversion depends on two or more mutations, a true revertant will require extended replication for exploration of sequence space (Chapter 3), and selection of compensatory mutations elsewhere in the genome (sometimes referred to as second site revertants) becomes an alternative for fitness gain. The term compensatory applies to mutations that compensate for the deleteriousness of other mutations. A typical example is a mutation that decreases the stability of a stem in an RNA stem-loop that functions as a *cis*-acting element. A compensatory mutation restores a stable stem needed for the activity. Transfection of cells by an engineered virus with some preselected genetic modification (produced either from cDNA copies of a viral genome or by chemical synthesis) may yield progeny genomes, which differ from the parent. If a substantial loss of replicative capacity is produced by a drastic genetic change (an *indel*, loss of a stem-loop structure, etc.) selection of a true revertant becomes extremely unlikely. The compensatory generation of alternative structures (or constellations of point mutations) that restores replication (partially or completely) becomes an interesting and informative possibility.

Procedures to copy an entire viral RNA genome into a cDNA for reverse genetics studies are now available (Fan and Di Bisceglie, 2010). If for technical reasons an infectious cDNA clone is constructed from several molecules, which were copied from different genomes present in the mutant spectrum, the ligation product may be transcribed into an RNA, which is not infectious. This is because, while some constellations of mutations may be compatible with infectivity, others may not, or may allow limited, suboptimal replication, thus favoring the selection of additional mutations or reversions. The same applies to a synthetic genome based on one of the multiple genomic sequences from a viral isolate. Individual mutations may be detrimental either per se, or by the combined presence of other mutations. The joint effect of different mutations in the same genome is often referred to as epistasis. Mutations that reinforce each other with regard to a viral function are said to produce positive epistasis, and those that interfere with each other produce negative epistasis (also mentioned in Section 2.3). Epistasis in RNA viruses may be blurred by the weight of mutant spectra in determining viral behavior through intergenomic interactions (Chapter 3).

An interesting contrast that recapitulates concepts given in Sections 2.5 and 2.6 is the effect of an active proofreading-repair activity in maintaining the infectivity of a viral genome upon its extended replication in vitro (in a test tube, in the absence of cellular extracts). The 19,285 bp bacteriophage φ29 DNA can be amplified at least 4000-fold without detectable loss of infectivity due to the fidelity of φ29 DNA polymerase conferred by a 3′−5′ proofreading-repair exonuclease activity (Bernad et al., 1989). Engineered φ29 DNA polymerases provide a powerful amplification tool in genomics (de Vega et al., 2010). In contrast, the 4220 nucleotides long Qβ RNA rapidly loses its infectivity when replicated by Qβ replicase in vitro due to the accumulation of mutations and deletions in the viral RNA (Mills et al., 1967; Sabo et al., 1977). The error-prone Qβ replicase is not adequate to amplify infectious viral RNA, but it was at the origin of the quasispecies concept to be discussed in Chapter 3. Mutagenic DNA polymerases (generally those involved in DNA repair) are an alternative to mutagenic PCR (Section 2.7) to produce randomly mutated collections of nucleic acid molecules (Forloni et al., 2018).

## 2.9 Recombination in DNA and RNA viruses

Recombination is the formation of a new genome by covalent linkage of genetic material from two or more different parental genomes (Fig. 2.7). Recombination can also involve different sites of the same genome to yield insertions or deletions, such as in the formation of defective interfering (DI) genomes. It is a widespread mechanism of genetic variation in all biological systems, and in cells, it underlies critical physiological and developmental processes (splicing, generation of diversity in

FIGURE 2.7    RNA recombination and segment reassortment. (A) Scheme of replicative and nonreplicative RNA recombination. RNA polarity is indicated by +, − symbols. Replicative recombination is displayed as the result of template switching during minus-strand RNA synthesis. Nonreplicative recombination is depicted as the outcome of breakage and ligation (joining) of fragments of plus-strand RNA. (B) An example of genome segment reassortment with the formation of a new segment constellation in which six genomic segments originate from one parent (*blue*) and two from the other (*gold*). Influenza virus is the best-known example (see text).

immunoglobulin genes and T cell receptors, transposition events, phase variation in bacteria, repair pathways that promote postreplicative error correction, etc.). Cellular DNA recombination relates to replication, repair, and completion of DNA replication, operations that involve multiple proteins displaying a variety of activities (Smith and Jones, 1999; Alberts et al., 2002; Nimonkar and Boehmer, 2003; Friedberg et al., 2006).

Recombination occurs both with DNA and RNA viruses, often with the participation of the virus replication machinery. Several types of recombination have been distinguished in viruses: homologous versus nonhomologous recombination, according to the extent of nucleotide sequence identity around the recombination (crossover) site, and replicative versus nonreplicative recombination, according to the requirement of viral genome replication for recombination to occur (Kirkegaard and Baltimore, 1986; King, 1988; Lai, 1992; Nagy and Simon, 1997; Plyusnin et al., 2002; Boehmer and Nimonkar, 2003; Gmyl et al., 2003; Chetverin et al., 2005; Agol, 2010; Simmonds, 2010; Bujarski, 2013; Perez-Losada et al., 2015; Agol and Gmyl, 2018; Bentley and Evans, 2018).

As in the case of cells, homologous recombination in double-stranded DNA viruses is intimately connected with DNA replication and repair. It implicates multiple viral gene products (DNA polymerase, single-stranded DNA-binding proteins, processivity factors, helicase-primase, eukaryotic topoisomerase I, etc.), and a succession of protein-catalyzed steps (Czarnecki and Traktman, 2017). In the copy choice (or template switching) mechanism, the nascent DNA switches from one template molecule to another, resulting in the synthesis of recombinant, daughter DNAs. In its basic form, recombination by breakage and rejoining starts with the introduction of a nick at one of the strands of each parental DNA, strand invasion of one parental DNA by the other, branch migration, ligation at the nicks (linking DNA

strands from the two parents), and further isomerization and cleavage reactions. DNA recombination is responsible for the endonuclease-mediated isomerization of herpesvirus genomes [four isomers defined by the orientation of the long (L) and short (S) regions of the viral genome]. During the late phase of herpes simplex virus-1 replication, the frequency of recombination has been estimated at 0.6% per Kb of the genome (Boehmer and Nimonkar, 2003). Integration or excision of proviral DNA or temperate bacteriophage DNA, are examples of site-specific recombination that involves specific enzyme activities (i.e., retroviral integrases), and requires a short stretch of nucleotide sequence identity.

The copy choice mechanism of homologous RNA recombination is also associated with genome replication. An RNA polymerase molecule with its nascent RNA product jumps into the corresponding position of another template molecule, to complete synthesis of the RNA product (Fig. 2.7). Given the large numbers of viral genomes often present in replication complexes (also termed replication factories) in each infected cell, it is not surprising that this mechanism may give rise to frequent recombinant progeny genomes. Formation of mosaic genomes has long been recognized as an essential feature of the genetics of some retroviruses and plant RNA viruses. For HIV-1 and some plant RNA viruses recombination frequencies have been estimated at 2%—10% of progeny per 100 nucleotides; for picornaviruses and coronaviruses the number of recombinants amounts to 10%—20% of the progeny (King, 1988; Lai, 1992; Nagy and Simon, 1997; Levy et al., 2004; Urbanowicz et al., 2005; Sztuba-Solinska et al., 2011). Using a phylogenetic approach the average recombination rate of HIV-1 in vivo was estimated in $1.4 \times 10^{-4}$ recombination events/site/generation, which is about fivefold greater than the average point mutation rate (Shriner et al., 2004). A ten-fold lower value of $1.4 \times 10^{-5}$ recombination events/site/

generation was estimated from the changes in the genetic composition of HIV-1 within single patients (Neher and Leitner, 2010). Recombination is required for HIV-1 replication and genome integrity (Rawson et al., 2018). In negative-strand RNA viruses recombination may be inefficient or absent, but some of them can display homologous recombination (Plyusnin et al., 2002), and a high rate of generation of DI RNAs and other types of defective genomes (Roux et al., 1991; Rezelj et al., 2018).

Recombination frequency may be altered by environmental factors that affect viral replication. A decrease of intracellular nucleotide levels as a result of treatment of cells with hydroxyurea may favor template switching reflected in an increase of intra and intermolecular recombination (Pfeiffer et al., 1999; Svarovskaia et al., 2000). Homologous RNA recombination can also be influenced by amino acid substitutions in the polymerase, the primary sequence in the RNA (i.e., high frequency of template switching in AU-rich regions), the sequence identity between the nascent strand and acceptor template, and secondary structures at or around the crossover sites, among other influences (Nagy and Simon, 1997; Alejska et al., 2005; Agol, 2010; Agol and Gmyl, 2018). Since recombination necessitates coinfections of the same cell by at least two parental genomes, the persistence of a viral genome in a cell increases the likelihood of sequential coinfections, unless some reinfection or superinfection exclusion mechanism operates [(Webster et al., 2013) and references therein]. Without such restrictions, persistently infected cells may be an environment with a higher probability of recombination than transiently infected cells, assuming comparable genome loads at the sites of replication.

Nonreplicative recombination does not require replication of the viral genome, and has been described upon cotransfection of cells with viral RNA fragments that could not replicate by themselves (Gmyl et al., 2003; Gallei et al., 2004; Agol, 2010; Agol and Gmyl, 2018; Bentley and Evans, 2018). It appears to be a promiscuous event with a required 3'-phosphate in the 5' partner RNA and a 5'-hydroxyl residue in the 3' partner RNA mediated by cellular components whose mechanisms of activity are not understood.

## 2.9.1 Molecular occurrence versus observed recombination

The emerging picture is that the frequency of recombination varies among viruses, and that as new tools for genome analyses have become available, recombination has been detected in an increasing number of viruses. Recognition of recombination in a viral system is facilitated when a cell culture system is available. Controlled infection of cells with genetically marked parental viruses has been essential to estimate recombination frequencies, and to distinguish true recombination from mutation-reversion events that may mimic the formation of recombinants. As with the concept of high genetic variation in RNA viruses, recombination has often gone from being considered marginal to prominent and relevant; HCV is a typical example (Galli and Bukh, 2014), with presently at least one chimera established as a circulating recombinant form in the field.

Viral replicative machineries may be endowed with features that influence the occurrence of recombination. One such feature is processivity of the viral polymerase (capacity of continued copying of the same template molecule). Genome detachment of the polymerase complex from one genome to bind either to a different genome or to a distant site of the same genome is part of the standard replicative cycle of viruses, such as retroviruses and coronaviruses. Reverse transcriptase participates in strand transfer during DNA synthesis, and coronavirus polymerase switches from one template site to another during discontinuous RNA synthesis. It may be significant that they belong to viral families displaying high recombination frequencies (Makino et al., 1986). Thus, here again, we encounter a "molecular instruction" that evolved as an essential feature of viral genome replication, and that can be exploited to generate variation, and permit new genomic forms to undergo the scrutiny of selection (Neher and Leitner, 2010) (Chapter 3).

Recombination is diagnosed by discordant positions of different genes or genomic regions in phylogenetic trees, as a result of the transfer of part of a viral genome from representatives of one lineage to representatives of another lineage (Chapter 7). A commonly used procedure measures similarity values between sequences using a sliding-window scanning method. The recombination crossover point (where the two parental sequences meet) is identified by the point (or region) where the similarity plot crosses from one sequence into another (Salemi and Vandamme, 2004; Martin et al., 2005; Kosakovsky Pond et al., 2006, Perez-Losada et al., 2015). Crossover points along a viral genome are not distributed at random, either because polymerase detachment from the template is sequence-dependent or because many of the recombination events do not lead to viable progeny. Absence of recombinant viability introduces a parallel with the distinction between mutation rate and mutation frequency (Section 2.5); that is, a difference between what does occur at the biochemical level during replication and what is subsequently observed upon analysis of the replication products. Newly arising recombinants may be subjected to negative selection, and only a viable subset might be detectable in the progeny virus (King, 1988; Lai, 1992). An elegant study by D. J. Evans and colleagues documented "imprecise" enterovirus recombinant intermediates that were lost upon serial virus passage (Lowry et al., 2014). Recombination is viewed as a biphasic process consisting of initial imprecise events followed by a stage of resolution in favor of fit recombinants.

The distinction between generation and resolution events that applies both to mutants and recombinants has yet another implication for RNA virus genetics. Some mutants or recombinants that in isolation do not exhibit sufficient replicative fitness to acquire dominance in a population may nevertheless persist as minority genomes. They may display low-level replication or be maintained by complementation by partner genomes (as in the case of two FMDV genome segments that are described in Section 2.11). As minority genomes, they may engage in modulatory activities (Chapter 3).

## 2.10 Genome segment reassortment

In viruses whose genomes are composed of two or more RNA or DNA segments, genome segment reassortment consists in the formation of new constellations of viral genomic segments from two or more parental genomes (McDonald et al., 2016) (Fig. 2.7). Reassortment can produce new phenotypic traits. It is the main mechanism of antigenic shift of influenza A viruses—often associated with new influenza pandemics (Webster et al., 1992; Morse, 1994; Gibbs et al., 1995; Domingo et al., 2001)—as opposed to antigenic drift, which is mediated by amino acid substitutions in the surface proteins hemagglutinin and neuraminidase (Barbezange et al., 2018). Reassortments occur among the 9−12 double-stranded RNA segments of the widespread *Reoviridae* family (Tanaka et al., 2012). Fitness differences among all possible segment combinations ($2^n$, for two types of coinfecting particles with n genome segments) determine the types of genome segment groupings that dominate subsequent rounds of infection. In the laboratory, analysis of reassortant viruses has been applied to map a viral function into one segment or a combination of segments.

Genomic segments can be encapsidated either into a single virus particle (as in Orthomyxoviruses or Arenaviruses) or into separate particles (as in multipartite plant viruses). A multipartite virus can have either RNA or DNA as genetic material. The plant Nanoviruses have 6-8 molecules of single-stranded circular DNA of about 900—1000 nucleotides, and each segment encodes a single protein. In the case of the nanovirus Faba bean necrotic stunt virus, its eight segments vary in frequency in a host-dependent manner (Sicard et al., 2013). This observation led the authors to propose a "set-point genome formula," which may reflect the control of segment (gene) copy number that may provide some still unrecognized benefit to the multipartite phenotype (see next Section 2.11). In principle, replication of multipartite viruses requires that each cell be coinfected by at least one of each type of particle harboring a different genome type, which in fact represents a remarkable cost for replicative efficiency. The fact that unsegmented and segmented RNA viruses are well represented in our biosphere suggests that neither of the two organizations confers a definitive and general advantage for long-term survival.

## 2.11 Transition toward viral genome segmentation: implications for general evolution

The origin of viral genome segmentation is a debated issue, although there is general agreement that it may confer adaptive flexibility to viruses. Most proposals have been based on theoretical studies. Segmentation has been viewed as a form of sex that facilitates genomic exchanges to counteract the effect of deleterious mutations (Chao, 1988; Szathmary, 1992). An alternative, not mutually exclusive model is that segmentation confers an advantage because replication of shorter RNA molecules is completed earlier than the unsegmented counterparts (Nee, 1987). Yet another possibility is that the lifestyle of a virus (in particular, the particle yield in connection with the number of surrounding susceptible cells), shaped over long evolutionary periods, may favor segmentation over intactness of a genetic message or vice versa.

An experimental system of genome segmentation is available with the picornavirus foot-and-mouth disease virus (FMDV). Its single-stranded RNA genome underwent a modification akin to genome segmentation when the standard virus was subjected to 260 passages in BHK-21 cells at high multiplicity of infection. The experiment was originally intended to investigate the limits of fitness gain following prolonged multiplication in a defined environment, in this case, BHK-21 cells in culture. The starting FMDV had not been well adapted to the BHK-21 cell culture environment since it derived from a diseased swine during a disease outbreak, and it was minimally propagated in BHK-21 cells to obtain a biological clone by plaque isolation. Upon extensive replication of the clone in BHK-21 cells, the virus evolved toward a bipartite genome (García-Arriaza et al., 2004). Each of the two pieces of RNA that composed the bipartite (or segmented) genome version contained in-frame deletions affecting *trans*-acting proteins (Fig. 2.8). Each segment in isolation could not infect cells productively, but, when present together, they were infectious by complementation, and killed cells in the absence of standard FMDV. A low multiplicity of infection rapidly selected the full-length genome as a result of recombination of the two parental, defective segments (García-Arriaza et al., 2004). The particles containing the shortened RNA were thermally more stable than the standard particles (Ojosnegros et al., 2011), but this difference did not explain the initial trigger of the segmentation event. The solution to this question came with the demonstration that the transition toward genome segmentation was possible because of an extensive exploration of the mutational sequence space by the standard virus. Indeed, the mutations that accumulated during serial passages enhanced the fitness of the segmented

**FIGURE 2.8** Evolution toward RNA genome segmentation in the laboratory. The monopartite, standard FMDV genome (clone C-S8c1 or pMT28, top) was subjected to 260 passages in BHK-21 cells. The resulting population p260 lacked detectable standard genome that could be rescued by low MOI passages. The evolved C-S8p260 accumulated 30 point mutations (depicted as *vertical lines* on the genome at the bottom) and consisted in two segments that were infections by complementation: Δ417, that lacked most of the L protease-coding region, and Δ999 that lacked most of the capsid proteins VP3, VP1-coding region. See text for further details and references.

genome version to a much higher extent than the fitness of the standard genome (Moreno et al., 2014) (Fig. 2.8). Thus, gradual evolution (drift in sequence space) was a requirement for the major transition toward segmentation, thus adding reassortment to mutation and recombination as potential mechanisms of genetic variation of this laboratory-adapted picornavirus. Cooperation and complementation are discussed in Section 3.8 of Chapter 3.

It should be noted that segmented forms of RNA viruses have been engineered, but little is known of the incurred fitness cost; it is relevant to have shown that segmentation was possible upon unperturbed replication of a virus with an unsegmented RNA genome. The experimental result suggests that in evolution there is no unsurmountable barrier that allows the conversion between intact and split forms of the same genome, reflecting remarkable genome flexibility that will be emphasized in Chapter 7 in the context of the relevance of virus variation in the emergence of viral pathogens.

## 2.12 Mutation, recombination, and reassortment as individual and combined evolutionary forces

Mutation, recombination, and segment reassortment contribute to the evolution of most DNA and RNA viruses. Sometimes one form of genetic change appears to be more prominent than another, and sometimes the concerted action of recombination or reassortment with the mutation is apparent [i.e., antigenic drift in influenza virus, following the origin of a new antigenic type through reassortment (Ghedin et al., 2005)]. A mutation is a universal form of genetic change. It underlies numerous adaptive responses and critical biological transitions in viruses, and it is a prerequisite for recombination and reassortment to have a biological impact. If mutations were not present in different template molecules during replication, recombinants with the crossover point at equivalent positions of the parental genomes would be "silent," and display the same behavior as the parental genomes. Apparently, "silent" recombination events may take place within replicative units; even if some mutations distinguished individual genomes of the same quasispecies swarm, a recombinant would not be distinguished from a mutant genome. The frequency of recombination in HIV-1 was noticed only when the acquired immune deficiency syndrome (AIDS) pandemic had advanced, and the virus had diversified through the accumulation of mutations. Similar arguments apply to segment reassortment. Genomes necessitate mutation-driven diversification for reassortment to provide a biological difference; detection of a reassortant will be easier the larger the replicative advantage it confers to the virus (see also Chapter 10).

The evolutionary significance of recombination has been viewed in two opposite ways: as a means to rescue fit genomes from less fit parents (a conservative force that eliminates deleterious mutations), or as a means to explore new genomic forms for adaptive potential (a vast substrate for the exploration of sequence space; Chapter 3) [reviewed in (Zimmern, 1988; Lai, 1992; Worobey and Holmes, 1999; Simmonds, 2010; Perez-Losada et al., 2015)]. Recombination has been probably at the origin of new viruses that presently occupy a well-established niche, and it is also at play today to expand diversity during the spread of viruses. As a historical event, the coronavirus mouse hepatitis virus appears to have acquired its hemagglutinin-esterase gene by recombination with an influenza C virus. The alphavirus Western equine encephalitis virus originated probably by recombination between Sindbis-like and Eastern equine encephalitis-like viruses [reviewed in different chapters of (Domingo et al., 2008)].

Several recent poliomyelitis outbreaks have been associated with recombinants between oral poliovirus vaccine (OPV) viruses and other circulating enteroviruses (Gavrilin et al., 2000; Kew et al., 2002; Oberste et al., 2004; Muslin et al., 2015). Intersubtype HIV-1 recombinants play a key role in current HIV-1 diversification, with around 100 circulating recombinant forms (and the number is growing) displaying complex mosaic structures (multiple crossover sites) (Thomson et al., 2002; Gerhardt et al., 2005). In addition, other HIV-1 recombinants have been characterized that are not established epidemiologically. Fewer HCV recombinants have been identified, but the number is likely to increase as the virus diversifies in nature. Positive selection of HIV-1 recombinants that unite different drug-resistant mutations in the same genome offers an example of the conservative force of recombination to rescue fit viruses in the face of a strong selective constraint (Menéndez-Arias, 2002). Recombination is expected to play an increasing role in the spread of drug resistance among viruses for which new antiviral agents are in use, such as HBV and HCV.

Some defective DNA and RNA genomes that include *indels*, notably DI RNAs, which originate from recombination events can play an important role in the establishment and maintenance of persistent infections in cell culture, and can modulate viral infections in vivo (Holland and Villarreal, 1974; Roux et al., 1991; Rezelj et al., 2018). Detailed genetic and biochemical analyses by A. Huang, J.J. Holland and their colleagues on the generation of VSV DI's and their interplay with the standard, infectious VSV contributed to unveil a continuous dynamics of genetic variation, competition, and selection, observable within short time intervals, a hallmark of RNA genetics (Palma and Huang, 1974; Holland et al., 1982), fully confirmed by application of new sequencing techniques. DI particles and defective genomes are present in populations of positive- and negative-strand RNA viruses as they multiply in their natural hosts (Nüesch et al., 1989; Drolet et al., 1995; Li et al., 2011; Saira et al., 2013; Ke et al., 2013; Rezelj et al., 2018). Their widespread presence in vivo may mean that they are an unavoidable side-product of the replication machineries (i.e., instruction to recombine) or that selection might have favored their generation. Both possibilities are compatible. An instruction whose result is a means to modulate replication of the corresponding standard viruses or the antiviral immune response will be selected as a consequence of its biological effects.

DI RNAs can be regarded as the tip of the iceberg of many classes of defective genomes with a range of interfering or potentiating capacities that may coexist with standard animal, plant, insect, and bacterial viruses, and that may facilitate persistence and modulate disease symptoms (Holland et al., 1982; Vogt and Jackson, 1999; López-Ferber et al., 2003; Rosario et al., 2005; Sachs and Bull, 2005; Villarreal, 2005; Aaskov et al., 2006; Rezelj et al., 2018). Noncytopathic coxsackievirus B3 (CVB3) variants with deletions at the untranslated 5'-genomic region were isolated from hearts of mice inoculated with CVB3. The variants replicated in vivo and were associated with long-term viral persistence (Kim et al., 2005).

Despite the continuous dynamics of the escape of infectious virus to the interfering activities of DIs, some authors consider DIs as potential antiviral agents (Dimmock and Easton, 2014). If defective genomes are competent in RNA (or DNA) synthesis or are complemented to replicate, they can act as dominant-negative swarms, provided they reach a sufficient load. In this manner, they may underlie the suppressive effects of mutant spectra of viral quasispecies. Intramutant spectrum decrease of replicative capacity due to the presence of defective genomes is one of the mechanisms of virus extinction evoked by enhanced mutagenesis (Chapters 3 and 9).

Recombination events must have been the last step in ancestral processes of horizontal gene transfer that mediated the incorporation of host genes (or gene segments) into viral genomes, and vice versa. Host genes related to immune responses were probably captured by complex DNA viruses at early stages of their evolution (Alcami, 2003; McFadden, 2005). Mosaicism associated with nonhomologous recombination events is the norm among tailed bacteriophages (Canchaya et al., 2003). Nonhomologous recombination can give rise to genomic sequences with a viral and a nonviral moiety. They include DI RNAs of Sindbis virus-containing cellular RNA sequences at their 5' ends, some cytopathic forms of bovine viral diarrhea virus, RNA of potato leafroll virus-containing tobacco chloroplast

RNA, or an influenza virus with an insertion of ribosomal RNA into the hemagglutinin gene, mentioned in Chapter 5 regarding transient, high fitness levels [reviewed in (Domingo et al., 2001)].

Phylogenetic analyses have suggested that recombination between RNA and DNA viruses might have occurred to give rise to some present-day single-stranded DNA viruses (Stedman, 2013). However, the evidence for this attractive possibility is indirect and, to my knowledge, no experimental evidence of viral RNA-viral DNA recombination in cell culture or in vivo has been reported. Viability of mutant and recombinant viral genomes is severely constrained by the evolutionary history of the virus that has shaped viral genomes as coordinated sets of modules (Botstein, 1980, 1981; Zimmern, 1988; Koonin and Dolja, 2014). Experimental studies with engineered recombinant viruses have shown that modularity can restrict recombination (Martin et al., 2005).

The three molecular mechanisms of viral genome variation (mutation, recombination, and reassortment) are not incompatible, although it may sometimes be difficult to discern their occurrence (Varsani et al., 2018). It would be truly remarkable if a viral system could be proven to be totally devoid of one of the mechanisms of genetic variation. It would imply that there are powerful molecular reasons to dispense with an effective adaptive mechanism. Absence of a mechanism is extremely difficult to demonstrate but, if we could, its basis would open a new chapter of molecular virology.

## 2.12.1 Mechanistically unavoidable versus evolutionarily relevant genetic variation

There is an ongoing controversy regarding clonality versus nonclonality in biological evolution, particularly concerning the evolution of cellular parasites (Heitman, 2010; Tibayrenc and Ayala, 2012, 2014; Ramirez and Llewellyn, 2014; Hauser and Cushion, 2018). Clonality means asexual progeny from a single ancestor. In the case of viruses, clonal evolution emphasizes reproduction without the exchange of genetic material among two or more parental genomes. Sexual reproduction necessarily involves the exchange of genetic material. The question for viruses is interesting because we would be inclined to propose clonal evolution despite considerable promiscuity of recombination and reassortment. A tentative solution was offered based on one assumption and some experimental observations. The assumption is that recombination is not a requirement for viruses to complete their infection cycles. Despite the possibility that recombination might be imprinted in the replication apparatuses (rendering it inherent to replication), its occurrence is not a necessity. The experimental observations are of two sorts. One is that historically recombination and reassortment have been at the origin of the emergence and reemergence of viral pathogens [western equine encephalitis virus, pandemic influenza viruses, emergent circulating poliovirus, HIV-1 and HCV recombinants, etc. (Section 2.12)]. The second observation is that recombination-based genome segmentation can occur given the adequate population dynamics and competitive environment, as documented with FMDV (Section 2.11). In consequence, the distinction was made between mechanistically unavoidable but biologically irrelevant, and meaningful recombination (Perales et al., 2015). The latter form of recombination requires prior diversification of parental genomes by mutation and a number of cellular and epidemiological conditions. Despite its relevance to evolutionary transitions and viral emergence, it is not a requirement for virus survival, propagation, and evolution. This marks a contrast with the genomic exchanges associated with sexual reproduction. The proposal is that viruses evolve clonally at widely different time scales (intrahost or within-host evolution vs. long-term evolution at the epidemiological

level). Similar arguments apply to mutation. This point will be addressed again in the closing Chapter 10, concerning implications of clonality.

## 2.13  Overview and concluding remarks

All forms of genetic variation of viruses must be viewed essentially as blind processes despite preferences of nucleotide sequences or structures for mutation and recombination events: *hot spots* with higher than average rate values, and *cold spots* with lower than average rate values. Mutation originates largely in fluctuations of electronic structure that modify base-pairing properties, and from features of polymerase-template interactions, not subject to regulation, in the sense that we understand the regulation of gene expression or enzymatic activity. Absence of regulation is not incompatible with long-term evolution having shaped the molecular interactions that yield a level of mutagenesis compatible with survival and adaptability. Given the biological consequences of mutation rates, many additional studies are needed for the biological phyla, to quantity not only basal mutation rates but also the possible presence of mutator strains.

Similar arguments can be used for recombination and reassortment. The number of segments that enter a new genomic constellation may be regulated but not which variant forms of the individual segments will make up the new viral particles. It is short-term selection acting at the very center of replication and recombinant complexes that preserves some mutant and recombinant forms in detriment of others. Subsequent levels of selection occur when variant forms expand in multiple rounds of infection first within cells, then within an organism and then at the epidemiological level. The very nature of life in our planet has been built upon an inherent tendency to instruct variation in an incessant fashion, as necessary and unavoidable as the physical principles that dictate the behavior of our universe.

The net result of all mechanisms of genetic variation available to a virus is the generation of repertoires of variant genomes for random drift and selective forces to act upon. In other terms, genetic variation sets the scene for the actors of evolution to play their roles, and secure a continuous input of new forms despite subtle or catastrophic environmental perturbations. The same forces that drive general evolution have produced the dominant virus forms we see in nature, with all their nuances in the interaction with cell components. The adaptation of viruses to participate in intracellular processes with cells dictates that genetic variation of viruses has its limits to prevent deleteriousness. This is currently exemplified by the effects of amino acid substitutions in viral polymerases that either increase or decrease template-copying fidelity. Viruses have reached a compromise between the stability of core information and flexibility for adaptability. Although not yet treated in this chapter, viral population numbers are a key parameter in the evolutionary events. Next chapters address some of these questions, not only in general conceptual terms but also in the way evolution affects our daily confrontation with viral disease (see Summary Box).

---

## Summary Box

- Mutation, recombination, and genome segment reassortment are the mechanisms of genetic variation used by DNA and RNA viruses. Mutations are due mainly to changes in the electronic distributions of the standard nucleotides, to damage of nucleotides by external influences, and by alignment alterations of the template relative to product polynucleotide chains. The effect of a mutation can range from being well-tolerated to highly detrimental or lethal.

- Mutation frequencies are only an indirect consequence of mutation rates. Their values for viruses whose replication is catalyzed by polymerases devoid of proofreading-repair activity are $10^5$- to $10^6$-fold higher than those displayed by replicative cellular DNA polymerases. Error-prone replication is a hallmark of RNA viruses and some DNA viruses. The larger the amount of genetic information encoded in a viral genome, the lower the mutation rate must be to maintain the genetic message.

- Several mechanisms of genetic recombination have been described for DNA and RNA viruses. The best characterized is homologous recombination whose frequency of occurrence is dependent on the replicative machinery, in particular, polymerase processivity. Genome segment reassortment is operative in segmented genomes, and it gives rise to biologically relevant changes, such as an antigenic shift in the influenza type A viruses.

- Studies with foot-and-mouth disease virus have shown that extensive evolution of an unsegmented RNA genome has the potential to undergo a recombination-mediated transition akin to genome segmentation. Therefore, segmented and unsegmented forms of RNA viruses need not be considered as completely unrelated classes of genome organization.

- Recombination and genome segment reassortment have been viewed as conservative forces to rescue viable genomes from a damaged pool, and also as a means to explore new genomic compositions that deviate from their parents. All forms of genetic variation give rise to repertoires of variant genomes on which selection and random drift act to produce the viral forms that we isolate and study.

# References

Aaskov, J., Buzacott, K., Thu, H.M., Lowry, K., Holmes, E.C., 2006. Long-term transmission of defective RNA viruses in humans and Aedes mosquitoes. Science 311, 236–238.

Achaz, G., 2009. Frequency spectrum neutrality tests: one for all and all for one. Genetics 183, 249–258.

Agol, V.I., 2010. Picornaviruses as a model for studying the nature of RNA recombination. In: Ehrenfeld, E., Domingo, E., Roos, R.P. (Eds.), The Picornaviruses. ASM Press, Washington, DC, pp. 239–252.

Agol, V.I., Gmyl, A.P., 2018. Emergency services of viral RNAs: repair and remodeling. Microbiol. Mol. Biol. Rev. 82 e00067-17.

Aguilera, A., Garcia-Muse, T., 2013. Causes of genome instability. Annu. Rev. Genet. 47, 1–32.

Akashi, H., 2001. Gene expression and molecular evolution. Curr. Opin. Genet. Dev. 11, 660–666.

Alberts, B., Johnson, A., Lewis, J., Raff, M., Roberts, K., et al., 2002. Molecular Biology of the Cell. Garland Science, New York, N.Y.

Alcami, A., 2003. Viral mimicry of cytokines, chemokines and their receptors. Nat. Rev. Immunol. 3, 36–50.

Alejska, M., Figlerowicz, M., Malinowska, N., Urbanowicz, A., Figlerowicz, M., 2005. A universal BMV-based RNA recombination system—how to search for general rules in RNA recombination. Nucleic Acids Res. 33, e105.

Aragones, L., Guix, S., Ribes, E., Bosch, A., Pinto, R.M., 2010. Fine-tuning translation kinetics selection as the driving force of codon usage bias in the hepatitis a virus capsid. PLoS Pathog. 6, e1000797.

Arias, A., Arnold, J.J., Sierra, M., Smidansky, E.D., Domingo, E., et al., 2008. Determinants of RNA-dependent RNA polymerase (in)fidelity revealed by kinetic analysis of the polymerase encoded by a foot-and-mouth disease virus mutant with reduced sensitivity to ribavirin. J. Virol. 82, 12346–12355.

Arnold, J.J., Vignuzzi, M., Stone, J.K., Andino, R., Cameron, C.E., 2005. Remote site control of an active site fidelity checkpoint in a viral RNA-dependent RNA polymerase. J. Biol. Chem. 280, 25706–25716.

Barbezange, C., Jones, L., Blanc, H., Isakov, O., Celniker, G., et al., 2018. Seasonal genetic drift of human influenza A virus quasispecies revealed by deep sequencing. Front. Microbiol. https://doi.org/10.3389/fmicb.2018.02596.

Barrett, J.W., McFadden, G., 2008. Origin and evolution of poxviruses. In: Domingo, E., Parrish, C.R., Holland, J.J. (Eds.), Origin and Evolution of Viruses, second ed. Elsevier, Oxford, pp. 431–446.

Batschelet, E., Domingo, E., Weissmann, C., 1976. The proportion of revertant and mutant phage in a growing population, as a function of mutation and growth rate. Gene 1, 27–32.

Beaucourt, S., Vignuzzi, M., 2014. Ribavirin: a drug active against many viruses with multiple effects on virus replication and propagation. Molecular basis of ribavirin resistance. Curr. Opin. Virol. 8C, 10–15.

Bebenek, A., Ziuzia-Graczyk, I., 2018. Fidelity of DNA replication – a matter of proofreading. Curr. Genet. 64, 985–996.

Bentley, K., Evans, D.J., 2018. Mechanisms and consequences of positive-strand RNA virus recombination. J. Gen. Virol. 99, 1345–1356.

Bernad, A., Blanco, L., Lazaro, J.M., Martin, G., Salas, M., 1989. A conserved 3′- 5′ exonuclease active site in prokaryotic and eukaryotic DNA polymerases. Cell 59, 219–228.

Bloomfield, V.A., Crothers, D.M., Tinoco Jr., I., 2000. Nucleic Acids. Structures, Properties, and Functions. University Science Books, Sausalito, CA.

Boehmer, P.E., Nimonkar, A.V., 2003. Herpes virus replication. IUBMB Life 55, 13–22.

Borderia, A.V., Rozen-Gagnon, K., Vignuzzi, M., 2016. Fidelity variants and RNA quasispecies. Curr. Top. Microbiol. Immunol. 392, 303–322.

Botstein, D., 1980. A theory of modular evolution for bacteriophages. Ann. NY Acad. Sci. 354, 484–491.

Botstein, D., 1981. A modular theory of virus evolution. In: Fields, B.N., Jaenisch, R., Fox, C.F. (Eds.), Animal Virus Genetics. Academic Press, New York, pp. 363–384.

Bradwell, K., Combe, M., Domingo-Calap, P., Sanjuan, R., 2013. Correlation between mutation rate and genome size in riboviruses: mutation rate of bacteriophage Qbeta. Genetics 195, 243–251.

Brest, P., Lapaquette, P., Souidi, M., Lebrigand, K., Cesaro, A., et al., 2011. A synonymous variant in IRGM alters a binding site for miR-196 and causes deregulation of IRGM-dependent xenophagy in Crohn's disease. Nat. Genet. 43, 242–245.

Brule, C.E., Grayhack, E.J., 2017. Synonymous codons: choose wisely for expression. Trends Genet. 33, 283–297.

Bujarski, J.J., 2013. Genetic recombination in plant-infecting messenger-sense RNA viruses: overview and research perspectives. Front. Plant Sci. 4, 68.

Bushman, F., 2002. Lateral DNA Transfer. Mechanisms and Consequences. Cold Spring Harbor Laboratory Press, Cold Spring Harbor, New York.

Campagnola, G., McDonald, S., Beaucourt, S., Vignuzzi, M., Peersen, O.B., 2015. Structure-function relationships underlying the replication fidelity of viral RNA-dependent RNA polymerases. J. Virol. 89, 275–286.

Canchaya, C., Fournous, G., Chibani-Chennoufi, S., Dillmann, M.L., Brussow, H., 2003. Phage as agents of lateral gene transfer. Curr. Opin. Microbiol. 6, 417–424.

Castro, C., Arnold, J.J., Cameron, C.E., 2005. Incorporation fidelity of the viral RNA-dependent RNA polymerase: a kinetic, thermodynamic and structural perspective. Virus Res. 107, 141–149.

Cattaneo, R., Billeter, M.A., 1992. Mutations and A/I hypermutations in measles virus persistent infections. Curr. Top. Microbiol. Immunol. 176, 63–74.

Chao, L., 1988. Evolution of sex in RNA viruses. J. Theor. Biol. 133, 99–112.

Chetverin, A.B., Kopein, D.S., Chetverina, H.V., Demidenko, A.A., Ugarov, V.I., 2005. Viral RNA-directed RNA polymerases use diverse mechanisms to promote recombination between RNA molecules. J. Biol. Chem. 280, 8748–8755.

Cheynier, R., Kils-Hutten, L., Meyerhans, A., Wain-Hobson, S., 2001. Insertion/deletion frequencies match those of point mutations in the hypervariable regions of the simian immunodeficiency virus surface envelope gene. J. Gen. Virol. 82, 1613–1619.

Coffey, L.L., Beeharry, Y., Borderia, A.V., Blanc, H., Vignuzzi, M., 2011. Arbovirus high fidelity variant loses fitness in mosquitoes and mice. Proc. Natl. Acad. Sci. U.S.A. 108, 16038–16043.

Coffin, J.M., 1990. Genetic variation in retroviruses. Appl. Virol. Res. 2, 11–33.

Combe, M., Sanjuan, R., 2014. Variation in RNA virus mutation rates across host cells. PLoS Pathog. 10, e1003855.

Crandall, K.A., Kelsey, C.R., Imamichi, H., Lane, H.C., Salzman, N.P., 1999. Parallel evolution of drug resistance in HIV: failure of nonsynonymous/synonymous substitution rate ratio to detect selection. Mol. Biol. Evol. 16, 372–382.

Czarnecki, M.W., Traktman, P., 2017. The vaccinia virus DNA polymerase and its processivity factor. Virus Res. 234, 193–206.

Czech, A., Fedyunin, I., Zhang, G., Ignatova, Z., 2010. Silent mutations in sight: co-variations in tRNA abundance as a key to unravel consequences of silent mutations. Mol. Biosyst. 6, 1767–1772.

de Vega, M., Lazaro, J.M., Mencia, M., Blanco, L., Salas, M., 2010. Improvement of phi29 DNA polymerase amplification performance by fusion of DNA binding motifs. Proc. Natl. Acad. Sci. U.S.A. 107, 16506–16511.

Denny, S.K., Greenleaf, W.J., 2018. Linking RNA sequence, structure, and function on massively parallel high-throughput sequences. Cold Spring Harb. Perspect. Biol. https://doi.org/10.1101/cshperspect.a 032300.

Dimmock, N.J., Easton, A.J., 2014. Defective interfering influenza virus RNAs: time to reevaluate their clinical potential as broad-spectrum antivirals? J. Virol. 88, 5217–5227.

Domingo, E., 2005. Virus entry into error catastrophe as a new antiviral strategy. Virus Res. 107 (Special Issue), 115–228.

Domingo, E., Perales, C., 2019. Viral Quasispecies. PLoS Genetics in press.

Domingo, E., Sabo, D., Taniguchi, T., Weissmann, C., 1978. Nucleotide sequence heterogeneity of an RNA phage population. Cell 13, 735–744.

Domingo, E., Mateu, M.G., Martínez, M.A., Dopazo, J., Moya, A., et al., 1990. Genetic variability and antigenic diversity of foot-and-mouth disease virus. In: Kurkstak, E., Marusyk, R.G., Murphy, S.A., Van-Regenmortel, M.H.V. (Eds.), Applied Virology Research. Plenum Publishing Co., New York, pp. 233–266.

Domingo, E., Holland, J.J., Biebricher, C., Eigen, M., 1995. Quasispecies: the concept and the word. In: Gibbs, A., Calisher, C., García-Arenal, F. (Eds.), Molecular Evolution of the Viruses. Cambridge University Press, Cambridge, pp. 171–180.

Domingo, E., Biebricher, C., Eigen, M., Holland, J.J., 2001. Quasispecies and RNA Virus Evolution: Principles and Consequences. Landes Bioscience, Austin.

Domingo, E., Escarmis, C., Baranowski, E., Ruiz-Jarabo, C.M., Carrillo, E., et al., 2003. Evolution of foot-and-mouth disease virus. Virus Res. 91, 47–63.

Domingo, E., Escarmís, C., Menéndez-Arias, L., Perales, C., Herrera, M., et al., 2008. Viral quasispecies: dynamics, interactions and pathogenesis. In: Domingo, E., Parrish, C., Holland, J.J. (Eds.), Origin and Evolution of Viruses. Elsevier, Oxford, pp. 87–118.

Drake, J.W., 1991. A constant rate of spontaneous mutation in DNA-based microbes. Proc. Natl. Acad. Sci. U.S.A. 88, 7160–7164.

Drake, J.W., Holland, J.J., 1999. Mutation rates among RNA viruses. Proc. Natl. Acad. Sci. U.S.A. 96, 13910–13913.

Drolet, B.S., Chiou, P.P., Heidel, J., Leong, J.A., 1995. Detection of truncated virus particles in a persistent RNA virus infection *in vivo*. J. Virol. 69, 2140–2147.

Earl, D.J., Deem, M.W., 2004. Evolvability is a selectable trait. Proc. Natl. Acad. Sci. U.S.A. 101, 11531–11536.

Eckerle, L.D., Lu, X., Sperry, S.M., Choi, L., Denison, M.R., 2007. High fidelity of murine hepatitis virus replication is decreased in nsp14 exoribonuclease mutants. J. Virol. 81, 12135–12144.

Eckerle, L.D., Becker, M.M., Halpin, R.A., Li, K., Venter, E., et al., 2010. Infidelity of SARS-CoV Nsp14-exonuclease mutant virus replication is revealed by complete genome sequencing. PLoS Pathog. 6, e1000896.

Eigen, M., 2002. Error catastrophe and antiviral strategy. Proc. Natl. Acad. Sci. U.S.A. 99, 13374–13376.

Eigen, M., 2013. From Strange Simplicity to Complex Familiarity. Oxford University Press, Oxford.

Eigen, M., Biebricher, C.K., 1988. Sequence space and quasi-species distribution. In: Domingo, E., Ahlquist, P., Holland, J.J. (Eds.), RNA Genetics. CRC Press Inc, Boca Raton, FL, pp. 211–245.

Elena, S.F., Sanjuan, R., 2005. Adaptive value of high mutation rates of RNA viruses: separating causes from consequences. J. Virol. 79, 11555–11558.

Escarmís, C., Dávila, M., Charpentier, N., Bracho, A., Moya, A., et al., 1996. Genetic lesions associated with Muller's ratchet in an RNA virus. J. Mol. Biol. 264, 255–267.

Escarmís, C., Dávila, M., Domingo, E., 1999. Multiple molecular pathways for fitness recovery of an RNA virus debilitated by operation of Muller's ratchet. J. Mol. Biol. 285, 495–505.

Fan, X., Di Bisceglie, A.M., 2010. RT-PCR amplification and cloning of large viral sequences. Methods Mol. Biol. 630, 139–149.

Faure, G., Ogurtsov, A.Y., Shabalina, S.A., Koonin, E.V., 2017. Adaptation of mRNA structure to control protein folding. RNA Biol. 14, 1649–1654.

Ferrer-Orta, C., Arias, A., Escarmis, C., Verdaguer, N., 2006. A comparison of viral RNA-dependent RNA polymerases. Curr. Opin. Struct. Biol. 16, 27–34.

Ferrer-Orta, C., Agudo, R., Domingo, E., Verdaguer, N., 2009. Structural insights into replication initiation and elongation processes by the FMDV RNA-dependent RNA polymerase. Curr. Opin. Struct. Biol. 19, 752–758.

Forloni, M., Liu, A.Y., Wajapeyee, N., 2018. Random mutagenesis using error-prone DNA polymerases. Cold Spring Harb. Protoc. https://doi.org/10.1101/pdb.prot097741.

Friedberg, E.C., Wagner, R., Radman, M., 2002. Specialized DNA polymerases, cellular survival, and the genesis of mutations. Science 296, 1627–1630.

Friedberg, E.C., Walker, G.C., Siede, W., Wood, R.D., Schultz, R.A., et al., 2006. DNA Repair and Mutagenesis. American Society for Microbiology, Washington, DC.

Fu, Y.X., 1997. Statistical tests of neutrality of mutations against population growth, hitchhiking and background selection. Genetics 147, 915–925.

Gago, S., Elena, S.F., Flores, R., Sanjuan, R., 2009. Extremely high mutation rate of a hammerhead viroid. Science 323, 1308.

Gallei, A., Pankraz, A., Thiel, H.J., Becher, P., 2004. RNA recombination in vivo in the absence of viral replication. J. Virol. 78, 6271–6281.

Galli, A., Bukh, J., 2014. Comparative analysis of the molecular mechanisms of recombination in hepatitis C virus. Trends Microbiol. 22, 354–364.

Ganai, R.A., Johansson, E., 2016. DNA replication-a matter of fidelity. Mol. Cell 62, 745–755.

García-Arriaza, J., Manrubia, S.C., Toja, M., Domingo, E., Escarmís, C., 2004. Evolutionary transition toward defective RNAs that are infectious by complementation. J. Virol. 78, 11678–11685.

Gatenby, R.A., Frieden, B.R., 2004. Information dynamics in carcinogenesis and tumor growth. Mutat. Res. 568, 259–273.

Gavrilin, G.V., Cherkasova, E.A., Lipskaya, G.Y., Kew, O.M., Agol, V.I., 2000. Evolution of circulating wild poliovirus and of vaccine-derived poliovirus in an immunodeficient patient: a unifying model. J. Virol. 74, 7381–7390.

Gerhardt, M., Mloka, D., Tovanabutra, S., Sanders-Buell, E., Hoffmann, O., et al., 2005. In-depth, longitudinal analysis of viral quasispecies from an individual triply infected with late-stage human immunodeficiency virus type 1, using a multiple PCR primer approach. J. Virol. 79, 8249–8261.

Ghedin, E., Sengamalay, N.A., Shumway, M., Zaborsky, J., Feldblyum, T., et al., 2005. Large-scale sequencing of human influenza reveals the dynamic nature of viral genome evolution. Nature 437, 1162–1166.

Gibbs, A.J., Calisher, C.H., García-Arenal, F. (Eds.), 1995. Molecular Basis of Virus Evolution. Cambridge University Press, Cambridge.

Gmyl, A.P., Pilipenko, E.V., Maslova, S.V., Belov, G.A., Agol, V.I., 1993. Functional and genetic plasticities of the poliovirus genome: quasi- infectious RNAs modified in the 5′-untranslated region yield a variety of pseudorevertants. J. Virol. 67, 6309–6316.

Gmyl, A.P., Korshenko, S.A., Belousov, E.V., Khitrina, E.V., Agol, V.I., 2003. Nonreplicative homologous RNA recombination: promiscuous joining of RNA pieces? RNA 9, 1221–1231.

Gnadig, N.F., Beaucourt, S., Campagnola, G., Borderia, A.V., Sanz-Ramos, M., et al., 2012. Coxsackievirus B3 mutator strains are attenuated in vivo. Proc. Natl. Acad. Sci. U.S.A. 109, E2294–E2303.

Graci, J.D., Cameron, C.E., 2004. Challenges for the development of ribonucleoside analogues as inducers of error catastrophe. Antivir. Chem. Chemother. 15, 1–13.

Hamano, T., Matsuo, K., Hibi, Y., Victoriano, A.F., Takahashi, N., et al., 2007. A single-nucleotide synonymous mutation in the gag gene controlling human immunodeficiency virus type 1 virion production. J. Virol. 81, 1528–1533.

Hauser, P.M., Cushion, M.T., 2018. Is sex necessary for the proliferation and transmission of *Pneumocystis*? PLoS Pathog. 14, e1007409.

Heitman, J., 2010. Evolution of eukaryotic microbial pathogens via covert sexual reproduction. Cell Host Microbe 8, 86–99.

Hirose, Y., Onuki, M., Tenjimbayashi, Y., Mori, S., Ishii, Y., et al., 2018. Within-host variations of human papillomaviruses reveal APOBEC signature mutagenesis in the viral genome. J. Virol. 92, e00017–18.

Hirsch, N., Birnbaum, R.Y., 2015. Dual function of DNA sequences: protein-coding sequences function as transcriptional enhancers. Perspect. Biol. Med. 58, 182–195.

Holland, J.J., Villarreal, L.P., 1974. Persistent noncytocidal vesicular stomatitis virus infections mediated by defective T particles that suppress virion transcriptase. Proc. Natl. Acad. Sci. U.S.A. 71, 2956–2960.

Holland, J.J., Spindler, K., Horodyski, F., Grabau, E., Nichol, S., et al., 1982. Rapid evolution of RNA genomes. Science 215, 1577–1585.

Holland, J.J., de la Torre, J.C., Steinhauer, D.A., Clarke, D., Duarte, E., et al., 1989. Virus mutation frequencies can be greatly underestimated by monoclonal antibody neutralization of virions. J. Virol. 63, 5030–5036.

Jin, Z., Leveque, V., Ma, H., Johnson, K.A., Klumpp, K., 2013. NTP-mediated nucleotide excision activity of hepatitis C virus RNA-dependent RNA polymerase. Proc. Natl. Acad. Sci. U.S.A. 110, E348–E357.

Joyce, G.F., 2004. Directed evolution of nucleic acid enzymes. Annu. Rev. Biochem. 73, 791–836.

Ke, R., Aaskov, J., Holmes, E.C., Lloyd-Smith, J.O., 2013. Phylodynamic analysis of the emergence and epidemiological impact of transmissible defective dengue viruses. PLoS Pathog. 9, e1003193.

Kew, O., Morris-Glasgow, V., Landaverde, M., Burns, C., Shaw, J., et al., 2002. Outbreak of poliomyelitis in Hispaniola associated with circulating type 1 vaccine-derived poliovirus. Science 296, 356–359.

Kim, K.S., Tracy, S., Tapprich, W., Bailey, J., Lee, C.K., et al., 2005. 5′-Terminal deletions occur in coxsackievirus B3 during replication in murine hearts and cardiac myocyte cultures and correlate with encapsidation of negative-strand viral RNA. J. Virol. 79, 7024–7041.

Kimura, M., 1983. The Neutral Theory of Molecular Evolution. Cambridge University Press, Cambridge, UK.

Kimura, M., 1989. The neutral theory of molecular evolution and the world view of the neutralists. Genome 31, 24–31.

King, A.M.Q., 1988. Genetic recombination in positive strand RNA viruses. In: Domingo, E., Holland, J.J., Ahlquist, P. (Eds.), RNA Genetics. CRC Press Inc, Boca Raton, FL, pp. 149–165.

King, J.L., Jukes, T.H., 1969. Non-Darwinian evolution. Science 164, 788–798.

Kirkegaard, K., Baltimore, D., 1986. The mechanism of RNA recombination in poliovirus. Cell 47, 433–443.

Kolakofsky, D., Roux, L., Garcin, D., Ruigrok, R.W., 2005. Paramyxovirus mRNA editing, the "rule of six" and error catastrophe: a hypothesis. J. Gen. Virol. 86, 1869–1877.

Koonin, E.V., Dolja, V.V., 2014. Virus world as an evolutionary network of viruses and capsidless selfish elements. Microbiol. Mol. Biol. Rev. 78, 278–303.

Kosakovsky Pond, S.L., Posada, D., Gravenor, M.B., Woelk, C.H., Frost, S.D., 2006. GARD: a genetic algorithm for recombination detection. Bioinformatics 22, 3096–3098.

Kryazhimskiy, S., Plotkin, J.B., 2008. The population genetics of dN/dS. PLoS Genet. 4, e1000304.

Kunkel, T.A., Erie, D.A., 2005. DNA mismatch repair. Annu. Rev. Biochem. 74, 681–710.

Lafforgue, G., Martinez, F., Sardanyes, J., de la Iglesia, F., Niu, Q.W., et al., 2011. Tempo and mode of plant RNA virus escape from RNA interference-mediated resistance. J. Virol. 85, 9686–9695.

Lai, M.M.C., 1992. Genetic recombination in RNA viruses. Curr. Top. Microbiol. Immunol. 176, 21–32.

Lenski, R.E., Winkworth, C.L., Riley, M.A., 2003. Rates of DNA sequence evolution in experimental populations of *Escherichia coli* during 20,000 generations. J. Mol. Evol. 56, 498–508.

Levy, D.N., Aldrovandi, G.M., Kutsch, O., Shaw, G.M., 2004. Dynamics of HIV-1 recombination in its natural target cells. Proc. Natl. Acad. Sci. U.S.A. 101, 4204–4209.

Li, D., Lott, W.B., Lowry, K., Jones, A., Thu, H.M., Aaskov, J., 2011. Defective interfering viral particles in acute dengue infections. PLoS One 6, e19447.

Ling, J., Reynolds, N., Ibba, M., 2009. Aminoacyl-tRNA synthesis and translational quality control. Annu. Rev. Microbiol. 63, 61–78.

Liu, D., Keijzers, G., Rasmussen, L.J., 2017. DNA mismatch repair and its many roles in eukaryotic cells. Mutat. Res. 773, 174–187.

Liu, W., Shi, X., Gong, P., 2018. A unique intra-molecular fidelity-modulating mechanism identified in a viral RNA-dependent RNA polymerase. Nucleic Acids Res. 46, 10840–10854.

López-Ferber, M., Simon, O., Williams, T., Caballero, P., 2003. Defective or effective? Mutalistic interactions between virus genotypes. Proc. R. Soc. Lond. B Biol. Sci. 270, 2249–2255.

Lowry, K., Woodman, A., Cook, J., Evans, D.J., 2014. Recombination in enteroviruses is a biphasic replicative process involving the generation of greater-than genome length 'imprecise' intermediates. PLoS Pathog. 10, e1004191.

Maas, S., Rich, A., Nishikura, K., 2003. A-to-I RNA editing: recent news and residual mysteries. J. Biol. Chem. 278, 1391–1394.

Makhoul, C.H., Trifonov, E.N., 2002. Distribution of rare triplets along mRNA and their relation to protein folding. J. Biomol. Struct. Dyn. 20, 413–420.

Makino, S., Keck, J.G., Stohlman, S.A., Lai, M.M., 1986. High-frequency RNA recombination of murine coronaviruses. J. Virol. 57, 729–737.

Malpica, J.M., Fraile, A., Moreno, I., Obies, C.I., Drake, J.W., et al., 2002. The rate and character of spontaneous mutation in an RNA virus. Genetics 162, 1505–1511.

Mansky, L.M., Temin, H.M., 1995. Lower *in vivo* mutation rate of human immunodeficiency virus type 1 than that predicted from the fidelity of purified reverse transcriptase. J. Virol. 69, 5087–5094.

Mansky, L.M., Le Rouzic, E., Benichou, S., Gajary, L.C., 2003. Influence of reverse transcriptase variants, drugs, and Vpr on human immunodeficiency virus type 1 mutant frequencies. J. Virol. 77, 2071–2080.

Martin, D.P., Williamson, C., Posada, D., 2005. RDP2: recombination detection and analysis from sequence alignments. Bioinformatics 21, 260–262.

Martínez, M.A., Jordan-Paiz, A., Franco, S., Nevot, M., 2016. Synonymous viral genome recoding as a tool to impact viral fitness. Trends Microbiol. 24, 134–147.

McDonald, S.M., Nelson, M.I., Turner, P.E., Patton, J.T., 2016. Reassortment in segmented RNA viruses: mechanisms and outcomes. Nat. Rev. Microbiol. 14, 448–460.

McFadden, G., 2005. Poxvirus tropism. Nat. Rev. Microbiol. 3, 201–213.

McGeoch, D.J., Davison, A.J., Dolan, A., Gatherer, D., Sevilla-Reyes, E.E., 2008. Molecular evolution of the *Herpesvirales*. In: Domingo, E., Parrish, C.R., Holland, J.J. (Eds.), Origin and Evolution of Viruses, second ed. Elsevier, Oxford, pp. 447–476.

Menéndez-Arias, L., 2002. Molecular basis of fidelity of DNA synthesis and nucleotide specificity of retroviral reverse transcriptases. Prog. Nucleic Acid Res. Mol. Biol. 71, 91–147.

Meng, T., Kwang, J., 2014. Attenuation of human enterovirus 71 high-replication-fidelity variants in AG129 mice. J. Virol. 88, 5803–5815.

Methot, S.P., Di Noia, J.M., 2017. Molecular mechanisms of somatic hypermutation and class switch recombination. Adv. Immunol. 133, 37–87.

Meyer, P.R., Matsuura, S.E., So, A.G., Scott, W.A., 1998. Unblocking of chain-terminated primer by HIV-1 reverse transcriptase through a nucleotide-dependent mechanism. Proc. Natl. Acad. Sci. U.S.A. 95, 13471–13476.

Meyerhans, A., Vartanian, J.-P., 1999. The fidelity of cellular and viral polymerases and its manipulation for hypermutagenesis. In: Domingo, E., Webster, R.G., Holland, J.J. (Eds.), Origin and Evolution of Viruses. Academic Press, San Diego, pp. 87–114.

Miller, R.V., Day, M.J., 2004. Microbial Evolution. Gene Establishment, Survival and Exchange. ASM Press, Washington, D.C.

Mills, D.R., Peterson, R.L., Spiegelman, S., 1967. An extracellular Darwinian experiment with a self-duplicating nucleic acid molecule. Proc. Natl. Acad. Sci. U.S.A. 58, 217–224.

Minskaia, E., Hertzig, T., Gorbalenya, A.E., Campanacci, V., Cambillau, C., et al., 2006. Discovery of an RNA virus $3' \rightarrow 5'$ exoribonuclease that is critically involved in coronavirus RNA synthesis. Proc. Natl. Acad. Sci. U.S.A. 103, 5108–5113.

Mittal, P., Brindle, J., Stephen, J., Plotkin, J.B., Kudla, G., 2018. Codon usage influences fitness through RNA toxicity. Proc. Natl. Acad. Sci. U.S.A. 115, 8639–8644.

Moreno, E., Ojosnegros, S., Garcia-Arriaza, J., Escarmis, C., Domingo, E., et al., 2014. Exploration of sequence space as the basis of viral RNA genome segmentation. Proc. Natl. Acad. Sci. U.S.A. 111, 6678–6683.

Moreno, E., Gallego, I., Gregori, J., Lucía-Sanz, A., Soria, M.E., et al., 2017. Internal disequilibria and phenotypic diversification during replication of hepatitis C virus in a noncoevolving cellular environment. J. Virol. 91, e02505–e02516.

Morse, S.S. (Ed.), 1994. The Evolutionary Biology of Viruses. Raven Press, New York.

Muslin, C., Joffret, M.-L., Pelletier, I., Blondel, B., Delpeyroux, F., 2015. Evolution and emergence of enteroviruses through intra- and inter-species recombination: plasticity and phenotypic impact of modular genetic exchanges in the 5′untranslated region. PLoS Pathog. 11, e1005266.

Naegeli, H., 1997. Mechanisms of DNA Damage Recognition in Mammalian Cells. Landes Bioscience, Austin, Texas.

Nagy, P.D., Simon, A.E., 1997. New insights into the mechanisms of RNA recombination. Virology 235, 1–9.

Nee, S., 1987. The evolution of multicompartmental genomes in viruses. J. Mol. Evol. 25, 277–281.

Neher, R.A., Leitner, T., 2010. Recombination rate and selection strength in HIV intra-patient evolution. PLoS Comput. Biol. 6, e1000660.

Nei, M., Gojobori, T., 1986. Simple methods for estimating the numbers of synonymous and nonsynonymous nucleotide substitutions. Mol. Biol. Evol. 3, 418–426.

Nevot, M., Martrus, G., Clotet, B., Martinez, M.A., 2011. RNA interference as a tool for exploring HIV-1 robustness. J. Mol. Biol. 413, 84–96.

Nevot, M., Jordan-Paiz, A., Martrus, G., Andrés, C., García-Cehic, D., et al., 2018. HIV-1 protease evolvability is affected by synonymous nucleotide recoding. J. Virol. 92 e00777-18.

Nimonkar, A.V., Boehmer, P.E., 2003. Reconstitution of recombination-dependent DNA synthesis in herpes simplex virus 1. Proc. Natl. Acad. Sci. U.S.A. 100, 10201–10206.

Nishikura, K., 2010. Functions and regulation of RNA editing by ADAR deaminases. Annu. Rev. Biochem. 79, 321–349.

Nishimura, I., Kurokawa, M., Liu, L., Ying, B.W., 2017. Coordinated changes in mutation and growth rates induced by genome reduction. mBio 8 e00676-17.

Novella, I.S., 2004. Negative effect of genetic bottlenecks on the adaptability of vesicular stomatitis virus. J. Mol. Biol. 336, 61–67.

Novella, I.S., Zarate, S., Metzgar, D., Ebendick-Corpus, B.E., 2004. Positive selection of synonymous mutations in vesicular stomatitis virus. J. Mol. Biol. 342, 1415–1421.

Nüesch, J.P.F., De Chastonay, J., Siegl, G., 1989. Detection of defective genomes in hepatitis A virus particles present in clinical specimens. J. Gen. Virol. 70, 3475–3480.

Oberste, M.S., Maher, K., Pallansch, M.A., 2004. Evidence for frequent recombination within species human enterovirus B based on complete genomic sequences of all thirty-seven serotypes. J. Virol. 78, 855–867.

Ochman, H., Elwyn, S., Moran, N.A., 1999. Calibrating bacterial evolution. Proc. Natl. Acad. Sci. U.S.A. 96, 12638–12643.

Ojosnegros, S., Garcia-Arriaza, J., Escarmis, C., Manrubia, S.C., Perales, C., et al., 2011. Viral genome segmentation can result from a trade-off between genetic content and particle stability. PLoS Genet. 7, e1001344.

Palma, E.L., Huang, A.S., 1974. Cyclic production of vesicular stomatitis virus caused by defective interfering particles. J. Infect. Dis. 126, 402–410.

Parmley, J.L., Chamary, J.V., Hurst, L.D., 2006. Evidence for purifying selection against synonymous mutations in mammalian exonic splicing enhancers. Mol. Biol. Evol. 23, 301–309.

Pathak, V.H., Temin, H.M., 1992. 5-Azacytidine and RNA secondary structure increase the retrovirus mutation rate. J. Virol. 66, 3093–3100.

Perales, C., Moreno, E., Domingo, E., 2015. Clonality and intracellular polyploidy in virus evolution and pathogenesis. Proc. Natl. Acad. Sci. U.S.A. 112, 8887–8892.

Pérez-Losada, M., Arenas, M., Galán, J.C., Palero, F., González-Candelas, F., 2015. Recombination in viruses: mechanisms, methods of study, and evolutionary consequences. Infect. Genet. Evol. 30, 296–307.

Pfaller, C.K., Donohue, R.C., Nersisyan, S., Brodsky, L., Cattaneo, R., 2018. Extensive editing of cellular and viral double-stranded RNA structures accounts for innate immunity suppression and the proviral activity of ADAR1$^{P150}$. PLoS Biol. 16, e2006577.

Pfeiffer, J.K., Kirkegaard, K., 2003. A single mutation in poliovirus RNA-dependent RNA polymerase confers resistance to mutagenic nucleotide analogs via increased fidelity. Proc. Natl. Acad. Sci. U.S.A. 100, 7289–7294.

Pfeiffer, J.K., Kirkegaard, K., 2005. Ribavirin resistance in hepatitis C virus replicon-containing cell lines conferred by changes in the cell line or mutations in the replicon RNA. J. Virol. 79, 2346–2355.

Pfeiffer, J.K., Topping, R.S., Shin, N.H., Telesnitsky, A., 1999. Altering the intracellular environment increases the frequency of tandem repeat deletion during Moloney murine leukemia virus reverse transcription. J. Virol. 73, 8441–8447.

Plyusnin, A., Kukkonen, S.K., Plyusnina, A., Vapalahti, O., Vaheri, A., 2002. Transfection-mediated generation of functionally competent Tula hantavirus with recombinant S RNA segment. EMBO J. 21, 1497–1503.

Poirier, E.Z., Mounce, B.C., Rozen-Gagnon, K., Hooikaas, P.J., Stapleford, K.A., et al., 2015. Low-fidelity polymerases of alphaviruses recombine at higher rates to overproduce defective interfering particles. J. Virol. 90, 2446–2454.

Preston, B.D., Dougherty, J.P., 1996. Mechanisms of retroviral mutation. Trends Microbiol. 4, 16–21.

Ramirez, J.D., Llewellyn, M.S., 2014. Reproductive clonality in protozoan pathogens – truth or artefact? Mol. Ecol. 23, 4195–4202.

Rawson, J.M.O., Nicolaitchik, O.A., Keele, B.F., Pathak, V.K., Hu, W.S., 2018. Recombination is required for efficient HIV-1 replication and the maintenance of the viral genome integrity. Nucleic Acids Res. 46, 10535–10545.

Resch, A.M., Carmel, L., Marino-Ramirez, L., Ogurtsov, A.Y., Shabalina, S.A., et al., 2007. Widespread positive selection in synonymous sites of mammalian genes. Mol. Biol. Evol. 24, 1821–1831.

Rezelj, V.V., Levi, L.I., Vignuzzi, M., 2018. The defective component of viral populations. Curr. Opin. Virol. 33, 74–80.

Richmond, R.C., 1970. Non-Darwinian evolution: a critique. Nature 225, 1025–1028.

Rocha, E.P., 2004. Codon usage bias from tRNA's point of view: redundancy, specialization, and efficient decoding for translation optimization. Genome Res. 14, 2279–2286.

Rosario, D., Perez, M., de la Torre, J.C., 2005. Functional characterization of the genomic promoter of borna disease virus (BDV): implications of 3'-terminal sequence heterogeneity for BDV persistence. J. Virol. 79, 6544–6550.

Roux, L., Simon, A.E., Holland, J.J., 1991. Effects of defective interfering viruses on virus replication and pathogenesis in vitro and in vivo. Adv. Virus Res. 40, 181–211.

Rozen-Gagnon, K., Stapleford, K.A., Mongelli, V., Blanc, H., Failloux, A.B., et al., 2014. Alphavirus mutator variants present host-specific defects and attenuation in mammalian and insect models. PLoS Pathog. 10, e1003877.

Sabo, D.L., Domingo, E., Bandle, E.F., Flavell, R.A., Weissmann, C., 1977. A guanosine to adenosine transition in the 3' terminal extracistronic region of bacteriophage Qβ RNA leading to loss of infectivity. J. Mol. Biol. 112, 235–252.

Sachs, J.L., Bull, J.J., 2005. Experimental evolution of conflict mediation between genomes. Proc. Natl. Acad. Sci. U.S.A. 102, 390–395.

Saira, K., Lin, X., DePasse, J.V., Halpin, R., Twaddle, A., et al., 2013. Sequence analysis of in vivo defective interfering-like RNA of influenza A H1N1 pandemic virus. J. Virol. 87, 8064–8074.

Salemi, M., Vandamme, A.M., 2004. The Phylogenetic Handbook. A Practical Approach to DNA and Protein Phylogeny. Cambridge University Press, Cambridge.

Sanjuan, R., Nebot, M.R., Chirico, N., Mansky, L.M., Belshaw, R., 2010. Viral mutation rates. J. Virol. 84, 9733–9748.

Santiago, M.L., Greene, W.C., 2008. The role of the APOBEC3 family of cytidine deaminase in innate immunity, G-to-A hypermutation, and evolution of retroviruses. In: Domingo, E., Webster, R.G., Holland, J.J. (Eds.), Origin and Evolution of Viruses, second ed. Elsevier, Oxford, pp. 183–206.

Sarisky, R.T., Nguyen, T.T., Duffy, K.E., Wittrock, R.J., Leary, J.J., 2000. Difference in incidence of spontaneous mutations between Herpes simplex virus types 1 and 2. Antimicrob. Agents Chemother. 44, 1524–1529.

Sheehy, A.M., Gaddis, N.C., Choi, J.D., Malim, M.H., 2002. Isolation of a human gene that inhibits HIV-1 infection and is suppressed by the viral Vif protein. Nature 418, 646–650.

Shriner, D., Rodrigo, A.G., Nickle, D.C., Mullins, J.I., 2004. Pervasive genomic recombination of HIV-1 in vivo. Genetics 167, 1573–1583.

Sicard, A., Yvon, M., Timchenko, T., Gronenborn, B., Michalakis, Y., et al., 2013. Gene copy number is differentially regulated in a multipartite virus. Nat. Commun. 4, 2248.

Simmonds, P., 2010. Recombination in the evolution of picornaviruses. In: Ehrenfeld, E., Domingo, E., Roos, R.P. (Eds.), The Picornaviruses. ASM Press, Whashington, DC, pp. 229–237.

Smith, E.C., Denison, M.R., 2013. Coronaviruses as DNA wannabes: a new model for the regulation of RNA virus replication fidelity. PLoS Pathog. 9, e1003760.

Smith, D.B., Inglis, S.C., 1987. The mutation rate and variability of eukaryotic viruses: an analytical review. J. Gen. Virol. 68, 2729–2740.

Smith, P.J., Jones, C.J., 1999. DNA Recombination and Repair. Oxfird University Press, Oxford.

Smith, E.C., Blanc, H., Vignuzzi, M., Denison, M.R., 2013. Coronaviruses lacking exoribonuclease activity are susceptible to lethal mutagenesis: evidence for proofreading and potential therapeutics. PLoS Pathog. 9, e1003565.

Stadler, P.F., 2016. Evolution of RNA-based networks. Curr. Top. Microbiol. Immunol. 392, 43–59.

Stavrou, S., Crawford, D., Blouch, K., Browne, E.P., Kohli, R.M., et al., 2014. Different modes of retrovirus restriction by human APOBEC3A and APOBEC3G in vivo. PLoS Pathog. 10, e1004145.

Stedman, K., 2013. Mechanisms for RNA capture by ssDNA viruses: grand theft RNA. J. Mol. Evol. 76, 359–364.

Steinhauer, D.A., Holland, J.J., 1986. Direct method for quantitation of extreme polymerase error frequencies at selected single base sites in viral RNA. J. Virol. 57, 219–228.

Steinhauer, D.A., Domingo, E., Holland, J.J., 1992. Lack of evidence for proofreading mechanisms associated with an RNA virus polymerase. Gene 122, 281–288.

Steitz, T.A., 1999. DNA polymerases: structural diversity and common mechanisms. J. Biol. Chem. 274, 17395–17398.

Stratton, M.R., Campbell, P.J., Futreal, P.A., 2009. The cancer genome. Nature 458, 719–724.

Suarez, P., Valcárcel, J., Ortín, J., 1992. Heterogeneity of the mutation rates of influenza A viruses: isolation of mutator mutants. J. Virol. 66, 2491–2494.

Supek, F., 2016. The code of silence: widespread associations between synonymous codon biases and gene function. J. Mol. Evol. 82, 65–73.

Suzuki, T., Moriyama, K., Otsuka, C., Loakes, D., Negishi, K., 2006. Template properties of mutagenic cytosine analogues in reverse transcription. Nucleic Acids Res. 34, 6438–6449.

Svarovskaia, E.S., Delviks, K.A., Hwang, C.K., Pathak, V.K., 2000. Structural determinants of murine leukemia virus reverse transcriptase that affect the frequency of template switching. J. Virol. 74, 7171–7178.

Swetina, J., Schuster, P., 1982. Self-replication with errors. A model for polynucleotide replication. Biophys. Chem. 16, 329–345.

Szathmary, E., 1992. Viral sex, levels of selection, and the origin of life. J. Theor. Biol. 159, 99–109.

Sztuba-Solinska, J., Urbanowicz, A., Figlerowicz, M., Bujarski, J.J., 2011. RNA-RNA recombination in plant virus replication and evolution. Annu. Rev. Phytopathol. 49, 415–443.

Tanaka, T., Eusebio-Cope, A., Sun, L., Suzuki, N., 2012. Mycoreovirus genome alterations: similarities to and differences from rearrangements reported for other reoviruses. Front. Microbiol. 3, 186.

Thomson, M.M., Perez-Alvarez, L., Najera, R., 2002. Molecular epidemiology of HIV-1 genetic forms and its significance for vaccine development and therapy. Lancet Infect. Dis. 2, 461–471.

Tibayrenc, M., Ayala, F.J., 2012. Reproductive clonality of pathogens: a perspective on pathogenic viruses, bacteria, fungi, and parasitic protozoa. Proc. Natl. Acad. Sci. U.S.A. 109, E3305–E3313.

Tibayrenc, M., Ayala, F.J., 2014. Cryptosporidium, Giardia, Cryptococcus, Pnemocystis genetic variability: cryptic biological species or clonal near-clades? PLoS Pathog. 10, e1003908.

Upton, D.C., Gregory, B.L., Arya, R., Unniraman, S., 2011. AID: a riddle wrapped in a mystery inside an enigma. Immunol. Res. 49, 14–24.

Urbanowicz, A., Alejska, M., Formanowicz, P., Blazewicz, J., Figlerowicz, M., et al., 2005. Homologous crossovers among molecules of brome mosaic bromovirus RNA1 or RNA2 segments in vivo. J. Virol. 79, 5732–5742.

Valcarcel, J., Ortin, J., 1989. Phenotypic hiding: the carryover of mutations in RNA viruses as shown by detection of *mar* mutants in influenza virus. J. Virol. 63, 4107–4109.

Valente, L., Nishikura, K., 2005. ADAR gene family and A-to-I RNA editing: diverse roles in posttranscriptional gene regulation. Prog. Nucleic Acid Res. Mol. Biol. 79, 299–338.

Varela-Echavarria, A., Garvey, N., Preston, B.D., Dougherty, J.P., 1992. Comparison of Moloney murine leukemia virus mutation rate with the fidelity of its reverse transcriptase in vitro. J. Biol. Chem. 267, 24681–24688.

Varsani, A., Lefeuvre, P., Roumagnac, P., Martin, D., 2018. Notes on recombination and reassortment in multipartite/segmented viruses. Curr. Opin. Virol. 33, 156–166.

Vartanian, J.P., Henry, M., Wain-Hobson, S., 2001. Simulating pseudogene evolution in vitro: determining the true number of mutations in a lineage. Proc. Natl. Acad. Sci. U.S.A. 98, 13172–13176.

Velicer, G.J., Raddatz, G., Keller, H., Deiss, S., Lanz, C., et al., 2006. Comprehensive mutation identification in an evolved bacterial cooperator and its cheating ancestor. Proc. Natl. Acad. Sci. U.S.A. 103, 8107–8112.

Venkatesan, S., Rosenthal, R., Kanu, N., McGranahan, N., Barket, J., et al., 2018. Perspective: APOBEC mutagenesis in drug resistance and immune escape in HIV and cancer evolution. Ann. Oncol. 29, 563–572.

Vignuzzi, M., Andino, R., 2010. Biological implications of picornavirus fidelity mutants. In: Ehrenfeld, E., Domingo, E., Roos, R.P. (Eds.), The Picornaviruses. ASM Press, Washington DC, pp. 213–228.

Vignuzzi, M., Stone, J.K., Arnold, J.J., Cameron, C.E., Andino, R., 2006. Quasispecies diversity determines pathogenesis through cooperative interactions in a viral population. Nature 439, 344–348.

Viguera, E., Canceill, D., Ehrlich, S.D., 2001. Replication slippage involves DNA polymerase pausing and dissociation. EMBO J. 20, 2587–2595.

Vijg, J., Dong, X., Milholland, B., Zhang, L., 2017. Genome instability: a conserved mechanism of ageing? Essays Biochem. 61, 305–315.

Villarreal, L.P., 2005. Viruses and the Evolution of Life. ASM Press, Washington, D.C.

Satellites and defective viral RNAs. In: Vogt, P.K., Jackson, A.O. (Eds.), 1999. Current Topics in Microbiol. and Immunol., vol. 239. Springer, Berlin.

Wainberg, M.A., Drosopoulos, W.C., Salomon, H., Hsu, M., Borkow, G., et al., 1996. Enhanced fidelity of 3TC-selected mutant HIV-1 reverse transcriptase. Science 271, 1282–1285.

Wang, Y., Schmitt, K., Guo, K., Santiago, M.L., Stephens, E.B., 2016. Role of the single deaminase domain APOBEC3A in virus restriction, retrotranscription, DNA damage and cancer. J. Gen. Virol. 97, 1–17.

Ward, C.D., Flanegan, J.B., 1992. Determination of the poliovirus RNA polymerase error frequency at eight sites in the viral genome. J. Virol. 66, 3784–3793.

Ward, C.D., Stokes, M.A., Flanegan, J.B., 1988. Direct measurement of the poliovirus RNA polymerase error frequency in vitro. J. Virol. 62, 558–562.

Webster, R.G., Bean, W.J., Gorman, O.T., Chambers, T.M., Kawaoka, Y., 1992. Evolution and ecology of influenza A viruses. Microbiol. Rev. 56, 152–179.

Webster, B., Ott, M., Greene, W.C., 2013. Evasion of superinfection exclusion and elimination of primary viral RNA by an adapted strain of hepatitis C virus. J. Virol. 87, 13354–13369.

Westrich, J.A., Warren, C.J., Klausner, M.J., Guo, K., Liu, C.W., et al., 2018. Human papilloma 16 E7 stabilizes APOBEC3A protein by inhibiting cullin 2-dependent protein degradation. J. Virol. 92 e01318-17.

Winters, M.A., Merigan, T.C., 2005. Insertions in the human immunodeficiency virus type 1 protease and reverse transcriptase genes: clinical impact and molecular mechanisms. Antimicrob. Agents Chemother. 49, 2575–2582.

Worobey, M., Holmes, E.C., 1999. Evolutionary aspects of recombination in RNA viruses. J. Gen. Virol. 80 (Pt 10), 2535–2543.

Wu, J., Gong, P., 2018. Visualizing the nucleotide addition cycle of viral RNA-dependent RNA polymerase. Viruses 10, E24.

Yamaguchi, T., Yamashita, Y., Kasamo, K., Inoue, M., Sakaoka, H., et al., 1998. Genomic heterogeneity maps to tandem repeat sequences in the herpes simplex virus type 2 UL region. Virus Res. 55, 221–231.

Yang, Z., Bielawski, J.P., 2000. Statistical methods for detecting molecular adaptation. Trends Ecol. Evol. 15, 496–503.

Zaher, H.S., Green, R., 2009. Fidelity at the molecular level: lessons from protein synthesis. Cell 136, 746–762.

Zimmern, D., 1988. Evolution of RNA viruses. In: Domingo, E., Holland, J.J., Ahlquist, P. (Eds.), RNA Genetics. CRC Press Inc, Boca Raton, FL, pp. 211–240.

# Darwinian principles acting on highly mutable viruses

## Abbreviations

**A** adenine
**AD** average distance
**AZC** azacytidine
**C** cytosine
**cDNA** complementary deoxyribonucleic acid
**CTL** cytotoxic T lymphocyte
**DI** defective interfering
**E. coli** Escherichia coli
**FMDV** foot-and-mouth disease virus
**FU** 5-fluorouracil
**G** guanine
**HBV** hepatitis B virus
**HIV-1** human immunodeficiency virus type 1
**HSV-1** herpes simplex virus type 1
**HTLV-1** human T-lymphotropic virus type 1
**HTLV-2** human T-lymphotropic virus type 2
**Kb** kilobase
**LCMV** lymphocytic choriomeningitis virus
**MARLS** monoclonal antibody resistant FMDV
   mutant
**MOI** multiplicity of infection
**MV** measles virus
**PAQ** partition analysis of quasispecies
**PCR** polymerase chain reaction
**PV** poliovirus
**R** ribavirin
**RdRp** RNA-dependent RNA polymerase
**RNase** ribonuclease
**RT-PCR** reverse transcription followed by
   polymerase chain reaction
**s/nt** substitutions per nucleotide
**s/s/y** substitutions per site and year
**U** uracil
**VSV** vesicular stomatitis virus

## 3.1 Theoretical frameworks to approach virus evolution

Population thinking is the term to refer to a view of biology that started at the beginning of the 19th century, and that grew as the result of observations on differences among individuals of the same biological group that were made by collectors of natural specimens and by animal and plant breeders. The accumulated evidence prepared a scientific environment that was ready to react in 1859 to Darwin's Origin of Species. As described by E. Mayr, the contribution of Darwin was threefold: (i) He provided evidence that evolution indeed occurred, (ii) He proposed natural selection as a mechanism for evolutionary change, and (iii) He replaced "typological" thinking by "population" thinking (Mayr, 1959). Concerning the third point, the traditional Platonic view of reality conferred to fixed "ideas" or "types" an essential value in the description of nature. Variations of the "types" were considered secondary, when not irrelevant, and not worthy of study. Plato had an impact on subsequent philosophical currents and permeated scientific thinking. The lesser value of "variations" of the standard types also impacted microbiology and virology (Domingo, 1999). As an extension of what E. Mayr explains for general evolution, virology is now

*Virus as Populations*
https://doi.org/10.1016/B978-0-12-816331-3.00003-9

confronted with the evidence that a virus "type" (we may call it the "wild type") is an abstraction, while the variations are real.

A major breakthrough in clarifying evolutionary thinking was the formulation known as *the modern synthesis* through work led mainly by S.S. Chetverikov, R.A. Fisher, J.B.S. Haldane, and S. Wright. The "synthesis" recognized the role of Darwinian natural selection acting on phenotypic variation, itself the result of genetic variation. It represented a unification of natural selection with Mendelian genetics, and identified genetic change and natural selection as the major forces of evolution (Huxley, 1942). The concepts contributed by the "synthesis" were in agreement with the molecular mechanisms of heredity, the details of which were unveiled in subsequent decades and buried other appealing but incorrect views, such as Lamarckism (the inheritance of modifications introduced in the body of organisms) (Mayr and Provine, 1980).

Since the first nucleotide and amino acid sequences of biological systems were obtained, procedures to establish evolutionary relationships among them were developed. The extent of sequence similarity indicated how closely homologous genes (those with recognized common ancestry) were. Phylogenetic trees were invented by W.M. Fitch and E. Margoliash in the middle of the 20th century (Fitch and Margoliash, 1967). They calculate ancestry relationships and quantify the relatedness among multiple traits, mainly nucleotide or amino acid sequences. Phylogenetic procedures are currently abundantly used to approach virus evolution, and to estimate the time at which the most common ancestor of a set of genomes under comparison existed. Relatively conserved viral genes (typically those encoding nonstructural proteins) are useful to establish connections—order of ancestry and degree of genetic diversification—among relatively distant viruses, while variable genes (typically those encoding capsid and surface proteins) can define short-term relationships among epidemiologically related isolates

(Chapter 7). There are excellent accounts of phylogenetic procedures that expand on the concepts and procedures sketched here (Salemi and Vandamme, 2004; Holmes, 2008, 2009; St. John, 2017; Baele et al., 2018). In particular, Chapter 1 of the M. Salemi and A.M. Vandamme book, written by A.M. Vandamme, is a concise and insightful introduction to key concepts that are relevant to virus evolution.

The extremely high mutation rate of RNA viruses and also many DNA viruses (Chapter 2), introduces a new element to be considered in the studies of evolution since events can be observed within short time intervals (in some cases even in hours and days). A relevant and debated question is whether biological features, which are dependent on high mutability of viruses, are amenable to the description by classic population genetics or some new paradigm is needed. One such paradigm that has permeated virology is quasispecies theory, one of the mathematical formulations of molecular evolution that has exerted great impact in the understanding of RNA viruses (Section 3.6). However, a unanimous view is that no matter which theoretical framework is adopted, the fundamental mechanisms involved in viral evolution are presided by general Darwinian principles. They can be succinctly expressed as the succession of genetic variation, competition among the generated variants, selection of the fittest, and perturbations due to random events, particularly population bottlenecks.

Darwinian principles were verified experimentally in early studies with *Drosophila* (Dobzhansky et al., 1977), as well as with rapidly replicating bacterial and animal cells and organisms. Due to their mutability and rapid replication, viruses have become accepted systems to address experimentally basic questions of general evolution (Morse, 1994; Gibbs et al., 1995; Moya et al., 2000; Domingo et al., 2001; Domingo and Perales, 2019). In fact, viruses have gone from being ignored by the most influential population geneticists in the middle of the 20th century to being recognized as ideal systems to address problems in evolution that are very difficult to approach with differentiated organisms. [The anecdote goes that in the last century when a prominent population geneticist (whose name I omit) was asked to comment on microbial evolution, the response was: "Next question please"]. One of the reasons to ignore viruses as systems to explore evolution could have been a lack of understanding of the contribution of viruses to the biological world. For many decades it was not clear what the influence of virus replication was on their host cells and organisms, except for being disease agents. There was no overview of what viruses might have represented either historically (in the early phases of evolution of life, Chapter 1) or currently (in present-day biology). Once the replicative mechanisms displayed by viruses were put into the perspective of general evolution, it became clear that at least part of the fundamental questions that have been addressed in basic general genetics can be studied experimentally with viruses.

Not a single theory has been able to capture the multiple features of virus evolution, and not a single formulation of population dynamics is totally independent and unrelated to others that approach variation of biological entities. K.M. Page and M.A. Nowak explained the relationships among several mathematical equations that describe evolutionary and ecological dynamics (i.e., replicator-mutator, Price, Lotka-Volterra, game dynamical, and quasispecies) (Page and Nowak, 2002) (Fig. 3.1). The Price equation (from the American geneticist G. Price) is considered key to describe the behavior of evolving populations in general (for review of theoretical aspects, see Okasha, 2013). However, because of the conspicuous participation of mutation in virus evolution and the extreme heterogeneity (or "polyploidy" using a classical term) of RNA viruses within infected cells, quasispecies theory is suitable for RNA viruses that display elevated mutation rates

**FIGURE 3.1** Correspondence among different equations that describe evolutionary dynamics, with or without the participation of mutation. The middle row indicates the equivalence of three important equations for general evolutionary dynamics: the replicator-mutator, Price, and replicator-mutator-Price equations. The mathematical equivalences were described by K.M. Page and M.A. Nowak. From the replicator-mutator equation, the quasispecies equation is obtained when mutation is present, while the game-dynamical and Lotka-Volterra equations do not consider mutation. In the quasispecies equation, $dx_i/dt$ describes the concentration of mutant $i$ as a function of time; $x_k$ describes the concentration of mutant k; the replication, degradation and the synthesis of $i$ from $k$ are described by $A_i$, $D_i$, and $W_{ik}$, respectively. $Q_i$ expresses the fraction of correct copies of $i$ produced as a result of copying $i$. Finally, $\phi_i$ corresponds to the flux of molecule $i$ that in the most simple case, it will be proportional to $x_i$. Concerning the Lotka-Volterra equation that describes the predator-prey relationship, $x$ is the number of prey, $y$ is the number of predators, $dx/dt$ and $dy/dt$ their variation as a function of time, and α, β, γ, and δ are parameters that describe the predator-prey interactions. *Figure based in data published in Page, K.M., Nowak, M.A., 2002. Unifying evolutionary dynamics. J. Theor. Biol. 219, 93–98; Domingo, E., Sheldon, J., Perales, C., 2012. Viral quasispecies evolution. Microbiol. Mol. Biol. Rev. 76, 159–216; Domingo, E., Schuster, P., 2016. Quasispecies: From Theory to Experimental Systems. Curr. Top. Microbiol. Immunol. vol. 392. Springer, Switzerland.*

(Domingo and Schuster, 2016). Other equations of evolutionary dynamics may suffice to describe the population dynamics of DNA-based organisms, viruses displaying lower average mutation rates than RNA viruses, or competition dynamics when the mutation rate is not a relevant parameter (prey-predator relationship expressed by the Lotka-Volterra equation in Fig. 3.1). The theoretical concepts contributed by classic genetics and the new quasispecies-based formulations have inspired experimental designs that, facilitated by the impulse of deep sequencing, are currently broadening our understanding of evolutionary mechanisms in general, not only for viruses (Chapter 10).

### 3.1.1 Theory and experiment

The connections between experimental studies and theoretical formulations (models of population dynamics with virtual replicons, in silico simulations that mimic virus replication, phase transitions in complex adaptive systems, bioinformatic approaches for the analysis of massive numbers of sequences, metagenomic investigations, machine learning approaches, etc.) are increasingly important in virology. The input of theory has been greatly aided by the availability of powerful computers, together with developments in bioinformatics, including extensive information repositories in data banks (Chapter 7). Theory has inspired experiments and vice versa. For theoretical studies to help in understanding virus behavior, it is important that the relevant algorithms include parameters based on solid experimental values for mutation rates, rounds of replication, population size, frequency of advantageous and deleterious mutations, and distributions of fitness values in mutant spectra. Very often, predictions of virus behavior based upon simulations with virtual replicons cannot take into consideration the array of selective pressures and population size variations that intervene in their infectious cycles; they are difficult to identify, and thus, to mimic. Keeping such limitations in mind, complementation between experiment and theory (using experimental data-driven models, in combination with exploratory models and predictive algorithms) has produced remarkable new insights on virus behavior as described in several chapters of this book.

Theoretical predictions based on unrealistic parameters are a continuous temptation to theoretical biologists eager to anticipate the behavior of real objects, in this case, viruses. This is legitimate, provided predictions do not contradict experimental findings. The excessive use of theory in detriment of experimentation has created skepticism among experimentalists regarding the value of theoretical models. Excesses have to be avoided since, as M. Eigen puts it, at least in biology, "pure theory" may be "poor theory" (Eigen, 2013). In particular, and to give a specific example, when a theoretical formulation includes the term "infinite" ($\infty$), either referring to size or time, strange conclusions may be reached. Extreme numbers (zero or $\infty$) are justified as a tool, but predictions derived from their use in theoretical models should be subjected to the scrutiny of experimentation. Parallel arguments have been expressed regarding the importance of models derived from machine learning to fit the intended clinical objectives through validation procedures [(Nevin, 2018) and articles in the same PLoS Medicine issue].

As clarifying as it was, the new theoretical framework provided by the "modern synthesis" did not consider a major concept that having originated in physics, has pervaded modern biology: complexity (Solé and Goodwin, 2000; Albergante et al., 2016). Complexity is a characteristic of systems with many interacting components, that give rise to an emergent result that cannot be explained as the mere sum of contributions of the parts involved. Important issues pertinent to the study of viruses at the population level, what we may term "population virology," relate to complexity, and some of them are discussed in Section 3.9. It is not obvious how to establish a connection between the theoretical treatment of complexity and experimental findings on virus evolution, but such a connection would provide a fresh approach to important aspects of virus evolution and viral pathogenesis (see Sections 3.8 and 3.9).

# 3.2 Genetic variation, competition, and selection

Viruses undergo genetic variation through mutation, recombination, and genome segment reassortment, as summarized in Chapter 2. Variant forms of the same virus are continuously generated and compete with each other in changing environments. As a result of such competitions, and also as a consequence of random events, many (probably most) viral variants that are generated during replication are extinguished, while others continue to multiply in their host organisms with different replication efficiencies. The long-term reiteration of such processes has given rise to the multiple pathogenic and nonpathogenic viruses we isolate and study.

Replication with the production of error copies in natural hosts has been extensively documented by infecting or transfecting cells or organisms with either biological or molecular clones (DNA or RNA transcripts from molecular cDNA clones that encode information to express an infectious virus). Transfection is the penetration of viral genetic material into cells rendered transiently permeable by physical or chemical means. Because mutations are produced during preparation of a biological clone from a single genome, and transcription has also limited fidelity (Fong et al., 2013; Xu et al., 2014; Gamba and Zenkin, 2018), infection or transfection are necessarily performed with multiple variant genomes present at low frequencies, and therefore, are intrinsically inaccurate processes regarding which precise genomic sequences enter the cells. New repertoires of variants that differ from the input are unavoidably generated in the infected cells, and they spread in the organism. Consensus sequences determined for entire populations may give the appearance of invariance between the infecting and progeny population, but mutant spectra must be analyzed to derive a reasonable picture of events at the molecular level (Section 3.6.3). Competition among variants has been profusely documented in vivo and in cell culture. Competition is a continuous and unavoidable process due to the limitation of resources. The latter include, in its broadest sense, host cell numbers, space confinements in the compartments of differentiated organisms, intracellular structures, biomolecules, and metabolites, among others. The limitation of resources is a key determinant of differential multiplication.

Following the competition, some variants become dominant over others due to positive selection, unless random events intervene (Fig. 3.2). While genetic variation takes place in genomes and affects genotypes, selection acts on phenotypes, the traits that result from the expression of genomes that have acquired the relevant mutations. Different forms of selection (Section 3.4) together with random sampling

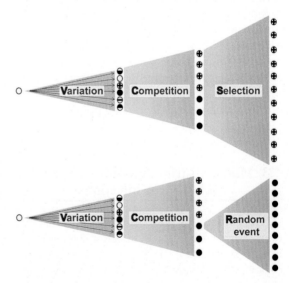

**FIGURE 3.2** Random events may prevent a potentially advantageous mutant swarm from reaching dominance. Top: variation of an initial population, followed by competition and selection results in the dominance of genomes represented by circles with a cross. Bottom: a random event may modify the outcome of selection and result in the dominance of genomes represented by a filled circle.

are continuously and transiently shaping viral populations. The experiments designed to test relative viral fitness (an important parameter that captures the overall replicative capacity of a virus in a given environment) are based on competition between variants under a given set of environmental conditions (Chapter 5).

Populations of viruses, and also microbes and unicellular organisms that can grow in culture, have a tremendous advantage that experiments to test Darwinian principles can be designed in the laboratory and the results recorded in days or even hours. The field of experimental evolution is particularly adequate for microbial entities. A round of viral replication can be completed in minutes, and changes in viral genomes can be recorded in real time. Viruses provide scientists with an accelerated version of those basic molecular events that have presided long-term evolution of life on Earth. Although it is rather ambiguous to refer to numbers of "generations" of a virus (because viral genome replication is not merely a regular process of genome division), simple laboratory experiments can involve hundreds of viral passages, each one comprising several rounds of genome copying in thousands or millions of cells. For comparison, our human ancestors living 300 generations ago were in the preagricultural era!

## 3.3 Mutant distributions during DNA and RNA virus infections

The continuous presence of mutant distributions during RNA virus infections in vivo defines a key feature of viral populations, which is not only relevant to interpret viral evolution but also viral pathogenesis. Although we refer generically to RNA viruses, the presence of mutant distributions of different amplitude (different average number of mutations per genome within a viral sample) is general for viruses. Studies on drug- and monoclonal antibody-escape mutants, as well as nucleotide

sequence surveys of molecular and biological clones from virus isolates, indicate the presence of heterogeneous populations during infections by DNA viruses, including those with complex genomes that encode many proteins (Smith and Inglis, 1987; Sarisky et al., 2000; Domingo et al., 2001; López-Bueno et al., 2003; Quinlivan et al., 2004; Domingo, 2006; Ba Abdullah et al., 2017). In particular, a similar mutant spectrum divergence is observed in single-stranded DNA and in RNA viruses. Well-documented cases are those of parvoviruses, and the plant geminiviruses with the presence of multiple variants in one individual host (Isnard et al., 1998; Sanz et al., 1999; Battilani et al., 2006; López-Bueno et al., 2006; Ge et al., 2007; Sánchez-Campos et al., 2018). Mutant distributions have also been characterized in populations of polyoma virus JC, both in patients with progressive multifocal leukoencephalopathy and in asymptomatic individuals (Van Loy et al., 2015). The relevance of viral variation and population complexity in virus behavior has been established by many studies, and different aspects can be found in specific articles and reviews (Holland et al., 1982; Kimata et al., 1999; Farci and Purcell, 2000; Domingo et al., 2001, 2008; Figlerowicz et al., 2003; Pfeiffer and Kirkegaard, 2005; Vignuzzi et al., 2006; Lauring and Andino, 2010; Perales et al., 2010; Domingo and Perales, 2012; Domingo and Schuster, 2016; Marz et al., 2014; Seifert and Beerenwinkel, 2016; Berkhout et al., 2017).

The genetic heterogeneity of populations of DNA viruses with large genomes has been studied in far less detail than that of RNA viruses. Heterogeneity levels may depend partly on the DNA polymerase (cellular or virus-coded) that catalyzes their replication. In eukaryotes several replicative, reparative and damage response DNA polymerases have been characterized that differ in template copying fidelity. While the error rate of replicative DNA polymerases is in the range of $10^{-7}$ to $10^{-10}$ errors per nucleotide copied, the corresponding values for reparative polymerases is of $10^{-1}$ to $10^{-7}$, depending on the polymerase family (Hubscher et al., 2002; McCulloch and Kunkel, 2008) (see also Chapter 2, Section 2.6).

Complex DNA viruses (carrying a large amount of encoded information) may have limitations to exploit generalized high mutation rates due to the lethal character of accumulation of mutations in large genomes, in agreement with the inverse correlation between genome size and mutation tolerance (Chapter 2). However, variations at specific *loci* (such as at recombination or mutation hot spots, gain or loss of tandem repeats, copy number expansions, etc.) may occur frequently and may have an adaptive value (Elde et al., 2012). In an experiment intended to examine the genetic and phenotypic consequences of repeated bottleneck events, the complex DNA virus herpes simplex type 1 (HSV-1) behaved very similarly to RNA viruses (Jaramillo et al., 2012; further discussed in Chapter 6). Thus, as our capacity to penetrate into the composition of viral populations improves, nucleotide sequence heterogeneity of genomes during replication is increasingly seen as a shared feature of all viruses.

The average genetic distance among genomes of a population is expected to increase with the number of rounds of replication, following an infection initiated by one or few founder genomes (see also Section 4.1.2 in Chapter 4 on gain of mutant spectrum complexity following a population bottleneck). It is often difficult to know whether in vivo an initial founder infection (an infection started by one or a limited number of infectious particles) has occurred close to or far from the time at which the heterogeneity of the viral population is examined. This is one of several reasons that limit our capacity to interpret comparative heterogeneity levels among natural viral isolates. In the case of retroviruses, another factor that influences heterogeneity is the time that the virus spends as a proviral element integrated into host DNA versus the time during which viral sequences

experience transcription from DNA (proviral DNA copied into RNA) and reverse transcription (viral RNA copied into DNA). Once viral DNA has been integrated, and the integrated DNA is not expressed (silent integration which is a form of latency), the provirus adheres to the evolutionary stasis typical of cellular DNA, with million-fold lower mutation rates than during reverse transcription. The two Deltaretroviruses human T-lymphotropic virus types 1 and 2 (HTLV-1 and HTLV-2), appear to be less variable than other retrovirus genera because of their extended latency periods as integrated proviruses. Their population dynamics is characterized by clonal expansions of the immune cells they infect (Chapter 7). Concerning Gammaretroviruses, it was calculated that the *v-mos* gene of Moloney murine sarcoma virus evolved at a rate of $1.3 \times 10^{-3}$ substitutions per site and year (s/s/y) while its cellular homolog *c-mos* evolved at the million-fold lower rate of $1.7 \times 10^{-9}$ s/s/y (Gojobori and Yokoyama, 1985). Several early studies confirmed a very different evolutionary rate of RNA viruses versus cells, an extremely important point first emphasized by J.J. Holland and his colleagues (Holland et al., 1982; Steinhauer and Holland, 1987). The biological consequences of divergent evolutionary tempos among components of our biosphere are probably still underappreciated.

## 3.4 Positive versus negative selection: two sides of the same coin

Two types of selection act on viruses: positive (also termed Darwinian, linear, dynamic, or directional) and negative (also termed purifying, normalizing, or stabilizing selection). Positive selection results in the survival of subsets of variants because their replication and capacity to persist are superior to other variants of the same entourage. In the opinion of M. Eigen, "selection is the fixation of the genotype by the continual evaluation of the phenotype"

(Eigen, 1992). A drug-resistant viral mutant is selected because its capacity to replicate in the presence of the drug is higher than that of other components of the viral population; the fixed genetic change is the nucleotide that encodes the amino acid that determines the resistance phenotype. Important examples of medically relevant, positive selection events are provided by the frequent occurrence in the viral quasispecies of mutants with amino acid substitutions that mediate escape from antibodies, cytotoxic T lymphocytes (CTLs), or antiviral inhibitors (reviewed in Domingo et al., 2001; Figlerowicz et al., 2003; Gerrish and Garcia-Lerma, 2003; Perales, 2018). In other terms, the escape mutants display higher replicative fitness than other members of the same mutant distribution in the presence of the selective agent (see Chapter 5).

The level of heterogeneity of the parental viral population prior to the application of selection will be generally higher than that of the selected population because the latter has been founded by a minority of genomes that will participate in the ensuing rounds of replication in the presence of the selective agent. These types of selection events that lead to a decrease in the heterogeneity of viral populations or to a decrease in the polymorphism in host organisms have been termed *sweeping* selection events. A selection sweep can produce the hitchhiking of those mutations that accompany the selected mutation in the same genome. If hitchhiked mutations are independent of the selected mutation, their chance occurrence in the selected genomes can influence the future course of the virus evolution. This is one of the mechanisms of antigenic variation in the absence of immune selection described in Chapter 4. Sweeping events may have an effect similar to a population bottleneck regarding a reduction of population complexity (Section 3.8.3).

Positive selection can reduce the diversity of a viral population by focusing the population toward one or a few genome types, but it may also result in a repertoire of variants harboring different genetic changes compatible with replication in the face of the selective constraint (Ciurea et al., 2000; Perales et al., 2005; Martin and Domingo, 2008). During the evolution of parallel viral lineages, identical adaptive mutations may be found (de la Torre et al., 1992; Borrego et al., 1993; Couderc et al., 1994; Chumakov et al., 1994; Martín Hernández et al., 1994; Mateu et al., 1994; Lu et al., 1996; Brown et al., 2001; Ruiz-Jarabo et al., 2003; Herrera et al., 2008) while, in other cases, multiple molecular pathways for fitness gain can be observed (Escarmís et al., 1999; Wichman et al., 1999; Domingo et al., 2001; Tsibris et al., 2009). Deep sequencing of viral populations as they respond to a selection event often reveals multiple, transient episodes of selection until a particularly well-suited (in terms of fitness) subpopulation overtakes the population. It would be interesting to explore whether an increasing complexity (in the sense of genetic information content) of the virus genome may limit the number of alternative molecular pathways to respond to a selective constraint. Positive selection can contribute to biological novelty, and it is an integral part of the establishment of emergent or reemergent virus in a host population (Chapter 7).

Negative selection acts to eliminate or to diminish the frequency of subsets of variants. Mutations that abolish regulatory elements or result in inactivation of an essential viral enzyme will generally be lethal. Lethality can be regarded as the extreme manifestation of negative selection. When genomic RNA acquires mutations that modify the secondary structure of an RNA motif involved in replication, the mutant genomes will be subjected to negative selection. Not all mutations specifically inflict a disadvantage to the genomes that harbor them. When an amino acid substitution diminishes by a factor of two the catalytic efficiency of a viral polymerase (or polymerization complex), the RNAs whose replication is catalyzed by the altered polymerase will not compete favorably with other genomes of the same replicative ensemble whose replication is catalyzed by a fully functional polymerase. Deleterious mutations in *cis*-acting functional elements—those that affect the same genome that harbors the element but no other genomes of the population—will result in negative selection acting on the genomes that include the deleterious mutations, but not others. In contrast, deleterious mutations that alter *trans*-acting functions—those that are expressed by one genome and which influence the replication of other genomes of the same population—may impair the replication of many genomes, not only those that include the mutation. The negative effect of *trans*-acting gene products is difficult to quantify since it depends on the dose of the product and its relative affinity for the sites whose recognition determines the relevant biological activity. The proportion of *cis*-versus *trans*-acting functions in viruses, together with their capacity to accept nonlethal deleterious mutations, may influence the generation and maintenance of the defective interfering (DI) particles in a virus-host system. The *cis* versus *trans* proportion is also pertinent to the understanding of internal interactions within mutant spectra (Section 3.8), and to the interpretation of virus extinction by lethal mutagenesis through lethal defection mediated by a class of defective genomes (Chapter 9). Deleterious mutations affecting *trans*-acting functions may be maintained in the viral population by genetic complementation, rendering negative selection dependent on the population context. A mutation in a genome that expresses a phenotype, which is selected against when the genome replicates in isolation, may be perpetuated because the corresponding function is expressed by other genomes of the same replicative ensemble.

Despite general agreement in the distinction between positive and negative selection, there might be instances during virus evolution in which difference between the two modes of selection is not obvious. Suppression by negative selection of all but a limited subset of variant viruses in a population will have the same effect as positive selection acting on the spared subset (Domingo and Holland, 1994). Thus, we accept that a monoclonal antibody-resistant mutant of a virus has been positively selected because replication of all other viruses that did not include the required escape mutation(s) was suppressed (subjected to negative selection).

Positive and negative selection can be viewed as two facets of the same phenomenon, and in some theoretical treatments, no distinction is made between these two modes of selection. The intensity of selection, expressed as a selection coefficient, can be measured through competition between two populations in the absence and presence of a selective constraint, often in the form of a fitness assay (Chapter 5).

Positive selection can be regarded both as a response for survival and as an exploratory process. Virus variants selected for a specific trait may, unpredictably, manifest an unexpected attribute, therefore contributing to biological novelty (Baranowski et al., 2003) (Chapter 7). In contrast, negative selection is basically a conservative force that maintains the biological identity of viruses despite a dissipative mutational pressure, particularly accentuated in the case of RNA and DNA viruses, which display error-prone replication.

## 3.5 Selection and random drift

Viral populations are of finite size, and the size undergoes large modifications in the course of the natural history of any virus. It is very often reduced during transmission from infected hosts into uninfected recipients, or when the virus colonizes new tissues or organs. The viral population size may decrease as a result of an immune response that fails to clear the virus. In this case, a subset of the initial viral population may be perpetuated by positive selection, for example, because it included key amino acid substitutions in B-cell or T-cell epitopes that conferred an advantage in the face of the immune response. Subsets of genomes need not continue replicating only because the others have been selected against, but merely by their chance survival during a random event that entails population size reduction. Therefore, the mutant spectrum subsets that go on to reproduce and populate new niches are not necessarily those dictated purely by selection. In an infected organism, a virus with high replicative capacity may multiply mainly in an internal organ such as the liver, causing liver injury. Yet, a particle that is not necessarily among the best adapted to multiply in the internal organ may surface the infected host by chance and be transmitted to a susceptible individual through contact with secretions, excretions, or blood. Successive transmissions of this sort introduce an additional element of randomness in virus evolution (see also Section 3.2 and Chapter 4).

As a brief digression from viral populations, and to emphasize the role of contingency in natural events, we are all aware that our own personal existence is due to a succession of chance events that need not be detailed here. As science has progressed, we have learned that not only our existence but also that of the universe we inhabit is probably the result of chance episodes, such as the generation of a particular pocket universe (one of the "megaverse") in which matter could be made, and life thrive (according to recent cosmological theories). Even if this cosmological argument is not accepted, one should consider that mammals might not have had a chance to evolve toward hominids if the mass extinction 65 million years ago—with more than 50% of existing species, including dinosaurs, gone extinct—had not occurred. That particular mass extinction was a key event for the future of our biosphere, and it was likely produced by the chance collision of a large asteroid with Earth. Thus, random events are rooted in the way nature works, virus evolution included (see Chapter 1 for a discussion of the evolution of life and mass extinctions in Earth history).

A point of interest and controversy in evolutionary virology, as well as in general population genetics is the relative contribution of random events versus positive selection in shaping the genetic composition of viral populations. The random drift of sequences harboring neutral or quasi-neutral mutations (defined in Chapter 2), independent of positive selection, was proposed as the main mechanism of molecular evolution in the neutral theory championed by M. Kimura and his school (Kimura, 1983). In the second half of the 20th century, the term "nonDarwinian evolution" was coined to refer to evolutionary changes due to random processes rather than to natural selection; the term has lost much of its initial impact.

Not only chance events are not incompatible with the ensemble of processes that we call "Darwinian," but it is also apparent that random evolutionary events act on biological objects that at some point must have been selected. The nature of the viral replication cycle, involving overtly competitive steps, suggests that before reaching a privileged survival position through random drift, a virus (or virus variant) had to face the scrutiny of selection. Viruses that survive by chance (i.e., the transmitted subset vs. the majority present in a donor infected host) will meet a situation in which they will compete with other variants. Extending the example of a virus replicating in the liver mentioned in the previous paragraph, if the liver is the major site of replication, the viral particles that circulate in the blood—and therefore, can be transmitted through blood-to-blood contact—are likely to reflect those particles that replicated best in the liver. M. Eigen disclosed a conversation in which M. Kimura proposed to reformulate the Darwinian principle as "survival of the luckiest" (also published in Kimura, 1989). Eigen agreed but with the addition that the luckiest "always has to be a member of the elite club of the fittest" (Eigen, 2013). J. Holland and I used similar terms: "In RNA virus evolution it is not exactly the luckiest that perpetuates itself, it is the selected luckiest that does" (Domingo and Holland, 1994). I do not know if M. Eigen read our publication, but I take this coincidence to mean that J.J. Holland and I were on the right track in this issue.

Despite these arguments, the neutralist school of molecular evolution exerted a great influence in the understanding of the mechanisms that underlie biological evolution and enriched the way we think about the operation of selection, and the multiple factors that determine evolutionary outcomes. As we learn more about virus evolution, it appears that both, positive selection and random drift of viral genomes occur regularly during viral multiplication and spread, with the prominent negative selection accepted by all schools. Viral populations continuously face episodes of selection and random drift, acting either conjointly or successively.

## 3.6 Viral quasispecies

RNA viruses and some DNA viruses have high mutation rates as one of their distinctive features (Chapter 2), with continuous production of variant genomes that provide a rich substrate for selection and random drift to act upon. Viral quasispecies are defined as collections of related but nonidentical viral genomes subjected to a continuous process of genetic variation, competition among variant forms, and that can act as a unit of selection. The term describes the continuously changing mutant clouds that compose replicating viral populations. Quasispecies theory has become the most adequate framework for RNA virus dynamics because it had limited copying fidelity as a key parameter in its mathematical formulation (Fig. 3.1), and emphasized the presence of mutant distributions during replication. No other models of population biology dealt with mutant spectra. A mutation is parallel to correct replication (Domingo and Schuster, 2016; Domingo and Perales, 2018).

### 3.6.1 The origins of quasispecies theory

The term quasispecies was first proposed by M. Eigen and P. Schuster to describe error-prone replication, and the self-organization potential of primitive replicons postulated to have populated the Earth at the onset of life, some $4 \times 10^9$ years ago (Eigen and Schuster, 1979) (Chapter 1). Quasispecies theory was built on a first theoretical treatment by M. Eigen that described a system characterized by the regular generation of error copies during replication with limited template-copying fidelity (Eigen, 1971) (Fig. 3.1). As explained by M. Eigen in his pioneer publication (this seminal paper was not accepted for publication in several journals because of its length; personal communication from C. Biebricher), his interest was stimulated by S. Spiegelman's experiments on in vitro replication of Qβ RNA—which evidenced production and selection of mutant RNAs—as well as by a "breakfast discussion" with F. Crick, in which Crick expressed the interest of developing a mathematical description of replication with the production of mutants. The so-called *Spiegelman experiments* referred to by Eigen involved serial amplification passages of bacteriophage Qβ RNA in the test tube. The basic design preceded many subsequent experiments on RNA evolution in vitro, the "serial passage" experiments of viruses and bacteria that today are familiar to virologists and bacteriologists (Fig. 3.3), and even the protein misfolding cyclic amplification in the case of prions. The origin of serial passages may well go back to the attenuating passages carried out by L. Pasteur to develop the first human vaccines. Serial passages of viruses in cell culture or live animal and plant hosts constitute one of the basic designs for experimental studies on virus evolution discussed in Chapter 6.

The experiments by S. Spiegelman and colleagues did not involve infection of living cells, but amplification of bacteriophage Qβ RNA (or its low molecular mass RNA derivatives) by purified Qβ replicase—the RNA-dependent RNA

**FIGURE 3.3**    Serial passages with viral nucleic acid or with the virus. Schematic depiction of serial passages as performed in experimental evolution. Top: The Spiegelman-Weissmann passages of Qβ RNA replicated by the enzyme Qβ replicase. Tube 1 contains the complete reaction mixture (Qβ RNA, Qβ replicase, host factor, nucleoside-triphosphates, and buffer with $Mg^{2+}$); after a chosen time of reaction, an aliquot from tube 1 is transferred to tube 2 which contains the complete reaction mixture except for Qβ RNA. The process is repeated the desired number of times ($n$). Analogous designs have been implemented for the amplification of prions. Bottom: Serial passages of the virus in cell monolayers or cells in suspension. Once an initial infection is completed in plate 1, the virus is titrated and the desired amount is used to infect the cells in plate 2. The process can be repeated n times. Several examples with cytolytic and persistent infections are described in connection with experimental evolution designs in Chapter 6.

polymerase (RdRp) of the virus—in the test tube (Mills et al., 1967) (Fig. 3.3). Bacteriophage Qβ belongs to a group of RNA-containing bacterial viruses that infect *E. coli*, and that was discovered by N. Zinder and his colleagues (Zinder, 1975). RNA bacteriophages were instrumental in the development of molecular biology, as a follow-up of the fundamental work of M. Delbrück and his school with DNA bacteriophages. Their RNA genomes furnished the first natural heteropolymeric RNAs for studies on in vitro protein synthesis. Moreover, genomic Qβ RNA can be copied into a full length complementary (minus) strand by the enzyme Qβ replicase in the test tube, and continuous production of plus and minus strands can be achieved in the presence of the replicase, a host factor, nucleotide substrates, and a suitable ionic environment (Weissmann, 1974; Gytz et al., 2015). To avoid the nucleotide substrates being exhausted, an aliquot of the reaction mixture is transferred into a fresh mixture containing the necessary ingredients for synthesis to proceed. Many transfers can be performed attaining remarkable

amplifications of RNA in the test tube so that the initial template molecules are no longer present at the end of the transfer series (representative publications of Spiegelman's group on in vitro RNA evolution are Haruna et al., 1963; Mills et al., 1967; Saffhill et al., 1970).

The origin of self-organization of primitive replicons, one of the main aims of M. Eigen's treatment, represented a link between the principles of Darwinian evolution and classical information theory (Eigen, 1971). "Self-instructive" behavior was distinguished from "general autocatalytic" behavior, and it was pointed out that "self-instruction" is required for "template" activity. A quality factor was defined that determines the fraction of copying processes that leads to an exact copy of the template (the relevant equation is given in Fig. 3.1). The "master copy" will produce "error copies" which will occur with a certain probability distribution. The error copy distribution is what today we term *mutant spectrum* or *mutant cloud*, but "comet tail" of error copies was the term used by Eigen (1971).

In subsequent elaborations of the theory by M. Eigen, P. Schuster, and their colleagues, quasispecies typified infinite, steady-state distributions of related genomes in equilibrium (Eigen and Schuster, 1979; Eigen and Biebricher, 1988; Eigen, 1992; Schuster and Stadler, 1999). The equilibrium is usually metastable: it will be replaced by a new one represented by another mutant distribution when an advantageous mutant appears in the first distribution. A treatment that considers a mutant spectrum of infinite size in equilibrium implies evolution as a deterministic process. This means that whenever a mutant with a selective advantage is produced, it will inevitably outgrow the former distribution, while generating a new mutant cloud. A deterministic model has a mathematical solution, and it is a frequent first formulation, to be followed by more realistic extensions that include stochastic (chance) components. This is a common way to proceed in theoretical biology, not only in the progression of quasispecies theory.

## 3.6.2 Deterministic versus stochastic quasispecies

In considering an application of quasispecies theory to the dynamics of viral populations, it was obvious that viruses (as opposed to theoretical mutant spectra) are not steady-state populations of infinite size in equilibrium when they replicate in their natural hosts or in cell culture. Neither were the primitive replicons in equilibrium at the onset of life and whose behavior was addressed by quasispecies theory (unless the Earth 4 billion years ago included oceans filled with million-trillions of replicons under steady-state conditions, a situation that defies our imagination). A deterministic model is not realistic for RNA viruses or for any other natural replicative entity, a limitation that has been amply recognized by theoreticians and virologists. The nondeterministic nature of mutant generation has deep roots in how mutant genomes are generated, and how they may reach dominance in a viral population. First, mutagenesis results from atomic fluctuations subject to quantum-mechanical uncertainty; for example, those fluctuations that determine the presence of one tautomeric form or another in a template nucleotide, precisely at the time at which an incoming nucleotide reaches the catalytic site of the polymerase elongation complex (Chapter 2). The protein conformation at the catalytic site and its surrounding residues is also dynamic and can contribute to incorporation uncertainties. Second, any growth-competition process is also exposed to statistical fluctuations, which may alter the outcome anticipated by a deterministic behavior. Third, we know now that a new ingredient has to be introduced in the way selection operates: a newly arising mutant that shows a selective advantage need not outgrow the population; its dominance is dependent on the mutant spectrum context, a key issue in viral quasispecies which is addressed in Section 3.8. Results on the long-term dynamics of hepatitis C virus (HCV) suggest that a true population equilibrium may not be reached even in a constant cell culture environment (at least within the usual time frames of nucleotide sequence determinations), as evidenced by continuous mutational waves (Moreno et al., 2017).

An important consideration is whether the generation of mutant distributions in viral populations can adhere to the quasi-equilibrium model as defined in theoretical biology (Domingo and Schuster, 2016). That is, whether the formation of mutants and their equilibration occurs faster than the environmental changes that the mutant distributions must confront. This might be fulfilled by the RNA replicons displaying high mutation rates in the microenvironment of a single replicative unit inside a cell, but a quasi-equilibrium will generally not be perceived by current intracellular RNA sequencing potentialities. If such short equilibrium steps are assumed, it would be legitimate to view the modification of mutant spectra as a succession of short equilibrium steps, each of them following the requirements of the deterministic quasispecies theory. In fact, such an abstraction provides an intuitively useful way to view virus evolution as a function of time (Fig. 3.4). Abstractions of this kind are not unusual in biological studies. In a discussion of virulence models for host-parasite relationships, successive equilibrium steps were proposed in relation to numbers of uninfected hosts, parasite virulence levels, and variation in the optimal transmission rate (Lenski and May 1994). Even more fundamental, the Hardy-Weinberg law of population genetics states the constancy of the frequency of alleles in a randomly mating, large population (also termed *panmictic* population), as a direct consequence of Mendelian heredity. It has attached to it a formula that predicts the frequency of alternative alleles in a randomly mating diploid population. Yet, as emphasized

by E.D. Wilson, there are "some large conditions attached to the use of the Hardy-Weinberg formula in the real world" (Wilson, 1998). The same limitations have been expressed by other authors. A number of simplifications—including finite population sizes—must be conceded to approximate the behavior of real populations to the predictions of the Hardy-Weinberg law. This does not diminish its value as a theoretical framework to understand the behavior of mating diploid populations.

The criticism that a theoretical model does not reflect real situations in a precise manner could be made virtually of any model, including those of population genetics. Also significant in this regard is a statement by M. Eigen when referring to specific conditions in the experimental implementation of the theoretical treatment of selection: "Important as these specifications of defined conditions are for an understanding of the principles of evolution and for a quantitative

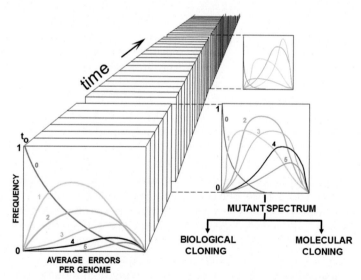

FIGURE 3.4   A representation of temporal evolution of a viral quasispecies as a succession of very short equilibrium steps. At any time point in the evolution of a viral quasispecies (here represented by a "slice" of time) a mutant distribution can be considered a "frozen" distribution of genomes with 0, 1, 2, 3, 4, and 5, etc. mutations at the frequencies given in ordinate. As time elapses, the different genome categories vary in frequency, and the process of change can be considered as consisting of a succession of equilibrium steps ("slices" of time). Molecular and biological cloning provide a snapshot of mutant distributions that, in reality, are changing continuously.

evaluation of experimental data, it is not in the least necessary that any real evolutionary process in nature should have taken place under these special conditions—just as no steam engine ever had to work under the exactly specified thermodynamic equilibrium conditions of the Carnot cycle" (Eigen, 1971). Thus, based on these several scientific precedents, as well as many others one could draw from connections between theoretical principles and their experimental implementation in physics, chemistry, and biology, it is not justified to raise the nonequilibrium argument against viral quasispecies.

The issue has also been brought up whether quasispecies is a necessary concept to understand virus evolution or not [for this controversy the reader is referred to (Holmes and Moya, 2002; Domingo, 2002; Geoghegan and Holmes, 2018)]. C.O. Wilke emphasized that there is no conflict between quasispecies and classical Fisher–Wright models (Wilke, 2005). A reason why virologists did not adhere to classical population genetics models of evolution might be that in the middle of the 20th-century microbes were not considered as satisfactory systems to approach evolution. In addition, most virologists did not find in the formulations of population genetics a means to grasp the key observations related to the presence of mutant spectra and the potential of the rapid evolution of RNA viruses. Population geneticists ignored viruses, and virologists did not establish the connections between observations on viral populations and the body of conceptual and mathematical tools available from population genetics. In contrast, since the very beginning, in the 1970s, actors involved in the development of quasispecies theory, and those that contributed the first experimental observations on RNA virus heterogeneity, made a point of connecting the two areas of scientific activity (Domingo and Wain-Hobson, 2009). Perhaps the key issue can be summarized with the following statement which was addressed to authors who object to the term "quasispecies": "Whether they are termed

quasispecies or not, we are dealing with a very unique population structure that has defied antiviral interventions despite huge economic investments, and that, presently, is helping to interpret virus behavior and to design new means to control viral disease. As Shakespeare said: A rose by any other name would smell as sweet" (quoted from Perales et al., 2010).

### 3.6.3 Mutant spectra, master genomes, and consensus sequences

Extensions of quasispecies theory to finite populations and changing environments (variable fitness landscapes) (Chapter 5) have been developed (Eigen, 2000; Wilke et al., 2001; Nowak, 2006; Saakian and Hu, 2006; Saakian et al., 2006, 2009; Park et al., 2010; Schuster, 2010; Domingo et al., 2012; Domingo and Schuster, 2016). In the way virologists view quasispecies, a mutant distribution may be dominated by one or several master sequences whose preponderance may be transient or extended. The master genome dominates because it displays the highest fitness value in the mutant spectrum. A dominant master and the consensus sequence may or may not be identical (Fig. 3.5). Box 3.1 summarizes definitions of quasispecies from the perspective of physics, chemistry, biology or medicine. Some terms used in quasispecies descriptions.

Recombination (Chapter 2) is an added element of stochasticity and a factor of population disequilibrium that has been also integrated into quasispecies theory (Boerlijst et al., 1996; Park and Deem, 2007; Muñoz et al., 2008; Wagner et al., 2016). Thus, quasispecies theory has incorporated environmental and molecular events that perturb equilibrium, as well as epigenetic effects or RNA interference in the process of template copying. Overall, these additions have reinforced the adequacy of quasispecies as a framework to understand virus evolution. The key concept for virology is that viral populations consist of complex and dynamic distributions of related

## QUASISPECIES

FIGURE 3.5 Two ways of representing a viral quasispecies. The symbols in the population depicted on the left mean different mutants expressing a relevant phenotypic trait (drug-resistance or other); identical symbols express the same phenotype, but they may differ in genomic nucleotide sequence. The representation on the right illustrates individual genomes as horizontal lines and mutations as symbols on the lines. Note that the master (dominant) sequence does not coincide with the consensus, as explained in the text. The two types of representation are used in this book, depending on the concept to be conveyed.

genomes and that the ensemble often plays a role in determining virus behavior.

The first experimental evidence that an RNA virus population consisted of a spectrum of mutants was obtained in C. Weissmann's laboratory in Zürich, working with bacteriophage Qβ replicating in its natural host *E. coli*. The studies included the calculation of a mutation rate for a specific nucleotide transition at the 3′-extracistronic region of Qβ RNA (Batschelet et al., 1976), and the demonstration that biological clones (the progeny of a single viral genome) evolved rapidly to form complex spectra of mutants. A key sentence summarized the results: "A Qβ phage population is in a dynamic equilibrium, with viable mutants arising at a high rate on the one hand, and being strongly selected against on the other. The genome of Qβ phage

cannot be described as a defined unique structure, but rather as a weighted average of a large number of different individual sequences" (Domingo et al., 1978). Subsequent studies with other RNA viruses, some DNA viruses, and subviral replicons revealed that the same statement applies to their population structure in cell culture and in vivo (Holland et al., 1982; Eigen and Biebricher, 1988; Martell et al., 1992; Duarte et al., 1994; Morse, 1994; Coffin, 1995; Gibbs et al., 1995; Isnard et al., 1998; Vogt and Jackson, 1999; Gavrilin et al., 2000; Sarisky et al., 2000; Domingo et al., 2001, 2012; López-Bueno et al., 2003; Domingo, 2005; Domingo and Perales, 2018; Filippou et al., 2018); for historical accounts of the early studies on theoretical and viral quasispecies and their impact, see Domingo and Wain-Hobson (2009) and Domingo and Schuster (2016). The experimental results with viruses provided the biological definition of quasispecies (Box 3.1) (Eigen and Biebricher, 1988; Domingo et al., 2001; Domingo and Schuster, 2016). The quasispecies concept has extended its applicability to nonviral, cellular, and macromolecular assemblies, and such interesting ramifications are treated in Chapter 10.

The mutant spectrum nature of viral populations is full of implications. Modifications of host cell tropism, failures of synthetic vaccines to produce solid and durable protection, and the very frequent selection of virus mutants resistant to antiviral agents, are some of the important events now readily understood on the basis of the dynamic mutant spectra on which selection acts (Fig. 3.6). Indeed, swarms of mutant genomes are continuously being placed under the test of selection, and mutants resistant to antibodies or inhibitors can arise and be maintained as minority components in the population until they increase in frequency in the presence of the relevant antibodies or inhibitors. Even when replicating in a stable cellular environment, mutant spectra may be perturbed, as reflected in mutational waves that may or may not alter the consensus sequence (Moreno et al., 2017). This renders it difficult to establish a connection between biological behavior and a consensus nucleotide

BOX 3.1

### Definitions and terms related to viral quasispecies

- Physical definition of quasispecies
  Cloud in sequence space.
- Chemical definition of quasispecies
  Rated distribution of related but nonidentical nucleic acid sequences.
- Biological definition of quasispecies
  Target of selection.
- Medical significance of quasispecies
  The many drug resistant mutants present or arising in infected patients that may result in treatment failure.
- Usual definition by virologists
  An ensemble of viral genomes subjected to genetic variation, competition, and selection and that can act as a unit of selection.
- Master sequence in a viral quasispecies

The genomic nucleotide sequence that is present at the highest frequency in a mutant distribution because of its higher fitness than the other genomes. The master sequence may still be a minority relative to the ensemble of low frequency variants.

- Consensus sequence
  The sequence that has at each position (nucleotide or amino acid) the residue, which is most abundant at the corresponding position in a set of aligned sequences. The consensus sequence need not coincide with the master sequence.
- Mutant spectrum of a viral quasispecies
  The ensemble of mutant genomes that make up a viral quasispecies. Alternative terms are mutant distributions, clouds or swarms.

*Partial resistance to inhibitors, antibodies, CTLs*

*Cyclic adaptation*

*Molecular memory*

*Altered cell tropism or host range (receptor recognition, other)*

*Switch between acute and persistent infection*

*Altered IFN induction or sensitivity*

*Altered capsid stability, pH sensitivity*

*Altered virulence*

**FIGURE 3.6** Mutant spectra are not neutral spaces. Arrows point to phenotypes that have been associated with minority components of mutant spectra. See text for specific examples and references.

control of viral disease and different possible strategies to overcome the adaptive capacity of mutant spectra are treated in Chapters 8 and 9. Lethal mutagenesis is based on one of the corollaries of quasispecies theory that asserts that in any replication system there is a maximum information content which can be maintained for a given average copying fidelity (termed by M. Eigen the "recognition parameter" $q$). Here again, the concept of error threshold has been extended to finite populations and variable fitness landscapes (Nowak and Schuster, 1989; Alves and Fontanari, 1998; Hogeweg and Takeuchi, 2003; Ochoa, 2006; Schuster, 2016).

### 3.6.4 Measurement of quasispecies complexity

At any given time during replication, a viral quasispecies consists of a complex mutant spectrum, from which a consensus sequence can be derived either experimentally (e.g., by Sanger sequencing of an entire genome population) or

sequence, as evidenced in the case of neurovirulence of mumps virus (Sander et al., 2006). A consensus sequence is a minimal and often insufficient description of a viral genome. The problem that quasispecies dynamics poses for the from a mutant spectrum sampled by biological or molecular cloning. A consensus sequence is determined by assigning to each position the nucleotide which is the most represented at that position among the sampled sequences (Fig. 3.5 and Box 3.1). Contrary to the term "consensus" applied to molecular biology (consensus promoter sequence, TATA box, polyadenylation signal, nuclear localization signal, etc.), in the case of viral quasispecies, the use of consensus does not imply universality. Independent or even sequential populations of the same virus can be characterized by different consensus sequences. An identical consensus may characterize mutant spectra of very different complexity or that display disparate behavior. Even if aggrandized by its inclusion in data banks, a consensus sequence may have as transient an existence as the mutant spectrum it intends to represent. The composition of a mutant spectrum can be sampled by biological or molecular cloning, followed by Sanger sequencing of individual clones (Fig. 3.7), or by subjecting the sample to deep sequencing. Each of these procedures has advantages and limitations. When biological clones (virus from individual viral plaques formed on a cell monolayer, or virus from end-point dilution series) are the starting materials, the genomic nucleic acid from each clone or clonal population is extracted. Then, the viral DNA or RNA is subjected to amplification by PCR (polymerase chain reaction) or RT-PCR (reverse transcription followed by polymerase chain reaction), respectively, and the amplified material is sequenced. The result is an average or consensus sequence of the sampled population of genomes that are part of each biological clone. As expected from viral mutation rates and from the rounds of viral amplification needed for plaque development, a viral biological clone is generally not homogeneous (Yin, 1993), let alone viral plaques derived from clinical isolates (Sukkaew et al., 2018). The nucleotide sequence obtained cannot be affected by possible misincorporations in individual copies introduced as a consequence of the limited copying fidelity of the enzymes used for the in vitro amplification, unless the amount

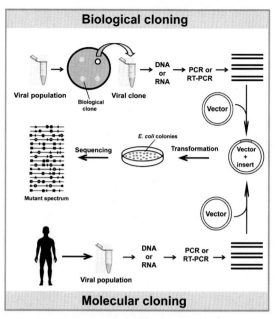

FIGURE 3.7 A schematic representation of steps involved in biological and molecular cloning for the analysis of viral quasispecies. Top: Biological cloning. A viral population is plated on a cell monolayer to obtain the virus from an individual plaque (biological clone). DNA or RNA is extracted and the viral sequences amplified by PCR or RT-PCR, respectively. The amplified DNA is ligated to a vector, and the product used to transform E. coli. Viral DNA inserted into individual bacterial colonies is further amplified and sequenced to provide a sample of genomic sequences present in the initial viral population. Bottom: Molecular cloning. DNA or RNA is extracted from a biological sample and the viral sequences amplified by PCR or RT-PCR, respectively. The products are ligated to vector DNA and the procedure completed as indicated for biological cloning. See text for controls and possible biases in mutant representation.

of template used for the amplification reaction is very small (amounting to a few molecules). Misincorporations can alter a genomic position in one of the amplified genomes, but not the average or consensus sequence representative of the clone (Fig. 3.5). The analysis of biological clones biases the sequence information in favor of genomes that are more infectious in the cells used for the biological cloning. New plaque detection and isolation assays may facilitate unbiased plaque scoring (Katzelnick et al., 2018; Wen et al., 2019). Genomes from noninfectious viral particles are excluded from the analysis unless complemented within a growing plaque, a possibility that cannot be excluded.

The alternative procedure, also using Sanger sequencing, is to obtain molecular clones from total viral DNA or RNA (or from an end-point dilution series of the nucleic acid preparation), followed by PCR or RT-PCR amplification and cloning of individual DNA or cDNA copies in a vector (often an *E. coli* plasmid) (Fig. 3.7). This method does not depend on infectivity of viral particles or genomes, and therefore, can analyze defective, noninfectious genomes, or probe the composition of viral preparations independently of their infectivity. A bias may be produced by the limited copying fidelity of enzymes used in the in vitro amplification. The problem can be largely solved using high-fidelity polymerases and correct amplification conditions, together with control experiments to determine the basal error rate inherent to the in vitro amplification protocol. When a population of viral DNA or RNA is used as the starting material, sufficient template DNA or RNA must be present to ensure that the sequence repertoire obtained approximates that of the viral genomes present in the population under study (Arias et al., 2001; Airaksinen et al., 2003; Polyak et al., 2005).

Despite their value in revealing the complexity of viral populations, cloning methods are limited by the number of sequences that are obtained, often in the range of 20−100 from a viral sample, due to the work involved. The advent of deep sequencing with its several platforms has represented important progress in general and viral genetics (Mardis, 2008; Fox et al., 2009; Quiñones-Mateu et al., 2014; Matranga et al., 2016). Some of the procedures can provide ten thousand to several hundred *thousand* sequences from the same sample, be it a biological specimen (tissue, secretion, excretion, tumor, entire animal, or plant) or an environmental sample (sea or lake water, soil, or toothbrush). Putative new viruses and others related to established viral pathogens have

been identified by deep sequencing, attesting of the ubiquity of viruses in our biosphere (Delwart, 2013; Drexler et al., 2013; Zhang et al., 2018) (Chapter 1).

The increased penetration into the composition of mutant spectra has amply confirmed the extreme genetic heterogeneity of viral populations, with myriad low-frequency variants that could not be detected by classic clonal surveys. Molecular amplification and bioinformatic procedures have been developed that allow reliable assignment of mutations present at frequencies in the range of 0.01%−1% (depending on the genome copying and amplification protocols, as well as the correcting operations and controls performed) (Zagordi et al., 2010; Jabara et al., 2011; Vandenbroucke et al., 2011; Chen-Harris et al., 2013; Poh et al., 2013; Acevedo et al., 2014; Gregori et al., 2014). Some studies have compared deep sequencing with molecular cloning-Sanger sequencing to characterize mutant spectra (Eriksson et al., 2008; Kuroda et al., 2010; Zagordi et al., 2010; Liang et al., 2011; Vandenbroucke et al., 2011; Wright et al., 2011; Gregori et al., 2014; Moreno et al., 2017). Although the mutant repertoire and the frequency of individual mutations often differ depending on the method used, the presence of dynamic mutant spectra has been evidenced for all RNA viruses which have been analyzed with attention to technical details (Domingo and Perales, 2018; Gire et al., 2014; Guedj et al., 2018). To limit the number of artifactual mutations, R. Andino and colleagues devised a method that consists in fragmenting and circularizing the RNA genomes to be sequenced and synthesizing tandem cDNA repeats (redundant copies) by rolling circle replication. Only those mutations present in each of the repeats of the same original templates are *bona fide* mutations existing in the RNA population (Acevedo et al., 2014). Applied to poliovirus (PV), this CirSeq procedure using the Illumina platform has confirmed the presence of multitudes of PV variants at frequencies of $10^{-3}$ to $10^{-5}$, giving an average of one−two mutations per genome. Curiously, but perhaps not coincidental, this estimate of heterogeneity is exactly the same obtained for bacteriophage Qβ 40 years ago using T1-oligonucleotide fingerprinting of genomic RNA (Domingo et al., 1978).

Although generally less informative than sequencing, alternative, rapid sampling procedures have been successfully used to compare relative genetic heterogeneities of viral populations: restriction enzyme site polymorphisms, RNase A mismatch cleavage (Sánchez-Palomino et al., 1993), DNA heteroduplex mobility assay, oligonucleotide mapping, walking within mutant spectra by successive, specific PCR amplifications (Garcia-Arriaza et al., 2007), among others. DNA microarrays have become a promising technology for diagnostic virology and to characterize viral populations, with prospects of an application to quantify the complexity of mutant spectra (Amexis et al., 2001; Wang et al., 2002; Mount, 2004; Briones and Domingo, 2008; Lauring and Andino, 2011; Martín et al., 2016). These methods have the advantage of comparing very large numbers of samples by automated procedures. If not coupled with sequence determinations, however, they lack the precision of information conveyed by sets of nucleotide sequences.

### 3.6.4.1 Mutant spectrum depth and dynamics

Comparative massive sequencing allows a molecular dissection of quasispecies dynamics during cross-species transmissions, and the identification of minority mutations that become dominant at later stages of evolution (Borucki et al., 2013) (see the concept of "harbinger" mutations in Chapter 5). Viral populations include "underground" swarms of low frequency genomes whose fitness, despite being very low, is not zero. Deep sequencing has acted as a magnifying glass that visualizes genomes that did not exist according to former techniques; it is like the underground police that watches events occurring at a previously hidden level. The once considered "lethal class" of mutations may actually be present in poorly replicating genomes,

thus forcing a reconsideration of the meaning of lethality (Domingo and Perales, 2018).

The penetration in the composition of mutant spectra has clinical implications for patient management. It allows the quantification of drug-resistance mutations in virus from treated or untreated infected patients, and of patients prior to treatment (Wang et al., 2007; Mitsuya et al., 2008; Solmone et al., 2009; Tsibris et al., 2009; Hedskog et al., 2010; Svicher et al., 2011; Quer et al., 2017). The methodology permits detection of mixed infections with different genotypes of the same virus that would go unnoticed by standard cloning analyses (Margeridon-Thermet et al., 2009; Wu et al., 2013; Quer et al., 2014). The capacity to interrogate the dynamics of minority quasispecies components has also revealed multiple antibody- or CTL-escape routes, some of them only transient, others successful, during episodes of immune selection (Fischer et al., 2010; Cale et al., 2011) (Fig. 3.8). Studies comparing rhesus macaques that controlled simian immunodeficiency virus infection suggest that a broadening of the CD8$^+$T cell response to target several viral proteins may be a predictor of control failure (Nomura et al., 2015). These important implications for the control of viral disease are further treated in Chapters 8 and 9.

The extremely complex and dynamic mutant "mixtures" render the current data banks a minimal and insufficient representation of virus identity (Domingo et al., 2006). The real composition of viral populations also demands new classification approaches to group components of mutant spectra (either from one isolate, from sequential isolates from one infected host, or from different hosts). Clustering methods, such as partition analysis of quasispecies (PAQ) (Baccam et al., 2001) and quasispecies evolution network-based transmission inference (QEN-TIN) (Skums et al., 2018) may help in establishing relationships among related genomes and identification of transmission clusters. In a pictorial representation of PAQ, a center genotype

**FIGURE 3.8** Transient selection events. In the upper part, genomes with a mutation represented by a red asterisk are directly selected because it confers an adequate phenotype for the existing environment. In the bottom part, genomes with a mutation represented by a *green asterisk* are transiently selected, until they are overgrown by genomes with the alternative mutation represented by the *red asterisk*. Application of deep sequencing to viral populations undergoing selection is revealing large numbers of transient and unsuccessful selection pathways that were previously unnoticed due to insufficient penetration into the mutant spectrum. See text for references and operating mechanisms.

nucleates a group of sequences within a circle with a previously set radius. In this fashion, multiple, overlapping or nonoverlapping groups of sequences can be defined within mutant spectra of viral quasispecies (Fig. 3.9). The procedure has been applied to the analysis of subpopulations of equine infectious anemia virus (Baccam et al., 2003), regulatory sequences of bovine viral diarrhea virus (Jones et al., 2002), sequential hepatitis A virus isolates (Costa-Mattioli et al., 2006), and mutant spectra of foot-and-mouth disease virus (FMDV) subjected to lethal mutagenesis (Ojosnegros et al., 2008). Other algorithms to organize complex mutant spectra have been developed (Prosperi et al., 2011; Poh et al., 2013; Gregori et al., 2014; Mangul et al., 2014;

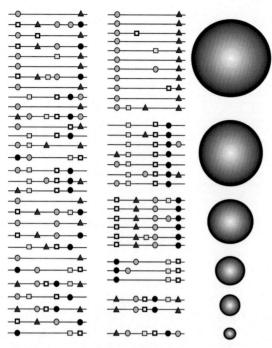

**FIGURE 3.9** Representation of viral quasispecies as a number of subpopulations, inspired in partition analysis of quasispecies developed by P. Baccam and colleagues. A seemingly random distribution of genomes (left) may actually include distinct groups according to mutation types (second distribution). On the right the mutant classes are represented as spheres whose size is proportional to the number of genomes in the class they represent. *The figure is modified from Domingo, E., Sheldon, J., Perales, C., 2012. Viral quasispecies evolution. Microbiol. Mol. Biol. Rev. 76, 159–216, with permission from the American Society for Microbiology, Washington DC, USA.*

Topfer et al., 2014; Marz et al., 2014; Jayasundara et al., 2015; Karagiannis et al., 2017). Elucidating which information is biologically and medically relevant, and worthy of inclusion in data banks is an important challenge.

### 3.6.4.2 Diversity indices

Once a mutant spectrum has been characterized and possible differentiated subpopulations sorted out, there are several parameters and

**TABLE 3.1**  Diversity indices to characterize mutant spectra.

| Index category | Specific parameter |
|---|---|
| Incidence at the entity level | Number of haplotypes |
| | Hill numbers, for $q = 0$ |
| | Chao-1 |
| | Number of polymorphic sites |
| | Number of mutations |
| Abundance at the molecular level | Shannon entropy |
| | Simpson index |
| | Gini-Simpson index |
| | Hill numbers, for $q > 0$ |
| Functional differences among haplotypes | Incidence at the entity level: Mf min, Mfe, FAD, $\pi_e$ |
| | Abundance at the molecular level: Mf max, $\pi$ |

Incidence-based indices are those that correspond to counts of entities in the alignment of haplotypes. Abundance-based indices consider the entities and their frequency in the population. Functional indices are based on differences between haplotypes, and may include (incidence) or not (abundance) the frequency of each haplotype. Meaning of symbols: q, order of Hill numbers; Mf min, minimum mutation frequency; Mfe, mutation frequency by entity; $\pi$, nucleotide diversity; $\pi_e$, nucleotide diversity by entity; FAD, functional attribute diversity; Mf max, mutation frequency by molecule. See Gregory et al., 2016 for the formulae to calculate each index, software packages, and examples. See text for the context of quasispecies dynamics.

diversity indices that can be used to quantify their complexity (Box 3.2 and Table 3.1). The traditional indices used in quasispecies research are mutation frequency (maximum and minimal), average genetic (or Hamming) distances, and Shannon entropy (Gregori et al., 2014; Seifert and Beerenwinkel, 2016). Some additional indices have been adopted from ecology, and have been tested with laboratory and clinical samples of RNA viruses (Gregori et al., 2016). The adequacy of an index depends on the objective of the determination; for example, the rare haplotype load is an index that proved adequate to quantify complexity of mutant spectra of viruses on the way toward extinction by lethal mutagenesis (see Chapter 9).

---

**BOX 3.2**

---

**Parameters used to estimate the complexity of mutant spectra**

- Minimum mutation frequency
  The number of different mutations scored in a set of sequences divided by the total number of nucleotides sequenced (expressed as substitutions/nucleotide (s/nt)).
- Maximum mutation frequency
  The total number of mutations (be they repeated or not) in a set of sequences divided by the total number of nucleotides sequenced (expressed as substitutions/nucleotide (s/nt)).
- When mutation frequency refers to a specific mutation it is called *mutant frequency*.
- Shannon entropy

- The proportion of different sequences in a set of sequences. A value of 0 means that all compared sequences are identical; a value of 1 means that each compared sequence is different from the others.
- Average genetic (or Hamming) distance
  The average number of mutations that distinguishes any two sequences in the set under study.
- Parameters also obtained from deep sequencing: number of different mutations, number of polymorphic sites, number of haplotypes, among other diversity indices (Table 3.1).

---

## 3.6.5 Some key points on the impact of quasispecies in virology

As a general note, the quasispecies concept served as a first indication of the complexity inherent to biological entities. The term complexity is used here to mean differences among individuals of a biological community that can be genetic or epigenetic. Genomic complexity has been unveiled by metagenomic surveys but its level was unsuspected at the pre-sequencing times, when experimental quasispecies was documented.

Quasispecies insights have been instrumental for the understanding of the nature of viral populations, and to interpret intrahost evolution and pathogenesis of RNA viruses for the following reasons:

- Quasispecies provides a new definition of wild type

One of the most inspiring departures from classic concepts contributed by quasispecies theory is that the wild type is no longer a defined genome with a precise nucleotide sequence, but a collection of related genomes. This means that forces that promote evolution [selection

acting on phenotypes and random drift acting on genotypes (Sections 3.5 and 3.6)] can operate on collections of genomes. By using viral quasispecies reconstructed with minority genomes expressing a similar phenotype (resistance to a neutralizing antibody) it was shown that what is selected is indeed a cloud of mutants (Perales et al., 2005) (Section 3.8.3). The new structure of the wild-type influences several aspects of virus biology, as summarized in the bullet statements that follow, and in subsequent chapters.

- Mutant spectra increase the adaptive potential of viruses because phenotypic variants are present in the population or can be rapidly generated

The adaptedness conferred by mutant spectra was elegantly proven by a decreased pathogenic potential for mice of a poliovirus mutant whose polymerase displayed higher copying fidelity than the wild-type enzyme. The virus that harbored such mutant polymerase formed a narrower mutant spectrum (a lower average number of mutations per genome) than the virus with the wild-type polymerase (Pfeiffer and Kirkegaard, 2005; Vignuzzi et al., 2006) (Chapter 2).

Several fidelity mutants of other viruses have been subsequently characterized, and they offer a clear (but by no means the only) example of the selective advantage conferred by high mutation rates, and of the involvement of quasispecies in viral pathogenesis and persistence (Borderia et al., 2016). Mutant spectrum complexity may influence the outcome of a viral infection, and the response to antiviral treatments (Chapters 8 and 9).

- The information provided by consensus sequences may be insufficient to understand virus behavior

Consensus sequences are commonly determined for virus isolates, and they provide adequate information for purposes such as virus identification and classification, structural determinations, or studies on molecular epidemiology. However, for an interpretation of viral evolution, pathogenesis, and in general, what we may call the genotype-phenotype relationship, it is essential to consider that a consensus sequence is an average of many different sequences, and that a genome identical with the consensus may not exist in the population (Fig. 3.5). It is sometimes wrongly believed that the consensus sequence is the sequence of the major component of the population. This is not necessarily the case. The consensus sequence derived for a viral population may not even exist in that population. Following a reflection by M. Eigen, the consensus of the numbers of a telephone directory is no telephone number (Eigen, 2013). Furthermore, the most abundant genomes at one-time point need not be the ancestors of the genomes that dominate the population at a subsequent time point. The latter may have been derived from a hidden minority at an earlier phase of the evolution of the same lineage (Briones and Domingo, 2008). One or a few point mutations may not affect the phylogenetic position of a virus, and yet they may modify some of its relevant biological traits. This comprises

even a single synonymous mutation within a protein-coding region, which may control virus production (Novella and Ebendick-Corpus, 2004; Hamano et al., 2007). The recent literature includes many additional cases, too numerous to describe here, of single nucleotide or amino acid substitutions that produce salient phenotypic modifications [example in (Young et al., 2019)]. One of the most serious mistakes of viral evolutionary genetics has been to ignore mutant spectra.

- The behavior of individual genomes that compose a mutant spectrum may depend on other genomes

Either positive interactions (through cooperation or complementation), or negative interactions (of a suppressive or interfering nature), can modulate the behavior of the ensemble, and this important aspect is detailed in Section 3.8. Profound fitness differences can follow from alterations in the composition of the mutant spectrum, without modification of the consensus nucleotide sequence (González-López et al., 2004, 2005; Novella and Ebendick-Corpus, 2004; Grande-Pérez et al., 2005), including very drastic events, such as the transition toward extinction by lethal mutagenesis (Chapter 9).

- Uses of the term quasispecies

Consideration of quasispecies theory prompts to reserve the use of the term quasispecies for viruses to the ensemble of genomes of a single replicative unit. Focusing on an infected cell, it would mean the genome ensemble in one of several membrane-anchored replication complexes. This strict use has been relaxed to mean sets of related genomes found in a cell, in a group of cells (in culture or within an organism), tissues, organs, or even in entire organisms. Such extensions can be criticized to the extent that they represent a departure from the initial conceptual origins, but they are justified because it helps in understanding the viral dynamics in

**(A)**

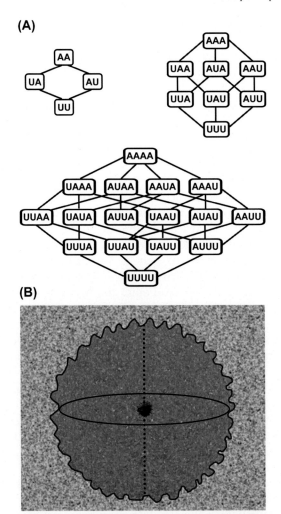

**(B)**

FIGURE 3.10 Representations of sequence space. (A) Connections between hypothetical genomes consisting of only two nucleotides (A,U) of length (*v*) two (top left), three (top right), and four (middle). The diagrams are intended to illustrate the structure of simple sequence spaces, the connectivity among points of the space, and the increase of complexity with genome length. The sequence space for a real virus is impossible to depict in the terms used in these three diagrams (Figure inspired and modified from several articles by M. Eigen and colleagues that can be found in references quoted in this and other chapters of the book). (B) A useful surrogate representation of the viable (replication competent) sequence space of a virus (four nucleotides with *v* = 3000−32,000 for RNA genomes) is in the form of spheres of size proportional to sequence space occupation. The outer square in *gray* represents a huge theoretical multidimensional

realistic scenarios, and its consequences. Viruses can be subjected to selective constraints that do not act on intracellular replicative units. The virus that reaches the blood from different sources may be differentially neutralized by circulating antibodies. This implies competition and selection acting on a "meta-quasispecies" that will result in survival of the fittest viruses in the blood environment, independently of the specific cell where the relevant mutations occurred. Parallel arguments apply to mutants displaying drug resistance or altered cell tropism that may compete in a body compartment. Many important medical implications of viral quasispecies can be apprehended with this "meta-quasispecies" extension.

Quasispecies has also reached environmental microbiology as a diversity concept, for example, in aquatic ecosystems in which viral heterogeneity captures production of variant genomes by hosts in changing habitats (López-Bueno et al., 2015; Cavicchioli and Endmann, 2015).

As the complexity of mutant spectra of RNA viruses is unveiled, additional implications of quasispecies become apparent, and the reader will find several of them in the following chapters.

## 3.7 Sequence space and state transitions

"Sequence space" applied to genomes refers to a theoretical representation of all possible variants of any nucleotide sequence (Fig. 3.10).

space that includes all possible viable and nonviable genomes. The outer *blue* and internal *red spheres* have been drawn with a 10-fold difference in radius that results in a 100-fold difference in surface area. The population represented by the large sphere has access to many more contacts with other genomes present in the sphere surface than the population represented by the small sphere. See text for the relevance of movements in sequence space in virus evolution. *Part (B) of the figure is reproduced from Domingo, E., Sheldon, J., Perales, C., 2012. Viral quasispecies evolution. Microbiol. Mol. Biol. Rev. 76, 159−216, with permission from the American Society for Microbiology, Washington DC, USA.*

In the language of physics, it is a "hypercube" (a geometrical representation that gathers multiple cubic forms) or a multidimensional set of points, analogous to a cube in a three-dimensional space (Eigen, 1992). "Space" is a concept that originated in information theory (Hamming, 1980), and has found broad application to indicate the range of positions or values for a given parameter or alternative states for physical or biological objects. Experts refer to the "structural space" of viral particles, or the "conformation space" of a given protein. Coexistence of minority conformers, that is, the presence of multiple points in the space of conformations within a protein population has been identified as a factor of prion adaptability to different host species (Chapter 10). In pharmacology, exploration of the "chemical space" of drugs with some desired biological activity is often used to describe the synthesis of new molecules by computer-assisted design. The following modest digression in psychology is justified because it relates to "quasispecies memory" of viral populations treated in Chapter 5. "Psychological space" was proposed in an attempt to apply physical principles to psychological science (Shepard, 1987). An individual's psychological space was mapped to a physical space that defined metric distances between brain stimuli able to elicit a response. The probability that a response to a stimulus extends to another stimulus will depend on the distance between them. In this view, upon removal of an external stimulus, its memory trace may undergo "diffusion" in psychological space (Staddon, 2001). The term "diffusion" will later be used as a metaphor to refer to delocalizations in sequence space undergone by viruses (Chapter 9), as well as quasispecies memory loss viewed as "diffusion" in sequence space (Chapter 5). For biological macromolecules, protein sequence space (Maynard Smith, 1970) and nucleic acid sequence space (Rechenberg, 1973; Eigen and Biebricher, 1988) are commonly used.

### 3.7.1 Virus evolution as a movement in sequence space

At any given time a virus occupies a cloud (multiple points, each corresponding to a genome with a nucleotide sequence) in a region of sequence space. A selective pressure or a bottleneck event acting on the replicating cloud reorganizes and relocalizes the cloud. Mutagenesis expands at least transiently the occupation of sequence space by a virus, sometimes with irreversible loss of genetic information when the system is driven toward nonfunctional points of the space. Lethal mutagenesis (virus extinction by an excess of mutations) can be regarded as a drift in sequence space until nonviable areas for the virus are reached; this view reconciles some theoretical models of lethal mutagenesis with the theoretical concept of error catastrophe and experimental observations (Chapter 9). A key feature of sequence space for virus adaptation or de-adaptation is its high connectivity (Eigen, 1992). Points separated by one mutation are linked by a line (Fig. 3.10). In the gigantic hypercube that represents the entire sequence space for a genome of $v$ nucleotides, the minimal distance between any two points is never larger than $v$. Fitness differences guide movements of mutant clouds through sequence space and allow distant points to be reached, with relevant consequences for virus adaptation (Chapter 5). Once the virus is in a distant point, the cloud produced by mutational pressure has a different domain of exploration than the previous cloud, and reiteration of this process allows the virus to increase its fitness in each specific environment. Physics notions of sequence space help in visualizing increased or decreased adaptation of a virus to an environment. The mathematics behind this and other abstractions should not deter us from the usage of physics concepts since mathematics is a formal representation of a concept, as commented with the equations that describe Darwinian evolution (Fig. 3.1).

Another, more every day type of image for a virus reaching its optimal local fitness through successions of exploratory mutations is to think how you reach a specific word in a classic printed dictionary: you open the dictionary at random, and then you approach the target page by sometimes going a few times to pages that precede or that follow the target, until you reach and stay in the correct page to read what you wanted. The major difference between a virus searching an ideal position in sequence space and your searching of a world is that while neighbor positions in sequence space often have related fitness (with exceptions though), words next to each other do not need to have related meanings (again, with exceptions). Thus, for highly mutable entities, adaptation is a gradual approximation to the desired endpoint; for reaching the objective, fitness is the guiding force, and mutation is the major toolbox.

At any time in an evolutionary process multiple points in sequence space are present in "one" viral population. Even if many such points are incompatible with genome replication, those that maintain some replicative ability can still guide the genome ensemble toward regions of sequence space which are compatible with replication and survival. Striking virus "resurrections" are sometimes observed, with viruses that have nearly lost their infectivity due to the action of a mutagenic agent, and that can nevertheless regain fitness when the mutagen is removed (Chapter 9). This "Lazarus effect" in virology is reminiscent of the "Lazarus phenomenon" in physiology (spontaneous return of blood circulation). In patients considered "cured" of an infection, on occasions, the virus may reappear after days or months.

Replication-competent viruses occupy only a tiny fraction of the immense theoretical sequence space: $4^{10,000}$ for a viral genome of 10,000 nucleotides! A palpably unimaginable number. Obviously, only very restricted subsets of genomic sequences are compatible with virus viability no matter which the biological environment might be (Eigen and Biebricher, 1988; Domingo et al., 2001). Importantly, the functional sequence space can be subdivided broadly into fully and conditionally functional subsets of genomes. Such subdivision is of relevance for interpreting internal interactions among components of a mutant spectrum, one of the events which determines that a viral quasispecies acts as a "unit of selection," meaning that what is selected and evolves may be an ensemble of genomes, not individual genomes (Section 3.8).

### 3.7.2 Exploration of sequence space and the sampling problem: viral population size as a key parameter

It may be thought that genomic diversity being a general feature of all replicating entities, be they viruses or cells, there are no distinctive features in RNA viruses (or the highly variable DNA viruses) that can justify the emphasis placed on their genetic heterogeneity. This is incorrect on several grounds, and here we consider one that relates sequence space occupation with adaptability.

The capacity to explore a large proportion of the theoretically available sequence space is a determinant of adaptability. In this respect, the impact of existing as quasispecies to permit adaptation is far greater for viruses than for cells. This is best understood when four parameters are considered: genome size, population size, the average number of mutations per genome, fecundity, and number of mutations needed for a phenotypic change (explained in Box 3.3). The connection between genome size and virus population size can be illustrated with the following arguments. A genome of 10,000 nucleotides can produce a total of $3 \times 10^4$ possible single mutants, although many will not survive or contribute to the next generation due to low

---

**BOX 3.3**

---

## Some parameters that influence the adaptability of viral quasispecies

- Genome size

  For RNA viruses, typically 3—33 Kb (see Chapter 1 for comparative genome size of viruses and cellular organisms). The limited genome size render RNA viruses more effective explorers of sequence space than cells.

- Population size

  It is highly variable. An acutely infected organism may harbor up to $10^{12}$ viral particles at a given time, each with a certain probability of productive infection. A viral plaque (the progeny of a single genome) on a cell monolayer can include $10^3$ to $10^{10}$ viral particles. (See Chapter 4 for comparisons of population size of viruses and host organisms.)

- Average number of mutations per genome calculated for the components of a mutant spectrum.

It is a wide range (that can be placed at 1—100 mutations per genome) that depends on error rate and proximity to a clonal origin. Application of deep sequencing has generally increased the estimates first obtained by cloning-Sanger sequencing.

- Fecundity or average yield of viral particles per infected cell

  Again, a broad range of values ($<1-10^6$ particles produced per cell) have been reported for infections in nature. High fecundity promotes exploration of sequence space.

- Number of mutations needed for a phenotypic change

  Critical to the pertinence of quasispecies to understand viruses is that one or few mutations are often sufficient to modify a relevant phenotypic trait.

- See text for illustration of the interplay among these parameters.

---

fitness. The potential number of single mutants is below the size of most viral populations that multiply in differentiated host organisms, and often even in a single cell from an organism. For a mammalian genome the potential number of single mutants is of the order of $10^{10}$, which is well above the population size of mammalian species (values of the population size of some biological groups are given in Chapter 4). This comparison illustrates that the capacity to explore sequence space (and to find adaptive pathways) is vastly superior for simple replicons displaying high mutation rates than for complex, cellular genomes (Eigen and Biebricher, 1988; Eigen, 1992; Domingo et al., 2001; Domingo, 2006). Many nucleotide positions in a large mammalian genome will remain invariant, no matter what the environmental demands are.

A mammalian genome can explore adaptive solutions through recombination and with a limited number of all possible mutations that it could theoretically undergo. In contrast, a small viral genome has the potential to explore the adaptive effect of changes in each of its nucleotide positions. Exploration of sequence space that leads to adaptation is facilitated by two additional parameters that relate to the complexity of mutant spectra: the great fecundity of most viruses, mainly those associated with lytic infections or with acute disease, and the number of mutations needed for a phenotypic change (Box 3.3; see also Section 3.7). If the number of mutations needed for an adaptive change of a virus was in the order of hundreds, the biological relevance of quasispecies with the present replicative and mutational

TABLE 3.2  Mutant distributions in viral quasispecies.

| Number of mutations in the genome | Number of virus particles with each number of mutations | |
|---|---|---|
| | With $1 \times 10^8$ total particles | With $2 \times 10^{11}$ total particles |
| 0 | $6.7 \times 10^5$ | $6.7 \times 10^8$ |
| 1 | $3.3 \times 10^6$ | $3.3 \times 10^9$ |
| 2 | $8.4 \times 10^6$ | $8.4 \times 10^9$ |
| 3 | $1.4 \times 10^7$ | $1.4 \times 10^{10}$ |
| 4 | $1.7 \times 10^7$ | $1.7 \times 10^{10}$ |
| 5 | $1.7 \times 10^7$ | $1.7 \times 10^{10}$ |
| 6 | $1.4 \times 10^7$ | $1.4 \times 10^{10}$ |
| — | — | — |
| 20 | $2.6 \times 10^1$ | $2.6 \times 10^3$ |

This calculation is based on a Poisson distribution of mutations among genomes, ignoring fitness effects of mutations. The formula is: $P(K) = m^k e^{-m}/K!$ where $P(K)$ is the frequency of a genome with K mutations, and m is the average number of mutations per genome. In the present example $m = 5$.

parameters would be minimal, and the emphasis on viral quasispecies would not be justified.

A numerical example of how the nonmutated class of genomes and single and multiple mutants may be distributed in a viral population is given in Table 3.2. Several conclusions can be drawn from examining the table: (i) genomes without mutations $(K = 0)$ are 25-fold less abundant than genomes with four or five mutations; (ii) genomes with 20 mutations are a million-fold less abundant than genomes with three to six mutations, and (iii) an increase in population size from $1 \times 10^8$ to $2 \times 10^{11}$ individuals entails a correspondingly larger presence of both nonmutated and mutated genomes. From what we know of the negative impact of mutations on fitness, real mutant distributions will be skewed toward a lower proportion of genomes with the highest numbers of mutations. Yet, current estimates either from the sequencing of molecular or biological clones or from deep sequencing reveal profound mutant spectra with many single and multiple mutants in viral

populations. The concept that mutants are a minority and that the nonmutated class dominates the population is obsolete in RNA virology. Viruses "are" collections of mutants.

Regarding measurements of population complexity, a sampling limitation must be commented. The available methodology (including deep sequencing) only allows for the analysis of an extremely tiny proportion of all genomic sequences present in many of the viral quasispecies of biological interest. Typically, an infected organism, or a laboratory dish with a monolayer of infected cells, can include or produce $10^{10}$ to $10^{12}$ infectious particles, measured at a given time point. These numbers exclude noninfectious viruses (or viral genomes) that may outnumber their infectious counterparts, and that may play important biological roles (Section 3.8). Altogether, we may be talking about $10^{11}$ to $10^{16}$ viral genomes in many infected hosts. The standard biological or molecular cloning procedures, followed by Sanger sequencing can realistically be used to provide sequences of about 100 (at most 1000) out of the total $10^{11}$ to $10^{16}$ genomes present in a host (i.e., 0.000,001% −0.00,000,000,001% of the genomes!). Deep sequencing can increase such proportion to 0.0001%−0.000,000,001%, still a tiny minority of the population. Both procedures offer complementary sampling methods to represent a far more complex reality (Ortega-Prieto et al., 2013; Gregori et al., 2014) (see also Section 3.6.4 and Fig. 3.6). There might even be additional "underground" levels of minority genomes in mutant spectra awaiting discovery when methodology will allow.

We could assume that if the sampling of 100 genomes is performed adequately (i.e., the 100 genomes are an unbiased representation of the ranking of genomes in the entire genome population in the sense that no minority genomes are overrepresented and no dominant genomes are underrepresented), a 100 genomes sample may be sufficient to characterize the population. Indeed, the great majority of analyses of mutant spectra carried out until 2010 were based on molecular and biological cloning and Sanger sequencing of not more than 20 genomes per sample. To the extent that this procedure unveiled the complexity of viral populations, it can be regarded as extremely successful. However, in addition to the limitations of biological and molecular cloning discussed in Section 3.6.4, it is often assumed that the PCR amplification procedures using standard protocols (at annealing and extension temperatures suitable to the primers available) will produce an unbiased representation of sequences. This assumption is not justified, as documented by studies that have allowed accession to new regions of sequence space in viral populations using PCR at low denaturing temperatures, a technique termed 3D PCR, developed by J.P. Vartanian, S. Wain-Hobson, and their colleagues at the Pasteur Institute in Paris (Suspène et al., 2005, 2008; Perales et al., 2011). The results obtained with 3D PCR cannot be regarded as a mere extension of the results using standard RT-PCR amplification. When applied to parallel FMDV populations passaged in the absence and presence of the mutagenic nucleoside analog ribavirin, AU-rich sequences were about as complex in the ribavirin-treated and untreated populations. However, the analysis unveiled that the major effect of ribavirin was to accelerate the occupation of the AU-rich portion of sequence space by the virus in its way toward extinction (Perales et al., 2011) (Chapter 9). The sampling representativeness is unavoidably conditioned by methodological filters.

PCR methodologies to explore either AU- or GC-enriched viral subpopulations, together with deep sequencing and complete viral genomes sequencing, combined with microarray-based methodologies, should permit increasingly detailed and accurate characterization of mutant spectra and their dynamics. It may become possible to define in quantitative terms changes in population structure that accompany selection and random drift episodes, or transitions toward increased or decreased fitness for viral survival or extinction. Transitions in population structure have a number of implications for medical practice that are discussed in Chapter 9.

## 3.8 Modulating effects of mutant spectra: interference, cooperation and complementation. An ensemble as the unit of selection

A major consequence of virus replicating as mutant spectra is that interactions among genomes are established, mainly through their expression products acting in *trans*. The result is a modulation of phenotypic traits displayed by an entire population that may not be attributable to specific single genomes that are part of the population. Observations both in vivo and in cell culture have substantiated that what could have been a dominant trait expressed by a genome in isolation (i.e., the same genome type filling an entire population) may be suppressed by the surrounding mutant spectrum. This important feature of viral quasispecies was predicted by computer simulation studies with short-chain replicons subjected to error-prone replication. The simulations suggested that a master sequence could dominate a fitter one by virtue of being surrounded by a better (more closely related) mutant spectrum (Swetina and Schuster, 1982; Eigen and Biebricher, 1988). Applied to virus behavior, the mutant spectrum of a virus can suppress replication of a mutant of superior fitness, as shown experimentally for the first time by J.C. de la Torre and J.J. Holland working with vesicular stomatitis virus (VSV) quasispecies (de la Torre and Holland, 1990), and then documented with different viruses in cell culture and in vivo (Box 3.4).

Interference among mutants has been documented experimentally (Crowder and Kirkegaard, 2005; Perales et al., 2007; Kirkegaard et al., 2016), and it may be accentuated when viral populations are mutagenized and approach extinction (González-López et al., 2004; Grande-Pérez et al., 2005). Studies with FMDV mutants showed that interference was exerted when their RNA replicated in the infected cells, irrespective of their ability to yield infectious particles (Perales et al., 2007). A single amino acid substitution in an interfering mutant abolished its interfering activity, implying that minimal movements in sequence space can generate or abrogate potentially modulating effects on mutant ensembles. A synergistic interference that slowed down the replication of the wild-type virus was produced by mixtures of FMDV capsid and polymerase mutants (Perales et al., 2007). Inhibition by

---

## BOX 3.4

### Examples of collective behavior, interference, cooperation and complementation among components of a mutant spectrum

- Biological clones of Qβ and VSV display higher fitness than the populations from which they were isolated (Domingo et al., 1978; Duarte et al., 1994).
- High fitness VSV clone was suppressed by a VSV mutant spectrum (de la Torre and Holland, 1990).
- Live-attenuated PV vaccines can suppress minority virulent PV (Chumakov et al., 1991).
- Antibody-resistant FMDV can suppress high fitness FMDV subpopulations (Borrego et al., 1993).
- LCMV that produces a growth hormone deficiency syndrome can be suppressed by non-pathogenic LCMV (Teng et al., 1996).
- Quasispecies memory affects the response to selective pressures (Ruiz-Jarabo et al., 2000; Briones and Domingo, 2008).
- Mutagenized FMDV populations interfere with replication of standard FMDV (González-López et al., 2004).
- Defector LCMV genomes interfere with replication of standard LCMV (Grande-Pérez et al., 2005).

- PV mutant spectra can suppress inhibitor-resistant PV mutants, and compounds directed to a dominant drug target can suppress selection of drug-resistant variants (Crowder and Kirkegaard, 2005; Tanner et al., 2014; Kirkegaard et al., 2016).
- A PV quasispecies was needed to mediate penetration of a PV mutant into the central nervous system (Pfeiffer and Kirkegaard, 2005; Vignuzzi et al., 2006).
- Two measles virus mutants were needed to mediate membrane fusion (Shirogane et al., 2012).
- Coexistence of HBV variants enhanced viral replication and immune response in a mouse model (Cao et al., 2014).
- Cooperation between two neuraminidase mutants in influenza virus (Xue et al., 2016).

Additional examples are given in the text. Quasispecies memory is discussed in Chapter 5, and its implications for antiviral treatments in Chapter 9.

several defective genomes, acting conjointly, probably mimics what is encountered with a mutagenized quasispecies in which a dispersed, abundant, and dynamic mutated swarm exerts a collective interfering activity (González-López et al., 2004). The dominant-negative effect of a mutagenized viral swarm is one of the molecular mechanisms which underlies lethal mutagenesis as an antiviral strategy (Chapter 9). The collective behavior of mutant spectra is a major departure from the classic mutation-selection equilibrium based on mere generation, competition, and coexistence of mutants. Components of a mutant spectrum are connected in such a manner that an ensemble can command behavior.

Inhibition of standard virus replication by DI particles can be viewed as a specific case of interference by heavily mutated genomes (with point mutations and deletions that render DI's noninfectious and dependent on their corresponding standard virus). Studies with VSV showed that the presence of DI resulted in a cyclic production of infectious virus (Palma and Huang, 1974). The observation that standard VSV and its DIs engaged in a continuous dynamics of mutation and competition led J.J. Holland and his colleagues to recognize the value of quasispecies dynamics to understand virus behavior (Holland et al., 1982; Holland, 1984). More generally, interference may be exerted by viable genomes that suboptimally replicate because they have acquired mutations by the action of mutagenic agents, as proposed for the first stages of a transition toward the error threshold (Chapter 9). When interference is prevalent (under increased mutagenesis) the average fitness of rescued viable biological clones tends to be higher than the fitness of the average parental population (Borrego et al., 1993; González-López et al., 2004) (Fig. 3.11). As a consequence, bottleneck events may serve to rescue higher fitness subpopulations (Bergstrom et al., 1999; Grande-Pérez et al., 2005; Domingo, 2006).

Genomes displaying various degrees of defectiveness are likely to play relevant roles as modulators of viral replication (Rezelj et al., 2018). In patients acutely infected with HIV-1, two-thirds of the infected cells included defective genomes, and early diversification from the founder virus resulted in populations of HIV-1 with about 30% of genomes with mutations that conferred defectiveness (Salazar-Gonzalez et al., 2009).

Opposite to suppression or interference, are interactions of cooperation or complementation among components of mutant spectra. Shirogane et al. (2012, 2016) distinguished cooperation (in which a new phenotype emerges as a result of interaction between variant proteins) from complementation (in which a functional protein encoded by a genome permits replication of another genome that encodes a defective form of the protein).

Cooperation is a frequent interaction in biological systems (Wingreen and Levin, 2006), including the division of tasks in bacterial biofilms (Dragos et al., 2018). Cooperation in measles virus was discovered in the course of studies of cell entry, which is mediated by the membrane fusion activity dependent on viral proteins H and F. A truncated form of protein H could not evoke cell to cell fusion, but the fusion activity was regained when truncated H was joined by two forms of protein F, and not by only one form of F. The two forms of F were the wild type and another with an amino acid substitution. Apparently, heterotrimers including mutant F exhibited higher fusion activity with wild type H than homotrimers composed of wild type F (Shirogane et al., 2012, 2016). In a human influenza virus H3N2 population, one type of neuraminidase mutant that was competent for cell entry cooperated with another type that was efficient in cell exit (Xue et al., 2016).

Complementation, also profusely studied in general genetics, also has some remarkable consequences for virus behavior. Among other

**(A)**

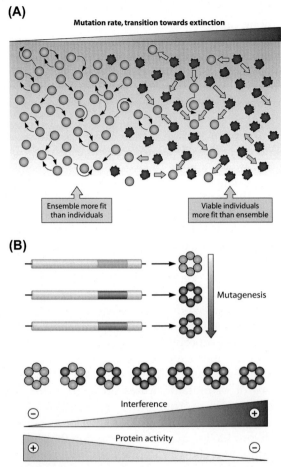

**Mutation rate, transition towards extinction**

Ensemble more fit than individuals

Viable individuals more fit than ensemble

**(B)**

Mutagenesis

Interference

⊖       ⊕

Protein activity

⊕       ⊖

**FIGURE 3.11** Representation of complementation and interference within the mutant spectrum of viral populations. (A) Representation of the transition from the dominance of complementation (left) toward dominance of interference (right). Standard genomes are drawn as blue circles and complementing interactions as *thin arrows*. Interfering genomes are depicted as rough *red symbols* and interference by large *yellow arrows*. Interference increases with the mutation rate operating during quasispecies replication. When complementation dominates, individual clones display lower relative fitness than the entire population; when interference dominates, individual viable genomes display higher fitness than the entire population. (B) A model of complementation and interference mediated by a *trans*-acting protein. The standard gene and encoded protein are represented in tan, and the increasingly mutagenized counterparts in *blue and red*. It is assumed that the protein is active as a hexameric complex, with a fully functional protein composed of six tan subunits,

examples in cell culture and in vivo, it is worth underlining the case of a marked, nonneurotropic poliovirus mutant that could not make its way into the brain of mice when inoculated alone, but reached the brain when coinoculated with neuropathogenic poliovirus populations (Vignuzzi et al., 2006) (discussed in Chapter 2 regarding PV fidelity mutants). In a mouse model, coexistence of hepatitis B virus (HBV) variants enhanced viral replication and evoked stronger intrahepatic CTL responses and surface protein-specific antibody responses (Cao et al., 2014). Thus, intra-quasispecies complementation may be not only a mechanism of enhanced viral replication but also of virus-induced immunopathogenesis.

The evolution of the standard FMDV genome into two segments that produced progeny (described in Section 2.11 of Chapter 2) was also possible by complementation once an adequate position in sequence space was reached. Many viral populations appear to contain high frequencies of noninfectious genomes (e.g., genomes with stop codons that truncate essential proteins), and some of them might be maintained at detectable levels by complementation. A defective dengue virus that included a nonsense mutation in its envelope protein could be transmitted and maintained in infected hosts (Aaskov et al., 2006). Genetic and phenotypic diversity of West Nile virus was maintained by coinfections and complementation among variants, and fitness of mutant ensembles exceeded that of individual

and decreasingly functional proteins composed of *blue and red* subunits. It should be noted that the blue protein can either decrease the activity of a functional (tan) ensemble or partially rescue the activity of a nonfunctional (*red*) ensemble, when present as a subunit component. Progressive decrease of protein activity is accompanied by increased interference. See text for references that justify this model. *Parts (A) and (B) of the figure are reproduced from Domingo, E., Sheldon, J., Perales, C., 2012. Viral quasispecies evolution. Microbiol. Mol. Biol. Rev. 76, 159–216, with permission from the American Society for Microbiology, Washington DC, USA.*

genome types (Ciota et al., 2012). Complementation may underlie a number of observations on genome ensembles that exhibit higher fitness than many of its individual components (Domingo et al., 1978; Duarte et al., 1994; Moreno et al., 1997; Novella, 2003; Wilke et al., 2004; Perales et al., 2007) (Fig. 3.11).

Transient maintenance of lethal or deleterious mutations has a parallel in deleterious recessive mutants of diploid organisms, which avoid negative selection when carried in a heterozygote with the dominant normal allele. The prevalence of complementation or interference depends on the quasispecies mutational status (Fig. 3.11).

Some theoretical models have introduced cooperation and complementation in quasispecies dynamics and the error threshold for the maintenance of genetic information (Arbiza et al., 2010; Sardanyés and Elena, 2010; Moreno et al., 2012). Here again, the interplay between theory and experiment may provide new insights into quasispecies behavior.

## 3.8.1 Molecular mechanisms of complementation and interference

Complementation and interference within viral populations (intracellular effects) can be viewed as two sides of the same coin, with a related molecular basis. A *trans*-acting functional protein expressed by a subset of genomes from the mutant spectrum can rescue a nonfunctional version of the same protein expressed by mutated genomes. The result is complementation. Yet, expression—above a certain critical level—of nonfunctional forms of the same protein encoded by mutated genomes may impede the protein fulfilling its functions. The result is interference. Malfunctioning due to incapacity to establish adequate protein-to-protein contacts can be easily visualized with proteins that act as homopolymeric and heteropolymeric ensembles during the virus life cycle (i.e., viral capsids

whose stability is based on intersubunit contacts among capsid proteins, or the case of replication complexes which include viral and host proteins). A defective subunit in these and other multisubunit complexes may compromise the function of the molecular assembly, leading to an unstable capsid, suboptimal replication complex, or other (Perales et al., 2007) (Fig. 3.11B).

Complementation and suppression are emerging (collective) properties that result from interactions among quasispecies components. Experimental results (Box 3.4) are coherent with theoretical predictions on emerging traits (Solé and Goodwin, 2000). The behavior of an ensemble of related mutants is not necessarily predicted from the behavior of the individual mutants (see Section 3.9). The influence that a mutant cloud exerts on an individual mutant establishes a conceptual link with interference exerted between different viruses, and with the effect of dominant-negative mutants, both well-established concepts in viral genetics.

Several biochemical mechanisms explain the inhibitory activities among distantly related viruses or of mutant viruses on their wild-type counterparts (Youngner and Whitaker-Dowling, 1999). They include interferon induction, production of a defective protein that enters a multimeric complex (termed the "rotten apple" hypothesis), a defective function that may sequester a factor necessary for the replication of the two viruses (the "road-block" hypothesis), imbalances in gene dosage (the "all things in moderation" hypothesis), or competition for host factors (direct competition hypothesis). Some of these mechanisms may also underlie the suppressive effect exerted by viral quasispecies, which can act as a "collective dominant-negative mutant." To establish the operation of such interference mechanisms, it is important to ascertain the specificity of interfering interactions. For example, in experiments with FMDV, altered capsid and polymerase mutants of

FMDV inhibited replication of FMDV but not of the other picornavirus encephalomyocarditis virus (Perales et al., 2007), thereby excluding some general inhibitory effect evoked by the presence of the mutants. Likewise, production of interferon type I or III (or other general antiviral activity associated with innate immunity) was excluded in the interference exerted by mutagenized FMDV on standard FMDV (González-López et al., 2004). Thus, despite interference being sometimes operational between distantly related viruses, the mechanisms (e.g., interferon production or competition for a cellular resource) are not the same as those operating in the interference that requires intimate protein-protein interaction during the viral life cycle (Fig. 3.11).

The negative effects of mutant spectra acting synergistically as a dominant-negative population aggregate are expected to be reinforced by the multifunctional nature of most viral proteins. A malfunction due to a defective protein will jeopardize any step in the virus life cycle in which the protein participates unless such participation involves distinct protein domains, and the protein lesion affects only one of the domains. The participation of one protein in multiple functions is increasingly documented in virology and cell biology. This fact, together with the presence of several isoforms of a protein, the discovery of the relevance of small proteins (expressed from open reading frames once thought to be meaningless or secondary), along with the realization that different domains of the same protein may be involved in distinct functions, complicates interpretation of protein-protein interaction maps. If clarification tools are developed, protein-protein interaction maps might provide insights into the mechanisms of intra-quasispecies interactions.

For a mutant cloud to interfere with a specific class of genomes, the cloud has to be generated either inside the same cell in which the specific class of genomes replicate or the cloud has to infect the same cell than the specific genome class. Intracellular generation of a mutant cloud occurs due to the basal mutation rates of RNA and some DNA viruses. Such rates are further increased during mutagenesis by nucleotide analogs. Since the cloud is generated intracellularly, it is one of the mechanisms of the transition of viruses toward extinction that does not require infection of the same cell by multiple viral particles (Chapter 9). In addition, infection of a cell with a mutant cloud may be simultaneous with infection by a specific mutant, or the two infections may occur successively, provided no superinfection exclusion mechanisms operate. Although double or multiple infections of the same cells in differentiated organisms may appear as unlikely, they may be more frequent than expected from dual hits occurring at random (Cicin-Sain et al., 2005; Chohan et al., 2005). In addition several mechanisms of bloc transmission of viruses have been documented [aggregates on the surface of by standing lipid structures, lipid vesicles, exosomes, cell to cell contacts, virological synapses, bacteria-mediated coinfections, etc. that add to high standard multiplicity of infection (MOI, or number of individual infectious particles per cell) when a large infectious yield is produced locally] (Chen et al., 2015; Altan-Bonnet et al., 2019). Bloc transmissions increase the initial mutant spectrum repertoire on which selective pressures act. They facilitate the presence of different genomes for the generation of new genome types by recombination and modulating effects among variant viruses observed in vivo (Box 3.4). Complementing interactions between mouse cytomegalovirus variants improved their dissemination and replication in the host, which agrees with poor prognosis in the case of patients infected with multiple variants of this virus (Cicin-Sain et al., 2005). In addition to intracellular complementation, altered expression of a diffusible regulatory factor, triggered by one class of virus variant, altered the behavior of another variant (Cicin-Sain et al., 2005; Sala et al., 2006). Thus, direct and indirect interactions among components of mutant spectra may result in an emergent behavior, with potentially important clinical implications. Complex viral quasispecies constitute inviting systems to explore experimentally molecular events behind properties that emerge from interacting individuals. Remarkably, influences among mutants and their gene products render viral quasispecies units of selection, one of the essential features of the theoretical quasispecies.

## 3.8.2 Individual versus group selection

Viral population numbers are very relevant on several grounds, and implications have been discussed in Section 3.7.2, and others are treated in Chapters 4 and 8. One of the influences of viral population numbers on evolutionary outcomes may result from the MOI. Its importance was documented for the first time by N. Sevilla and colleagues by showing that FMDV underwent adaptation to the MOI: a population with a passage history at low MOI overgrew a related population when the competition passages were carried out at low MOI but not at high MOI (Sevilla et al., 1998). Thus, the evolutionary history in terms of how massive have been the prior cellular invasions may influence an evolutionary outcome should a competition with other variants arise.

I.S. Novella and colleagues obtained evidence of density-dependent selection in VSV, also using the MOI as the key variable. In competition assays, the fitness of a deleterious mutant relative to the wild type increased as a function of the MOI. The results were interpreted in terms of increased complementation at high MOI, and were supported by a theoretical model that assumed that the two viruses shared their gene products (generalized *trans*-acting functions) during coinfections (Novella et al., 2004).

The evidence of interactions within quasispecies raises the issue of the target of selection being either an individual or a group of individuals. This has been a debated issue in general genetics with ramifications in social behavior (Wright, 1945; Maynard Smith, 1964; Lewontin, 1970; Williams, 1992; Okasha, 2013). In the case of viruses, this debate has not reached prominence yet, probably because the mechanisms that may favor group selection have been unveiled only recently.

The intramutant spectrum interactions (Sections 3.8 and 3.8.1) confer selective forces the potential to act at the level of a single viral genome (viewed as liberated from possible links with surrounding genomes) or a group of genomes (tied by molecular dependences). Here, I suggest a model according to which the preferential operation of individual or group selection in viral quasispecies may depend on the mutant spectrum context (i.e., how firmly have the internal interactions been established) as well as on the intensity and specificity of the selective constraint. An antiviral inhibitor directed to a viral protein, or a neutralizing monoclonal antibody directed to a single epitope within an antigenic site, will tend to select resistant "individuals." The selection process will be little affected by the surrounding community of viral genomes, unless the latter is strongly suppressive, for example due to low fitness (several examples are described in Section 3.8). In contrast, adaptation to a less focused multicomponent environment (e.g., the penetration of a virus into new tissue or organ) may be prone to modulation by the collectivity. In this view, still to be largely confirmed by experiments, the duality of individual versus group selection would in itself confer a selective advantage to viruses because of an inherent flexibility in the means to respond to selective constraints.

### 3.8.3 Stochastic effects in selected collectivities

The evidence put forward in previous sections suggests that selection may operate either at the level of a collectivity of related mutants or the level of an individual (or a very specific set of individuals). In the latter instance, the outcome (success or not) of the selected individual may depend on the progeny mutant spectrum generated, with the indeterminations inherent to mutagenesis (the atomic basis for the stochastic nature of the generation of mutations) (Chapter 2). The statement of G.C. Williams (Williams, 1992) "… a particular mutation in a particular individual at a particular time could have important effects on all descendants of the mutant individual and all others that interact with these descendants …" is very relevant to the development of mutant spectra as collections of interacting individuals. In the initial stages of the generation of a mutant spectrum, the stochastic nature of the mutations that are generated may alter the types of the successive mutations that will constitute the dynamic mutant spectrum (see Chapter 5 for the contrast between mutant generation and establishment of high fitness genomes facilitated by epistatic interactions).

Positive selection may lead to a decrease of heterogeneity in a viral population. A complex mutant spectrum may include variants that cover a range of phenotypic traits (cell tropism, antibody-escape, CTL-escape, drug-resistance, among others). When selection for one such trait occurs, the subset of genomes expressing the relevant trait will contribute to the following generations of genomes. The level of heterogeneity of the parental viral population prior to the application of selection will be generally higher than that of the selected population because the latter was founded by a minority of genomes that will participate in the ensuing rounds of replication in the presence of the selective agent.

These types of "sweeping" selection events may have an effect similar to a population bottleneck regarding a reduction of population complexity of the population (further discussed in Section 3.4). Fitness may also decrease if measured in the absence of the selective agent that was responsible for the selective sweep (Chapter 5).

When selection acts on a highly heterogeneous viral population, multiple, nonidentical genomes can possess different genomic sequences all of them amenable to selection. In a study with FMDV, a quasispecies was reconstructed with multiple antigenic variants resistant to a specific monoclonal antibody, each at a frequency that was very similar to their expected frequency in viral populations. When the reconstructed quasispecies was subjected to selection by the relevant antibody, a swarm of antibody-escape mutants was selected (Perales et al., 2005). This observation indicates that what could be regarded as a sweeping selection event does not necessarily lead to loss of heterogeneity since, depending on the mutant spectrum composition, multiple, nonidentical genomes may share a selectable trait and they will soon regenerate a complex mutant spectrum. The result also provided evidence that the target of selection in viral quasispecies can be a collection of related individuals, irrespective of their being linked by complementation.

## 3.9 Viral populations in connection with biological complexity

Complexity has at least two meanings in virology. It may denote the number of nonredundant information contents (in terms of regulatory regions and protein coding, open reading frames) embodied in a viral genome or the level of heterogeneity within a population. According to the first meaning of complexity, poxviruses—with up to 375,000 base pairs of a terminally covalently linked DNA encoding 200 to 300 proteins—are more complex than the RNA picornaviruses which typically contain between 7500 and 8000 nucleotides encoding about a dozen fully processed proteins. The difference in complexity is related to the strategies of virus-host interactions, in particular, the molecular mechanisms evolved to cope with the host immune response (Chapter 4). A compact RNA genome anticipates that evolution has selected a maximum number of functions associated with a small number of nucleotide residues, reflected in multifunctional proteins or protein precursors in the case of genome expression through polyproteins, a quite widespread strategy. The expectation, still to be confirmed, is that for the more compact (less complex from the point of view of the genetic information being carried), more extensive multifunctionality of virus-coded proteins may be found.

The second meaning of complexity, of relevance to the understanding of viruses as populations, relates to the variation in nucleotide sequence among individual genomes that compose a virus population and recapitulates features of viral quasispecies delineated in Section 3.6. All viruses possess core information required to complete their life cycle: penetration into the cell, uncoating and replication of the RNA or DNA genome, viral protein expression, particle assembly, and release from cells. The features embodied in this minimal central information largely determine virus identity; it is what allows a new viral isolate to be assigned to a group (influenza virus, papillomavirus, or other). Once the core information is secured, variations in nucleotide sequence (and in the corresponding amino acid sequence of the encoded proteins) are allowed among different isolates or the individual genomes within an isolate, without enfeeblement of virus identity. Despite not altering the core information, such variations are essential to the virus because they affect interactions of viruses with their physical and

biological environment that determine virus survival. Core information embodies minimal requirements for short-term survival, but variations of the core are essential for host interactions and long-term survival.

Suffice to mention as an example, the B-cell and T-cell epitopes in their interaction with components of the immune response. Variations in a few specific viral residues may have as the most extreme outcomes either virus evasion and survival or viral inactivation and extinction. From the point of view of viral pathogenesis, despite sharing the core information, only some influenza viruses are highly lethal, and only some papilloma-viruses are oncogenic.

We refer to "population complexity" to portray the numbers of different genomes that harbor mutations in viral populations. This meaning of complexity bears on the concept of complexity that originated in physics, and that has ramifications in disparate fields of science, such as neurobiology, astrophysics or even economics (Gell-Mann, 1994a,b; Simon, 1996; Albergante et al., 2016). No single parameter captures what we would like to convey to indicate that a viral population is more complex than another. Mutation frequencies, number of polymorphic sites, average genetic distances and other diversity indices used to describe quantitatively mutant spectra of viral quasispecies are only an approximation to the complexity of the population (Gregori et al., 2016) (Section 3.6.4.2).

Keeping in mind that we need to apprehend what may be more crucial in virus population complexity, it is relevant to examine the effective algorithmic complexity described by M. Gell-Mann. When his view is applied to viruses, a homogeneous collection of genomes (all identical) has the lowest complexity possible, as compared with any collection of genomes in which some of its components differ by some mutations. In the other end of the scale of homogeneity, a population consisting of random sequences, devoid of any information content in the considered environment, would not qualify as complex regarding the amount of information or the effective algorithmic complexity (Gell-Mann, 1994a), simply because it does not contain meaningful information. A degree of complexity must be defined relative to some information content since without information, complexity is devoid of meaning. According to the algorithmic complexity, population size (the number of genomes in the distribution) is not a significant variable. However, we know that the total viral population size is highly relevant to virus biological behavior, fundamentally because it determines the capacity to explore sequence space with its consequences (Section 3.7). The justification of such statement can be illustrated with the following example: if the frequency of mutations that can contribute to antibody-escape is $10^{-4}$ per nucleotide (not considering fitness effects), for a viral genome of $10^4$ nucleotides a population of at least $10^4$ is required to find one mutation relevant to antibody-escape. With the same mutation frequency, multiple antibody-escape mutations will be present in a $1.6 \times 10^5$ genomes population, and none in a 16 genomes population. Yet, the 16, 16,000, and 160,000 viral populations may share identical mutation frequency. Thus, in the case of mutant spectra of viral quasispecies, the statement of P. Anderson "more is different" (quoted in Levin, 2006) acquires its full validity. Following again physics language, properties of a viral population that do not depend on population size (e.g., the mutation frequency) have been termed "intrinsic" while properties that depend on the population size (e.g., the presence or not of a mutant present at a frequency of $10^{-4}$) have been termed *extrinsic* properties (Domingo and Perales, 2012).

Physics has contributed yet another meaning of complexity that relates to emerging properties, applicable also to viral quasispecies. Some biological phenomena are the result of multiple influences, and cannot be explained by the mere addition of the individual influences (Solé and Goodwin, 2000). The term "emergence" describes the inability to explain a whole as the sum of its parts, a concept not too appreciated by some schools of biochemistry and classical genetics. The rejection of "emergence" arises probably from the halo of mystery that it conveys, and that on occasions has been exploited to cast doubts on the capacity of science to explain the origin of some biological systems or mechanisms, and the proposal of "intelligent design" as a better alternative. Obviously, the term "emergence" does not imply any mysterious (scientifically unexplainable) intervention; it manifests that there may be transitional behaviors not directly explained by the parts involved considered in isolation, but explainable if we knew more. Mental activity, as an outcome of neuronal connections, is an example. This meaning of complexity bears on the understanding of virus evolution, pathogenesis, and disease emergence (Chapter 7) which are an outcome of sets of host-virus-environmental interactions. Only very recently we have begun to understand part of the molecular basis of some "emergent" properties of viral populations. Molecular virology is probably immersed in complexities of the types summarized here with us hardly noticing it.

## 3.10 Overview and concluding remarks

Experimental observations and theoretical treatments have contributed to our understanding of how Darwinian principles guide virus evolution. There are several mathematics equations that have been used to describe evolutionary dynamics in general. The equations are interconnected, and emphasize different aspects of the Darwinian processes. One of the equations corresponds to quasispecies theory, which includes mutation as a prominent event in the evolutionary behavior. The attention to error-prone replication renders quasispecies an adequate theoretical framework to describe the evolution of RNA viruses, and generally any highly variable virus. Quasispecies has impacted virology in several key aspects that range from a new concept of "wild type" as a collection of genomes to the evidence of a molecular mechanisms of adaptation to changing environments that was not obvious before quasispecies came into play.

An individual viral genome is characterized by a defined nucleotide sequence that has the information to generate infectious progeny in an infected cell. However, unavoidably the replication of such a genome generates a distribution of mutants that consists of multitudes of closely related, but different genomes. The "sequence" of the genome population is the average of a multitude of different sequences. Populations are often treated as if they had "a genomic sequence," but they have not. To assign "a sequence" to a viral population is a gross simplification perpetuated in textbooks and data banks. This seemingly inconsequential practice masks fundamental properties of viruses that cannot be anticipated by examining a consensus sequence. Deep sequencing methodologies have enhanced the potential to penetrate into the composition of mutant spectra and to dissect the molecular details of quasispecies dynamics. The new sequencing platforms have fully confirmed the extreme genetic heterogeneity of viral populations, with the presence of multitudes of low-frequency genomes whose role, if any, remains to be elucidated. Virus evolution consists fundamentally in movements in sequence space or the replacement of some viral subpopulations by others in response to selective constraints and random fluctuations in population size. It is easy for a virus to navigate in sequence space when the route is guided by a fitness gradient.

Mutant spectra are not a mere aggregate of mutants, which act independently one from the other. Rather, interactions of complementation or interference can be established within quasispecies that may dictate their behavior. The concept that the entire quasispecies can act as a unit of selection has been supported by many experimental observations in cell culture and in vivo, and it has found endorsement in theoretical models. The intramutant spectrum interactions raise the issue of selection acting at the level of the individual or ensembles of individuals. Current evidence suggests that viral quasispecies may be endowed with the dual potential of individual versus group selection, a feature that, if further substantiated by experimentation, would add to their repertoire of adaptive strategies.

Finally, this chapter has concluded with the concepts of complexity and emerging properties applied to viral populations. Despite being in its infancy, research on properties that have a multicomponent origin starts playing an important role in the understanding of viruses as populations. Emergent behavior will be encountered in further chapters (see Summary Box).

---

## Summary box

- Extended Darwinian principles consisting on generation of variant genomes, competition among them, selection of the fittest in the considered environment, and random drift of genomes preside virus evolution.

- The several mathematical equations of Darwinian formulations are related. Quasispecies emphasizes the effect of high mutation rates and, therefore, is adequate as

**Summary box** *(cont'd)*

the theoretical framework for highly variable viruses, in particular, the RNA viruses.

- Viral quasispecies is a complex and dynamic mutant distributions that has explained the adaptive potential of RNA viruses.
  Virus evolution can be regarded as a movement in sequence space, in which some genome subpopulations are replaced by others.
- Interactions among components of a mutant spectrum can influence the behavior of the ensemble. Selection can act either on individuals or group of individuals in viral quasispecies.
- Viruses are adequate experimental systems to approach complexity, in the sense of probing why the behavior of a genome collectivity may not be explained by the sum of contributions of the individual members of the collectivity.

# References

Aaskov, J., Buzacott, K., Thu, H.M., Lowry, K., Holmes, E.C., 2006. Long-term transmission of defective RNA viruses in humans and Aedes mosquitoes. Science 311, 236–238.

Acevedo, A., Brodsky, L., Andino, R., 2014. Mutational and fitness landscapes of an RNA virus revealed through population sequencing. Nature 505, 686–690.

Airaksinen, A., Pariente, N., Menendez-Arias, L., Domingo, E., 2003. Curing of foot-and-mouth disease virus from persistently infected cells by ribavirin involves enhanced mutagenesis. Virology 311, 339–349.

Albergante, L., Liu, D., Palmer, S., Newman, T.J., 2016. Insights into biological complexity from simple foundations. Adv. Exp. Med. Biol. 915, 295–305.

Altan-Bonnet, N., Perales, C., Domingo, E., 2019. Extracellular vesicles: Vehicles of *en block* viral transmission. Virus Res. 265, 143–149.

Alves, D., Fontanari, J.F., 1998. Error threshold in finite populations. Phys. Rev. E 57, 7008–7013.

Amexis, G., Oeth, P., Abel, K., Ivshina, A., Pelloquin, F., et al., 2001. Quantitative mutant analysis of viral quasispecies by chip-based matrix- assisted laser desorption/ionization time-of-flight mass spectrometry. Proc. Natl. Acad. Sci. U.S.A. 98, 12097–12102.

Arbiza, J., Mirazo, S., Fort, H., 2010. Viral quasispecies profiles as the result of the interplay of competition and cooperation. BMC Evol. Biol. 10, 137.

Arias, A., Lázaro, E., Escarmís, C., Domingo, E., 2001. Molecular intermediates of fitness gain of an RNA virus: characterization of a mutant spectrum by biological and molecular cloning. J. Gen. Virol. 82, 1049–1060.

Ba Abdullah, M.M., Palermo, R.D., Palser, A.L., Grayson, N.E., Kellam, P., et al., 2017. Heterogeneity of the Epstein-Barr virus (EBV) major internal repeat reveals evolutionary mechanisms of EBV and a functional defect in the prototype EBV strain B95-8. J. Virol. 91 pii: e00920-17.

Baccam, P., Thompson, R.J., Fedrigo, O., Carpenter, S., Cornette, J.L., 2001. PAQ: partition analysis of quasispecies. Bioinformatics 17, 16–22.

Baccam, P., Thompson, R.J., Li, Y., Sparks, W.O., Belshan, M., et al., 2003. Subpopulations of equine infectious anemia virus Rev coexist in vivo and differ in phenotype. J. Virol. 77, 12122–12131.

Baele, G., Dellicour, S., Suchard, M.A., Lemey, P., Vrancken, B., 2018. Recent advances in computational phylodynamics. Curr. Opin. Virol. 31, 24–32.

Baranowski, E., Ruiz-Jarabo, C.M., Pariente, N., Verdaguer, N., Domingo, E., 2003. Evolution of cell recognition by viruses: a source of biological novelty with medical implications. Adv. Virus Res. 62, 19–111.

Batschelet, E., Domingo, E., Weissmann, C., 1976. The proportion of revertant and mutant phage in a growing population, as a function of mutation and growth rate. Gene 1, 27–32.

Battilani, M., Scagliarini, A., Ciulli, S., Morganti, L., Prosperi, S., 2006. High genetic diversity of the VP2 gene of a canine parvovirus strain detected in a domestic cat. Virology 352, 22–26.

Bergstrom, C.T., McElhany, P., Real, L.A., 1999. Transmission bottlenecks as determinants of virulence in rapidly evolving pathogens. Proc. Natl. Acad. Sci. U.S.A. 96, 5095–5100.

Deep sequencing in virology. In: Berkhout, B., Beerenwinkel, N., Domingo, E. (Eds.), 2017. Virus Research, vol. 239. Elsevier.

Boerlijst, M.C., Boenhoefer, S., Nowak, M.A., 1996. Viral quasispecies and recombination. Proc. R. Soc. Lond. B 263, 1577–1584.

Borderia, A.V., Rozen-Gagnon, K., Vignuzzi, M., 2016. Fidelity variants and RNA quasispecies. Curr. Top. Microbiol. Immunol. 392, 303–322.

Borrego, B., Novella, I.S., Giralt, E., Andreu, D., Domingo, E., 1993. Distinct repertoire of antigenic variants of foot-and-mouth disease virus in the presence or absence of immune selection. J. Virol. 67, 6071–6079.

Borucki, M.K., Chen-Harris, H., Lao, V., Vanier, G., Wadford, D.A., et al., 2013. Ultra-deep sequencing of intra-host rabies virus populations during cross-species transmission. PLoS Negl. Trop. Dis. 7, e2555.

Briones, C., Domingo, E., 2008. Minority report: hidden memory genomes in HIV-1 quasispecies and possible clinical implications. AIDS Rev. 10, 93–109.

Brown, E.G., Liu, H., Kit, L.C., Baird, S., Nesrallah, M., 2001. Pattern of mutation in the genome of influenza A virus on adaptation to increased virulence in the mouse lung: identification of functional themes. Proc. Natl. Acad. Sci. U.S.A. 98, 6883–6888.

Cale, E.M., Hraber, P., Giorgi, E.E., Fischer, W., Bhattacharya, T., et al., 2011. Epitope-specific CD8+ T lymphocytes cross-recognize mutant simian immunodeficiency virus (SIV) sequences but fail to contain very early evolution and eventual fixation of epitope escape mutations during SIV infection. J. Virol. 85, 3746–3757.

Cao, L., Wu, C., Shi, H., Gong, Z., Zhang, E., et al., 2014. Coexistence of hepatitis B virus quasispecies enhances viral replication and the ability to induce host antibody and cellular immune responses. J. Virol. 88, 8656–8666.

Cavicchioli, R., Endmann, S., 2015. The discovery of Antarctic RNA viruses: a new game changer. Mol. Ecol. 24, 4809–4811.

Chen, Y.H., Du, W., Hagemeijer, M.C., Takvorian, P.M., Pau, C., et al., 2015. Phosphatidylserine vesicles enable efficient en bloc transmission of enteroviruses. Cell 160, 619–630.

Chen-Harris, H., Borucki, M.K., Torres, C., Slezak, T.R., Allen, J.E., 2013. Ultra-deep mutant spectrum profiling: improving sequencing accuracy using overlapping read pairs. BMC Genomics 14, 96.

Chohan, B., Lavreys, L., Rainwater, S.M., Overbaugh, J., 2005. Evidence for frequent reinfection with human immunodeficiency virus type 1 of a different subtype. J. Virol. 79, 10701–10708.

Chumakov, K.M., Powers, L.B., Noonan, K.E., Roninson, I.B., Levenbook, I.S., 1991. Correlation between amount of virus with altered nucleotide sequence and the monkey test for acceptability of oral poliovirus vaccine. Proc. Natl. Acad. Sci. U.S.A. 88, 199–203.

Chumakov, K.M., Dragunsky, E.M., Norwood, L.P., Douthitt, M.P., Ran, Y., et al., 1994. Consistent selection of mutations in the 5′-untranslated region of oral poliovirus vaccine upon passaging in vitro. J. Med. Virol. 42, 79–85.

Cicin-Sain, L., Podlech, J., Messerle, M., Reddehase, M.J., Koszinowski, U.H., 2005. Frequent coinfection of cells explains functional in vivo complementation between cytomegalovirus variants in the multiply infected host. J. Virol. 79, 9492–9502.

Ciota, A.T., Ehrbar, D.J., Van Slyke, G.A., Willsey, G.G., Kramer, L.D., 2012. Cooperative interactions in the West Nile virus mutant swarm. BMC Evol. Biol. 12, 58.

Ciurea, A., Klenerman, P., Hunziker, L., Horvath, E., Senn, B.M., et al., 2000. Viral persistence *in vivo* through selection of neutralizing antibody- escape variants. Proc. Natl. Acad. Sci. U.S.A. 97, 2749–2754.

Coffin, J.M., 1995. HIV population dynamics in vivo: implications for genetic variation, pathogenesis, and therapy. Science 267, 483–489.

Costa-Mattioli, M., Domingo, E., Cristina, J., 2006. Analysis of sequential hepatitis A virus strains reveals coexistence of distinct viral subpopulations. J. Gen. Virol. 87, 115–118.

Couderc, T., Guedo, N., Calvez, V., Pelletier, I., Hogle, J., et al., 1994. Substitutions in the capsids of poliovirus mutants selected in human neuroblastoma cells confer on the Mahoney type 1 strain a phenotype neurovirulent in mice. J. Virol. 68, 8386–8391.

Crowder, S., Kirkegaard, K., 2005. Trans-dominant inhibition of RNA viral replication can slow growth of drug-resistant viruses. Nat. Genet. 37, 701–709.

de la Torre, J.C., Holland, J.J., 1990. RNA virus quasispecies populations can suppress vastly superior mutant progeny. J. Virol. 64, 6278–6281.

de la Torre, J.C., Giachetti, C., Semler, B.L., Holland, J.J., 1992. High frequency of single-base transitions and extreme frequency of precise multiple-base reversion mutations in poliovirus. Proc. Natl. Acad. Sci. U.S.A. 89, 2531–2535.

Delwart, E., 2013. A roadmap to the human virome. PLoS Pathog. 9, e1003146.

Dobzhansky, T., Ayala, F.J., Stebbins, G.L., Valentine, J.W., 1977. Evolution. W.H. Freeman and Company, San Francisco.

Domingo, E., 1999. Vers une compréhension des virus comme systèmes complexes et dynamiques. Ann. Med. Vet. 143, 225–235.

Domingo, E., 2002. Quasispecies theory in virology. J. Virol. 76, 463–465.

Domingo, E., 2005. Virus entry into error catastrophe as a new antiviral strategy. Virus Res. 107, 115–228.

Domingo, E., Holland, J.J., 1994. Mutation rates and rapid evolution of RNA viruses. In: Morse, S.S. (Ed.), Evolutionary Biology of Viruses. Raven Press, New York, pp. 161–184.

Domingo, E., Perales, C., 2012. From quasispecies theory to viral quasispecies: how complexity has permeated virology. Math. Model. Nat. Phenom. 7, 32–49.

Domingo, E., Perales, C., 2018. Quasispecies and virus. Eur. Biophys. J. 47, 443–457.

Domingo, E., Perales, C., 2019. Virus evolution. Encyclopedia of microbiology. In: fourth ed. Microbial Ecology and Evolution Elsevier, Ámsterdam.

Domingo, E., Schuster, P., 2016. Quasispecies: From Theory to Experimental Systems. Curr. Top. Microbiol. Immunol., vol. 392. Springer, Switzerland.

Domingo, E., Wain-Hobson, S., 2009. The 30th anniversary of quasispecies. Meeting on "Quasispecies: past, present and future". EMBO Rep. 10, 444–448.

Domingo, E., Sabo, D., Taniguchi, T., Weissmann, C., 1978. Nucleotide sequence heterogeneity of an RNA phage population. Cell 13, 735–744.

Domingo, E., Biebricher, C., Eigen, M., Holland, J.J., 2001. Quasispecies and RNA Virus Evolution: Principles and Consequences. Landes Bioscience, Austin.

Domingo, E., Brun, A., Núñez, J.I., Cristina, J., Briones, C., et al., 2006. Genomics of viruses. In: Hacker, J., Dobrindt, U. (Eds.), Pathogenomics: Genome Analysis of Pathogenic Microbes. Wiley-VCH Verlag GmbH & Co. KGaA, Weinheim, pp. 369–388.

Domingo, E., Parrish, C., Holland, J.J.E., 2008. Origin and Evolution of Viruses, second ed. Elsevier, Oxford.

Domingo, E., Sheldon, J., Perales, C., 2012. Viral quasispecies evolution. Microbiol. Mol. Biol. Rev. 76, 159–216.

Dragos, A., Kiesewalter, H., Martin, M., Hsu, C.Y., Hartmann, R., et al., 2018. Division of labor during biofilm matrix production. Curr. Biol. 28, 1903–1913.

Drexler, J.F., Geipel, A., Konig, A., Corman, V.M., van Riel, D., et al., 2013. Bats carry pathogenic hepadnaviruses antigenically related to hepatitis B virus and capable of infecting human hepatocytes. Proc. Natl. Acad. Sci. U.S.A. 110, 16151–16156.

Duarte, E.A., Novella, I.S., Ledesma, S., Clarke, D.K., Moya, A., et al., 1994. Subclonal components of consensus fitness in an RNA virus clone. J. Virol. 68, 4295–4301.

Eigen, M., 1971. Self-organization of matter and the evolution of biological macromolecules. Naturwissenschaften 58, 465–523.

Eigen, M., 1992. Steps Towards Life. Oxford University Press, Oxford.

Eigen, M., 2000. Natural selection: a phase transition? Biophys. Chem. 85, 101–123.

Eigen, M., 2013. From Strange Simplicity to Complex Familiarity. Oxford University Press, Oxford.

Eigen, M., Biebricher, C.K., 1988. Sequence space and quasispecies distribution. In: Domingo, E., Ahlquist, P., Holland, J.J. (Eds.), RNA Genetics. CRC Press, Boca Raton, FL, pp. 211–245.

Eigen, M., Schuster, P., 1979. The Hypercycle. A Principle of Natural Self-Organization. Springer, Berlin.

Elde, N.C., Child, S.J., Eickbush, M.T., Kitzman, J.O., Rogers, K.S., et al., 2012. Poxviruses deploy genomic accordions to adapt rapidly against host antiviral defenses. Cell 150, 831–841.

Eriksson, N., Pachter, L., Mitsuya, Y., Rhee, S.Y., Wang, C., et al., 2008. Viral population estimation using pyrosequencing. PLoS Comput. Biol. 4, e1000074.

Escarmís, C., Dávila, M., Domingo, E., 1999. Multiple molecular pathways for fitness recovery of an RNA virus debilitated by operation of Muller's ratchet. J. Mol. Biol. 285, 495–505.

Farci, P., Purcell, R.H., 2000. Clinical significance of hepatitis C virus genotypes and quasispecies. Semin. Liver Dis. 20, 103–126.

Figlerowicz, M., Alejska, M., Kurzynska-Kokorniak, A., Figlerowicz, M., 2003. Genetic variability: the key problem in the prevention and therapy of RNA-based virus infections. Med. Res. Rev. 23, 488–518.

Filippou, C., Garrido-Jurado, I., Meyling, N.V., Quesada-Moraga, E., Coutts, R.H.A., et al., 2018. Mycoviral population dynamics in Spanish isolates of the entomopathogenic fungus *Beauveria bassiana*. Viruses 10 pii: E665.

Fischer, W., Ganusov, V.V., Giorgi, E.E., Hraber, P.T., Keele, B.F., et al., 2010. Transmission of single HIV-1 genomes and dynamics of early immune escape revealed by ultra-deep sequencing. PLoS One 5, e12303.

Fitch, W.M., Margoliash, E., 1967. Construction of phylogenetic trees. Science 155, 279–284.

Fong, Y.W., Cattoglio, C., Tjian, R., 2013. The intertwined roles of transcription and repair proteins. Mol. Cell 52, 291–302.

Fox, S., Filichkin, S., Mockler, T.C., 2009. Applications of ultra-high-throughput sequencing. Methods Mol. Biol. 553, 79–108.

Gamba, P., Zenkin, N., 2018. Transcription fidelity and its roles in the cell. Curr. Opin. Microbiol. 42, 13–18.

Garcia-Arriaza, J., Domingo, E., Briones, C., 2007. Characterization of minority subpopulations in the mutant spectrum of HIV-1 quasispecies by successive specific amplifications. Virus Res. 129, 123–134.

Gavrilin, G.V., Cherkasova, E.A., Lipskaya, G.Y., Kew, O.M., Agol, V.I., 2000. Evolution of circulating wild poliovirus and of vaccine-derived poliovirus in an immunodeficient patient: a unifying model. J. Virol. 74, 7381–7390.

Ge, L., Zhang, J., Zhou, X., Li, H., 2007. Genetic structure and population variability of tomato yellow leaf curl China virus. J. Virol. 81, 5902–5907.

Gell-Mann, M., 1994a. Complex adaptive systems. In: Cowan, G.A., Pines, D., Meltzer, D. (Eds.), Complexity. Metaphors, Models and Reality. Wesley Publishing Co., Reading, MA, pp. 17–45.

Gell-Mann, M., 1994b. The Quark and the Jaguar. Freeman, New York.

Gerrish, P.J., Garcia-Lerma, J.G., 2003. Mutation rate and the efficacy of antimicrobial drug treatment. Lancet Infect. Dis. 3, 28–32.

Geoghegan, J.L., Holmes, E.C., 2018. Evolutionary virology at 40. Genetics 210, 1151–1162.

Gibbs, A., Calisher, C., García-Arenal, F., 1995. Molecular Basis of Virus Evolution. Cambridge University Press, Cambridge.

Gire, S.K., Goba, A., Andersen, K.G., Sealfon, R.S., Park, D.J., et al., 2014. Genomic surveillance elucidates Ebola virus origin and transmission during the 2014 outbreak. Science 345, 1369–1372.

Gojobori, T., Yokoyama, S., 1985. Rates of evolution of the retroviral oncogene of Moloney murine sarcoma virus and of its cellular homologues. Proc. Natl. Acad. Sci. U.S.A. 82, 4198–4201.

González-López, C., Arias, A., Pariente, N., Gómez-Mariano, G., Domingo, E., 2004. Preextinction viral RNA can interfere with infectivity. J. Virol. 78, 3319–3324.

González-López, C., Gómez-Mariano, G., Escarmís, C., Domingo, E., 2005. Invariant aphthovirus consensus nucleotide sequence in the transition to error catastrophe. Infect. Genet. Evol. 5, 366–374.

Grande-Pérez, A., Lázaro, E., Lowenstein, P., Domingo, E., Manrubia, S.C., 2005. Suppression of viral infectivity through lethal defection. Proc. Natl. Acad. Sci. U.S.A. 102, 4448–4452.

Gregori, J., Perales, C., Rodríguez-Frías, F., Esteban, J.I., Quer, J., et al., 2016. Viral quasispecies complexity measures. Virology 493, 227–237.

Gregori, J., Salicru, M., Domingo, E., Sanchez, A., Esteban, J.I., et al., 2014. Inference with viral quasispecies diversity indices: clonal and NGS approaches. Bioinformatics 30, 1104–1111.

Guedj, J., Piorkowski, G., Jacquot, F., Madelain, V., Tram Nguyen, T.H., et al., 2018. Antiviral efficacy of favipiravir against Ebola virus: a translational study in cynomolgus macaques. PLoS Med. 15, e1002535.

Gytz, H., Mohr, D., Seweryn, P., Yoshimura, Y., Kutlubaeva, Z., et al., 2015. Structural basis for RNA–genome recognition during bacteriophage Qβ replication. Nucleic Acids Res. 43, 10893–10906.

Hamano, T., Matsuo, K., Hibi, Y., Victoriano, A.F., Takahashi, N., et al., 2007. A single-nucleotide synonymous mutation in the gag gene controlling human immunodeficiency virus type 1 virion production. J. Virol. 81, 1528–1533.

Hamming, R.W., 1980. Coding and Information Theory. Prentice-Hall, Englewood Cliffs.

Haruna, I., Nozu, K., Ohtaka, Y., Spiegelman, S., 1963. An RNA "Replicase" induced by and selective for a viral RNA: isolation and properties. Proc. Natl. Acad. Sci. U.S.A. 50, 905–911.

Hedskog, C., Mild, M., Jernberg, J., Sherwood, E., Bratt, G., et al., 2010. Dynamics of HIV-1 quasispecies during antiviral treatment dissected using ultra-deep pyrosequencing. PLoS One 5, e11345.

Herrera, M., Grande-Perez, A., Perales, C., Domingo, E., 2008. Persistence of foot-and-mouth disease virus in cell culture revisited: implications for contingency in evolution. J. Gen. Virol. 89, 232–244.

Hogeweg, P., Takeuchi, N., 2003. Multilevel selection in models of prebiotic evolution: compartments and spatial self-organization. Orig. Life Evol. Biosph. 33, 375–403.

Holland, J.J., 1984. Continuum of change in RNA virus genomes. In: Notkins, A.L., Oldstone, M.B.A. (Eds.), Concepts in Viral Pathogenesis. Springer-Verlag, New York.

Holland, J.J., Spindler, K., Horodyski, F., Grabau, E., Nichol, S., et al., 1982. Rapid evolution of RNA genomes. Science 215, 1577–1585.

Holmes, E.C., 2008. Comparative studies of RNA virus evolution. In: Domingo, E., Parrish, C.R., Holland, J.J. (Eds.), Origin and Evolution of Viruses, second ed. J.J. Elsevier, Oxford, pp. 119–134.

Holmes, E.C., 2009. The Evolution and Emergence of RNA Viruses. Oxford Series in Ecology and Evolution. Oxford University Press, New York.

Holmes, E.C., Moya, A., 2002. Is the quasispecies concept relevant to RNA viruses? J. Virol. 76, 460–465.

Hubscher, U., Maga, G., Spadari, S., 2002. Eukaryotic DNA polymerases. Annu. Rev. Biochem. 71, 133–163.

Huxley, J., 1942. Evolution: The Modern Synthesis. G. Allen and Unwin Ltd., London.

Isnard, M., Granier, M., Frutos, R., Reynaud, B., Peterschmitt, M., 1998. Quasispecies nature of three maize streak virus isolates obtained through different modes of selection from a population used to assess response to infection of maize cultivars. J. Gen. Virol. 79, 3091–3099.

Jabara, C.B., Jones, C.D., Roach, J., Anderson, J.A., Swanstrom, R., 2011. Accurate sampling and deep sequencing of the HIV-1 protease gene using a Primer ID. Proc. Natl. Acad. Sci. U.S.A. 108, 20166–20171.

Jaramillo, N., Domingo, E., Muñoz, M.C., Tabarés, E., Gadea, I., 2012. Evidence of Muller's ratchet in herpes simplex virus type 1. J. Gen. Virol. 94, 366–375.

Jayasundara, D., Saeed, I., Chang, B.C., Tang, S.L., Halgamuge, S.K., 2015. Accurate reconstruction of viral quasispecies spectra through improved estimation of strain richness. BMC Bioinformatics 16 (Suppl. 18), S3.

Jones, L.R., Zandomeni, R., Weber, E.L., 2002. Quasispecies in the 5′ untranslated genomic region of bovine viral diarrhoea virus from a single individual. J. Gen. Virol. 83, 2161–2168.

Karagiannis, K., Simonyan, V., Chumakov, K., Mazumder, R., 2017. Separation and assembly of deep sequencing data into discrete subpopulation genomes. Nucleic Acids Res. 45, 10989–11003.

Katzelnick, L.C., Coello Escoto, A., McElvany, B.D., Chávez, C., Salje, H., et al., 2018. Viridot: an automated virus plaque (immunofocus) counter for the measurement of serological neutralizing responses with application to dengue virus. PLoS Negl. Trop. Dis. 12, e0006862.

Kimata, J.T., Kuller, L., Anderson, D.B., Dailey, P., Overbaugh, J., 1999. Emerging cytopathic and antigenic simian immunodeficiency virus variants influence AIDS progression. Nat. Med. 5, 535–541.

Kimura, M., 1983. The Neutral Theory of Molecular Evolution. Cambridge University Press, Cambridge, UK.

Kimura, M., 1989. The neutral theory of molecular evolution and the world view of the neutralists. Genome 31, 24–31.

Kirkegaard, K., van Buuren, N.J., Mateo, R., 2016. My cousin, my enemy: quasispecies suppression of drug resistance. Curr. Opin. Virol. 210, 106–111.

Kuroda, M., Katano, H., Nakajima, N., Tobiume, M., Ainai, A., et al., 2010. Characterization of quasispecies of pandemic 2009 influenza A virus (A/H1N1/2009) by de novo sequencing using a next-generation DNA sequencer. PLoS One 5, e10256.

Lauring, A.S., Andino, R., 2010. Quasispecies theory and the behavior of RNA viruses. PLoS Pathog. 6, e1001005.

Lauring, A.S., Andino, R., 2011. Exploring the fitness landscape of an RNA virus using a universal barcode microarray. J. Virol. 85, 3780–3791.

Lenski, R.E., May, R.M., 1994. The evolution of virulence in parasites and pathogens: reconciliation between two competing hypotheses. J. Theor. Biol. 169, 253–265.

Levin, S.A., 2006. Fundamental questions in biology. PLoS Biol. 4, e300.

Lewontin, R.C., 1970. The units of selection. Annu. Rev. Ecol. Syst. 1, 1–18.

Liang, B., Luo, M., Scott-Herridge, J., Semeniuk, C., Mendoza, M., et al., 2011. A comparison of parallel pyrosequencing and sanger clone-based sequencing and its impact on the characterization of the genetic diversity of HIV-1. PLoS One 6, e26745.

López-Bueno, A., Mateu, M.G., Almendral, J.M., 2003. High mutant frequency in populations of a DNA virus allows evasion from antibody therapy in an immunodeficient host. J. Virol. 77, 2701–2708.

López-Bueno, A., Rastrojo, A., Peiró, R., Arenas, M., Alcamí, A., 2015. Ecological connectivity shapes quasispecies structure of RNA viruses in an Antarctic lake. Mol. Ecol. 24, 4812–4825.

López-Bueno, A., Rubio, M.P., Bryant, N., McKenna, R., Agbandje-McKenna, M., et al., 2006. Host-selected amino acid changes at the sialic acid binding pocket of the parvovirus capsid modulate cell binding affinity and determine virulence. J. Virol. 80, 1563–1573.

Lu, Z., Rezapkin, G.V., Douthitt, M.P., Ran, Y., Asher, D.M., et al., 1996. Limited genetic changes in the Sabin 1 strain of poliovirus occurring in the central nervous system of monkeys. J. Gen. Virol. 77 (Pt 2), 273–280.

Mangul, S., Wu, N.C., Mancuso, N., Zelikovsky, A., Sun, R., et al., 2014. Accurate viral population assembly from ultra-deep sequencing data. Bioinformatics 30, i329–i337.

Mardis, E.R., 2008. The impact of next-generation sequencing technology on genetics. Trends Genet. 24, 133–141.

Margeridon-Thermet, S., Shulman, N.S., Ahmed, A., Shahriar, R., Liu, T., et al., 2009. Ultra-deep pyrosequencing of hepatitis B virus quasispecies from nucleoside and nucleotide reverse-transcriptase inhibitor (NRTI)-treated patients and NRTI-naive patients. J. Infect. Dis. 199, 1275–1285.

Martell, M., Esteban, J.I., Quer, J., Genesca, J., Weiner, A., et al., 1992. Hepatitis C virus (HCV) circulates as a population of different but closely related genomes: quasispecies nature of HCV genome distribution. J. Virol. 66, 3225–3229.

Martín Hernández, A.M., Carrillo, E.C., Sevilla, N., Domingo, E., 1994. Rapid cell variation can determine the establishment of a persistent viral infection. Proc. Natl. Acad. Sci. U.S.A. 91, 3705–3709.

Martín, V., Domingo, E., 2008. Influence of the mutant spectrum in viral evolution: focused selection of antigenic variants in a reconstructed viral quasispecies. Mol. Biol. Evol. 25, 1544–1554.

Martín, V., Perales, C., Fernández-Algar, M., Dos Santos, H.G., Garrido, P., et al., 2016. An efficient microarray-based genotyping platform for the identification of drug-resistance mutations in majority and minority subpopulations of HIV-1 quasispecies. PLoS One 11, e0166902.

Marz, M., Beerenwinkel, N., Drosten, C., Fricke, M., Frishman, D., et al., 2014. Challenges in RNA virus bioinformatics. Bioinformatics 30, 1793–1799.

Mateu, M.G., Hernández, J., Martínez, M.A., Feigelstock, D., Lea, S., et al., 1994. Antigenic heterogeneity of a foot-and-mouth disease virus serotype in the field is mediated by very limited sequence variation at several antigenic sites. J. Virol. 68, 1407–1417.

Matranga, C.B., Gladden-Young, A., Qu, J., Winnicki, S., Nosamiefan, D., 2016. Unbiased deep sequencing of RNA viruses from clinical samples. J. Vis. Exp. 113 https://doi.org/10.3791/54117.

Maynard Smith, J., 1964. Group selection and kin selection. Nature 201, 1145—1147.

Maynard Smith, J.M., 1970. Natural selection and the concept of a protein space. Nature 225, 563—564.

Mayr, E., 1959. Evolution and the Diversity of Life. Harvard University Press, Cambridge, Massachussets.

Mayr, E., Provine, W.B., 1980. The Evolutionary Synthesis. University Press, Cambridge, MA.

Mills, D.R., Peterson, R.L., Spiegelman, S., 1967. An extracellular Darwinian experiment with a self-duplicating nucleic acid molecule. Proc. Natl. Acad. Sci. U.S.A. 58, 217—224.

McCulloch, S.D., Kunkel, T.A., 2008. The fidelity of DNA synthesis by eukaryotic replicative and translesion synthesis polymerases. Cell Res. 18, 148—161.

Mitsuya, Y., Varghese, V., Wang, C., Liu, T.F., Holmes, S.P., et al., 2008. Minority human immunodeficiency virus type 1 variants in antiretroviral-naive persons with reverse transcriptase codon 215 revertant mutations. J. Virol. 82, 10747—10755.

Moreno, E., Gallego, I., Gregori, J., Lucia-Sanz, A., Soria, M.E., et al., 2017. Internal disequilibria and phenotypic diversification during replication of hepatitis C virus in a noncoevolving cellular environment. J. Virol. 91 (10), e02505—e02516.

Moreno, I.M., Malpica, J.M., Rodriguez-Cerezo, E., Garcia-Arenal, F., 1997. A mutation in tomato aspermy cucumovirus that abolishes cell-to-cell movement is maintained to high levels in the viral RNA population by complementation. J. Virol. 71, 9157—9162.

Moreno, H., Tejero, H., de la Torre, J.C., Domingo, E., Martin, V., 2012. Mutagenesis-mediated virus extinction: virus-dependent effect of viral load on sensitivity to lethal defection. PLoS One 7, e32550.

Morse, S.S., 1994. The Evolutionary Biology of Viruses. Raven Press, New York.

Mount, D.W., 2004. Bioinformatics. Sequence and Genome Analysis. Cold Spring Harbor Laboratory Press, Cold Spring Harbor, New York.

Moya, A., Elena, S.F., Bracho, A., Miralles, R., Barrio, E., 2000. The evolution of RNA viruses: a population genetics view. Proc. Natl. Acad. Sci. U.S.A. 97, 6967—6973.

Muñoz, E., Park, J.M., Deem, M.W., 2008. Quasispecies theory for horizontal gene transfer and recombination. Phys. Rev. 78, 061921.

Nevin, L., 2018. Advancing the beneficial use of machine learning in health care and medicine: toward a community understanding. PLoS Med. 15, e1002708.

Nomura, T., Yamamoto, H., Ishii, H., Akari, H., Naruse, T.K., et al., 2015. Broadening of virus-specific CD8+T-cell responses is indicative of residual viral replication in aviremic SIV controllers. PLoS Pathog. 11 (11), e1005247.

Novella, I.S., 2003. Contributions of vesicular stomatitis virus to the understanding of RNA virus evolution. Curr. Opin. Microbiol. 6, 399—405.

Novella, I.S., Ebendick-Corpus, B.E., 2004. Molecular basis of fitness loss and fitness recovery in vesicular stomatitis virus. J. Mol. Biol. 342, 1423—1430.

Novella, I.S., Reissig, D.D., Wilke, C.O., 2004. Density-dependent selection in vesicular stomatitis virus. J. Virol. 78, 5799—5804.

Nowak, M.A., 2006. Evolutionary Dynamics. The Belknap Press of Harvard University Press, Cambridge, Massachusetts and London, England.

Nowak, M., Schuster, P., 1989. Error thresholds of replication in finite populations mutation frequencies and the onset of Muller's ratchet. J. Theor. Biol. 137, 375—395.

Ochoa, G., 2006. Error thresholds in genetic algorithms. Evol. Comput. 14, 157—182.

Ojosnegros, S., Agudo, R., Sierra, M., Briones, C., Sierra, S., et al., 2008. Topology of evolving, mutagenized viral populations: quasispecies expansion, compression, and operation of negative selection. BMC Evol. Biol. 8, 207.

Okasha, S., 2013. Evolution and the Levels of Selection. Clarendon Press, Oxford.

Ortega-Prieto, A.M., Sheldon, J., Grande-Pérez, A., Tejero, H., Gregori, J., et al., 2013. Extinction of hepatitis C virus by ribavirin in hepatoma cells involves lethal mutagenesis. PLoS One 8, e71039.

Page, K.M., Nowak, M.A., 2002. Unifying evolutionary dynamics. J. Theor. Biol. 219, 93—98.

Palma, E.L., Huang, A.S., 1974. Cyclic production of vesicular stomatitis virus caused by defective interfering particles. J. Infect. Dis. 126, 402—410.

Park, J.M., Deem, M.W., 2007. Phase diagrams of quasispecies theory with recombination and horizontal gene transfer. Phys. Rev. Lett. 98, 058101.

Park, J.M., Munoz, E., Deem, M.W., 2010. Quasispecies theory for finite populations. Phys. Rev. 81, 011902.

Perales, C., 2018. Quasispecies dynamics and clinical significance of hepatitis C virus (HCV) antiviral resistance. Int. J. Antimicrob. Agents. https://doi.org/10.1016/j.ijantimicag.2018.10.005.

Perales, C., Martin, V., Ruiz-Jarabo, C.M., Domingo, E., 2005. Monitoring sequence space as a test for the target of selection in viruses. J. Mol. Biol. 345, 451—459.

Perales, C., Mateo, R., Mateu, M.G., Domingo, E., 2007. Insights into RNA virus mutant spectrum and lethal mutagenesis events: replicative interference and complementation by multiple point mutants. J. Mol. Biol. 369, 985—1000.

Perales, C., Lorenzo-Redondo, R., López-Galíndez, C., Martínez, M.A., Domingo, E., 2010. Mutant spectra in virus behavior. Future Virol. 5, 679—698.

Perales, C., Henry, M., Domingo, E., Wain-Hobson, S., Vartanian, J.P., 2011. Lethal mutagenesis of foot-and-mouth disease virus involves shifts in sequence space. J. Virol. 85, 12227–12240.

Pfeiffer, J.K., Kirkegaard, K., 2005. Increased fidelity reduces poliovirus fitness under selective pressure in mice. PLoS Pathog. 1, 102–110.

Poh, W.T., Xia, E., Chin-Inmanu, K., Wong, L.P., Cheng, A.Y., et al., 2013. Viral quasispecies inference from 454 pyrosequencing. BMC Bioinformatics 14, 355.

Polyak, S.J., Sullivan, D.G., Austin, M.A., Dai, J.Y., Shuhart, M.C., et al., 2005. Comparison of amplification enzymes for Hepatitis C virus quasispecies analysis. J. Virol. 2, 41.

Prosperi, M.C., Prosperi, L., Bruselles, A., Abbate, I., Rozera, G., et al., 2011. Combinatorial analysis and algorithms for quasispecies reconstruction using next-generation sequencing. BMC Bioinformatics 12, 5.

Quer, J., Gregori, J., Rodriguez-Frias, F., Buti, M., Madejon, A., et al., 2014. High-resolution hepatitis C virus subtyping using NS5B deep sequencing and phylogeny, an alternative to current methods. J. Clin. Microbiol. 53, 219–226.

Quer, J., Rodríguez-Frias, F., Gregori, J., Tabernero, D., Soria, M.E., et al., 2017. Deep sequencing in the management of hepatitis virus infections. Virus Res. 239, 115–125.

Quinlivan, M.L., Gershon, A.A., Steinberg, S.P., Breuer, J., 2004. Rashes occurring after immunization with a mixture of viruses in the Oka vaccine are derived from single clones of virus. J. Infect. Dis. 190, 793–796.

Quiñones-Mateu, M.E., Avila, S., Reyes-Teran, G., Martínez, M.A., 2014. Deep sequencing: becoming a critical tool in clinical virology. J. Clin. Virol. 61, 9–19.

Rechenberg, I., 1973. Evolutionsstrategie. Problemata frommann-holzboog, Stuttgart-Bad Cannstatt.

Rezelj, V.V., Levi, L.I., Vignuzzi, M., 2018. The defective component of viral populations. Curr. Opin. Virol. 33, 74–80.

Ruiz-Jarabo, C.M., Arias, A., Baranowski, E., Escarmís, C., Domingo, E., 2000. Memory in viral quasispecies. J. Virol. 74, 3543–3547.

Ruiz-Jarabo, C.M., Miller, E., Gómez-Mariano, G., Domingo, E., 2003. Synchronous loss of quasispecies memory in parallel viral lineages: a deterministic feature of viral quasispecies. J. Mol. Biol. 333, 553–563.

Saakian, D.B., Hu, C.K., 2006. Exact solution of the Eigen model with general fitness functions and degradation rates. Proc. Natl. Acad. Sci. U.S.A. 103, 4935–4939.

Saakian, D.B., Munoz, E., Hu, C.K., Deem, M.W., 2006. Quasispecies theory for multiple-peak fitness landscapes. Phys. Rev. E 73, 041913.

Saakian, D.B., Biebricher, C.K., Hu, C.K., 2009. Phase diagram for the Eigen quasispecies theory with a truncated fitness landscape. Phys. Rev. 79, 041905.

Saffhill, R., Schneider-Bernloehr, H., Orgel, L.E., Spiegelman, S., 1970. In vitro selection of bacteriophage Q-beta ribonucleic acid variants resistant to ethidium bromide. J. Mol. Biol. 51, 531–539.

Sala, M., Centlivre, M., Wain-Hobson, S., 2006. Clade-specific differences in active viral replication and compartmentalization. Curr. Opin. HIV AIDS 1, 108–114.

Salazar-Gonzalez, J.F., Salazar, M.G., Keele, B.F., Learn, G.H., Giorgi, E.E., et al., 2009. Genetic identity, biological phenotype, and evolutionary pathways of transmitted/founder viruses in acute and early HIV-1 infection. J. Exp. Med. 206, 1273–1289.

Salemi, M., Vandamme, A.M. (Eds.), 2004. The Phylogeny Handbook. A Practical Approach to DNA and Protein Phylogeny. Cambridge University Press, Cambridge.

Sánchez-Campos, S., Dominguez-Huerta, G., Díaz-Martínez, L., Tomás, D.M., Navas-Castillo, J., et al., 2018. Differential shape of geminivirus mutant spectra across cultivated and wild hosts with invariant viral consensus sequences. Front. Plant Sci. 9, 932.

Sánchez-Palomino, S., Rojas, J.M., Martínez, M.A., Fenyo, E.M., Najera, R., et al., 1993. Dilute passage promotes expression of genetic and phenotypic variants of human immunodeficiency virus type 1 in cell culture. J. Virol. 67, 2938–2943.

Sanz, A.I., Fraile, A., Gallego, J.M., Malpica, J.M., García-Arenal, F., 1999. Genetic variability of natural populations of cotton leaf curl geminivirus, a single-stranded DNA virus. J. Mol. Evol. 49, 672–681.

Sardanyés, J., Elena, S.F., 2010. Error threshold in RNA quasispecies models with complementation. J. Theor. Biol. 265, 278–286.

Sarisky, R.T., Nguyen, T.T., Duffy, K.E., Wittrock, R.J., Leary, J.J., 2000. Difference in incidence of spontaneous mutations between Herpes simplex virus types 1 and 2. Antimicrob. Agents Chemother. 44, 1524–1529.

Sauder, C.J., Vandenburgh, K.M., Iskow, R.C., Malik, T., Carbone, K.M., et al., 2006. Changes in mumps virus neurovirulence phenotype associated with quasispecies heterogeneity. Virology 350, 48–57.

Schuster, P., 2010. Genotypes and phenotypes in the evolution of molecules. In: Caetono-Anolles, G. (Ed.), Evolutionary Genomics and Systems Biology. Wiley-Blackwell, Hoboken, New Jersey, pp. 123–152.

Schuster, P., 2016. Quasispecies on fitness landscapes. Curr. Top. Microbiol. Immunol. 392, 61–120.

Schuster, P., Stadler, P.F., 1999. Nature and evolution of early replicons. In: Domingo, E., Webster, R.G., Holland, J.J. (Eds.), Origin and Evolution of Viruses. Academic Press, San Diego, pp. 1–24.

Seifert, D., Beerenwinkel, N., 2016. Estimating fitness of viral quasispecies from next-generation sequencing data. Curr. Top. Microbiol. Immunol. 392, 181–200.

Sevilla, N., Ruiz-Jarabo, C.M., Gómez-Mariano, G., Baranowski, E., Domingo, E., 1998. An RNA virus can adapt to the multiplicity of infection. J. Gen. Virol. 79, 2971–2980.

Shepard, R.N., 1987. Toward a universal law of generalization for psychological science. Science 237, 1317–1323.

Shirogane, Y., Watanabe, S., Yanagi, Y., 2012. Cooperation between different RNA virus genomes produces a new phenotype. Nat. Commun. 3, 1235.

Shirogane, Y., Watanabe, S., Yanagi, Y., 2016. Cooperative interaction within RNA virus mutant spectra. Curr. Top. Microbiol. Immunol. 392, 219–229.

Simon, H.A., 1996. The Sciences of the Artificial, third ed. The MIT Press, Cambridge, Massachusetts.

Skums, P., Zelikovsky, A., Singh, R., Gussler, W., Dimitrova, Z., et al., 2018. QUENTIN: reconstruction of disease transmission from viral quasispecies genomic data. Bioinformatics 34, 163–170.

Smith, D.B., Inglis, S.C., 1987. The mutation rate and variability of eukaryotic viruses: an analytical review. J. Gen. Virol. 68, 2729–2740.

Solé, R., Goodwin, B., 2000. Signs of Life. How Complexity Pervades Biology. Basic Books, New York.

Solmone, M., Vincenti, D., Prosperi, M.C., Bruselles, A., Ippolito, G., et al., 2009. Use of massively parallel ultradeep pyrosequencing to characterize the genetic diversity of hepatitis B virus in drug-resistant and drug-naive patients and to detect minor variants in reverse transcriptase and hepatitis B S antigen. J. Virol. 83, 1718–1726.

St John, K., 2017. Review paper: the shape of phylogenetic treespace. Syst. Biol. 66, e83–e94.

Staddon, J.E.R., 2001. Adaptive Dynamics. The Theoretical Analysis of Behavior. The MIT Press, Cambridge, Massachusetts.

Steinhauer, D.A., Holland, J.J., 1987. Rapid evolution of RNA viruses. Annu. Rev. Microbiol. 41, 409–433.

Sukkaew, A., Thanagith, M., Thongsakulprasert, T., Mutso, M., Mahalingam, S., et al., 2018. Heterogeneity of clinical isolates of chikungunya virus and its impact on the responses of primary human fiboblast-like synoviocytes. J. Gen. Virol. 99, 525–535.

Suspène, R., Henry, M., Guillot, S., Wain-Hobson, S., Vartanian, J.P., 2005. Recovery of APOBEC3-edited human immunodeficiency virus G→A hypermutants by differential DNA denaturation PCR. J. Gen. Virol. 86, 125–129.

Suspène, R., Renard, M., Henry, M., Guetard, D., Puyraimond-Zemmour, D., et al., 2008. Inversing the natural hydrogen bonding rule to selectively amplify GC-rich ADAR-edited RNAs. Nucleic Acids Res. 36, e72.

Svicher, V., Balestra, E., Cento, V., Sarmati, L., Dori, L., et al., 2011. HIV-1 dual/mixed tropic isolates show different genetic and phenotypic characteristics and response to maraviroc in vitro. Antiviral Res. 90, 42–53.

Swetina, J., Schuster, P., 1982. Self-replication with errors. A model for polynucleotide replication. Biophys. Chem. 16, 329–345.

Tanner, E.J., Liu, H.M., Oberste, M.S., Pallansch, M., Collett, M.S., et al., 2014. Dominant drug targets suppress the emergence of antiviral resistance. eLife 3, e03830.

Teng, M.N., Oldstone, M.B., de la Torre, J.C., 1996. Suppression of lymphocytic choriomeningitis virus-induced growth hormone deficiency syndrome by disease-negative virus variants. Virology 223, 113–119.

Topfer, A., Marschall, T., Bull, R.A., Luciani, F., Schonhuth, A., et al., 2014. Viral quasispecies assembly via maximal clique enumeration. PLoS Comput. Biol. 10, e1003515.

Tsibris, A.M., Korber, B., Arnaout, R., Russ, C., Lo, C.C., et al., 2009. Quantitative deep sequencing reveals dynamic HIV-1 escape and large population shifts during CCR5 antagonist therapy in vivo. PLoS One 4, e5683.

Van Loy, T., Thys, K., Ryschkewitsch, C., Lagatie, O., Monaco, M.C., et al., 2015. JC virus quasispecies analysis reveals a complex viral population underlying progressive multifocal leukoencephalopathy and supports viral dissemination via the hematogenous route. J. Virol. 89, 1340–1347.

Vandenbroucke, I., Van Marck, H., Verhasselt, P., Thys, K., Mostmans, W., et al., 2011. Minor variant detection in amplicons using 454 massive parallel pyrosequencing: experiences and considerations for successful applications. Biotechniques 51, 167–177.

Vignuzzi, M., Stone, J.K., Arnold, J.J., Cameron, C.E., Andino, R., 2006. Quasispecies diversity determines pathogenesis through cooperative interactions in a viral population. Nature 439, 344–348.

Vogt, P.K., Jackson, A.O., 1999. Satellites and Defective Viral RNAs. Curr. Top. Microbiol. Immunol., vol. 239. Springer-Verlag, Berlin.

Wagner, N., Atsmon-raz, Y., Ashkenasy, G., 2016. Theoretical models of generalized quasispecies. Curr. Top. Microbiol. Immunol. 392, 141–159.

Wang, D., Coscoy, L., Zylberberg, M., Avila, P.C., Boushey, H.A., et al., 2002. Microarray-based detection and genotyping of viral pathogens. Proc. Natl. Acad. Sci. U.S.A. 99, 15687–15692.

Wang, C., Mitsuya, Y., Gharizadeh, B., Ronaghi, M., Shafer, R.W., 2007. Characterization of mutation spectra with ultra-deep pyrosequencing: application to HIV-1 drug resistance. Genome Res. 17, 1195–1201.

Weissmann, C., 1974. The making of a phage. FEBS Lett. 40 (Suppl.), S10–S18.

Wen, Z., Citron, M., Bett, A.J., Espeseth, A.S., Vora, K.A., 2019. Development and application of a higher throughput RSV plaque assay by immunofluorescent imaging. J. Virol. Methods 263, 88–95.

Wichman, H.A., Badgett, M.R., Scott, L.A., Boulianne, C.M., Bull, J.J., 1999. Different trajectories of parallel evolution during viral adaptation. Science 285, 422–424.

Wilke, C.O., 2005. Quasispecies theory in the context of population genetics. BMC Evol. Biol. 5, 44.

Wilke, C.O., Ronnewinkel, C., Martinetz, T., 2001. Dynamic fitness landscapes in molecular evolution. Phys. Rep. 349, 395–446.

Wilke, C.O., Reissig, D.D., Novella, I.S., 2004. Replication at periodically changing multiplicity of infection promotes stable coexistence of competing viral populations. Evolution 58, 900–905.

Williams, G.C., 1992. Natural Selection. Domains, Levels and Challenges. Oxford University Press, New York, Oxford.

Wilson, E.D., 1998. Consilience. Abacus. Little, Brown and Co., U.K.

Wingreen, N.S., Levin, S.A., 2006. Cooperation among microorganisms. PLoS Biol. 4, e299.

Wright, S., 1945. Tempo and mode of evolution. A critical review. Ecology 26, 415–419.

Wright, C.F., Morelli, M.J., Thebaud, G., Knowles, N.J., Herzyk, P., et al., 2011. Beyond the consensus: dissecting within-host viral population diversity of foot-and-mouth disease virus by using next-generation genome sequencing. J. Virol. 85, 2266–2275.

Wu, S., Kanda, T., Nakamoto, S., Jiang, X., Miyamura, T., et al., 2013. Prevalence of hepatitis C virus subgenotypes 1a and 1b in Japanese patients: ultra-deep sequencing analysis of HCV NS5B genotype-specific region. PLoS One 8, e73615.

Xu, L., Da, L., Plouffe, S.W., Chong, J., Kool, E., et al., 2014. Molecular basis of transcriptional fidelity and DNA lesion-induced transcriptional mutagenesis. DNA Repair 19, 71–83.

Xue, K.S., Hooper, K.A., Ollodart, A.R., Dingens, A.S., Bloom, J.D., 2016. Cooperation between distinct viral variants promotes growth of $H_3N_2$ influenza in cell culture. eLife 5, e13974.

Yin, J., 1993. Evolution of bacteriophage T7 in a growing plaque. J. Bacteriol. 175, 1272–1277.

Young, D.F., Wignall-Fleming, E.B., Busse, D.C., Pickin, M.J., Hankinson, J., et al., 2019. The switch between acute and persistent paramyxovirus infection caused by single amino acid substitutions in the RNA polymerase P subunit. PLoS Pathog. 15, e1007561.

Youngner, J.S., Whitaker-Dowling, P., 1999. Interference. In: Granoff, A., Webster, R.G. (Eds.), Encyclopedia of Virology. Academic Press, San Diego, California, pp. 850–854.

Zagordi, O., Klein, R., Daumer, M., Beerenwinkel, N., 2010. Error correction of next-generation sequencing data and reliable estimation of HIV quasispecies. Nucleic Acids Res. 38, 7400–7409.

Zhang, Y.Z., Shi, M., Holmes, E.C., 2018. Using metagenomics to characterize an expanding virosphere. Cell 172, 1168–1172.

Zinder, N.D., 1975. RNA Phages. Cold Spring Harbor Monograph Series. Cold Spring Harbor Laboratory, New York.

# Interaction of virus populations with their hosts

## Abbreviations

**A** Adenine
**ABV** Avian Bornavirus
**ACE two** Angiotensin-Converting Enzyme two
**AIDS** Acquired Immune Deficiency Syndrome
**Arg (R)** Arginine
**BDV** Borna Disease Virus
**C** Cytosine
**CAR** Coxsackievirus and Adenovirus Receptor
**CTL** Cytotoxic T Lymphocyte
**CVB** Group B Coxsackieviruses
**DAF** Decay-Accelerating Factor
**EIAV** Equine Infectious Anemia Virus
**FMDV** Foot-and-Mouth Disease Virus
**G** Guanine
**HA** Hemagglutinin
**HAV** Hepatitis A Virus
**HBV** Hepatitis B Virus
**HCV** Hepatitis C Virus
**HIV-1** Human Immunodeficiency Virus Type 1
**IFN** Interferon
**IV** Influenza Virus
**LCMV** Lymphocytic Choriomeningitis Virus
**LTR** Long Terminal Repeat
**MHC** Major Histocompatibility Complex
**MIC** Mutual Information Criterion
**MOI** Multiplicity of Iinfection
**mRNA** Messenger RNA
**N** Population Size
**Ne** Effective Population Size
**NK cells** Natural Killer Cells
**NTPase** Nucleoside-Triphosphatase
**PV** Poliovirus
**PVR** Poliovirus Receptor
**RGD** Arginine-Glycine-Aspartic Acid

**RNAi** RNA Interference
**SARS CoV** Severe Acute Respiratory Syndrome Coronavirus
**SIV** Simian Immunodeficiency Virus
**T** Thymine
**TCID$_{50}$** Tissue Culture Infectious Dose 50 (the amount of virus needed to infect 50% of a cell population)
**tRNA** Transfer RNA
**Trp (W)** Tryptophan
**U** Uracil
**VPg** Viral Protein Genome Linked
**VSV** Vesicular Stomatitis Virus

## 4.1 Contrasting viral and host population numbers

Although we often refer to viruses as "autonomous" genetic elements, their replication is dependent on host functions. This dependence occurs at two levels: at the stage of replication within individual cells and at the stage of dissemination among populations of susceptible cells, animals, or plants. Host cells and organisms have mediated survival of the viruses that we can isolate and study, but they have conditioned their persistence to the capacity to overcome selective constraints imposed by the cellular world. The paradox that represents that host functions are sometimes recruited by the virus

*Virus as Populations*
https://doi.org/10.1016/B978-0-12-816331-3.00004-0

to ensure its replication, while other times they become part of the innate immune response, can be interpreted as a result of long-term coevolution between viruses and cells, and the basic mechanisms that mediate biological evolution in our biosphere. The concept of evolutionary tinkering proposed by F. Jacob in the last century (Jacob, 1977) is adequate to interpret the paradoxical interplay of viral and cellular functions (Domingo, 2011). In this view, viruses and cells, in their coevolutionary race for survival, must have taken advantage, as needed, of what existed at any given stage of the evolutionary process. Here, again, viral population numbers and molecular instructions to ensure genome diversity have played essential roles.

The infection of a host by a virus can be viewed as a specific example of a predator (virus)-pray (host) relationship, with a very distinctive attribute: the disproportionate difference in population size between viruses and their hosts, dramatically in favor of viruses. The large population size of many viral populations (both, as replicative ensembles inside individual cells and as particles available for new rounds of infection and transmission in the field) favors their adaptability, as a consequence of the ease of exploration of sequence space (Section 3.7 in Chapter 3). It is well established in ecology that, except in the case of being a pathogenic agent, the predator cannot be more numerous than the prey (Remmert, 1980). A virus in an infected organism can reach $10^9$ to $10^{12}$ potentially infectious particles at a given point in time. Early calculations were made in cattle infected with foot-and-mouth disease virus (FMDV) (Sellers, 1971), and during togavirus infections (Halstead, 1980). Similar estimates have been obtained for human immunodeficiency virus type 1 (HIV-1), hepatitis B virus (HBV), and hepatitis C virus (HCV) (Wei et al., 1995; Nowak et al., 1996; Neumann et al., 1998) or for a tobacco leaf infected with tobacco mosaic virus (review in Gutierrez et al., 2012). Large amounts of virus can be present in excretions and secretions from infected

individuals, where the total amount of virus and its concentration is a factor of virus spread. In adult volunteers infected with influenza virus (IV) type A, maximum titers of $10^3$ to $10^7$ tissue culture infectious dose 50 per ml (TCID$_{50}$/ml) were determined in nasopharyngeal fluids 1 day after infection (Murphy and Webster, 1985). In children with type B IV, titers of $10^4$ TCID$_{50}$/ml were present in nasal washings also 1 day after infection (Hall et al., 1979). In the laboratory, for viruses that infect cells in culture, a visible viral plaque on a cell monolayer (or a focus formed in a cluster of infected cells) contains a variable number of infectious units that generally exceeds $10^3$. A range of $10^3$ to $10^9$ infectious units per plaque is quite frequent for cytopathic viruses, such as FMDV and VSV used in studies of experimental evolution (Chapter 6). The lytic plaque and infected cell focus size, and the number of viral particles in them, depends on the replication rate of the viral clone, the way of transmission from cell to cell (either exit into the intercellular medium prior to penetration into a neighbor cell, or direct cell-to-cell transmission), virus stability in the extracellular environment (agar overlay or other medium), and the host cell viability during the time of plaque or focus development. Culture medium added to confluent bacteriophage plaques on a bacterial lawn is an effective method to achieve high bacteriophage titers for viral purification and physical studies. For animal viruses that infect cells in culture, infections of cell monolayers or cells in suspension can be scaled up to produce viral populations of $10^{10}$ to $10^{11}$ particles in the case of the most fecund viruses.

The viral population numbers have been given to underline the sharp contrast with the population numbers of their hosts. Considering mammals, each of many primate species includes a total of hundreds to a few thousand individuals. Among the most abundant, the gibbon *Hylobates muelleri* is represented by $3 \times 10^5$ to $4 \times 10^5$ individuals, and humans by $7 \times 10^9$ individuals, with a projection of $9 \times 10^9$

individuals for 2040, values which are orders of magnitude lower than the total number of HCV particles in a liver acutely infected with HCV. Mammals are, however, modest in representation as compared with other types of organisms. The number of insect species is uncertain, with estimates broadly ranging from $1 \times 10^6$ to $20 \times 10^6$, with about $10^{18}$-$1^{19}$ individual insects alive in our planet at a given time. Despite these impressive numbers (imagine how many viruses might be hosted by insects that have never been analyzed!), the estimated number of individual insects is still $10^{13}$-fold lower than the total number of viral particles on Earth (compare with figures given in Chapter 1). Insects are only exceeded by zooplankton (about $10^{21}$ individuals) and nematodes (about $10^{22}$ individuals). To give some additional comparative figures that will become pertinent when dealing with zoonotic transmissions and viral disease emergence (Chapter 7), the total number of livestock is $2.4 \times 10^{10}$, and the total number of birds, mammals, reptiles, amphibians, or fish is $10^{10}$–$10^{13}$. Estimates of the number of biological species and of individuals within species are regularly published, and the reader will find numbers that are all extremely modest compared with the VIROME (Viral Informatics Resource for Metagenome Exploration) regarding virus diversity and anticipated number of individual viral particles per group (Wommack et al., 2012; Virgin, 2014; Zhang et al., 2018).

Long-term virus survival has been based not only in multiple strategies to cope with the host immune response but also in their life cycles generally having evolved to produce vast numbers of progeny. From the terminology of ecology, viruses are *r* strategists in the sense that they base their success in rapid reproduction to confront multiple habitats (intrahost compartmentalization and multiple selective constraints, as discussed in Section 4.2). In contrast, large animals are *K* strategists that produce limited progeny, have a long life span, and inhabit relatively stable environments (Remmert, 1980).

## 4.1.1 Productive power of some viral infections

The exploration of sequence space is commensurate with the number of newly synthesized viral genomes per unit time in infected organisms. Only for a few virus-host systems, the velocity of genome replication (number of nucleotides incorporated into a growing viral RNA or DNA genome per unit time) has been calculated. Early studies indicated that the average time needed to synthesize an entire plus strand of bacteriophage Qβ RNA (4220 nucleotides) in vivo was about 90 s (Robertson, 1975). For poliovirus (PV) it has been estimated that it takes about 1 min to synthesize a full-length genomic RNA (7440 nucleotides) and that when PV RNA synthesis reaches its maximum, 2000–3000 RNA molecules are produced per cell and minute (Richards and Ehrenfeld, 1990; Paul, 2002). HCV polymerase incorporates 5 to 20 nucleotides per second (reviewed in Fung et al., 2014) These values imply that with the mutation rates and frequencies typical of RNA viruses (Chapter 2) mutant distributions of $10^5$ to $10^7$ genomes can be produced in infected cell cultures or host organisms in minutes.

In the course of infections by HIV-1, it has been estimated that $10^{10}$ to $10^{11}$ new virions are produced each day (Coffin, 1995; Ho et al., 1995). The average life span of cells productively infected with HIV-1 has been estimated in 1–2.2 days, with a half-life ($t^1/_2$) of about 1.5 days. The average life span of HIV-1 virions in plasma is about 6 h, with a $t^1/_2$ of 2–4 h (Ho, 1995; Wei et al., 1995; Perelson et al., 1996; Markowitz et al., 2003). The half-life of HCV particles circulating in infected individuals is about 2.7 h, and about $10^{12}$ particles are produced and cleared every day (Neumann et al., 1998; Ramratnam et al., 1999). A typical active HBV infection can produce $10^{13}$ viral particles per day; with an average mutation rate of $10^{-4}$ mutations per nucleotide, $10^9$ new mutations can be tested every day in the 3200 bp HBV genome (Whalley et al., 2001) (review in Quer et al., 2008). Thus, replication of some important viral pathogens is extremely rapid; viruses undergo continuous genetic change and are constantly replaced by new variants (rapid turnover). Genetic and phenotypic diversification is fast and observable. This has been directly noticed in the case of HIV-1 upon the reconstruction of the genomic nucleotide sequence of the transmitted (or founder) virus in a number of patients (Keele et al., 2008; Salazar-Gonzalez et al., 2009). These studies revealed a rapid diversification of the founder, biologically active HIV-1 into multiple replication-competent and defective progeny. There is little question that quasispecies dynamics, as defined in Chapter 3, is operating in vivo, and it implies an effective exploration of the permissive area of sequence space. Rapid, error-prone replication is the basis of virus behavior as *r* strategists, an adaptation to their long-term survival in heterogeneous environments.

## 4.1.2 Population size limitations and the effect of bottlenecks: the effective population size

High viral yields are not universal during viral infections. Viral production can be very high in acute infections in vivo and in cytopathic infections in cell culture. However, viruses can also establish latent infections with intermittent periods of virus production and intervals without detection of infectious virus. In latent infections by DNA viruses or retroviruses, the virus can be undetectable or present in minimal quantities until recurrence of the infection by activation of the latent reservoir takes place. Latency can occur with or without the integration of viral DNA into the host DNA. Chronic infections involve continuous but variable production of infectious virus, with or without disease manifestations that may become apparent only after prolonged chronicity.

An acute infection can be followed by a persistent stage, sometimes producing highly mutated forms of the acute virus that give rise to new pathologies. This is the case of subacute sclerosing panencephalitis, a rare brain disease associated with hypermutated variants of measles virus (Chapter 2).

Persistent infections in cell culture have been divided into steady-state and carrier cell infections. The maintenance of a steady-state system depends on the continued division of cells carrying the replicating virus, and not on reinfection of a susceptible subset of uninfected cells. In contrast, in carrier cell cultures, there is a continuous supply of a small number of uninfected cells that engage in a sustained, low-level viral production. Persistent infections in cell culture have been instrumental to learn about the consequences of virus-host cell interactions, and they are studied in Chapter 6.

In the steady-state persistent infections, cells often produce and release a limited amount of virus, while cells divide with little metabolic affectation. Persistent infections by Borna disease virus (BDV) are particularly illustrative because as little as 0.01–0.05 infectious units are present per infected cell (Pauli and Ludwig, 1985). This behavior may relate to the frequent occurrence of asymptomatic infection of several animal species by this unique viral pathogen (de la Torre, 2002). Limited Bornavirus replication may also explain its relative evolutionary stasis in some hosts, while its rate of evolution and extent of diversification appear to be larger among newly described avian bornaviruses (ABVs) (Philadelpho et al., 2014), with 17-fold higher substitution rates for ABV than the mammal-infecting BDV (He et al., 2014). A requirement of rapid replication is not universal for the different lifestyles exhibited by viruses.

The maintenance of a large population size of viruses in cytolytic or acute infections in vivo is conditioned to the absence of population bottlenecks (strong reductions in population size that

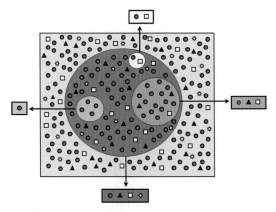

**FIGURE 4.1** A depiction of population bottlenecks of different intensity. The big rectangle represents a viral population with multiple variants distinguished by various symbols. Internal circles denote reductions in population size that result in the isolation of one, two, three, or four variant types (different external rectangles) to initiate the next round of infection. The evolutionary outcome will depend on the number of founder variants that participate in the infection. *The figure is modified from Domingo, E., Sheldon, J., Perales, C., 2012. Viral quasispecies evolution. Microbiol. Mol. Biol. Rev. 76, 159–216, with permission from the American Society for Microbiology, Washington DC, USA.*

can alter the course of selective events, as discussed in Chapter 3) (Fig. 4.1). Bottlenecks have been characterized during viral transmission from an infected into a susceptible host, and also within infected hosts (Foy et al., 2004; Scholle et al., 2004; Quer et al., 2005; Ali et al., 2006; Pfeiffer and Kirkegaard, 2006; Kuss et al., 2008; Smith et al., 2008; Haaland et al., 2009; Bull et al., 2011; Domingo et al., 2012; Gutierrez et al., 2012; Forrester et al., 2014). As emphasized by S. Gutierrez, S. Blanc, and colleagues, a variety of population bottleneck sizes is probably encountered during virus-host interactions (Gutierrez et al., 2012). Aerosols produced by an infected individual can spread a very small proportion of the total number of infectious particles, thus representing strong bottleneck restrictions (estimates were reported in Artenstein and Miller, 1966; Gerone et al., 1966; see also Clarke et al., 1994 and

Chapter 6 for the effect of serial bottlenecks, as assessed in laboratory experiments). Bottlenecks are also intense when an infection is initiated through contact with small volumes of secretions or excretions. Different evolutionary outcomes can be expected when a virus is transmitted through a small or large amount of infected fluids, for example, sharing a syringe versus a transfusion with contaminated blood in the case of HIV-1 or HCV when blood screening for these viruses had not been implemented (Shepard et al., 2005; Sharma and Sherker, 2009; De Cock et al., 2012). Not only the probability of infection is higher when a susceptible individual is exposed to a large amount of contaminated fluid, but a massive amount of initial virus facilitates the adaptation of the quasispecies to the recipient host, by providing multiple initial genomes from which to initiate adaptive walks.

Severe population bottlenecks occur during plant leave inoculation, and seed or aphid transmission of plant viruses, such as cucumber mosaic virus, tobacco mosaic virus, pea seed-borne mosaic virus, and potato virus Y variants, among others (Li and Roossinck, 2004; Ali et al., 2006; Moury et al., 2007; Betancourt et al., 2008; Sacristan et al., 2011; Fabre et al., 2014; reviewed in Roossinck, 2008; Gutierrez et al., 2012). Aphids transmit an average of 0.5-5 virus particles into the recipient plant host, a range of values that is very similar to that estimated for HIV-1, with evidence that in about 75% of HIV-1-infected patients a single founder genome initiated the infection while in the others a minimum of two to five viruses was involved (Keele et al., 2008). In other cases, no severe transmission bottlenecks or major barriers for dissemination seem to operate during intrahost virus expansion (Murcia et al., 2010; Erickson and Pfeiffer, 2013) (Fig. 4.2). Additional studies are needed to elucidate how variations in population size can affect evolutionary outcomes in viruses. Deep sequencing should provide new insights toward the consequence of bottlenecks for the genetic heterogeneity of viral populations

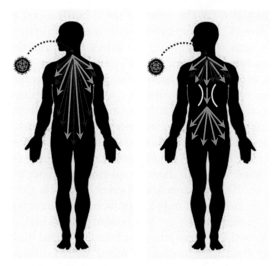

FIGURE 4.2    A simplified scheme of the consequences of a bottleneck event in vivo. On the left, a virus spreads in an infected individual without intervening bottlenecks. Any variant produced (here only two are represented by red and blue arrows) can reach any target organ. On the right, the occurrence of a bottleneck restricts the types of variants (here represented by a blue arrow) that can reach some target organs.

(see Section 4.10 on bottlenecks during the arbovirus life cycle).

Not all viral particles present at a given time in an infected organism are productively replicating. The term effective population size ($N_e$) has been used to indicate the fraction of the total viral population ($N$) that contributes to progeny. $N_e$ in virology was adopted from a concept initially introduced by S. Wright in population genetics to mean the number of breeding individuals, and therefore, the number of the total $N$ that contributes to the next generation. It is considered important to evaluate the variability in a population and the relative participation of selection and random drift in evolution (Charlesworth, 2009). $N_e$ is also relevant for conservation biology since it can predict the rate of inbreeding and the level of genetic diversity in wildlife (Palstra and Ruzzante, 2008).

The application of $N_e$ to viral populations is not easy and has not been without controversy.

Measurements of $N_e$ for HIV-1 have yielded values that range from $10^3$ to $10^6$ (Rouzine et al., 2014). There are at least two reasons to account for the fact that not all viral particles present in an infected individual are replicating to produce infectious progeny at any given time: (i) despite all infectious virus being potentially replication-competent, for stochastic reasons not all of them make their way into a susceptible cell. (ii) A proportion of viral particles that are counted in virus censuses based on quantification of genetic material are intrinsically noninfectious (they could not infect despite having available permissive host cells). The reasons for defectiveness are multiple, encompassing genetic lesions, and virion assembly defects. It is very difficult to estimate the proportion of defective viruses in the different host compartments in which replication takes place. Regarding transmission, it means that when one to four particles are transmitted (external rectangles in Fig. 4.1) some particles may be defective or unfit, thus contributing to possible unproductive infections. Viral $N$ and $N_e$ values may vary locally in different tissues and organs, rendering the interest of the actual $N_e$ value highly dependent on the purpose of the determination (i.e., during viremia to interpret the neutralizing activity of circulating antibodies, or in a specific host compartment to evaluate pathogenic consequences of the infection or the probability of generation of mutants resistant to an inhibitor). The distinction between $N$ and $N_e$ has been attempted with a few virus-host systems (Gutierrez et al., 2012; Rouzine et al., 2014, and references therein). In the cases in which estimates of $N_e$ have been obtained, the values are such that both selection and random drift can influence the evolutionary outcomes. For all these reasons, in the present book, we refer only to virus population size, without attempting a distinction between $N$ and $N_e$. The interpretation of several observations related to differences in population size that will be discussed in Chapters

5 and 6 is not critically dependent on such a distinction, further justifying our simplification, despite acknowledging the relevance of $N_e$ for nonviral systems, and in some particular cases for virus evolution (for example, the population size version that contributes to fitness gain during viral replication; Chapter 5).

Differences in population size and the severity and frequency of bottleneck events determine the contributions of random drift versus selection as evolutionary influences (Chapter 3). In addition, a bottleneck event, irrespective of its perturbing effect in the process of selection, will reduce the diversity of the post-bottleneck population. A few rounds of replication will generally be needed to restore the mutant spectrum amplitude of the initial populations (Fig. 4.3). It has been proposed that one of the driving forces for RNA viruses to maintain as a universal trait high mutation rates is to favor a rapid repertoire of mutants

to ensure adaptability following bottleneck events that they have to undergo as part of their life cycles (Vignuzzi and Andino, 2010).

Despite having population sizes far smaller than viruses, differentiated organisms can also undergo bottleneck events, for example, through geographical isolation of a subset of individuals of an animal species. Such founder events, by which progeny with different alleles from those dominant in a large parental population has a chance to proliferate, can contribute to geographical differentiation of host species, and it is one of the models proposed for the generation of new species (speciation).

## 4.2 Types of constraints and evolutionary *trade-offs* in virus-host interactions

Infection of host organisms by a vast amount of viruses must be necessarily limited to ensure

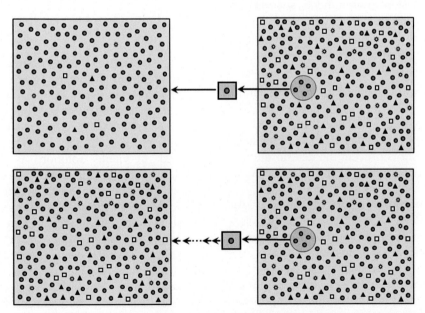

**FIGURE 4.3** A population bottleneck results in the reduction of population complexity, and complexity restoration may require multiple rounds of infection. A severe bottleneck is represented by a single type of virus (blue sphere in the small square). In the upper drawing, a single infection leads to limited relevant phenotypic diversification (although virtually any progeny virus can be a mutant). In the bottom drawing, multiple infective cycles have restored a phenotypic complexity similar to that of the parental population.

the long-term survival of both, viruses and hosts. The term, constraint, is used to refer to multiple mechanisms that prevent generalized and devastating infections by viruses, notably, barriers for virus penetration into organisms, and the host immune response. In Chapter 3, positive and negative selection acting on viruses were discussed in conceptual terms, using some specific examples, but without detailing types of constraints that viruses must face in their hosts. At the cellular level, superinfection exclusion provides a set of mechanisms for an infected cell to prevent the infection by other related viruses. However, virus mutants can be selected to partly overcome the exclusion mechanism (Webster et al., 2013; Campbell et al., 2014).

In addition to immune responses and exclusion mechanisms, viruses are continuously and unavoidably subjected to functional and structural constraints, even when replication occurs in relatively constant environments, such as those provided by established cell lines in culture. Several classes of selective constraints can be distinguished (Morse, 1994; Gibbs et al., 1995; Domingo et al., 2001; Simmonds et al., 2004):

- Basal constraints inherent to the virus, and to the requirements for viral genome replication that operate independently of the environment in which a virus is immersed. Such constraints prevent the deterioration of the "core information" that determines viral identity.
- Selective constraints are intrinsic to individual host cells. They include the availability of membrane structures, nucleotide pool levels, tRNA abundances, ionic composition, accessibility of proteins that can act as receptors for viruses, host factors that have to be incorporated into replication complexes or used for viral protein synthesis, presence of RNA or protein chaperones, etc. A balance of many host cell-dependent influences is

necessary for a virus to complete its replication cycle inside individual cells.
- Selective constraints internal to the host as an organism understood as an integrated aggregate of cells, tissues, and organs. These constraints include immune responses, metabolic alterations that may result in fever or abnormal concentrations of metabolites, increased levels of reactive oxygen species, phenotypic heterogeneity among cells of the tissues where the virus replicates, among other organism-dependent perturbations.
- Selective constraints due to external influences exerted on the host. They relate to human intervention and include administration of antiviral drugs or therapeutic antibodies, vaccination, immunosuppressant treatments, and others.

Next, the origin and features of the different classes of constraints are examined.

## 4.2.1 Long-term history dictates basal constraints

As expressed for the developmental program of an organism (Maynard Smith et al., 1985), present-day viruses can be viewed as the result of a long historical process. From the molecular mechanisms and the building blocks available at the different evolutionary periods, viruses have attained a viable solution as autonomous, cell-dependent, genetic elements. A viral genome was shaped by natural selection acting on phenotypes that were the result of transfers of functional modules, duplication of genomic regions, constellations of mutations acting in concert, and other molecular events that changed the genetic materials. It is not easy to partially dismantle the coordinated set of elements without destroying the replicative capacity of the construct. The requirement of a finely tuned assemblage of elements introduces what we refer to as basal constraints. Their existence at multiple structural and functional levels has become more and more evident as knowledge of the molecular biology of viruses has progressed. This is one of the reasons why synonymous mutations are not necessarily neutral, as justified in Chapter 2, and why manipulation of viral genomes in the laboratory (changes of gene order, construction of chimeric viruses, codon deoptimizations, etc.) generally results in fitness decrease or even lethality.

High-order RNA structures (such as the internal ribosome entry site for the initiation of protein synthesis of some viruses, or pseudoknot structures involved in protein recognition and ribosomal frameshifting) tolerate limited numbers of mutations, and when they occur, compensatory mutations are selected to maintain their folding (Olsthoorn et al., 1994; Pleij, 1994; Escarmís et al., 1995; Arora et al., 1996; Belsham and Sonenberg, 1996; Martinez-Salas et al., 1996). Open-reading frames may include regulatory signals that are independent of the protein-coding function of the same RNA region. Restrictions to the fixation of silent substitutions within open-reading frames may also be imposed by codon usage and translation efficiency (Britten, 1993; Eyre-Walker, 1996) (Section 4.3). The long evolutionary history of viruses (Chapter 1) has derived in the array of basal constraints that we are beginning to characterize at the molecular level.

## 4.2.2 Cell-dependent constraints: No free lunch

Other constraints for viruses are a consequence of the complete dependence of virus replication on the cell, and the need for the virus to be competent for transmission from a donor cell into a recipient cell. Constraints to variation of surface residues in the viral particle are one of the mechanisms proposed to explain antigenic stability of some viruses despite their displaying high mutation rates and frequencies (e.g., measles virus or rabies virus), or the presence of widely different numbers of serotypes among picornavirus genera (Chapter 7). Constraints introduce conflicts regarding the evolutionary change. A viral capsid must confer stability to virions outside the cell, but be labile enough (or capable of a transition toward a labile form) to permit uncoating and release of the viral genome as a result of some intracellular environmental change.

The integrin-recognition site of FMDV illustrates constraints acting on this rapidly evolving virus. Contrary to other picornaviruses, the capsid of FMDV lacks a canyon or pit where residues involved in cell receptor recognition lie. FMDV has a smooth surface with a protruding, mobile loop that has the dual activity of interacting with integrin receptors as a means to enter cells and binding to neutralizing antibodies that impair infectivity (Acharya et al., 1989; Verdaguer et al., 1995; Mateu, 2017). Thus, antibody escape through amino acid substitutions in this loop must be compatible with receptor recognition. This explains the limited repertoire of amino acid substitutions at this multifunctional loop among field isolates and laboratory populations of FMDV (Martínez et al., 1992; Borrego et al., 1993; Mateu, 1995) (Section 4.4).

Other types of conflicting requirements are becoming evident as we learn about the multiple molecular interactions between viruses and cells. They force viruses to evolutionary *trade-offs* by which some nucleotide or amino acid substitutions are introduced to confront a constraint, despite rendering the genome suboptimal for another trait. Suboptimality, however, has a limit, and any *trade-off* must fulfill a balance regarding how a genetic change is positive for a trait and negative for other traits. Most virus functions (replication rate, viral protein synthesis, protein processing, particle assembly, etc.) are unlikely to have been optimized for maximum rate; rather, they are the result of *trade-offs* to fulfill multiple requirements to complete the virus life cycle, and each individual function has a performance level, which is conditioned by the requirements of other viral functions.

In the context of *trade-offs*, the theorem of "no free lunch," amply used in economy and algorithm solution searching, applies. Stated in simple terms, a person or a society cannot obtain "something for nothing." There is always a cost, even if that cost is hidden. A virus will endure a fitness cost as a result of having confronted a selective constraint successfully (as illustrated with specific examples in Chapters 5 and 8), and a universal optimization is highly unlikely. Application of the "no free lunch" theorem to complexity theory is currently under investigation regarding its applicability to biological processes, such as genetic optimization algorithms (Whitley and Watson, 2005; Manning et al., 2013; Buenno et al., 2015; Popovici, 2018). Data of viral RNA and protein functions suggest that due to the extremely compact biological information imprinted into a physically small genome, each nucleotide is exploited for multiple functions, even those nucleotides that do not belong to overlapping genes (alternative open-reading frames with two different proteins encoded in part by the same nucleotide sequence). As a consequence, possible neutral sites ("neutral" meaning that their modification does not entail any functional difference) are probably very rare, as explained in Chapters 2 and 3. Some sites may be referred to as neutral only in the sense that a modification in them still allows the virus to survive. Few nucleotides of an RNA virus genome will conform to the definition of neutral alleles in the sense that "one could be substituted for the other …. without affecting the altered individual's prospect of survival and reproduction under any environmental circumstance" (Reeve et al., 1990). The scarcity of neutral mutations is one of the major reasons for fitness landscapes being extremely rugged for viruses (Chapter 5). Finding *trade-offs* is a way of life for viruses. No free lunch for individuals, societies, or viruses.

The constraints often cited as inherent to the host cells (supply of cellular membranes, components of the translation apparatus exploited by viruses for their own replication machinery, etc.) constitute only a minority of the cellular functions that are, one way or another, involved in virus replication. Application of microarray hybridization analysis to quantitate the alteration of host gene expression during viral infections typically shows upregulation or downregulation of hundreds of host genes. Testing the influence of cellular functions by RNAi (interference) screens often reveals unsuspected effects of host gene products in virus replication, even when appropriate controls to exclude off-target effects of the interfering RNAs are used. Part of the host modifications in gene expression observed in infected cells may be just the indirect consequence of other cellular perturbations, and thus, they may not be essential to the progression of the infection. However, the massive response of collectivities of genes illustrates the multiple connections that a replicating virus establishes with cells and suggests that disarrangements of some cellular functions lead to other (perhaps compensatory) changes in the cells. Viruses are structurally and functionally deeply integrated into the cells in which they replicate, in line with theories of virus origins that favor long-term coexistence of viral elements with precellular and primitive cell organizations (Chapter 1).

## 4.2.3 Constraints in host organisms: contrast with man-made antiviral interventions

Constraints internal to organisms have evolved to display at least three lines of defense against pathogens: intrinsic (preexisting factors that restrict virus replication), innate (activated when a virus enters the organism; i.e., several interferons, apoptosis, NK cells, etc.), and adaptive (immune interferon, T cells, B cells, that expand and evoke a specific cellular and antibody response against the invading virus, among other activities). Long-term coevolution of host organisms and their viruses must have contributed to the survival of both (Woolhouse et al., 2002; Switzer et al., 2005; Villarreal, 2005). Viruses may respond to short-term perturbations by the dominance of subsets of variants that, in addition to increasing fitness in the new environment, must not compromise viral survival in the presence or absence of the perturbation. Many examples can be cited: an amino acid substitution permitting antibody escape should be compatible with virion stability and receptor recognition; a substitution that decreases virus affinity for a soluble cellular receptor should still allow recognition and binding to the cell-anchored receptor for cell entry, or allow an alternative entry pathway, etc. They are changes that must conform to the *trade-off* concept described in Section 4.2.2.

The external influences exerted on host organisms with the aim of limiting viral replication can be divided into two major groups: those that consist of inducing or mimicking an immune response (vaccination or passive immune therapy, respectively), and those that consist of challenging the virus with antiviral inhibitors which are not among the metabolites of their hosts. Viruses do not have an evolutionary history of confrontation with man-made antiviral agents, and nevertheless, viruses can overcome the effect of this class of inhibitors. This reflects that viruses have evolved survival mechanisms of such a general nature (basically the different genetic variation strategies described in Chapter 2) that they constitute a flexible "tool-box" ready for contingencies. One of the major problems in antiviral therapy is the selection of viral mutants resistant to drugs used in therapy. There is virtually no antiviral agent for which no resistance mutations have been described. In fact, the capacity of a drug to select resistant mutants has been traditionally considered a proof of the selectivity of the drug (Herrmann and Herrmann, 1977). Although many studies on inhibitor-resistant mutants have involved HIV-1, HBV, and HCV

infections, all viruses respond to specific antiviral agents by selecting escape mutants, as described in Chapter 8. Remarkably, different mutations in the same viral genes may determine resistance to, or dependence on, an antiviral agent (de la Torre et al., 1990; Baldwin et al., 2004; Baldwin and Berkhout, 2007, and references therein). This adaptive flexibility is obviously of great practical relevance, and it has encouraged the exploration of new antiviral strategies that aim at avoiding selection of virus mutants resistant to antiviral agents (Chapter 9).

There are antiviral inhibitors that establish a connection between external and internal constraints. They target a cellular metabolic pathway whose inhibition results in a stimulation of innate immune response genes. Inhibitors of pyrimidine biosynthesis (for example, A3, DD264, and brequinar) may affect specific replicative steps of some viruses, and they trigger induction of components of the innate immune response. They confront the virus with a broad (multifactorial) antiviral response (Lucas-Hourani et al., 2013; Ortiz-Riano et al., 2014, and references therein). For this reason, this class of inhibitors are broad-spectrum antiviral agents, and the hope is that they may be less prone than standard inhibitors to select resistant mutants. If this expectation is confirmed, this class of inhibitors could be incorporated into new antiviral strategies to control viral quasispecies (Chapter 9).

## 4.3 Codon usage as a selective constraint: virus attenuation through codon and codon-pair deoptimization

Codon use bias in relation to cellular tRNA abundances was listed as one of the mechanisms by which synonymous mutations can affect virus behavior (Section 2.3 in Chapter 2), and the possibility of codon choice is part of the relevant information harbored by the genetic code (Maraia and Iben, 2014). Here we expand the discussion of codon usage as a constraint for viral infections and its implications for long-term virus-host interactions. Several computations have established that the choice among synonymous codons is not random in many biological systems, not only in viruses. Evolutionary events must have resulted in the preferential use of some codons over others, and several possibilities have been proposed. One is that misincorporation tendencies of nucleic acid polymerases may have led to the selection of codons rich in the bases preferentially introduced during genome replication. Long-term priority for some mutation types (because of the catalytic and fidelity properties of cellular or viral polymerases) may decant synonymous codons in favor of those containing the nucleotides that arise more frequently as a consequence of the mutational bias. An alternative, but not mutually exclusive possibility, is that codon bias might have been an evolutionary outcome for optimal RNA secondary structure or RNA-RNA interactions. RNA is rarely a linear unstructured polynucleotide (as usually drawn for simplicity), but rather a complex molecule with a number of high-order structures (stem-loops, pseudoknots, kissing loops, etc.) that play functional roles, and contribute to RNA stability. The appearance of an RNA molecule is certainly closer to that of a protein than to double-stranded DNA (as examples, see Cantara et al., 2014; de Borba et al., 2015 and references therein). Regulatory signals in viral RNA may be located either in untranslated regions or within open-reading frames. In the latter case, preservation of the higher-order structure may restrict the possibility of some triplets to mutate to synonymous ones. This occurrence would constitute an example of negative selection acting on synonymous mutations amply documented during RNA virus evolution (Chapter 2).

tRNAs are among the oldest biological molecules dating back to the time in which the genetic code was developed (Eigen, 1992). In accordance with their ancient nature and with the critical role they play in the transmission of information, tRNAs are extremely conserved among cellular organisms, and their sequences have served, to date, the origin of the genetic code (discussed in Chapter 1). tRNAs must have been extremely restricted with regard to nucleotide sequence changes, due to folding requirements and their role as "adaptor" molecules. Regulation of translation could not be achieved (at least in an effective manner) through tRNA sequence modifications. Instead, regulation could be attained through differences in the abundances of tRNAs that recognize different synonymous codons.

The accommodation of synonymous codon usage to the cellular tRNA pool is known as translational selection. Other mechanisms related to selection of base and dinucleotide frequencies, composition of enhancers of splicing, or translation kinetics have also been proposed as underlying the variation of codon usage (dos Reis et al., 2004; Chamary and Hurst, 2005; Lavner and Kotlar, 2005; Shackelton et al., 2006; Yang and Nielsen, 2008; Aragones et al., 2010). Picornaviruses have been studied regarding synonymous codon usage and the response of a virus when codon frequencies are artificially altered. Introduction of unpreferred synonymous codons in the capsid-coding region of PV resulted in fitness decrease, attributed to alteration of an early step in the virus replication cycle (Burns et al., 2006). The relative fitness of the modified virus, measured in HeLa cells, decreased in proportion to the number of replaced codons. Codon deoptimization resulted in reduced viral RNA yields and decreased specific infectivities of purified virus (Mueller et al., 2006). The specific infectivity is the ratio between infectious and physical particles, an important parameter that is further discussed in Chapter 9 because its decrease is associated with the extinction of viruses by lethal mutagenesis. Not only codon usage, but also codon-pair frequencies can affect PV fitness. It has been suggested that viruses deoptimized for codon pairs may open the way to a new generation of antiviral vaccines

(Coleman et al., 2008), and this approach is currently investigated with other viral systems [(Martrus et al., 2013; Le Nouen et al., 2014; Nogales et al., 2014; Cheng et al., 2015; Li et al., 2018), among other investigations]. These observations reflect once again the multiple ways in which a viral RNA per se can be part of the viral phenotype, independently of its protein-coding function.

There is no clear picture of the molecular mechanisms involved in fitness recovery of codon-deoptimized viruses. Evidence for HIV-1 suggests that phenotypic reversion can come about through true reversion of some of the initial modifications, together with additional mutations (Martrus et al., 2013). Information on the mutational pathways for fitness recovery of these highly debilitated viruses is essential to quantify the stability of the attenuated phenotype of candidate vaccines based on codon modifications.

In contrast to PV, hepatitis A virus (HAV) uses rare synonymous codons (those that correspond to tRNAs that are present at low concentrations in the cells it infects) to control the rate of translation. The adequate combination of common and rare codons allows HAV to regulate ribosome traffic and to slow down the synthesis of capsid proteins to facilitate their proper folding, thus contributing to capsid stability (Sánchez et al., 2003; Aragones et al., 2010; Costafreda et al., 2014). In an elegant study, HAV was replicated in cells treated with actinomycin D (a specific inhibitor of DNA-dependent RNA polymerases), which provided an altered cellular environment in which the tRNA pool available for translation of viral RNA was increased. HAV adaptation to the modified environment resulted in a new deoptimization of codon usage in the capsid-coding region, again supporting translation kinetics selection as the basis for biased codon usage by HAV. Proper protein folding may be essential for nonenveloped viruses that are transmitted via the fecal-oral route to survive for prolonged time periods in the external

environment (Aragones et al., 2010). For viruses whose infectivity is maintained in the external environment, the time-dependent difference between intra and interhost evolutionary rate might be accentuated (discussed in Chapter 7). The differences between PV and HAV regarding fitness effects of codon usage modification were reviewed by Bosch et al. (2010).

In the course of the studies with HAV, evidence of quasispecies memory (explained in Chapter 5) and of selection for fine-tuning translation kinetics acting on the mutant spectrum as a whole was obtained (Aragones et al., 2010). Also, the continuous re-deoptimization of HAV to the new environment to maintain fitness constitutes further support of the "Red Queen" hypothesis (Van Valen, 1973; Krakauer and Jansen, 2002), one of the concepts of population genetics shown to operate with RNA viruses. (Concepts first proposed in general population genetics and then shown to operate in viruses are treated in Chapter 6).

Codon usage also has biotechnological implications. When the codon frequencies in an expression system do not match the codon usage of the viral genomes to be expressed, viral yields may be diminished. This is a relevant factor, which is taken into consideration for the choice of expression systems, including the use of synthetic genes with optimized codon composition (Lanza et al., 2014 and references therein).

### 4.3.1 The synonymous codon space can affect an evolutionary outcome

It is worth expanding on the concept outlined in Chapter 2 that synonymous codons may lie at a different distance from a nonsynonymous codon in sequence space, and this may modify the extent of genetic change needed for adaptation to a new environment (e.g., to reach an amino acid substitution to confer resistance to an antiviral inhibitor). Some mutations have been termed *quasisynonymous* because, despite not leading to an amino acid replacement, they can affect the evolutionary course (Salemi and Vandamme, 2004). As an example, there are six triplets that encode the amino acid R (Arg), and they all have a G in the middle position (AGG, AGA, CGG, CGA, CGC, CGU); out of these, only two (AGG and CGG) are within a one nucleotide distance from W (Trp) (UGG); transversion A → U is required to change the triplet AGG (R) to UGG (W), and transition C → U is required to change CGG (R) to UGG (W). The only triplet encoding W (UGG) is within a single nucleotide distance from two termination codons (UGA and UAG) (Fig. 4.4). Thus, a difference in the R codons that a virus uses is not a neutral trait since two of them are within a one nucleotide distance of the codon for W, which in turn is at one nucleotide distance of two termination (stop) codons. The latter may be reached as a result of increased mutational pressure, either due to a decrease of polymerase fidelity or to antiviral lethal mutagenesis treatments (Chapter 9). A strategy to attenuate virus consists in engineering triplets at one mutation distance from a stop codon, or expressed more generally, in transferring genomes to unfavorable regions of sequence space (Moratorio et al., 2017; Moratorio and Vignuzzi, 2018).

In several viral populations that have been examined by deep sequencing, a remarkably high frequency of termination codons is observed, denoting the presence of defective genomes that represent either dead-end evolutionary pathways or genomes that are maintained by complementation during quasispecies replication (Rodriguez-Frías et al., 2012). The frequency of termination codons may increase as a result of treatment with antiviral agents and may decrease the fitness of the quasispecies ensemble.

The studies and concepts summarized in this section bear on the increasing realization of the potential biological impact of synonymous codon modifications (Hunt et al., 2014), and how movements in a neutral sequence space can actually affect the phenotypic space (Schuster, 2011). Position in sequence space is

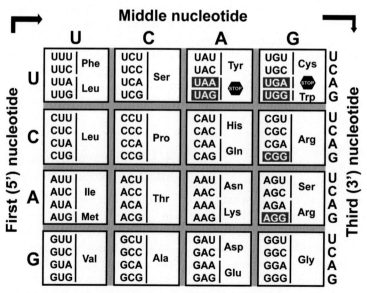

FIGURE 4.4    The genetic code with an indication of triplets and encoded amino acids. The triplets in red are those discussed in the text in connection with the phenotypic nonequivalence of synonymous codons, regarding the proximity to STOP (protein synthesis terminating) codons.

pertinent to the viral dynamics in the infected hosts.

## 4.4 Modifications of host cell tropism and host range

Any step in the replication cycle can determine the permissivity of a cell type to sustain the replication of a virus. However, recognition by viruses of cellular receptors is a key determinant of cell tropism. Macromolecules that act as viral receptors are diverse, and they include cell adhesion and cell-to-cell contact proteins, extracellular matrix components, sugar and lipid derivatives, chemokine and G-protein-coupled receptors, growth factor receptors, complement control protein superfamily, low- and high-density lipoprotein receptor, tumor necrosis factor-related proteins, and transporter proteins, among others (as review, see Baranowski et al., 2003; Bhella, 2015). The expression of a virus receptor is a necessary but not sufficient condition for the virus to infect a cell. The PV receptor (PVR or CD155) belongs to the immunoglobulin-like superfamily of cell adhesion molecules that constitute an important group of the cellular receptor for viruses (Mendelsohn et al., 1989; Bhella, 2015). Many tissues that express PVR are not infected by PV in vivo. Likewise, the sialic acids that act as IV receptors are common in cell surfaces, and the virus attaches to them, but the productive infection is highly restricted to the epithelial cells of the respiratory tract (Bergelson, 2010).

About $10^3$ to $10^4$ receptor molecules can be present per cell (Lonberg-Holm et al., 1976; Consigli et al., 1986; Thulke et al., 2006). Although the MOI during in vivo infections is difficult to estimate, it is unlikely that the number of receptor molecules represents a limitation to an infection. Bottlenecks are more probably due to restrictions of the number of infectious viruses that reaches a target tissue or organ than to an insufficient number of receptor molecules on the cell surface. The interaction of a virus with one or several receptors (or a receptor and a coreceptor) will generally allow virus entry, which is a multistep process that involves changes in virion structure, a succession of low- and high-affinity binding to one or more cellular proteins, and membrane fusion in the case of enveloped viruses (Verdaguer et al., 2014; Strauss et al., 2015).

Features of virus-receptor interactions relevant to viral evolution are summarized in Box 4.1, based on concepts reviewed in Baranowski et al. (2003). Of note is that differentiated organisms do no express the same set of cell surface macromolecules in different tissues and organs. This implies not only compartmentalization of viral infections but also the possibility of selection of viral subpopulations in specific host compartments, a translation in vivo of one of the tenets of quasispecies dynamics. A significant example is the cyclical adaptation of measles virus quasispecies to epithelial and lymphocytic cells (Donohue et al., 2019).

Members of the same virus family, or viruses associated with related disease manifestations, may use different receptors. Among other examples, consider the diversity among viruses that can cause brain or liver disease. Since tropism changes are one of the biologically most relevant consequences of viruses existing as mutant spectra, several studies have been listed in Table 4.1 to emphasize that tropism changes are not exceptional; the list is by no means exhaustive. Some residues of the IV hemagglutinin (HA) dictate the specificity for sialic acid linked to galactose by either an α-2,3 or an α-2,6 linkage, and linkage preference is a determinant of host specificity (a preference for α-2,6 linkage by human IVs, and for α-2,3 linkage by avian IVs) (Skehel and Wiley, 2000; Parrish and Kawaoka, 2005; Yang et al., 2015). In lymphocytic choriomeningitis virus (LCMV), high-affinity binding to its receptor α-dystroglycan is associated with immunosuppression and viral persistence in mice, whereas low-affinity binding results in

---

**BOX 4.1**

### Facts related to receptor usage by viruses

- Different compartments within an organism do not express the same surface macromolecules that can act as viral receptors.
- A virus may use different receptors and coreceptors.
- A receptor type can be shared by different viruses and other microbial pathogens.
- A phylogenetic position or biological features of a virus do not predict the use of some

receptor types. Members of the same virus family or viruses that are associated with similar disease manifestations may use different receptors.

- One or a few amino acid substitutions in capsid or surface proteins may modify receptor recognition, with consequences for viral pathogenesis.

---

**TABLE 4.1**   Examples of one or few amino acid substitutions that can modify virus cell tropism.

| Observation | References |
| --- | --- |
| Host restriction of avian polyomavirus budgerigar fledgling disease virus | Stoll et al. (1994) |
| Colonic tropism and persistence of murine norovirus | Nice et al. (2013) |
| N versus B tropism of murine leukemia virus | Jung and Kozak (2000) |
| Tissue tropism of adeno-associated viruses | Wu et al. (2006) |
| Cellular tropism of feline immunodeficiency virus | Verschoor et al. (1995), Vahlenkamp et al. (1997), Lerner and Elder (2000) |
| Loss of enteric tropism of transmissible gastroenteritis coronavirus by two amino acid substitutions in the spike protein | Ballesteros et al. (1997) |
| HIV-1 tropism by a single amino acid substitution in gp120 | Takeuchi et al. (1991), Boyd et al. (1993) |
| SARS coronavirus recognition of ACE2 receptor (see also text) | Li et al. (2005) |
| Receptor preferences of herpes simplex virus | Spear et al. (2000) |
| Hemagglutinin residues in influenza virus tropism (see also text) | Rogers et al. (1983), Connor et al. (1994) |
| Substitutions in poliovirus capsid expand receptor recognition | Colston and Racaniello (1995) |
| Conversion of encephalomyocarditis D into a diabetogenic variant through altered cell tropism | Bae and Yoon (1993) |
| High-affinity binding of measles virus to CD46 | Hsu et al. (1998) |
| Decreased neurovirulence of Sindbis virus through impaired receptor recognition in neural cells | Tucker and Griffin (1991), Lee et al. (2002) |
| Parvovirus host range (see also text and Fig. 4.5) | Hueffer et al. (2003) |
| Change in receptor recognition by foot-and-mouth disease virus in vivo and in cell culture (see also text and Fig. 4.6) | Baranowski et al. (2000, 2001), Ruiz-Jarabo et al. (2004) |

clearance of infection (Sevilla et al., 2000, 2002; Smelt et al., 2001).

One or few amino acid substitutions have been associated with changes in cell tropism and host range of parvoviruses (Hueffer and Parrish, 2003; Parrish and Kawaoka, 2005; Shackelton et al., 2005; Lopez-Bueno et al., 2006) (Fig. 4.5), adenoviruses (Huang et al., 1999), herpes simplex viruses (Spear et al., 2000), and lentiviruses. HIV-1 coreceptor usage varies in the course of infection in humans. Most primary HIV-1 isolates belong to the R5 receptor specificity (they use coreceptor CCR5), and as the infection progresses, dualtropic (R5X4) and X4 variants (that use coreceptor CXCR4) arise and often become dominant. The expansion by HIV-1 of coreceptor usage to include CXCR4 is associated with loss of $CD4^+T$ cells, and to progression to AIDS (Connor et al., 1997). The R5 to X4 transition in coreceptor usage constitutes an example of tropism change with a direct impact in viral pathogenesis, a change that is observed in about half of HIV-1-infected patients. Multiple, additional coreceptors can be used by HIV and simian immunodeficiency virus (SIV) variants, and transitions in receptor usage vary with virus types and subtypes. Two amino acids in the spike (S) protein of the SARS coronavirus (SARS CoV) modulate the binding to either human or palm civet angiotensin-converting enzyme 2 (ACE 2), a functional receptor for the virus (Li et al., 2005).

It must be clarified that the examples in Table 4.1 and those commented in the text are centered on the effects of a few amino acid substitutions on cell tropism. Some of these substitutions and additional ones may affect virus pathogenesis, which is often (but not always) related to tropism alterations. Moreover, other genetic lesions, such as insertions or deletions (not only point mutations) may also affect receptor recognition and pathogenic potential.

The selective forces that trigger a modification of cell tropism are not easy to identify although several observations suggest that the availability of the type of receptor may select a variant

subpopulation out of a mutant distribution. HIV-1 variants enter $CD8^+$ cells at late disease stages (Saha et al., 2001). In a mouse model, a modified form of RANTES, a natural ligand for CCR5, selected HIV-1 mutants that used CXCR4 as a coreceptor (Mosier et al., 1999) (also discussed in Chapter 9 in connection with antiviral agents directed to cellular targets). The bicyclam AMD3100, a selective antagonist of CXCR4, led to the suppression of X4 variants in cell culture and prevented the switch from R5 HIV-1 to X4 HIV-1 (Este et al., 1999). Group B coxsackieviruses (CVB) use the coxsackievirus and adenovirus receptor to infect cells (Bergelson, 2010). When a CVB was passaged in a cell line expressing a limited amount of CAR, the virus expanded its cell receptor specificity to bind multiple molecules including CAR and decay-accelerating factor (DAF). The modification involved a limited number of amino acid substitutions in the viral capsid (Carson et al., 2011).

Multiplication of a virus in a given cell line (or primary culture) may select virus subpopulations present as a minority in a biological sample from a naturally infected host. Passage of biological clone of FMDV in BHK-21 cells resulted in the dispensability of the RGD (integrin recognition Arg-Gly-Asp sequence) and expansion of cell tropism (Ruiz-Jarabo et al., 2004) (Fig. 4.6). Thus, paradoxically, repeated replication in a cell line may lead to the relaxation of virus specialization for that cell line. Application of deep sequencing to describe mutant spectra in vivo should help in establishing whether low-frequency mutants with altered receptor recognition sites might be present in evolving quasispecies. Their basal frequency will depend on their fitness relative to other components of the mutant spectrum. Here again, the relevance of virus population numbers is evident; the larger the population size, the higher the probability that variants with altered cell recognition are present as minority components of the mutant cloud.

**FIGURE 4.5** Amino acid substitutions found in the capsid protein of the prototypic strain of the parvovirus minute virus of mice (MVMp) upon the passage in immunodeficient mice. At the top, genomic residue numbers and amino acid substitutions (single letter code) found in virus from several organs are listed. At the bottom, a more detailed list of substitutions found in many individual clones is given. The numbers (*n*) indicate the number of clones analyzed, and the letters the organ from which the clones were isolated (B, brain; K, kidney; L, liver). *The figure is modified from Lopez-Bueno, A., Rubio, M.P., Bryant, N., McKenna, R., Agbandje-McKenna, M., et al., 2006. Host-selected amino acid changes at the sialic acid binding pocket of the parvovirus capsid modulate cell binding affinity and determine virulence. J. Virol. 80, 1563–1573, with permission from the American Society for Microbiology, Washington DC, USA.*

**FIGURE 4.6** Expansion of host cell tropism. FMDV biological clone C—S8c1 (filled square at the top) was passaged in BHK-21 cell at an MOI of 1—5 PFU per cell (open circles at the top). The resulting population included amino acid substitutions in the viral capsid and acquired the capacity to infect additional cell lines (large arrows below the capsid structures). *Scheme based on the results reported in Ruiz-Jarabo, C.M., Pariente, N., Baranowski, E., Dávila, M., Gómez-Mariano, G., et al., 2004. Expansion of host-cell tropism of foot-and-mouth disease virus despite replication in a constant environment. J. Gen. Virol. 85, 2289—2297.*

| FMDV | Lesion score | Amino acid VP1 (147) | 2C (248) | 3A (44) |
|---|---|---|---|---|
| C-S8c1 | | L | I | Q |
| V 2.1 | | L | I/T(10%) | Q |
| V 2.2 | + | L | T | Q/R (40%) |
| V 2.3 | + | L | T | R |
| V 2.4a | + | L/P(40%) | T | R |
| V 2.4b | ++ | L/P(20%) | T | R |
| V 2.5a | + | P | T | R |
| V 2.5b | + | L/P(50%) | T | R |
| V 2.6a | ++ | P | T | R |
| V 2.6a | + | P | T | R |
| V 2.6b | + | P | T | R |
| V 2.7b | ++ | P | T | R |
| V 2.7b | ++ | P | T | R |
| V 2.10b | ++ | P | T | R |

**FIGURE 4.7** Amino acid substitutions in proteins VP1, 2C, and 3A in the course of adaptation of swine FMDV C—S8c1 to the guinea pig. Sequential virus isolates from different animals are represented in the first column. The lesion score indicates the presence of a vesicle at the inoculation site (+) or the presence of vesicles at the inoculation site and other sites (secondary lesions) (++). The amino acid substitutions (single letter code) in VP1, 2C, and 3A (amino acid position given for each protein) are indicated in the three columns on the right. Results are based on consensus sequences; the presence of amino acid mixtures (percentage in parenthesis) is based on the peaks of the sequencing experiments. *The figure has been adapted from Núñez, J.I., Baranowski, E., Molina, N., Ruiz-Jarabo, C.M., Sánchez, C., et al., 2001. A single amino acid substitution in nonstructural protein 3A can mediate adaptation of foot-and-mouth disease virus to the Guinea pig. J. Virol. 75, 3977—3983, with permission from the American Society for Microbiology, Washington DC, USA.*

## 4.4.1 Nonstructural viral proteins and RNA in cell tropism and host range of viruses

The capacity of a virus to penetrate a new cell type has consequences for its host range, and the potential emergence of new viral pathogens (Chapter 7). Nonstructural proteins (those that are not present in the viral particles) and regulatory regions in viral genomes may also mediate cellular tropism and host range changes in viruses. Deletions and point mutations in the nonstructural protein 3A of FMDV have been associated with attenuation for cattle (reviewed in Baranowski et al., 2003). A single amino acid substitution in 3A was critical for the adaptation of a swine FMDV to guinea pig (Núñez et al., 2001) (Fig. 4.7). Despite the conservation of the LTR of a macrophage-tropic strain of the lentivirus equine infectious anemia virus (EIAV)

in vivo, the stepwise generation of a new transcription factor-binding motif within the enhancer element was associated with the expansion of tropism to endothelial cells and fibroblasts (Maury et al., 2005). A comparison of the minimal promoter-enhancer element of HIV-1 of clades B, C, and E, engineered into the same SIV genetic background, indicated that this element could modulate viral replication in different cell subsets in vivo (Centlivre et al., 2005; Sala et al., 2006). Polymorphism in the binding sites of transcription factors is probably involved in differential expression of HIV-1 clades in different tissues. The polymerase genes of IV may influence host range, and probably, gene constellations affect the relative replication capacity of viruses in different hosts (Parrish and Kawaoka, 2005).

The dominance of viral subpopulations in vivo can be due to cells offering higher permissivity to variant forms of a virus, or depletion of cells due to cytolytic infection, and survival of other cell types permissive to other viral variants (Centlivre et al., 2005; Sala et al., 2006). Cell dynamics can exert an important influence on short-term virus evolution and pathogenesis (See Chapter 6 for a description of the dependence on host cell variation for the initiation of persistent infection of FMDV in cell culture). Not only nonstructural proteins but modifications in viral RNA may also contribute to host-specific viral fitness. RNA structural elements and mutations within the elements may influence the adaptation of arboviruses to alternative hosts (Ventoso, 2012; Villordo et al., 2015; Filomatori et al., 2017).

From an evolutionary perspective, the capacity of viruses to use alternative receptors and to modify receptor specificity or intracellular preferences by modest genetic change—involving short distances in the genotypic sequence space (Chapter 3)—may manifest a necessity of viruses to parasitize increasingly differentiated organisms. Any virus whose capacity to expand cell tropism was limited by genetic constraints (e.g., the need to move to long distances in sequence space, incompatible with viral population numbers) would have a lower probability of long-term survival in an increasingly differentiated cellular world. A large virus population size is the key to provide a sufficient number of variants with new potential cell recognition specificities. That is, the application of the Darwinian principles to the long-term evolution of viruses is expected to have produced a flexible virus-host cell interaction in the sense that it may be modified by a limited genetic change in the virus.

## 4.5 Trait coevolution: mutual influences between antigenic variation and tropism change

Coevolution between organisms and their pathogenic agents means that reciprocal and adaptive genetic modifications have occurred in them because of their interaction as biological systems that have shared space and resources over prolonged time periods. Coevolution is a general concept that applies not only to host-parasite relationships but also to other interacting biological species that have shaped biological systems up to the present epoch (Futuyma and Slatkin, 1983; Woolhouse et al., 2002; Gomez et al., 2015). Long-term coevolution has probably led to increasingly subtle molecular mechanisms to deal with the host immune response, based on virus-coded proteins that interact with host proteins. In addition, mutual evolutionary influences may be exerted among different sites within a virus. Such influences are favored by overlaps between domains involved in distinct functions in viral nucleic acids or proteins that may also introduce additional evolutionary constraints. Coevolution of amino acids at or around important functional domains may contribute to functional stability (Gloor et al., 2005). Homologous structure analyses can be applied to identify functional residues in proteins (Du et al., 2016). Intramolecular coevolution can be measured for amino acid pairs by the probability of their occurrence at some defined positions, also termed mutual information criterion (MIC) (Korber et al., 1993).

The three-dimensional structure of several viruses has evidenced that frequently, there is an overlap between antibody and receptor recognition sites (reviewed in Baranowski et al., 2003) (Tables 4.2 and 4.3). As mentioned in Section 4.2.2, the FMDV capsid includes an Arg-Gly-Asp (RGD) triplet at an exposed mobile loop in protein VP1 (Acharya et al., 1989; Fry and

Stuart, 2010) which is involved in the binding of neutralizing antibodies and in recognition of integrin receptors (Verdaguer et al., 1995). Functional alterations in FMDV can be interpreted as a consequence of the overlap between a major antigenic site and the integrin recognition domain: variants selected with neutralizing monoclonal antibodies displayed altered integrin recognition (Martinez et al., 1997; Baranowski et al., 2000; Ruiz-Jarabo et al., 2004). Adaptation of FMDV to cell culture may result in antigenic variation and in the use of heparan sulfate as a molecule that facilities virus entry; some of the residues involved in heparin-binding map at antigenic sites (Curry et al., 1996; Sa-Carvalho et al., 1997; Fry et al., 1999; Baranowski et al., 2000). Cattle that were partially immunized with synthetic peptides, representing the VP1 loop sequence, selected FMDV mutants with substitutions within the RGD or at neighboring sites. The mutants showed altered host cell tropism (Taboga et al., 1997; Tami et al., 2003) (see also Chapter 8).

Studies with IVs have shown several consequences of the close connection between antigenic and receptor interaction sites. HA antigenic variants were selected upon egg adaptation of the virus (Robertson et al., 1987). Treatment with antibodies resulted in the selection of variants with altered receptor binding (Laeeq et al., 1997). The hemagglutinating activity of type C IV can be modulated by amino acid residues involved in antibody binding (Matsuzaki et al., 1992). Passage of this virus in MHV-II cells resulted in antigenic variants that displayed an advantage in receptor binding (Umetsu et al., 1992). The receptor-binding specificity of IV can modify the antigenic profile of the virus as analyzed by the reactivity of monoclonal antibodies with the HA (Yamada et al., 1984).

**TABLE 4.2**    Examples of overlap between antigenic sites and receptor recognition sites in DNA viruses.

| Virus | Observation | References |
|---|---|---|
| Adenovirus-3 | Fiber knob includes receptor-binding and antigenic sites | Liebermann et al. (1998) |
| Adeno-associated virus (serotypes 1 and 5) | Virus-antibody complex structures show epitopes at receptor recognition sites | Tseng et al. (2015) |
| Bovine herpesvirus-1 | Antiidiotypic antibodies bind to cellular receptors | Thaker et al. (1994), Varthakavi and Minocha (1996) |
| Human cytomegalovirus | Antiidiotypic antibodies bind to cellular receptors | Keay et al. (1989), Keay and Baldwin (1991) |
| Herpes simplex virus | Antiidiotypic antibodies bind to cellular receptors | Huang and Campadelli-Fiume (1996) |
| | Overlap between a receptor-binding domain in gD and an antigenic site | Whitbeck et al. (1999) |
| Hepatitis B virus | Antiidiotypic antibodies that mimic cellular structures bind to small HBV surface antigen. Synthetic peptide analog is recognized by antiHBV antibodies and cell receptors | Neurath et al. (1986) |
| | Antiidiotypic antibodies bind to cellular receptors | Petit et al. (1992), Hertogs et al. (1994), Budkowska et al. (1995) |
| Duck hepatitis B virus | Residues involved in the interaction with cells are also critical for virus neutralization | Tong et al. (1995), Li et al. (1996), Sunyach et al. (1999) |

**TABLE 4.3** Examples of overlap between antigenic sites and receptor recognition sites in RNA viruses.

| Virus | Observation | References |
|---|---|---|
| Poliovirus | Receptor recognition influenced by residues of antigenic sites | Murray et al. (1988), Harber et al. (1995) |
| | Critical role of VP1 BC loop in receptor interaction | Yeates et al. (1991) |
| Human rhinovirus | Neutralizing antibody to HRV14 penetrates the receptor-binding canyon | Smith et al. (1996) |
| | Exposed VP1 BC- and HI-loops covered by footprint of very low-density lipoprotein receptor | Hewat et al. (2000) |
| Theiler's encephalomyelitis virus | Neutralizing antibodies map close to putative receptor-binding site | Sato et al. (1996) |
| | Substitutions of adaptation to some cells map in antigenic sites | Jnaoui and Michiels (1998) |
| Foot-and-mouth disease virus | Overlap of integrin- and antibody-binding sites (Additional studies presented in the text) | Verdaguer et al. (1995) |
| Human influenza virus | Amino acid residues of the sialic acid-binding pocket are accessible to neutralizing antibodies (Additional studies presented in the text) | Stewart and Nemerow (1997) |
| Newcastle disease virus | Monoclonal antibodies to HN glycoprotein prevent virus attachment | Iorio et al. (1989) |
| Rabies virus | Antiidiotypic antibodies bind to cellular receptors | Hanham et al. (1993) |
| | Residues critical for neurotropism are involved in antibody binding | Coulon et al. (1998) |
| Bovine viral diarrhea virus | Antiidiotypic antibodies bind to cellular receptors | Xue and Minocha (1993), Minocha et al. (1997) |
| Dengue virus | Residues critical for mouse neurovirulence are involved in antibody binding | Hiramatsu et al. (1996) |
| Yellow fever virus | Residues critical for neurotropism are involved in antibody binding | Jennings et al. (1994) |
| Murine coronavirus | Overlap between epitopes and receptor-binding sites | Kubo et al. (1993, 1994) |
| Middle East respiratory syndrome (MERS) coronavirus | Antibodies bind to receptor recognition site | Ying et al. (2014) |
| Sindbis virus | Antiidiotypic antibodies bind to cellular receptors | Ubol and Griffin (1991), Wang et al. (1991), Strauss et al. (1994) |
| Ross River virus | Binding of antibodies to cell-receptor recognition regions | Smith et al. (1995) |
| Reovirus | Antiidiotypic antibodies bind to cellular receptors | Co et al. (1985), Gaulton et al. (1985), Williams et al. (1988, 1989, 1991) |
| Bluetongue virus | Antiidiotypic antibodies bind to cellular receptors | Xu et al. (1997) |

The mutual influence between antigenic variation and receptor site modifications has been extended to recent IV isolates (Koel et al., 2013, 2014). These summarized accounts for FMDV and IV, in addition to the studies listed in Tables 4.2 and 4.3 provide incontestable evidence that antigenic changes, which are extremely frequent because of the continuous confrontation of viruses with antibodies, may contribute to the modification of cell-binding preferences.

Two facts facilitate antigenicity-cell tropism coevolution: (i) the limited number of amino acid substitutions, and hence of mutations, that are needed for an antigenic change (one of the parameters that render the quasispecies nature of viruses biologically relevant; see Box 3.3 and text in Chapter 3). (ii) surface residues tend to be less constrained structurally than internal residues in viral capsids or envelopes. Their tolerance to accept amino acid substitutions is one of the mechanisms of antigenic variation in the absence of immune selection (Section 4.7). Surface residues appear as the most variable when all the historically recorded substitutions are depicted on the three-dimensional structures of viral particles. In the case of FMDV high variability of surface amino acids has been corroborated with several virion structures determined by D. Stuart, E. Fry, D. Rowlands and their colleagues (one of the comparisons between two isolates of a subtype and an antigenic variant is described in Lea et al., 1995). Conformational antigenic sites, similarly to internal capsid residues, are involved in interactions, which may be necessary for virion stability, and therefore, their tolerance of amino acid substitutions is more restricted than in linear, continuous epitopes. This is the case with discontinuous epitopes that have been characterized within antigenic site $D_2$ of FMDV. The conformation of this site in FMDV $O_1BFS$ and $C_1$ is conserved despite different primary sequences. The substitutions in antibody-escape mutants of $D_2$ map in amino acids which are not involved in interactions with surrounding residues. Significantly, the only substitution found in a residue involved in hydrogen bonding led to an amino acid that maintained the hydrogen bond with the same neighbor amino acid, according to the modeling of the change based on crystallographic data (Lea et al., 1994). Epitope plasticity may allow antibody escape while maintaining the interaction with cellular ligands (Lindesmith et al., 2019). It is expected that disordered protein regions free of structural constraints can tolerate amino acid substitutions. In some cases, surprisingly, even a viral polymerase may contain domains that can be extensively changed and remain functional (Gitlin et al., 2014).

## 4.6 Escape from antibody and cytotoxic T cell responses in viral persistence: fitness cost

Viruses use two major strategies to cope with the host response: modulation and escape. Modulation consists of the expression of viral gene products that by any mechanism can alter the components of the immune response. The consequence is to facilitate virus survival to increase the probability of transmission, or virus persistence in the infected organism. The escape strategy is based on the exploitation of mutations in the viral genome that confer the virus increased resistance to inactivation by components of the immune response. Amino acid substitutions may prevent neutralization of viral particles by antibodies or elimination of infected cells by specific cytotoxic T cells. Both modulation and escape strategies are exploited by DNA and RNA viruses, although complex DNA viruses encode several proteins whose primary function is to interfere with the host defense mechanisms. They include homologues of cytokines, chemokines, viral proteins that act as a decoy for antiviral antibodies, proteins that block complement activation, that suppress MHC class I and II molecules, that interfere with ubiquitin-dependent proteolysis, that induce or inhibit apoptosis, among other proteins and activities (Alcami,

2003; Seet et al., 2003; Rustagi and Gale, 2014). Viral proteins block interferon induction (i.e., influenza virus NS1, Ebola virus P35, and others) (Basler and Garcia-Sastre, 2002; Katze et al., 2002; Weber et al., 2004). Some of the proteins involved provide remarkable examples of protein multi-functionality. The leader L proteinase of FMDV catalyzes its own cleavage from the polyprotein, cleaves the host cell translation factor eIF4G—leading to the shutoff of host cell translation dependent on capped mRNAs—and inhibits IFN induction in the infected cells (de Los Santos et al., 2006). HIV-1 Nef contributes to HIV-1 evasion of immune surveillance through interaction with membrane traffic regulators (Pawlak and Dikeakos, 2015). Several additional multi-functional viral proteins one of whose roles has an impact on the host antiviral response have been described.

Despite expressing proteins that interfere with the immune response, RNA viruses exploit evasion through genetic variation as a major means to cope with host defenses. This is probably an evolutionary coadaptation of high mutability and genome compactness. Genomic compression is evidenced by overlapping reading frames, ambisense RNA genomes, RNA editing, partial read-through of termination codons, overlap between regulatory and protein-coding regions, leaky ribosome scanning with initiation of protein synthesis at two in-frame AUGs, ribosome frameshifting, hopping, shunting and bypassing, synthesis of polyproteins whose partial or complete processing leads to several functional proteins, etc (Domingo et al., 2001; Alberts et al., 2002). Evolution has offered high mutation rates, small genome size, and escape pathways as an alternative to the modulation strategy.

Mutations that mediate escape from neutralizing antibodies and from CTLs can be readily observed in vivo (Weiner et al., 1995; Borrow et al., 1997; McMichael and Phillips, 1997; Ciurea et al., 2000; Sevilla et al., 2002). Rather than being a secondary phenomenon in the course of viral infections, the generation of antibody- and CTL

escape mutants may indeed contribute to viral persistence (Gebauer et al., 1988; Ciurea et al., 2000; Richman et al., 2003; Domingo, 2006). Evasion of an immune response, as the outcome of a selection event in viruses, may entail a fitness cost. Such a cost may bring about reversion to the initial sequences when selective forces (antibodies or CTLs) are no longer present (Borrow et al., 1997). Viruses often display multiple antibody-escape routes, and the preferred pathway may be imposed by the number and concentration of the antibodies (Borrego et al., 1993; Keck et al., 2014).

There is an important difference between the selective constraint imposed by antiviral drugs and by the host immune response. While drugs inhibit a specific step of the virus replication cycle (or two or more steps if two or multiple drugs are administered simultaneously), the immune response gives rise to multiple constraints that act upon the virus. As an example, 100 HCV passages in cell culture in the presence of IFN-α were necessary to select HCV mutants displaying significant resistance to IFN-α (Perales et al., 2013). The resistant HCV exhibited higher fitness than the populations passaged in the absence of IFN-α when fitness was measured in the presence of IFN-α, but not in its absence. Sequence analysis documented that amino acid substitutions that contributed to resistance were present in most viral proteins, and many substitutions differed among parallel viral lineages (Perales et al., 2013). It is expected that viruses find higher genetic and phenotypic barriers for responding to multicomponent antiviral responses than to a single inhibitor with a defined target (Perales et al., 2014). A high genetic barrier is due to the requirement of multiple mutations, while a high phenotypic barrier reflects the fitness cost inflicted by the mutations once they have been selected. The fact that IFN-α resistance requires several amino acid substitutions implies a high barrier to resistance since each of the substitutions may affect a multifunctional protein. The observations with HCV in cell culture agree with multiple possible IFN-α-resistance mutations identified in clinical practice, thus providing an interpretation of the absence of a catalog of IFN-α-resistance mutations that are available for other HCV inhibitors [reviewed in (Perales et al., 2014; Perales, 2018)].

It is an open question whether the concept of high fitness cost to overcome an interferon response can be extended to other systems. I. S. Novella, J.J. Holland, and colleagues showed that passage of VSV in IFN-treated cells selected only variants of limited IFN resistance (Novella et al., 1996). However, field isolates of VSV appear to contain clones with different capacity to resist or to induce IFN, to the point that interferon induction was used as a quasispecies marker for the virus (Marcus et al., 1998).

Quantification of fitness cost is described in Chapters 5 and 8, and the use of broad-spectrum antiviral agents that promote a multicomponent antiviral state is discussed in Chapter 9.

## 4.7 Antigenic variation in the absence of immune selection

Several cases of antigenic variation of DNA and RNA viruses in the absence of immune selection have been described (Domingo et al., 1993, 2001, and references therein). Two possible underlying mechanisms have been proposed: (i) tolerance of antigenic sites to accept amino acid replacements by virtue of being relatively free of structural constraints (Section 4.5); fluctuations of mutant distributions (through the selection of an unrelated trait or through random drift) that may raise antigenic variants to dominance. (ii) Not mutually exclusive with the previous mechanism, selective forces other than an immune response may result in amino acid replacements at antigenic sites (as in coevolution of receptor recognition specificity and antigenicity, discussed in Section 4.5). Antigenic variation may follow from the hitchhiking of mutations that encode amino acid substitutions at antigenic sites, following selection sweeps (Chapter 3) (Fig. 4.8). Therefore, a consequence of quasispecies dynamics is that amino acid substitutions observed at the antigenic sites of viruses are not necessarily the result of immune selection.

## 4.8 Constraints as a demand on mutation rate levels

The nature of the selective constraint influences the requirements—in terms of mutation rate and complexity of the mutant spectrum—to be fulfilled by a virus to overcome that specific selective constraint. For a virus to lose sensitivity to an antiviral inhibitor directed to a viral protein, when such a loss depends on a single transition mutation (low genetic barrier), the standard mutation rate, or even a mutation rate lower than the standard, may suffice to generate a resistant mutant. In contrast, when a virus has to adapt to a complex environment, multiple mutations located at different genomic sites might be required for effective replication in the new environment.

The occurrence of multiple mutations in the same viral genome will be favored by low-fidelity polymerases because they generate mutant spectra characterized by a larger average number of mutations per genome (higher mutation frequency and nucleotide diversity) than an enzyme with standard copying fidelity. We refer to "broad" or "narrow" mutant spectra to indicate whether they include many types of variants or a restricted repertoire of variants, respectively (Fig. 4.9). A selective disadvantage due to a narrow mutant spectrum was elegantly documented with a high-fidelity PV mutant and its inability to cause neurological disease using a mouse model (Pfeiffer and Kirkegaard, 2005; Vignuzzi et al., 2006) (see also Chapter 3). Examples of complex environments for a virus—those that often require a broad mutant spectrum for a virus to confront them successfully—are the invasion of

**FIGURE 4.8** Antigenic variation in the absence of immune selection. The genomic region encoding an antigenic site is delimited by two discontinuous vertical lines. At the top, extensive replication of the virus leads to the accumulation of amino acid substitutions at the antigenic site because of its structural flexibility. At the bottom, a genetic bottleneck or a selection event results in the fixation (hitchhiking) of two amino acid substitutions (green triangle and asterisk) at the antigenic site. *The figure has been modified from Domingo, E., Díez, J., Martínez, M.A., Hernández, J., Holguín, A., et al., 1993. New observations on antigenic diversification of RNA viruses. Antigenic variation is not dependent on immune selection. J. Gen. Virol. 74, 2039–2045.*

tissues or organs to ensure extended replication (that may or may not lead to disease) or survival in face of the immune response of the host, with its multiple (humoral and cellular) branches (Section 4.6).

Arguments occasionally put forward by theoretical biologists that high mutation rates are a consequence of some other parameter [i.e., rate of RNA genome replication (Fitzsimmonds et al., 2018)] but not a trait directly selected for adaptation, has the problem that it is very difficult to demonstrate experimentally in a sound manner with several viral systems; substantiation of this proposal would require showing a fitness advantage of high fidelity mutants using a fitness measurement that does

not involve short-term replication rate. Even assuming that high replication rate was a general need for viruses, it would be a case of trait coadaptation. The necessity of viruses to periodically respond to complex (multiple) selective constraints suggests that high mutation rates have been positively selected (Pfeiffer and Kirkegaard, 2005; Vignuzzi et al., 2006). To my knowledge, there is no evidence that persistent viruses—that in most cases would benefit from a limited replication rate—have selected high fidelity mutants in nature or the laboratory. Error-prone replication is maintained after prolonged viral persistence despite mutation rate being an evolvable trait and that copying fidelity can be modulated by proteins other than the

**FIGURE 4.9** Two representations of viral populations with different amplitude and size of the mutant spectrum. At the top, a narrow mutant spectrum that cannot cope with a complex selective constraint is represented (left), in contrast with a more complex mutant distribution that can cope with a constraint (right). Genomes are depicted as horizontal lines and mutations as symbols on the lines. At the bottom, mutant distributions of different size (large in the blue and small in the red curves) and complexity (average number of mutations per genome given in the abscissa) are represented. *The bottom part is reproduced from Domingo, E., Sheldon, J., Perales, C., 2012. Viral quasispecies evolution. Microbiol. Mol. Biol. Rev. 76, 159–216 with permission from the American Society for Microbiology, Washington DC, USA.*

polymerase (Section 4.9). Following a population bottleneck, an event which from all evidence is very frequent (Section 4.1.2) a single infectious virus must repopulate a mutant spectrum to provide a range of phenotypes that can mediate adaptation. A selective advantage over mutants that cannot provide a comparable phenotypic repertoire can be anticipated (Vignuzzi and Andino, 2010). Mutant viruses displaying altered template-copying fidelity generally exhibit lower fitness than their standard parental genomes. Therefore, current evidence favors the view that the observed mutation rate values are of benefit to the survival strategies employed by viruses, and probably a consequence of their evolutionary adaptation to changing environments.

## 4.9 Multifunctional viral proteins in interaction with host factors: joker substitutions

Many viral proteins are multifunctional, a feature that may be regarded as an adaptation of compact viral genomes to exploit, as much as possible, the phenotypic possibilities of each coding nucleotide (a point mentioned in Section 4.6 in connection with functions exerted by proteins that modulate the host immune response). Multifunctionality of proteins is not unique to viruses since it is increasingly documented for cellular proteins. However, it is in the case of viruses in which the broadest functional repertories of a single protein have been documented. Of many examples, a few are briefly mentioned here.

Proteins NS1 and NS2 of orbiviruses play roles in replication, assembly, and morphogenesis. NS1 is synthesized abundantly in bluetongue virus-infected cells, the protein assembles as tubular structures, and a single-chain antibody against NS1 expressed intracellularly can lead to the reduction of cytopathology and an increase of virus budding and release from the cells (reviewed in Roy, 2008).

Picornavirus nonstructural protein 2C constitutes another dramatic example of multifunctionality that extends even to influence the fidelity properties of the viral polymerase. 2C includes nucleoside-triphosphatase (NTPase) and RNA-binding activities, acts as a chaperone during picornaviral replication, and participates in viral RNA encapsidation, the uncoating of viral particles, and host cell membrane rearrangements required for picornavirus replication (reviewed in Ehrenfeld et al., 2010). Amino acid substitutions in 2C may affect the killing of cells in culture (Herrera et al., 2007) or virulence in vivo (Sanz-Ramos et al., 2008). A substitution in 2C acted as a compensatory mutation of a defect in cytopathology displayed by deletion mutants of FMDV that lack two copies of 3B, the gene encoding VPg, which is the protein involved in the initiation of picornaviral RNA replication (Arias et al., 2010). Mutations in FMDV 2C can modulate the incorporation of nucleotides in the absence of mutations in the viral polymerase (Agudo et al., 2016). 2C is the target of guanidine hydrochloride, a protein-denaturing agent that at low concentrations is an inhibitor of picornavirus replication. This inhibitor has been instrumental to estimate mutation rates of picornaviruses and to understand the molecular basis of virus extinction by enhanced mutagenesis (Chapter 9).

A specific protein domain can participate in two or more functions. The FMDV polymerase 3D includes in its N-terminal region a nuclear localization signal that also acts to regulate nucleotide incorporation during RNA synthesis (Ferrer-Orta et al., 2015; de la Higuera et al., 2018). The nucleoprotein (NP) of arenaviruses is the main structural element of the viral ribonucleoprotein that directs viral RNA synthesis. In addition, NP counteracts the activity of type I IFN, and this function was mapped in the C-terminal region of the protein. The latter region folds in a way similar to the DEDDh family of 3−5′ exoribonucleases (named after a conserved core domain with amino acids DEDD), the type of activity that in replicative DNA polymerases and coronaviruses RNA polymerase is responsible for proofreading repair of mismatched nucleotides at the growing 3'-end of the genome (Chapter 2). Arenavirus NP mutants lacking this activity display remarkable decreases in viral fitness (Martinez-Sobrido et al., 2009; reviewed in Grande-Perez et al., 2016).

Multifunctionality of viral proteins has several implications for the modulation of host cell functions during infections, virus evolution, viral genome stability, and the response of viruses to selective constraints. An amino acid substitution in a viral protein needed to overcome a specific selective pressure directed to that protein may also affect unrelated functions performed by the protein. This may result in a fitness cost, frequently observed with viral mutants selected for their capacity to overcome an inhibitory activity. When the selected mutants are allowed to continue their replication, additional mutations may occur to increase fitness. Involvement in two or more functions may limit the number of possible amino acid substitutions that a protein can accept for fitness increase, because the same substitutions that may compensate for a limitation of one of the functions may adversely affect other functions exerted by the same protein. Again, *trade-offs* (the virus *modus vivendi*) enter the picture (Section 4.2). Restrictions imposed by multifunctionality may underlie the observation that a single amino acid replacement I248T in protein 2C was selected repeatedly and independently when FMDV had to respond to different environmental demands: in the course of adaptation of a biological clone of the virus to guinea pig (Núñez et al., 2001), upon replication of the same clone in mice (Sanz-Ramos et al., 2008), or to modulate the relative rate of incorporation of different nucleotides by the FMDV polymerase (Agudo et al., 2016). Substitutions that serve a virus to increase its fitness under different environmental circumstances are termed "joker" substitutions. They lead to a general increase in replication efficiency under different environmental demands.

The expectation that the genomic nucleotide sequences that belong to overlapping genes (i.e., polynucleotide stretches that contribute to the coding of more than one protein) would be much more conserved than standard open-reading frames for a single protein has not been confirmed, at least to the extent of being able to derive a general conclusion. A reason for the limited differences in conservation of these two categories of coding regions may lie in protein multifunctionality, in combination with the phenotypic involvement of the viral RNA itself, independent of its coding function (Section 4.3 and Chapter 2). Indeed, if the same protein performs different functions and the RNA is also involved in regulatory activities, each nucleotide may be subjected to constraints that may not differ significantly from the constraints operating on the products expressed from overlapping genes (Domingo et al., 2012).

## 4.10 Alternating selective pressures: the case of arboviruses

The selective pressures that viruses encounter when infecting their hosts in nature are difficult to identify, but those that have been recognized are rarely constant and uniform. The experimental designs in which a virus is subjected to a specific selective pressure in an established cell line in culture are a gross simplification of reality that has nevertheless allowed quantification of the effects of well-defined variables (Chapter 6). However, the first point to note in dealing with the interaction of viral populations with their hosts is the multitude of selective pressures, often conflicting, that viruses must confront. Selective pressures vary in kind and intensity in space and time.

An interesting, biologically relevant case of alternating selective pressures is provided by the animal and plant arboviruses which alternate replication in vertebrate animals or in plants and insect hosts, and have successfully persisted in nature as disease agents (reviewed in different chapters of Morse, 1994; Gibbs et al., 1995; Domingo, 2006). Three strategies have been distinguished regarding the part of the virus life cycle that elapses in the vector: (i) the virus attaches to the vector, usually at its external organs, but does not reach the internal milieu, and does not undergo replication. This style has been termed noncirculative. (ii) The virus enters the vector through specific receptors and multiplies in it. This style is often termed circulative-propagative. (iii) The virus cycles inside the vector, but does not propagate in it. This is the case of the plant nanovirus in its aphid vectors. Interactive styles (i) and (iii) must be distinguished from purely mechanical transmission because inside the vector the virus may meet conditions that alter particle stability relative to the outside environment, or it may induce changes in the insect expression profile. During the circulative-propagative style, the virus undergoes several bottleneck events. In mosquito vectors, the virus must transit from one compartment into another, and most of the compartments are separated by basal lamina that limits the penetration of viruses (Forrester et al., 2014). Bottlenecks may affect fitness, limit the number of particles that can be transmitted and, in consequence, contribute to the unpredictability of evolutionary outcomes (Hanley and Weaver, 2008; Gutierrez et al., 2012; Forrester et al., 2014). The consequences of stochasticity are accentuated by the cloud nature of viral populations since it is virtually impossible that two bottleneck events acting on the same virus population will isolate two genomes with an identical sequence. Quasispecies ensures diverse initial genomes to pursue evolutionary trajectories following bottleneck episodes.

Zoonotic vector-borne flaviviruses and alphaviruses have been extensively studied because of their pathogenic potential for humans. They include yellow fever virus, Dengue virus, Chikungunya virus, and Venezuelan equine encephalitis virus, among others.

Several types of insect-mammalian infection cycles have been characterized by these viruses. A sylvatic cycle is defined as the one that involves nonhuman animal host and insects. Phylogenetic evidence suggests that sylvatic Dengue was the precursor of Dengue viruses that infected humans and established enzootic and endemic cycles (Fig. 4.10). Here we return to the importance of virus and host population numbers that were discussed in the first section of this Chapter. In the case of Dengue virus, it has been estimated that efficient human-to-human transmission required a minimum human population size in the range of $10^4$ to $10^6$ individuals, a size that was attained only with the advent of urban life (Gubler, 1997). Critical population numbers of hosts and vectors, as well as numbers of infectious particles in viremic hosts, are necessary for the emergence and maintenance of viral diseases (Chapter 7). Dengue virus appears to be highly adaptable to new animal and vector hosts, notably the transition from *Aedes albopictus* to *Aedes aegypti* as mosquito vectors (reviewed in Hanley and Weaver, 2008). Replication in different environments is the rule for the great majority of viruses, even those that have a preferred host species. Perhaps because of this, there are no salient features in the patterns of evolution of different viruses (overall rate of evolution, proportion of conserved genomic residues or protein domains, frequencies of antibody- or inhibitor-escape mutants, etc.) that allow a clear, quantitative and general distinction between viral pathogens that are directly transmitted among individual vertebrate hosts and those that require an invertebrate vector.

## 4.10.1 The sophistication of pathogen-vector-host interactions in plant viruses

Since plants do not have an independent capacity to move, virus transmission from one individual plant to another or from a cultivar to

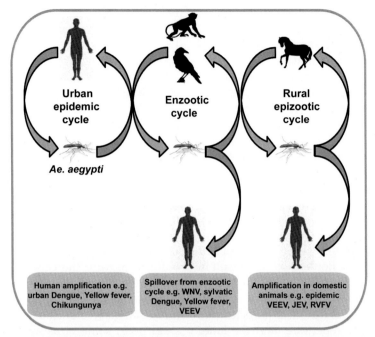

**FIGURE 4.10** Simplified representation of three types of infectious cycles (epidemic, enzootic, and rural epizootic) identified for zoonotic arboviruses. Host alternations between vertebrate and mosquito hosts are depicted by curved arrows. Examples of viruses documented to participate in these cycles are shown at the bottom (WNV, West Nile virus; VEEV, Venezuelan equine encephalitis virus; JEV, Japanese encephalitis virus; RVFV, Rift Valley fever virus). *Figure modified from Hanley, K.H., Weaver, S.C., 2008. Arbovirus evolution. In: Domingo, E., Parrish, C.R., Holland, J.J. (Eds.), Origin and Evolution of Viruses second ed. Elsevier, Oxford, pp. 351–392 with permission from the authors.*

another can rarely rely on direct interhost contacts and never on plant movements, as is the case with most animal viruses. Instead, viruses are transmitted among plants either via nematodes in the soil or flying insects [compilation in (Hogenhout et al., 2013)]. Virus particles have to survive during the process of insect-mediated transmission, an adaptation that has probably been shaped over the 300 million years since the first insects evolved. A surprising array of plant-insect-virus interactions has been described, with the exploitation of biochemical pathways that at the organism level rival with the adaptations attained by mutant spectra of the infecting viruses at the molecular level. In some cases, infected vectors are attracted to noninfected plants while uninfected vectors are attracted to infected plants, as a result of

molecular signals whose result is the enhancement virus transmission (Mauck et al., 2010; Carr et al., 2018). Virus factories and the transmission body of cauliflower mosaic virus can reversibly disassemble to enhance virion distribution, and favor aphid-mediated transmission (Bak et al., 2013). A true arms race and coevolutionary events at many levels are established among viruses, their host plants, and terrestrial or airborne insect vectors, to achieve the survival of all participants.

## 4.11 Overview and concluding remarks

Virus population numbers are orders of magnitude greater than the numbers of individuals in the host species that they infect. Large population numbers, however, are not constant during the virus life cycles. Viruses undergo periodic bottleneck events, that is, drastic reductions in population size as they spread within individual hosts and during host-to-host transmission. Bottlenecks accentuate the effect of chance (that promotes genetic drift) in virus evolution, made more severe by the mutant spectrum (cloud) nature of viral populations. Out of the large mutant repertoire that is produced during an infection, bottlenecks dictate that only a minority contributes to the next round of progeny production either in the same host (because of a change of compartment) or in a different host individual. It is unlikely that two bottleneck transmissions will involve the same genome repertoire, even when applied to the same population. Bottlenecks also result in transient reductions of population heterogeneity (a decrease in the amplitude of mutant spectra) because only a few founder viral particles will originate the progeny repertoire, and the latter necessitates some replication rounds to restore the population size and complexity that the population had prior to the bottleneck. It has been suggested that one of the reasons why high mutation rates are maintained in viruses is the need to reinstate population heterogeneity following a population bottleneck. This proposal is in line with the adaptive value of complex mutant spectra documented in Chapters 2 and 3. Despite several models on the evolutionary origins of high mutation rates, current evidence favors that high mutation rates of RNA viruses (and some DNA viruses) serve the adaptive needs of viruses; virtually, every chapter of this book includes examples.

Viruses are subjected to several classes of constraints during their multiplication in cellular hosts, an obvious condition for long-term virus-host coexistence given the disparity of their population numbers. Basal constraints are those imposed by the core replicative needs of viruses. Other constraints are due to responses of the individual cells or entire organisms in the event of an infection. Still, other constraints are the result of human intervention such as vaccination or the administration of antiviral drugs. Emphasis has been put on codon usage and codon-pair associations as a basal constraint for virus replication, for two reasons. Because it constitutes an example of the nonneutral character of synonymous mutations in viral RNAs, and because codon and codon-pair deoptimization have opened new prospects for the design of live-attenuated antiviral vaccines.

A major problem in evolutionary biology is the identification of the selective constraints that, together with the random drift of genomes, mold evolutionary outcomes. This is an immense problem often solved by shallow assumption. We often propose a selective advantage for the brain to acquire the capacity of consciousness, for the giraffe to have a long neck, or for a virus to assemble a stable capsid. Yet, successful creatures can lack consciousness and can have a short neck; some viruses that have survived to our times have a labile capsid. I suggest that the difficulty of identifying the selective constraints that viruses confront in their multiplication cycles is one of the major contributory factors to our fragmentary understanding of virus evolution.

Modulation and evasion are two distinct strategies used by viruses to overcome immune responses. Viral proteins that act as immune modulators are often multifunctional. The studies on evasion strategy, based on the selection of antibody- and CTL-resistant mutants, have been instrumental for the understanding of quasispecies dynamics. Viruses can use several receptors that belong to different families of cell surface proteins that mediate virus internalization. Viruses display remarkable flexibility to modify their cell target preference. Alterations of cell tropism and coevolution of antigenicity and receptor recognition specificity have been outlined as a consequence of quasispecies dynamics and the overlap between antigenic sites and receptor-recognition domains in viruses. The adaptability of a viral quasispecies is also reflected in the capacity of viruses to confront changing and alternating selective pressures. The arboviruses that alternate between insect and vertebrate or plant hosts constitute an example of adaptation to multiple hosts as a standard way of life, although to some extent any virus must confront different environments (see Summary Box).

## Summary Box

- Virus population numbers are several orders of magnitude larger than those of their host organisms.
- Viruses are subjected to constraints at multiple levels that limit their replication and contribute to maintaining their biological identity.
- Viruses exploit a variety of molecular mechanisms to overcome external constraints, such as therapeutic interventions.
- Codon usage and codon-pair frequencies represent constraints for viral gene expression that can be used to design new generation vaccines.
- Viruses can modify their cellular tropisms through amino acid substitutions in surface protein residues and other mechanisms. Overlap between antigenic sites and receptor recognition sites facilitates coevolution of antigenicity and cellular tropism.
- Viruses use two major strategies to cope with the host immune response: modulation and escape. Escape mutations generally imply a fitness cost for the virus.
- Viruses have evolved means to cope with multiple constraints, including remarkable environmental alternations. This is the case of pathogenic arboviruses whose natural infectious cycles involve replication in vertebrate and insect hosts.
- An important limitation to interpret evolutionary trajectories of viruses is that many selective constraints confronted by viruses in their life cycles have not been identified.

# References

Acharya, R., Fry, E., Stuart, D., Fox, G., Rowlands, D., et al., 1989. The three-dimensional structure of foot-and-mouth disease virus at 2.9 Å resolution. Nature 337, 709–716.

Agudo, R., de la Higuera, I., Arias, A., Grande-Pérez, A., Domingo, E., 2016. Involvement of a joker mutation in a polymerase-independent lethal mutagenesis escape mechanism. Virology 494, 257–266.

Alberts, B., Johnson, A., Lewis, J., Raff, M., Roberts, K., et al., 2002. Molecular Biology of the Cell. Garland Science, New York, N.Y.

Alcami, A., 2003. Viral mimicry of cytokines, chemokines and their receptors. Nat. Rev. Immunol. 3, 36–50.

Ali, A., Li, H., Schneider, W.L., Sherman, D.J., Gray, S., et al., 2006. Analysis of genetic bottlenecks during horizontal transmission of Cucumber mosaic virus. J. Virol. 80, 8345–8350.

Aragones, L., Guix, S., Ribes, E., Bosch, A., Pinto, R.M., 2010. Fine-tuning translation kinetics selection as the driving force of codon usage bias in the hepatitis a virus capsid. PLoS Pathog. 6, e1000797.

Arias, A., Perales, C., Escarmis, C., Domingo, E., 2010. Deletion mutants of VPg reveal new cytopathology determinants in a picornavirus. PLoS One 5, e10735.

Arora, R., Priano, C., Jacobson, A.B., Mills, D.R., 1996. cis-acting elements within an RNA coliphage genome: fold as you please, but fold you must!! J. Mol. Biol. 258, 433–446.

Artenstein, M.S., Miller, W.S., 1966. Air sampling for respiratory disease agents in army recruits. Bacteriol. Rev. 30, 571–572.

Bae, Y.S., Yoon, J.W., 1993. Determination of diabetogenicity attributable to a single amino acid, Ala776, on the polyprotein of encephalomyocarditis virus. Diabetes 42, 435–443.

Bak, A., Gargani, D., Macia, J.L., Malouvet, E., Vernerey, M.S., et al., 2013. Virus factories of cauliflower mosaic virus are virion reservoirs that engage actively in vector transmission. J. Virol. 87, 12207–12215.

Baldwin, C., Berkhout, B., 2007. HIV-1 drug-resistance and drug-dependence. Retrovirology 4, 78.

Baldwin, C.E., Sanders, R.W., Deng, Y., Jurriaans, S., Lange, J.M., et al., 2004. Emergence of a drug-dependent human immunodeficiency virus type 1 variant during therapy with the T20 fusion inhibitor. J. Virol. 78, 12428–12437.

Ballesteros, M.L., Sanchez, C.M., Enjuanes, L., 1997. Two amino acid changes at the N-terminus of transmissible gastroenteritis coronavirus spike protein result in the loss of enteric tropism. Virology 227, 378–388.

Baranowski, E., Ruíz-Jarabo, C.M., Sevilla, N., Andreu, D., Beck, E., et al., 2000. Cell recognition by foot-and-mouth disease virus that lacks the RGD integrin-binding motif: flexibility in aphthovirus receptor usage. J. Virol. 74, 1641–1647.

Baranowski, E., Ruiz-Jarabo, C.M., Domingo, E., 2001. Evolution of cell recognition by viruses. Science 292, 1102–1105.

Baranowski, E., Ruiz-Jarabo, C.M., Pariente, N., Verdaguer, N., Domingo, E., 2003. Evolution of cell recognition by viruses: a source of biological novelty with medical implications. Adv. Virus Res. 62, 19–111.

Basler, C.F., Garcia-Sastre, A., 2002. Viruses and the type I interferon antiviral system: induction and evasion. Int. Rev. Immunol. 21, 305–337.

Belsham, G.J., Sonenberg, N., 1996. RNA-protein interactions in regulation of picornavirus RNA translation. Microbiol. Rev. 60, 499–511.

Bergelson, J.M., 2010. Receptors. In: Ehrenfeld, E., Domingo, E., Roos, R.P. (Eds.), The Picornaviruses. ASM Press, Washington DC, pp. 73–86.

Betancourt, M., Fereres, A., Fraile, A., Garcia-Arenal, F., 2008. Estimation of the effective number of founders that initiate an infection after aphid transmission of a multipartite plant virus. J. Virol. 82, 12416–12421.

Bhella, D., 2015. The role of cellular adhesion molecules in virus attachment and entry. Philos. Trans. R. Soc. Lond. B Biol. Sci. 370.

Borrego, B., Novella, I.S., Giralt, E., Andreu, D., Domingo, E., 1993. Distinct repertoire of antigenic variants of foot-and-mouth disease virus in the presence or absence of immune selection. J. Virol. 67, 6071–6079.

Borrow, P., Lewicki, H., Wei, X., Horwitz, M.S., Peffer, N., et al., 1997. Antiviral pressure exerted by HIV-1-specific cytotoxic T lymphocytes (CTLs) during primary infection demonstrated by rapid selection of CTL escape virus. Nat. Med. 3, 205–211.

Bosch, A., Mueller, S., Pintó, R.M., 2010. Codon biases and viral fitness. In: Ehrenfeld, E., Domingo, E., Roos, R.P. (Eds.), The Picornaviruses. ASM Press, Washington, DC, pp. 271–283.

Boyd, M.T., Simpson, G.R., Cann, A.J., Johnson, M.A., Weiss, R.A., 1993. A single amino acid substitution in the V1 loop of human immunodeficiency virus type 1 gp120 alters cellular tropism. J. Virol. 67, 3649–3652.

Britten, R.J., 1993. Forbidden synonymous substitutions in coding regions. Mol. Biol. Evol. 10, 205–220.

Budkowska, A., Bedossa, P., Groh, F., Louise, A., Pillot, J., 1995. Fibronectin of human liver sinusoids binds hepatitis B virus: identification by an anti-idiotypic antibody bearing the internal image of the pre-S2 domain. J. Virol. 69, 840–848.

Buenno, L.H., Leme, J., Caricati, C.P., Tonso, A., Rocha, J.C., et al., 2015. Use of uniform designs in combination with neural networks for viral infection process development. Biotechnol. Prog. 31, 532–540.

Bull, R.A., Luciani, F., McElroy, K., Gaudieri, S., Pham, S.T., et al., 2011. Sequential bottlenecks drive viral evolution in early acute hepatitis C virus infection. PLoS Pathog. 7, e1002243.

Burns, C.C., Shaw, J., Campagnoli, R., Jorba, J., Vincent, A., et al., 2006. Modulation of poliovirus replicative fitness in HeLa cells by deoptimization of synonymous codon usage in the capsid region. J. Virol. 80, 3259–3272.

Campbell, C.L., Smith, D.R., Sanchez-Vargas, I., Zhang, B., Shi, P.Y., et al., 2014. A positively selected mutation in the WNV 2K peptide confers resistance to superinfection exclusion in vivo. Virology 464–465, 228–232.

Cantara, W.A., Olson, E.D., Musier-Forsyth, K., 2014. Progress and outlook in structural biology of large viral RNAs. Virus Res. 193, 24–38.

Carr, J.P., Donnelly, R., Tungadi, T., Murphy, A.M., Jiang, S., et al., 2018. Viral manipulation of plant stress responses and host interactions with insects. Adv. Virus Res. 102, 177–197.

Carson, S.D., Chapman, N.M., Hafenstein, S., Tracy, S., 2011. Variations of coxsackievirus B3 capsid primary structure, ligands, and stability are selected for in a coxsackievirus and adenovirus receptor-limited environment. J. Virol. 85, 3306–3314.

Centlivre, M., Sommer, P., Michel, M., Ho Tsong Fang, R., Gofflo, S., et al., 2005. HIV-1 clade promoters strongly influence spatial and temporal dynamics of viral replication in vivo. J. Clin. Investig. 115, 348–358.

Chamary, J.V., Hurst, L.D., 2005. Biased codon usage near intron-exon junctions: selection on splicing enhancers, splice-site recognition or something else? Trends Genet. 21, 256–259.

Charlesworth, B., 2009. Fundamental concepts in genetics: effective population size and patterns of molecular evolution and variation. Nat. Rev. 10, 195–205.

Cheng, B.Y., Ortiz-Riano, E., Nogales, A., de la Torre, J.C., Martinez-Sobrido, L., 2015. Development of live-attenuated arenavirus vaccines based on codon deoptimization. J. Virol. 89, 3523–3533.

Ciurea, A., Klenerman, P., Hunziker, L., Horvath, E., Senn, B.M., et al., 2000. Viral persistence *in vivo* through selection of neutralizing antibody-escape variants. Proc. Natl. Acad. Sci. U.S.A. 97, 2749–2754.

Clarke, D.K., Duarte, E.A., Elena, S.F., Moya, A., Domingo, E., et al., 1994. The red queen reigns in the kingdom of RNA viruses. Proc. Natl. Acad. Sci. U.S.A. 91, 4821–4824.

Co, M.S., Gaulton, G.N., Fields, B.N., Greene, M.I., 1985. Isolation and biochemical characterization of the mammalian reovirus type 3 cell-surface receptor. Proc. Natl. Acad. Sci. U.S.A. 82, 1494–1498.

Coffin, J.M., 1995. HIV population dynamics in vivo: implications for genetic variation, pathogenesis, and therapy. Science 267, 483–489.

Coleman, J.R., Papamichail, D., Skiena, S., Futcher, B., Wimmer, E., et al., 2008. Virus attenuation by genome-scale changes in codon pair bias. Science 320, 1784–1787.

Colston, E.M., Racaniello, V.R., 1995. Poliovirus variants selected on mutant receptor-expressing cells identify capsid residues that expand receptor recognition. J. Virol. 69, 4823–4829.

Connor, R.J., Kawaoka, Y., Webster, R.G., Paulson, J.C., 1994. Receptor specificity in human, avian, and equine H2 and H3 influenza virus isolates. Virology 205, 17–23.

Connor, R.I., Sheridan, K.E., Ceradini, D., Choe, S., Landau, N.R., 1997. Change in coreceptor use correlates with disease progression in HIV-1-infected individuals. J. Exp. Med. 185, 621–628.

Consigli, R.A., Griffith, G.R., Marriott, S.J., Ludlow, J.W., 1986. Biochemical characterization of poliomavirus-receptor interactions. In: Crowell, R.L., Lonberg-Holm, K. (Eds.), Virus Attachment and Entry into Cells. American Society for Microbiology, Washington, DC, pp. 44–53.

Costafreda, M.I., Perez-Rodriguez, F.J., D'Andrea, L., Guix, S., Ribes, E., et al., 2014. Hepatitis A virus adaptation to cellular shutoff is driven by dynamic adjustments of codon usage and results in the selection of populations with altered capsids. J. Virol. 88, 5029–5041.

Coulon, P., Ternaux, J.P., Flamand, A., Tuffereau, C., 1998. An avirulent mutant of rabies virus is unable to infect motoneurons in vivo and in vitro. J. Virol. 72, 273–278.

Curry, S., Fry, E., Blakemore, W., Abu-Ghazaleh, R., Jackson, T., et al., 1996. Perturbations in the surface structure of A22 Iraq foot-and-mouth disease virus accompanying coupled changes in host cell specificity and antigenicity. Structure 4, 135–145.

de Borba, L., Villordo, S.M., Iglesias, N.G., Filomatori, C.V., Gebhard, L.G., et al., 2015. Overlapping local and long range RNA-RNA interactions modulate dengue virus genome cyclization and replication. J. Virol. 89, 3430–3437.

De Cock, K.M., Jaffe, H.W., Curran, J.W., 2012. The evolving epidemiology of HIV/AIDS. AIDS 26, 1205–1213.

de la Torre, J.C., 2002. Molecular biology of Borna disease virus and persistence. Front. Biosci. 7, d569–d579.

de la Higuera, I., Ferrer-Orta, C., Moreno, E., de Avila, A.I., Soria, M.E., et al., 2018. Contribution of a multifunctional polymerase region of foot-and-mouth disease virus to lethal mutagenesis. J. Virol. 92 e01119-18.

de la Torre, J.C., Wimmer, E., Holland, J.J., 1990. Very high frequency of reversion to guanidine resistance in clonal pools of guanidine-dependent type 1 poliovirus. J. Virol. 64, 664–671.

de Los Santos, T., de Avila Botton, S., Weiblen, R., Grubman, M.J., 2006. The leader proteinase of foot-and-mouth disease virus inhibits the induction of beta interferon mRNA and blocks the host innate immune response. J. Virol. 80, 1906–1914.

Domingo, E. (Ed.), 2006. Quasispecies: Concepts and Implications for Virology. Curr. Top. Microbiol. Immunol., vol. 299. Springer, Berlin.

Domingo, E., 2011. Paradoxical interplay of viral and cellular functions. Viruses 3, 272–277.

Domingo, E., Díez, J., Martínez, M.A., Hernández, J., Holguín, A., et al., 1993. New observations on antigenic diversification of RNA viruses. Antigenic variation is not dependent on immune selection. J. Gen. Virol. 74, 2039–2045.

Domingo, E., Biebricher, C., Eigen, M., Holland, J.J., 2001. Quasispecies and RNA Virus Evolution: Principles and Consequences. Landes Bioscience, Austin.

Domingo, E., Sheldon, J., Perales, C., 2012. Viral quasispecies evolution. Microbiol. Mol. Biol. Rev. 76, 159–216.

Donohue, R.C., Pfaller, C.K., Cattaneo, R., 2019. Cyclical adaptation of measles virus quasispecies to epithelial and lymphocytic cells: to V, or not to V. PLoS Pathog. 15 (2), e1007605.

dos Reis, M., Savva, R., Wernisch, L., 2004. Solving the riddle of codon usage preferences: a test for translational selection. Nucleic Acids Res. 32, 5036–5044.

Du, Y., Wu, N.C., Jiang, L., Zhang, T., Gong, D., et al., 2016. Annotating protein functional residues by coupling high-throughput fitness profile and homologous-structure analysis. mBio 7 (6), e01801–e01816.

Ehrenfeld, E., Domingo, E., Ross, R.P., 2010. The Picornaviruses. ASM Press, Washington, DC.

Eigen, M., 1992. Steps towards Life. Oxford University Press, Oxford.

Erickson, A.K., Pfeiffer, J.K., 2013. Dynamic viral dissemination in mice infected with yellow fever virus strain 17D. J. Virol. 87, 12392–12397.

Escarmís, C., Dopazo, J., Dávila, M., Palma, E.L., Domingo, E., 1995. Large deletions in the 5′-untranslated region of foot-and-mouth disease virus of serotype C. Virus Res. 35, 155–167.

Este, J.A., Cabrera, C., Blanco, J., Gutierrez, A., Bridger, G., et al., 1999. Shift of clinical human immunodeficiency virus type 1 isolates from X4 to R5 and prevention of emergence of the syncytium-inducing phenotype by blockade of CXCR4. J. Virol. 73, 5577–5585.

Eyre-Walker, A., 1996. Synonymous codon bias is related to gene length in *Escherichia coli*: selection for translational accuracy? Mol. Biol. Evol. 13, 864–872.

Fabre, F., Moury, B., Johansen, E.I., Simon, V., Jacquemond, M., et al., 2014. Narrow bottlenecks affect Pea seedborne mosaic virus populations during vertical seed transmission but not during leaf colonization. PLoS Pathog. 10, e1003833.

Ferrer-Orta, C., De la Higuera, I., Caridi, F., Sanchez-Aparicio, M.T., Moreno, E., et al., 2015. Multifunctionality of a picornavirus polymerase domain: nuclear localization signal is involved in nucleotide recognition. J. Virol. 89, 6849–6859.

Filomatori, C.V., Carballeda, J.M., Villordo, S.M., Aguirre, S., et al., 2017. Dengue virus genomic variation associated with mosquito adaptation defines the pattern of viral non-coding RNAs and fitness in human cells. PLoS Pathog. 13, e1006265.

Fitzsimmonds, W.M., Woods, R.J., McCrone, J.T., Woodman, A., Arnold, J.J., et al., 2018. A speed-fidelity trade-off determines the mutation rate and virulence of an RNA virus. PLoS Biol. 16, e2006459.

Forrester, N.L., Coffey, L.L., Weaver, S.C., 2014. Arboviral bottlenecks and challenges to maintaining diversity and fitness during mosquito transmission. Viruses 6, 3991–4004.

Foy, B.D., Myles, K.M., Pierro, D.J., Sanchez-Vargas, I., Uhlirova, M., et al., 2004. Development of a new Sindbis virus transducing system and its characterization in three Culicine mosquitoes and two Lepidopteran species. Insect Mol. Biol. 13, 89–100.

Fry, E.E., Stuart, D.I., 2010. Virion structure. In: Ehrenfeld, E., Domingo, E., Roos, R.P. (Eds.), The Picornaviruses. ASM Press, Washington, DC, pp. 59–72.

Fry, E.E., Lea, S.M., Jackson, T., Newman, J.W., Ellard, F.M., et al., 1999. The structure and function of a foot-and-mouth disease virus- oligosaccharide receptor complex. EMBO J. 18, 543–554.

Fung, A., Jin, Z., Dyatkina, N., Wang, G., Beigelman, L., Deval, J., 2014. Efficiency of incorporation and chain termination determines the inhibition potency of 2′-modified nucleotide analogs against hepatitis C virus polymerase. Antimicrob. Agents Chemother. 58, 3636–3645.

Futuyma, D.J., Slatkin, M. (Eds.), 1983. Coevolution. Sinauer Associates Inc., Sunderland, Massachusetts.

Gaulton, G., Co, M.S., Greene, M.I., 1985. Anti-idiotypic antibody identifies the cellular receptor of reovirus type 3. J. Cell. Biochem. 28, 69–78.

Gebauer, F., de la Torre, J.C., Gomes, I., Mateu, M.G., Barahona, H., et al., 1988. Rapid selection of genetic and antigenic variants of foot-and-mouth disease virus during persistence in cattle. J. Virol. 62, 2041–2049.

Gerone, P.J., Couch, R.B., Keefer, G.V., Douglas, R.G., Derrenbacher, E.B., et al., 1966. Assessment of experimental and natural viral aerosols. Bacteriol. Rev. 30, 576–588.

Gibbs, A.J., Calisher, C.H., García-Arenal, F. (Eds.), 1995. Molecular Basis of Virus Evolution. Cambridge University Press, Cambridge.

Gitlin, L., Hagai, T., LaBarbera, A., Solovey, M., Andino, R., 2014. Rapid evolution of virus sequences in intrinsically disordered protein regions. PLoS Pathog. 10, e1004529.

Gloor, G.B., Martin, L.C., Wahl, L.M., Dunn, S.D., 2005. Mutual information in protein multiple sequence alignments reveals two classes of coevolving positions. Biochemistry 44, 7156–7165.

Gomez, P., Ashby, B., Buckling, A., 2015. Population mixing promotes arms race host-parasite coevolution. Proc. Biol. Sci. 282, 20142297.

Grande-Perez, A., Martin, V., Moreno, H., de la Torre, J.C., 2016. Arenavirus quasispecies and their biological implications. Curr. Top. Microbiol. Immunol. 392, 231–276.

Gubler, D.J., 1997. Dengue and dengue hemorrhagic fever: its history and resurgence as a global public health problem. In: Gubler, D.J., Kuno, G. (Eds.), Dengue and Dengue Hemorrhagic Fever. CAB International, New York.

Gutierrez, S., Michalakis, Y., Blanc, S., 2012. Virus population bottlenecks during within-host progression and host-to-host transmission. Curr. Opin. Virol. 2, 546–555.

Haaland, R.E., Hawkins, P.A., Salazar-Gonzalez, J., Johnson, A., Tichacek, A., et al., 2009. Inflammatory genital infections mitigate a severe genetic bottleneck in heterosexual transmission of subtype A and C HIV-1. PLoS Pathog. 5, e1000274.

Hall, C.B., Douglas Jr., R.G., Geiman, J.M., Meagher, M.P., 1979. Viral shedding patterns of children with influenza B infection. J. Infect. Dis. 140, 610–613.

Halstead, S.B., 1980. Immunological parameters of Togavirus disease syndromes. In: Schlesinger, R.W. (Ed.), The Togaviruses. Biology, Structure, Replication. Academic Press, New York, pp. 107–173.

Hanham, C.A., Zhao, F., Tignor, G.H., 1993. Evidence from the anti-idiotypic network that the acetylcholine receptor is a rabies virus receptor. J. Virol. 67, 530–542.

Hanley, K.H., Weaver, S.C., 2008. Arbovirus evolution. In: Domingo, E., Parrish, C.R., Holland, J.J. (Eds.), Origin and Evolution of Viruses, second ed. Elsevier, Oxford, pp. 351–392.

Harber, J., Bernhardt, G., Lu, H.H., Sgro, J.Y., Wimmer, E., 1995. Canyon rim residues, including antigenic determinants, modulate serotype-specific binding of polioviruses to mutants of the poliovirus receptor. Virology 214, 559–570.

He, M., An, T.Z., Teng, C.B., 2014. Evolution of mammalian and avian bornaviruses. Mol. Phylogenetics Evol. 79, 385–391.

Herrera, M., Garcia-Arriaza, J., Pariente, N., Escarmis, C., Domingo, E., 2007. Molecular basis for a lack of correlation between viral fitness and cell killing capacity. PLoS Pathog. 3, e53.

Herrmann Jr., E.C., Herrmann, J.A., 1977. A working hypothesis—virus resistance development as an indicator of specific antiviral activity. Ann. N. Y. Acad. Sci. 284, 632–637.

Hertogs, K., Depla, E., Crabbe, T., De Bruin, W., Leenders, W., et al., 1994. Spontaneous development of anti-hepatitis B virus envelope (anti-idiotypic) antibodies in animals immunized with human liver endonexin II or with the F(ab')2 fragment of anti-human liver endonexin II immunoglobulin G: evidence for a receptor-ligand-like relationship between small hepatitis B surface antigen and endonexin II. J. Virol. 68, 1516–1521.

Hewat, E.A., Neumann, E., Conway, J.F., Moser, R., Ronacher, B., et al., 2000. The cellular receptor to human rhinovirus 2 binds around the 5-fold axis not in the canyon: a structural view. EMBO J. 19, 6317–6325.

Hiramatsu, K., Tadano, M., Men, R., Lai, C.J., 1996. Mutational analysis of a neutralization epitope on the dengue type 2 virus (DEN2) envelope protein: monoclonal antibody resistant DEN2/DEN4 chimeras exhibit reduced mouse neurovirulence. Virology 224, 437–445.

Ho, D.D., 1995. Time to hit HIV, early and hard. N. Engl. J. Med. 333, 450–451.

Ho, D.D., Neumann, A.U., Perelson, A.S., Chen, W., Leonard, J.M., et al., 1995. Rapid turnover of plasma virions and CD4 lymphocytes in HIV-1 infection. Nature 373, 123–126.

Hogenhout, S.A., Mitchum, M., Smant, G., 2013. Focus on molecular plant-nematode and plant-insect interactions. Mol. Plant Microbe Interact. 26 https://doi.org/10.1094/MPMI-26-01-0008.

Hsu, E.C., Sarangi, F., Iorio, C., Sidhu, M.S., Udem, S.A., et al., 1998. A single amino acid change in the hemagglutinin protein of measles virus determines its ability to bind CD46 and reveals another receptor on marmoset B cells. J. Virol. 72, 2905–2916.

Huang, T., Campadelli-Fiume, G., 1996. Anti-idiotypic antibodies mimicking glycoprotein D of herpes simplex virus identify a cellular protein required for virus spread from cell to cell and virus-induced polykaryocytosis. Proc. Natl. Acad. Sci. U.S.A. 93, 1836–1840.

Huang, S., Reddy, V., Dasgupta, N., Nemerow, G.R., 1999. A single amino acid in the adenovirus type 37 fiber confers binding to human conjunctival cells. J. Virol. 73, 2798–2802.

Hueffer, K., Parrish, C.R., 2003. Parvovirus host range, cell tropism and evolution. Curr. Opin. Microbiol. 6, 392–398.

Hueffer, K., Parker, J.S., Weichert, W.S., Geisel, R.E., Sgro, J.Y., et al., 2003. The natural host range shift and subsequent evolution of canine parvovirus resulted from virus-specific binding to the canine transferrin receptor. J. Virol. 77, 1718–1726.

Hunt, R.C., Simhadri, V.L., Iandoli, M., Sauna, Z.E., Kimchi-Sarfaty, C., 2014. Exposing synonymous mutations. Trends Genet. 30, 308–321.

Iorio, R.M., Glickman, R.L., Riel, A.M., Sheehan, J.P., Bratt, M.A., 1989. Functional and neutralization profile of seven overlapping antigenic sites on the HN glycoprotein of Newcastle disease virus: monoclonal antibodies to some sites prevent viral attachment. Virus Res. 13, 245–261.

Jacob, F., 1977. Evolution and tinkering. Science 196, 1161–1166.

Jennings, A.D., Gibson, C.A., Miller, B.R., Mathews, J.H., Mitchell, C.J., et al., 1994. Analysis of a yellow fever virus isolated from a fatal case of vaccine-associated human encephalitis. J. Infect. Dis. 169, 512–518.

Jnaoui, K., Michiels, T., 1998. Adaptation of Theiler's virus to L929 cells: mutations in the putative receptor binding site on the capsid map to neutralization sites and modulate viral persistence. Virology 244, 397–404.

Jung, Y.T., Kozak, C.A., 2000. A single amino acid change in the murine leukemia virus capsid gene responsible for the Fv1(nr) phenotype. J. Virol. 74, 5385–5387.

Katze, M.G., He, Y., Gale Jr., M., 2002. Viruses and interferon: a fight for supremacy. Nat. Rev. Immunol. 2, 675–687.

Keay, S., Baldwin, B., 1991. Anti-idiotype antibodies that mimic gp86 of human cytomegalovirus inhibit viral fusion but not attachment. J. Virol. 65, 5124–5128.

Keay, S., Merigan, T.C., Rasmussen, L., 1989. Identification of cell surface receptors for the 86-kilodalton glycoprotein of human cytomegalovirus. Proc. Natl. Acad. Sci. U.S.A. 86, 10100–10103.

Keck, Z.Y., Angus, A.G., Wang, W., Lau, P., Wang, Y., et al., 2014. Non-random escape pathways from a broadly neutralizing human monoclonal antibody map to a highly conserved region on the hepatitis C virus E2 glycoprotein encompassing amino acids 412–423. PLoS Pathog. 10, e1004297.

Keele, B.F., Giorgi, E.E., Salazar-Gonzalez, J.F., Decker, J.M., Pham, K.T., et al., 2008. Identification and characterization of transmitted and early founder virus envelopes in primary HIV-1 infection. Proc. Natl. Acad. Sci. U.S.A. 105, 7552–7557.

Koel, B.F., Burke, D.F., Bestebroer, T.M., van der Vliet, S., Zondag, G.C., et al., 2013. Substitutions near the receptor binding site determine major antigenic change during influenza virus evolution. Science 342, 976–979.

Koel, B.F., van der Vliet, S., Burke, D.F., Bestebroer, T.M., Bharoto, E.E., et al., 2014. Antigenic variation of clade 2.1 H5N1 virus is determined by a few amino acid substitutions immediately adjacent to the receptor binding site. mBio 5 e01070–14.

Korber, B.T., Farber, R.M., Wolpert, D.H., Lapedes, A.S., 1993. Covariation of mutations in the V3 loop of human immunodeficiency virus type 1 envelope protein: an information theoretic analysis. Proc. Natl. Acad. Sci. U.S.A. 90, 7176–7180.

Krakauer, D.C., Jansen, V.A., 2002. Red queen dynamics of protein translation. J. Theor. Biol. 218, 97–109.

Kubo, H., Takase-Yoden, S., Taguchi, F., 1993. Neutralization and fusion inhibition activities of monoclonal antibodies specific for the S1 subunit of the spike protein of neurovirulent murine coronavirus JHMV cl-2 variant. J. Gen. Virol. 74 (Pt 7), 1421–1425.

Kubo, H., Yamada, Y.K., Taguchi, F., 1994. Localization of neutralizing epitopes and the receptor-binding site within the amino-terminal 330 amino acids of the murine coronavirus spike protein. J. Virol. 68, 5403–5410.

Kuss, S.K., Etheredge, C.A., Pfeiffer, J.K., 2008. Multiple host barriers restrict poliovirus trafficking in mice. PLoS Pathog. 4, e1000082.

Laeeq, S., Smith, C.A., Wagner, S.D., Thomas, D.B., 1997. Preferential selection of receptor-binding variants of influenza virus hemagglutinin by the neutralizing antibody repertoire of transgenic mice expressing a human immunoglobulin mu minigene. J. Virol. 71, 2600–2605.

Lanza, A.M., Curran, K.A., Rey, L.G., Alper, H.S., 2014. A condition-specific codon optimization approach for improved heterologous gene expression in Saccharomyces cerevisiae. BMC Syst. Biol. 8, 33.

Lavner, Y., Kotlar, D., 2005. Codon bias as a factor in regulating expression via translation rate in the human genome. Gene 345, 127–138.

Le Nouen, C., Brock, L.G., Luongo, C., McCarty, T., Yang, L., et al., 2014. Attenuation of human respiratory syncytial virus by genome-scale codon-pair deoptimization. Proc. Natl. Acad. Sci. U.S.A. 111, 13169–13174.

Lea, S., Hernández, J., Blakemore, W., Brocchi, E., Curry, S., et al., 1994. The structure and antigenicity of a type C foot-and-mouth disease virus. Structure 2, 123–139.

Lea, S., Abu-Ghazaleh, R., Blakemore, W., Curry, S., Fry, E., et al., 1995. Structural comparison of two strains of foot-and-mouth disease virus subtype O1 and a laboratory antigenic variant, G67. Structure 3, 571–580.

Lee, P., Knight, R., Smit, J.M., Wilschut, J., Griffin, D.E., 2002. A single mutation in the E2 glycoprotein important for neurovirulence influences binding of sindbis virus to neuroblastoma cells. J. Virol. 76, 6302–6310.

Lerner, D.L., Elder, J.H., 2000. Expanded host cell tropism and cytopathic properties of feline immunodeficiency virus strain PPR subsequent to passage through interleukin-2-independent T cells. J. Virol. 74, 1854–1863.

Li, H., Roossinck, M.J., 2004. Genetic bottlenecks reduce population variation in an experimental RNA virus population. J. Virol. 78, 10582–10587.

Li, J.S., Tong, S.P., Wands, J.R., 1996. Characterization of a 120-Kilodalton pre-S-binding protein as a candidate duck hepatitis B virus receptor. J. Virol. 70, 6029–6035.

Li, P., Ke, X., Wang, T., Tan, Z., Luo, D., et al., 2018. Zika virus attenuation by codon pair deoptimization induces sterilizing immunity in mouse models. J. Virol. 92, e00701–e00718.

Li, W., Zhang, C., Sui, J., Kuhn, J.H., Moore, M.J., et al., 2005. Receptor and viral determinants of SARS-coronavirus adaptation to human ACE2. EMBO J. 24, 1634–1643.

Liebermann, H., Mentel, R., Bauer, U., Pring-Akerblom, P., Dolling, R., et al., 1998. Receptor binding sites and antigenic epitopes on the fiber knob of human adenovirus serotype 3. J. Virol. 72, 9121–9130.

Lindesmith, L.C., Brewer-Jensen, P.D., Mallory, M.L., Yount, B., Collins, M.H., et al., 2019. Human norovirus epitope D plasticity allows escape from antibody immunity without loss of capacity for binding cellular ligands. J. Virol. 93 (2), e01813–e01818.

Lonberg-Holm, K., Gosser, L.B., Shimshick, E.J., 1976. Interaction of liposomes with subviral particles of poliovirus type 2 and rhinovirus type 2. J. Virol. 19, 746–749.

Lopez-Bueno, A., Rubio, M.P., Bryant, N., McKenna, R., Agbandje-McKenna, M., et al., 2006. Host-selected amino acid changes at the sialic acid binding pocket of the parvovirus capsid modulate cell binding affinity and determine virulence. J. Virol. 80, 1563–1573.

Lucas-Hourani, M., Dauzonne, D., Jorda, P., Cousin, G., Lupan, A., et al., 2013. Inhibition of pyrimidine biosynthesis pathway suppresses viral growth through innate immunity. PLoS Pathog. 9, e1003678.

Manning, T., Sleator, R.D., Walsh, P., 2013. Naturally selecting solutions: the use of genetic algorithms in bioinformatics. Bioengineered 4, 266–278.

Maraia, R.J., Iben, J.R., 2014. Different types of secondary information in the genetic code. RNA 20, 977–984.

Marcus, P.I., Rodriguez, L.L., Sekellick, M.J., 1998. Interferon induction as a quasispecies marker of vesicular stomatitis virus populations. J. Virol. 72, 542–549.

Markowitz, M., Louie, M., Hurley, A., Sun, E., Di Mascio, M., et al., 2003. A novel antiviral intervention results in more accurate assessment of human immunodeficiency virus type 1 replication dynamics and T-cell decay in vivo. J. Virol. 77, 5037–5038.

Martínez, M.A., Dopazo, J., Hernandez, J., Mateu, M.G., Sobrino, F., et al., 1992. Evolution of the capsid protein genes of foot-and-mouth disease virus: antigenic variation without accumulation of amino acid substitutions over six decades. J. Virol. 66, 3557–3565.

Martinez, M.A., Verdaguer, N., Mateu, M.G., Domingo, E., 1997. Evolution subverting essentiality: dispensability of the cell attachment Arg-Gly-Asp motif in multiply passaged foot-and-mouth disease virus. Proc. Natl. Acad. Sci. U.S.A. 94, 6798–6802.

Martinez-Salas, E., Regalado, M.P., Domingo, E., 1996. Identification of an essential region for internal initiation of translation in the aphthovirus internal ribosome entry site and implications for viral evolution. J. Virol. 70, 992–998.

Martinez-Sobrido, L., Emonet, S., Giannakas, P., Cubitt, B., Garcia-Sastre, A., et al., 2009. Identification of amino acid residues critical for the anti-interferon activity of the nucleoprotein of the prototypic arenavirus lymphocytic choriomeningitis virus. J. Virol. 83, 11330–11340.

Martrus, G., Nevot, M., Andres, C., Clotet, B., Martinez, M.A., 2013. Changes in codon-pair bias of human immunodeficiency virus type 1 have profound effects on virus replication in cell culture. Retrovirology 10, 78.

Mateu, M.G., 1995. Antibody recognition of picornaviruses and escape from neutralization: a structural view. Virus Res. 38, 1–24.

Mateu, M.G., 2017. The foot-and-mouth disease virion: structure and function. In: Sobrino, F., Domingo, E. (Eds.), Foot-and-Mouth Disease Virus: Current Research and Emerging Trends. Caister Academic Press, Norfolk, UK, pp. 61–105.

Matsuzaki, M., Sugawara, K., Adachi, K., Hongo, S., Nishimura, H., et al., 1992. Location of neutralizing epitopes on the hemagglutinin-esterase protein of influenza C virus. Virology 189, 79–87.

Mauck, K.E., De Moraes, C.M., Mescher, M.C., 2010. Deceptive chemical signals induced by a plant virus attract insect vectors to inferior hosts. Proc. Natl. Acad. Sci. U.S.A. 107, 3600–3605.

Maury, W., Thompson, R.J., Jones, Q., Bradley, S., Denke, T., et al., 2005. Evolution of the equine infectious anemia virus long terminal repeat during the alteration of cell tropism. J. Virol. 79, 5653–5664.

Maynard Smith, J., Burian, R., Dauffman, S., Alberch, P., Campbell, J., 1985. Developmental constraints and evolution. Q. Rev. Biol. 60, 265–287.

McMichael, A.J., Phillips, R.E., 1997. Escape of human immunodeficiency virus from immune control. Annu. Rev. Immunol. 15, 271–296.

Mendelsohn, C.L., Wimmer, E., Racaniello, V.R., 1989. Cellular receptor for poliovirus: molecular cloning, nucleotide sequence, and expression of a new member of the immunoglobulin superfamily. Cell 56, 855–865.

Minocha, H.C., Xue, W., Reddy, J.R., 1997. A 50 kDa membrane protein from bovine kidney cells is a putative receptor for bovine viral diarrhea virus (BVDV). Adv. Exp. Med. Biol. 412, 145–148.

Moratorio, G., Henningsson, R., Barbezange, C., Carrau, L., Borderia, A.V., et al., 2017. Attenuation of RNA viruses by redirecting their evolution in sequence space. Nat Microbiol 2, 17088.

Moratorio, G., Vignuzzi, M., 2018. Monitoring and redirecting virus evolution. PLoS Pathog. 14, e1006979.

Morse, S.S., 1994. The Evolutionary Biology of Viruses. Raven Press, New York.

Mosier, D.E., Picchio, G.R., Gulizia, R.J., Sabbe, R., Poignard, P., et al., 1999. Highly potent RANTES analogues either prevent CCR5-using human immunodeficiency virus type 1 infection in vivo or rapidly select for CXCR4-using variants. J. Virol. 73, 3544–3550.

Moury, B., Fabre, F., Senoussi, R., 2007. Estimation of the number of virus particles transmitted by an insect vector. Proc. Natl. Acad. Sci. U.S.A. 104, 17891–17896.

Mueller, S., Papamichail, D., Coleman, J.R., Skiena, S., Wimmer, E., 2006. Reduction of the rate of poliovirus protein synthesis through large-scale codon deoptimization causes attenuation by lowering specific infectivity. J. Virol. 80 (19), 9687–9696.

Murcia, P.R., Baillie, G.J., Daly, J., Elton, D., Jervis, C., et al., 2010. Intra- and interhost evolutionary dynamics of equine influenza virus. J. Virol. 84, 6943–6954.

Murphy, B.R., Webster, R.G., 1985. Influenza viruses. In: Fields, B.N. (Ed.), Virology. Raven Press, New York, pp. 1179–1239.

Murray, M.G., Bradley, J., Yang, X.F., Wimmer, E., Moss, E.G., et al., 1988. Poliovirus host range is determined by a short amino acid sequence in neutralization antigenic site I. Science 241, 213–215.

Neumann, A.U., Lam, N.P., Dahari, H., Gretch, D.R., Wiley, T.E., et al., 1998. Hepatitis C viral dynamics in vivo and the antiviral efficacy of interferon-alpha therapy. Science 282, 103–107.

Neurath, A.R., Kent, S.B., Strick, N., Parker, K., 1986. Identification and chemical synthesis of a host cell receptor binding site on hepatitis B virus. Cell 46, 429–436.

Nice, T.J., Strong, D.W., McCune, B.T., Pohl, C.S., Virgin, H.W., 2013. A single-amino-acid change in murine norovirus NS1/2 is sufficient for colonic tropism and persistence. J. Virol. 87, 327–334.

Nogales, A., Baker, S.F., Ortiz-Riano, E., Dewhurst, S., Topham, D.J., et al., 2014. Influenza A virus attenuation by codon deoptimization of the NS gene for vaccine development. J. Virol. 88, 10525–10540.

Novella, I.S., Cilnis, M., Elena, S.F., Kohn, J., Moya, A., et al., 1996. Large-population passages of vesicular stomatitis virus in interferon-treated cells select variants of only limited resistance. J. Virol. 70, 6414–6417.

Nowak, M.A., Bonhoeffer, S., Hill, A.M., Boehme, R., Thomas, H.C., et al., 1996. Viral dynamics in hepatitis B virus infection. Proc. Natl. Acad. Sci. U.S.A. 93, 4398–4402.

Núñez, J.I., Baranowski, E., Molina, N., Ruiz-Jarabo, C.M., Sánchez, C., et al., 2001. A single amino acid substitution in nonstructural protein 3A can mediate adaptation of foot-and-mouth disease virus to the Guinea pig. J. Virol. 75, 3977–3983.

Olsthoorn, R.C., Licis, N., van Duin, J., 1994. Leeway and constraints in the forced evolution of a regulatory RNA helix. EMBO J. 13, 2660–2668.

Ortiz-Riano, E., Ngo, N., Devito, S., Eggink, D., Munger, J., et al., 2014. Inhibition of arenavirus by A3, a pyrimidine biosynthesis inhibitor. J. Virol. 88, 878–889.

Palstra, F.P., Ruzzante, D.E., 2008. Genetic estimates of contemporary effective population size: what can they tell us about the importance of genetic stochasticity for wild population persistence? Mol. Ecol. 17, 3428–3447.

Parrish, C.R., Kawaoka, Y., 2005. The origins of new pandemic viruses: the acquisition of new host ranges by canine parvovirus and influenza A viruses. Annu. Rev. Microbiol. 59, 553–586.

Paul, A.V., 2002. Possible unifying mechanism of picornavirus genome replication. In: Semler, B.L., Wimmer, E. (Eds.), Molecular Biology of Picornaviruses. ASM Press, Washington, DC, pp. 227–246.

Pauli, G., Ludwig, H., 1985. Increase of virus yields and releases of Borna disease virus from persistently infected cells. Virus Res. 2, 29–33.

Pawlak, E.N., Dikeakos, J.D., 2015. HIV-1 Nef: a master manipulator of the membrane trafficking machinery mediating immune evasion. Biochim. Biophys. Acta 1850, 733–741.

Perales, C., 2018. Quasispecies dynamics and clinical significance of HCV antiviral resistance. Int. J. Antimicrob. Agents S0924–8579 (18), 30297–30298.

Perales, C., Beach, N.M., Gallego, I., Soria, M.E., Quer, J., et al., 2013. Response of hepatitis C virus to long-term passage in the presence of alpha interferon: multiple mutations and a common phenotype. J. Virol. 87, 7593–7607.

Perales, C., Beach, N.M., Sheldon, J., Domingo, E., 2014. Molecular basis of interferon resistance in hepatitis C virus. Curr. Opin. Virol. 8, 38–44.

Perelson, A.S., Neumann, A.U., Markowitz, M., Leonard, J.M., Ho, D.D., 1996. HIV-1 dynamics in vivo: virion clearance rate, infected cell life-span, and viral generation time. Science 271, 1582–1586.

Petit, M.A., Capel, F., Dubanchet, S., Mabit, H., 1992. PreS1-specific binding proteins as potential receptors for hepatitis B virus in human hepatocytes. Virology 187, 211–222.

Pfeiffer, J.K., Kirkegaard, K., 2005. Increased fidelity reduces poliovirus fitness under selective pressure in mice. PLoS Pathog. 1, 102–110.

Pfeiffer, J.K., Kirkegaard, K., 2006. Bottleneck-mediated quasispecies restriction during spread of an RNA virus from inoculation site to brain. Proc. Natl. Acad. Sci. U.S.A. 103, 5520–5525.

Philadelpho, N.A., Rubbenstroth, D., Guimaraes, M.B., Piantino Ferreira, A.J., 2014. Survey of bornaviruses in pet psittacines in Brazil reveals a novel parrot bornavirus. Vet. Microbiol. 174, 584–590.

Pleij, C.W.A., 1994. RNA pseudoknots. Curr. Opin. Struct. Biol. 4, 337–344.

Popovici, E., 2018. Co-optimization free lunches: tractability of optimal black-box algorithms for maximizing expected utility. Evol. Comput. 26, 145–175.

Quer, J., Esteban, J.I., Cos, J., Sauleda, S., Ocana, L., et al., 2005. Effect of bottlenecking on evolution of the nonstructural protein 3 gene of hepatitis C virus during sexually transmitted acute resolving infection. J. Virol. 79, 15131–15141.

Quer, J., Martell, M., Rodriguez, A., Bosch, A., Jardi, R., et al., 2008. The impact of rapid evolution of hepatitis viruses. In: Domingo, E., Parrish, C., Holland, J.J. (Eds.), Origin and Evolution of Viruses. Elsevier, Oxford, pp. 303–350.

Ramratnam, B., Bonhoeffer, S., Binley, J., Hurley, A., Zhang, L., et al., 1999. Rapid production and clearance of HIV-1 and hepatitis C virus assessed by large volume plasma apheresis. Lancet 354, 1782–1785.

Reeve, R., Smith, E., Wallace, B., 1990. Components of fitness become effectively neutral in equilibrium populations. Proc. Natl. Acad. Sci. U.S.A. 87, 2018–2020.

Remmert, H., 1980. Ecology. Springer-Verlag, Berlin Heidelberg.

Richards, O.C., Ehrenfeld, E., 1990. Pliovirus RNA replication. Curr. Top. Microbiol. Immunol. 161, 90–119.

Richman, D.D., Wrin, T., Little, S.J., Petropoulos, C.J., 2003. Rapid evolution of the neutralizing antibody response to HIV type 1 infection. Proc. Natl. Acad. Sci. U.S.A. 100, 4144–4149.

Robertson, H.D., 1975. Functions of replicating RNA in cells infected by RNA bacteriophages. In: RNA Phages. Cold Spring Harbor Laboratory, New York, pp. 113–115.

Robertson, J.S., Bootman, J.S., Newman, R., Oxford, J.S., Daniels, R.S., et al., 1987. Structural changes in the haemagglutinin which accompany egg adaptation of an influenza A(H1N1) virus. Virology 160, 31–37.

Rodriguez-Frías, F., Tabernero, D., Quer, J., Esteban, J.I., Ortega, I., et al., 2012. Ultra-deep pyrosequencing detects conserved genomic sites and quantifies linkage of drug-resistant amino acid changes in the hepatitis B virus genome. PLoS One 7, e37874.

Rogers, G.N., Paulson, J.C., Daniels, R.S., Skehel, J.J., Wilson, I.A., et al., 1983. Single amino acid substitutions in influenza haemagglutinin change receptor binding specificity. Nature 304, 76–78.

Roossinck, M.J., 2008. Mutant clouds and bottleneck events in plant virus evolution. In: Domingo, E., Parrish, C., Holland, J.J. (Eds.), Origin and Evolution of Viruses. Elsevier, Oxford, pp. 251–258.

Rouzine, I.M., Coffin, J.M., Weinberger, L.S., 2014. Fifteen years later: hard and soft selection sweeps confirm a large population number for HIV in vivo. PLoS Genet. 10, e1004179.

Roy, P., 2008. Molecular dissection of bluetongue virus. In: Mettenleiter, T.C., Sobrino, F. (Eds.), Animal Viruses. Molecular Biology. Cister Academic Press, Norfolk, U.K., pp. 305–353

Ruiz-Jarabo, C.M., Pariente, N., Baranowski, E., Dávila, M., Gómez-Mariano, G., et al., 2004. Expansion of host-cell tropism of foot-and-mouth disease virus despite replication in a constant environment. J. Gen. Virol. 85, 2289–2297.

Rustagi, A., Gale Jr., M., 2014. Innate antiviral immune signaling, viral evasion and modulation by HIV-1. J. Mol. Biol. 426, 1161–1177.

Sa-Carvalho, D., Rieder, E., Baxt, B., Rodarte, R., Tanuri, A., et al., 1997. Tissue culture adaptation of foot-and-mouth disease virus selects viruses that bind to heparin and are attenuated in cattle. J. Virol. 71, 5115–5123.

Sacristan, S., Diaz, M., Fraile, A., Garcia-Arenal, F., 2011. Contact transmission of Tobacco mosaic virus: a quantitative analysis of parameters relevant for virus evolution. J. Virol. 85, 4974–4981.

Saha, K., Zhang, J., Gupta, A., Dave, R., Yimen, M., et al., 2001. Isolation of primary HIV-1 that target CD8+ T lymphocytes using CD8 as a receptor. Nat. Med. 7, 65–72.

Sala, M., Centlivre, M., Wain-Hobson, S., 2006. Clade-specific differences in active viral replication and compartmentalization. Curr. Opin. HIV AIDS 1, 108–114.

Salazar-Gonzalez, J.F., Salazar, M.G., Keele, B.F., Learn, G.H., Giorgi, E.E., et al., 2009. Genetic identity, biological phenotype, and evolutionary pathways of transmitted/ founder viruses in acute and early HIV-1 infection. J. Exp. Med. 206, 1273–1289.

Salemi, M., Vandamme, A.M., 2004. The Phylogenetic Handbook. A Practical Approach to DNA and Protein Phylogeny. Cambridge University Press, Cambridge.

Sánchez, G., Bosch, A., Pinto, R.M., 2003. Genome variability and capsid structural constraints of hepatitis a virus. J. Virol. 77, 452–459.

Sanz-Ramos, M., Diaz-San Segundo, F., Escarmis, C., Domingo, E., Sevilla, N., 2008. Hidden virulence determinants in a viral quasispecies in vivo. J. Virol. 82, 10465–10476.

Sato, S., Zhang, L., Kim, J., Jakob, J., Grant, R.A., et al., 1996. A neutralization site of DA strain of Theiler's murine encephalomyelitis virus important for disease phenotype. Virology 226, 327–337.

Scholle, F., Girard, Y.A., Zhao, Q., Higgs, S., Mason, P.W., 2004. Trans-packaged West Nile virus-like particles: infectious properties in vitro and in infected mosquito vectors. J. Virol. 78, 11605–11614.

Schuster, P., 2011. Mathematical modeling of evolution. Solved and open problems. Theor. Biosci. 130, 71–89.

Seet, B.T., Johnston, J.B., Brunetti, C.R., Barrett, J.W., Everett, H., et al., 2003. Poxviruses and immune evasion. Annu. Rev. Immunol. 21, 377–423.

Sellers, R.F., 1971. Quantitative aspects of the spread of foot-and-mouth disease. Vet. Bull. 41, 431–439.

Sevilla, N., Kunz, S., Holz, A., Lewicki, H., Homann, D., et al., 2000. Immunosuppression and resultant viral persistence by specific viral targeting of dendritic cells. J. Exp. Med. 192, 1249–1260.

Sevilla, N., Domingo, E., de la Torre, J.C., 2002. Contribution of LCMV towards deciphering biology of quasispecies in vivo. Curr. Top. Microbiol. Immunol. 263, 197–220.

Shackelton, L.A., Parrish, C.R., Truyen, U., Holmes, E.C., 2005. High rate of viral evolution associated with the emergence of carnivore parvovirus. Proc. Natl. Acad. Sci. U.S.A. 102, 379–384.

Shackelton, L.A., Parrish, C.R., Holmes, E.C., 2006. Evolutionary basis of codon usage and nucleotide composition bias in vertebrate DNA viruses. J. Mol. Evol. 62, 551–563.

Sharma, N.K., Sherker, A.H., 2009. Epidemiology, risk factors, and natural history of chronic hepatitis C. In: Shetty, K., Wu, G.Y. (Eds.), Clinical Gastroenterology. Chronic Viral Hepatitis. Diagnosis and Therapeutics. Humana Press, Springer Science Business media, LLC, New York, pp. 33–70.

Shepard, C.W., Finelli, L., Alter, M.J., 2005. Global epidemiology of hepatitis C virus infection. Lancet Infect. Dis. 5, 558–567.

Simmonds, P., Tuplin, A., Evans, D.J., 2004. Detection of genome-scale ordered RNA structure (GORS) in genomes of positive-stranded RNA viruses: implications for virus evolution and host persistence. RNA 10, 1337–1351.

Skehel, J.J., Wiley, D.C., 2000. Receptor binding and membrane fusion in virus entry: the influenza hemagglutinin. Annu. Rev. Biochem. 69, 531–569.

Smelt, S.C., Borrow, P., Kunz, S., Cao, W., Tishon, A., et al., 2001. Differences in affinity of binding of lymphocytic choriomeningitis virus strains to the cellular receptor alpha-dystroglycan correlate with viral tropism and disease kinetics. J. Virol. 75, 448–457.

Smith, T.J., Cheng, R.H., Olson, N.H., Peterson, P., Chase, E., et al., 1995. Putative receptor binding sites on alphaviruses as visualized by cryoelectron microscopy. Proc. Natl. Acad. Sci. U.S.A. 92, 10648–10652.

Smith, T.J., Chase, E.S., Schmidt, T.J., Olson, N.H., Baker, T.S., 1996. Neutralizing antibody to human rhinovirus 14 penetrates the receptor- binding canyon. Nature 383, 350–354.

Smith, D.R., Adams, A.P., Kenney, J.L., Wang, E., Weaver, S.C., 2008. Venezuelan equine encephalitis virus in the mosquito vector Aedes taeniorhynchus: infection initiated by a small number of susceptible epithelial cells and a population bottleneck. Virology 372, 176–186.

Spear, P.G., Eisenberg, R.J., Cohen, G.H., 2000. Three classes of cell surface receptors for alphaherpesvirus entry. Virology 275, 1–8.

Stewart, P.L., Nemerow, G.R., 1997. Recent structural solutions for antibody neutralization of viruses. Trends Microbiol. 5, 229–233.

Stoll, R., Hobom, G., Muller, H., 1994. Host restriction in the productive cycle of avian polyomavirus budgerigar fledgling disease virus type 3 depends on a single amino acid change in the common region of structural proteins VP2/VP3. J. Gen. Virol. 75 (Pt 9), 2261–2269.

Strauss, J.H., Wang, K.S., Schmaljohn, A.L., Kuhn, R.J., Strauss, E.G., 1994. Host-cell receptors for Sindbis virus. Arch. Virol. 9, 473–484.

Strauss, M., Filman, D.J., Belnap, D.M., Cheng, N., Noel, R.T., et al., 2015. Nectin-like interactions between poliovirus and its receptor trigger conformational changes associated with cell entry. J. Virol. 89, 4143–4157.

Sunyach, C., Rollier, C., Robaczewska, M., Borel, C., Barraud, L., et al., 1999. Residues critical for duck hepatitis B virus neutralization are involved in host cell interaction. J. Virol. 73, 2569–2575.

Switzer, W.M., Salemi, M., Shanmugam, V., Gao, F., Cong, M.E., et al., 2005. Ancient co-speciation of simian foamy viruses and primates. Nature 434, 376–380.

Taboga, O., Tami, C., Carrillo, E., Núñez, J.I., Rodríguez, A., et al., 1997. A large-scale evaluation of peptide vaccines against foot-and-mouth disease: lack of solid protection in cattle and isolation of escape mutants. J. Virol. 71, 2606–2614.

Takeuchi, Y., Akutsu, M., Murayama, K., Shimizu, N., Hoshino, H., 1991. Host range mutant of human immunodeficiency virus type 1: modification of cell tropism by a single point mutation at the neutralization epitope in the env gene. J. Virol. 65, 1710–1718.

Tami, C., Taboga, O., Berinstein, A., Nuñez, J.I., Palma, E.L., et al., 2003. Evidence of the coevolution of antigenicity and host cell tropism of foot-and-mouth disease virus in vivo. J. Virol. 77, 1219–1226.

Thaker, S.R., Stine, D.L., Zamb, T.J., Srikumaran, S., 1994. Identification of a putative cellular receptor for bovine herpesvirus 1. J. Gen. Virol. 75 (Pt 9), 2303–2309.

Thulke, S., Radonic, A., Nitsche, A., Siegert, W., 2006. Quantitative expression analysis of HHV-6 cell receptor CD46 on cells of human cord blood, peripheral blood and G-CSF mobilised leukapheresis cells. Virol. J. 3, 77.

Tong, S., Li, J., Wands, J.R., 1995. Interaction between duck hepatitis B virus and a 170-kilodalton cellular protein is mediated through a neutralizing epitope of the pre-S region and occurs during viral infection. J. Virol. 69, 7106–7112.

Tseng, Y.S., Gurda, B.L., Chipman, P., McKenna, R., Afione, S., et al., 2015. Adeno-associated virus serotype 1 (AAV1)- and AAV5-antibody complex structures reveal evolutionary commonalities in parvovirus antigenic reactivity. J. Virol. 89, 1794–1808.

Tucker, P.C., Griffin, D.E., 1991. Mechanism of altered Sindbis virus neurovirulence associated with a single-amino-acid change in the E2 Glycoprotein. J. Virol. 65, 1551–1557.

Ubol, S., Griffin, D.E., 1991. Identification of a putative alphavirus receptor on mouse neural cells. J. Virol. 65, 6913–6921.

Umetsu, Y., Sugawara, K., Nishimura, H., Hongo, S., Matsuzaki, M., et al., 1992. Selection of antigenically distinct variants of influenza C viruses by the host cell. Virology 189, 740–744.

Vahlenkamp, T.W., Verschoor, E.J., Schuurman, N.N., van Vliet, A.L., Horzinek, M.C., et al., 1997. A single amino acid substitution in the transmembrane envelope glycoprotein of feline immunodeficiency virus alters cellular tropism. J. Virol. 71, 7132–7135.

Van Valen, L., 1973. A new evolutionary law. Evol. Theory 1, 1–30.

Varthakavi, V., Minocha, H.C., 1996. Identification of a 56 kDa putative bovine herpesvirus 1 cellular receptor by anti-idiotype antibodies. J. Gen. Virol. 77 (Pt 8), 1875–1882.

Ventoso, I., 2012. Adaptive changes in alphavirus mRNA translation allowed colonization of vertebrate hosts. J. Virol. 86, 9484–9494.

Verdaguer, N., Mateu, M.G., Andreu, D., Giralt, E., Domingo, E., et al., 1995. Structure of the major antigenic loop of foot-and-mouth disease virus complexed with a neutralizing antibody: direct involvement of the Arg-Gly-Asp motif in the interaction. EMBO J. 14, 1690–1696.

Verdaguer, N., Ferrero, D., Murthy, M.R., 2014. Viruses and viral proteins. IUCrJ 1, 492–504.

Verschoor, E.J., Boven, L.A., Blaak, H., van Vliet, A.L., Horzinek, M.C., et al., 1995. A single mutation within the V3 envelope neutralization domain of feline immunodeficiency virus determines its tropism for CRFK cells. J. Virol. 69, 4752–4757.

Vignuzzi, M., Andino, R., 2010. Biological implications of picornavirus fidelity mutants. In: Ehrenfeld, E., Domingo, E., Roos, R.P. (Eds.), The Picornaviruses. ASM Press, Washington, DC, pp. 213–228.

Vignuzzi, M., Stone, J.K., Arnold, J.J., Cameron, C.E., Andino, R., 2006. Quasispecies diversity determines pathogenesis through cooperative interactions in a viral population. Nature 439, 344–348.

Villarreal, L.P., 2005. Viruses and the Evolution of Life. ASM Press, Washington, DC.

Villordo, S.M., Filomatori, C.V., Sanchez-Vargas, I., Blair, C.D., Gamarnik, A.V., 2015. Dengue virus RNA structure specialization facilitates host adaptation. PLoS Pathog. 11, e1004604.

Virgin, H.W., 2014. The virome in mammalian physiology and disease. Cell 157, 142–150.

Wang, K.S., Schmaljohn, A.L., Kuhn, R.J., Strauss, J.H., 1991. Antiidiotypic antibodies as probes for the Sindbis virus receptor. Virology 181, 694–702.

Weber, F., Kochs, G., Haller, O., 2004. Inverse interference: how viruses fight the interferon system. Viral Immunol. 17, 498–515.

Webster, B., Ott, M., Greene, W.C., 2013. Evasion of superinfection exclusion and elimination of primary viral RNA by an adapted strain of hepatitis C virus. J. Virol. 87, 13354–13369.

Wei, X., Ghosh, S.K., Taylor, M.E., Johnson, V.A., Emini, E.A., et al., 1995. Viral dynamics in human immunodeficiency virus type 1 infection. Nature 373, 117–122.

Weiner, A., Erickson, A.L., Kansopon, J., Crawford, K., Muchmore, E., et al., 1995. Persistent hepatitis C virus infection in a chimpanzee is associated with emergence of a cytotoxic T lymphocyte escape variant. Proc. Natl. Acad. Sci. U.S.A. 92, 2755–2759.

Whalley, S.A., Murray, J.M., Brown, D., Webster, G.J., Emery, V.C., et al., 2001. Kinetics of acute hepatitis B virus infection in humans. J. Exp. Med. 193, 847–854.

Whitbeck, J.C., Muggeridge, M.I., Rux, A.H., Hou, W., Krummenacher, C., et al., 1999. The major neutralizing antigenic site on herpes simplex virus glycoprotein D overlaps a receptor-binding domain. J. Virol. 73, 9879–9890.

Whitley, D., Watson, J.P., 2005. Complexity theory and the no free lunch theorem. In: Burke, E.K., Kendall, G. (Eds.), Search Methodologies. Introductory Tutorials in Optimization and Decision Support Techniques. Springer Science + Business Media, LLC, New York, pp. 317–339.

Williams, W.V., Guy, H.R., Rubin, D.H., Robey, F., Myers, J.N., et al., 1988. Sequences of the cell-attachment sites of reovirus type 3 and its anti-idiotypic/antireceptor antibody: modeling of their three-dimensional structures. Proc. Natl. Acad. Sci. U.S.A. 85, 6488–6492.

Williams, W.V., Weiner, D.B., Greene, M.I., 1989. Development and use of antireceptor antibodies to study interaction of mammalian reovirus type 3 with its cell surface receptor. Methods Enzymol. 178, 321–341.

Williams, W.V., Kieber-Emmons, T., Weiner, D.B., Rubin, D.H., Greene, M.I., 1991. Contact residues and predicted structure of the reovirus type 3-receptor interaction. J. Biol. Chem. 266, 9241–9250.

Wommack, K.E., Bhavsar, J., Polson, S.W., Chen, J., Dumas, M., et al., 2012. VIROME: a standard operating procedure for analysis of viral metagenome sequences. Stand. Genomic Sci. 6, 427–439.

Woolhouse, M.E.J., Webster, J.P., Domingo, E., Charlesworth, B., Levin, B.R., 2002. Biological and biomedical implications of the coevolution of pathogens and their hosts. Nat. Genet. 32, 569–577.

Wu, Z., Asokan, A., Grieger, J.C., Govindasamy, L., Agbandje-McKenna, M., et al., 2006. Single amino acid changes can influence titer, heparin binding, and tissue tropism in different adeno-associated virus serotypes. J. Virol. 80, 11393–11397.

Xu, G., Wilson, W., Mecham, J., Murphy, K., Zhou, E.M., et al., 1997. VP7: an attachment protein of bluetongue virus for cellular receptors in Culicoides variipennis. J. Gen. Virol. 78 (Pt 7), 1617–1623.

Xue, W., Minocha, H.C., 1993. Identification of the cell surface receptor for bovine viral diarrhoea virus by using anti-idiotypic antibodies. J. Gen. Virol. 74 (Pt 1), 73–79.

Yamada, A., Brown, L.E., Webster, R.G., 1984. Characterization of H2 influenza virus hemagglutinin with monoclonal antibodies: influence of receptor specificity. Virology 138, 276–286.

Yang, Z., Nielsen, R., 2008. Mutation-selection models of codon substitution and their use to estimate selective strengths on codon usage. Mol. Biol. Evol. 25, 568–579.

Yang, H., Carney, P.J., Chang, J.C., Guo, Z., Villanueva, J.M., et al., 2015. Structure and receptor binding preferences of recombinant human A(H3N2) virus hemagglutinins. Virology 477C, 18–31.

Yeates, T.O., Jacobson, D.H., Martin, A., Wychowski, C., Girard, M., et al., 1991. Three-dimensional structure of a mouse-adapted type 2/type 1 poliovirus chimera. EMBO J. 10, 2331–2341.

Ying, T., Du, L., Ju, T.W., Prabakaran, P., Lau, C.C., et al., 2014. Exceptionally potent neutralization of Middle East respiratory syndrome coronavirus by human monoclonal antibodies. J. Virol. 88, 7796–7805.

Zhang, Y.-Z., Shi, M., Holmes, E.C., 2018. Using metagenomics to characterize an expanding virosphere. Cell 172, 1168–1172.

# Viral fitness as a measure of adaptation

## Abbreviations

**A** Adenine
**AIDS** Acquired Immune Deficiency Syndrome
**BHK** Baby Hamster Kidney
**CHO** Chinese Hamster Oovary
**CVB** Coxsackievirus Group B
**DENV** Dengue Virus
**FMDV** Foot-and-Mouth Disease Virus
**G** Guanine
**HAV** Hepatitis A Virus
**HBV** Hepatitis B Virus
**HCV** Hepatitis C Virus
**HIV-1** Human Immunodeficiency Virus Type 1
**IV** Influenza Virus
**LCMV** Lymphocytic Choriomeningitis Virus
**MOI** Multiplicity of Infection
**PBMC** Peripheral Blood Mononuclear Cell
**PV** Poliovirus
**RTP** Ribavirin-Triphosphate
**RT-PCR** Reverse Transcription Followed by the Polymerase Chain Reaction
**TCID50** Tissue Culture Infectious Dose Needed to kill 50% of cells
**tRNA** Transfer RNA
**U** Uracil

## 5.1 Origin of the fitness concept and its relevance to viruses

*Viral fitness*, also termed *selective* or *adaptive value*, is an application of the concept of Darwinian fitness to viruses. From the early foundations of evolutionary theory, genetic variation, fitness, and evolutionary change were considered to be connected concepts. The relative reproductive efficiency that we recapitulate with the term "fitness," correlates with the rate of evolutionary change promoted by natural selection. Fisher's fundamental theorem of natural selection asserts that the rate of fitness increase of a population equates with the genetic variance of fitness at that time (Fisher, 1930). Fitness measurements go back to the early developments of population genetics, using *Drosophila* as a model organism (Dobzhansky et al., 1977; Spiess, 1977) with observations that were later reexamined with viruses, and provided a molecular interpretation of some fitness variations [reviewed in (Domingo et al. 2019)]. F.J. Ayala showed that mixed *Drosophila* populations or populations that were artificially diversified by limited radiation displayed higher fitness than the undiversified populations, thus providing experimental support to Fisher's theorem (Ayala, 1965, 1969). In practical terms, fitness refers to the relative contributions to the next generation made by the different individuals that compose a population. This definition is adequate for viruses because it can be adopted to mean the overall replication capacity of a mutant spectrum, despite possible modulation by intrapopulation interactions of cooperation, complementation, or interference (Chapter 3). Fitness quantifies the ability of a virus population (irrespective of its complexity) to produce infectious progeny, usually compared with a

*Virus as Populations*
https://doi.org/10.1016/B978-0-12-816331-3.00005-2

reference viral clone, in a defined environment (Domingo and Holland, 1997; Quinones-Mateu and Arts, 2006; Martinez-Picado and Martinez, 2008; Wargo and Kurath, 2012; Domingo et al., 2019). There is no general agreement on how should fitness be measured. Some authors prefer quantifying fitness as the contribution of progeny to the next generation, while others favor the number of progeny in the second or following generations. In the case of viruses, by emphasizing viral production in a specific environment, it is assumed that fitness measured as the progeny in infection will be maintained in the next generations (next rounds of infection in serial passages) provided the environment remains invariant. This assumption should, however, be taken with caution because of the continuous variation of mutant spectrum composition due to the stochastic occurrence of mutations, and consequent uncertainties inherent to quasispecies evolution, even in a constant environment (Section 5.1.3). Different points of view have been expressed regarding whether fitness should be considered an absolute or relative parameter, and if fitness is exactly equivalent to a selection coefficient (Maree et al., 2000; Wargo and Kurath, 2011). We will not dwell on these issues, we will use the term fitness rather than the selection coefficient, and for convenience, we consider that viral fitness is a relative value. This solution facilitates the comparison and interpretation of fitness values and fitness differences, as determined by growth-competition experiments between two viruses, the standard approach to quantifying fitness.

## 5.1.1 Measurement of viral fitness

Competition aimed at determining a relative fitness value is achieved by coinfecting a host (either a primary cell culture, an established cell line, an explanted tissue, an organoid, an animal, or a plant), with a mixture containing known proportions of the virus to be tested

and a reference variant or population of the same virus. For higher reliability, different initial proportions of the two competitors should be assayed in triplicate. The progeny virus obtained in the first infection is then used to infect fresh cells (or the chosen host) under the same conditions, and the proportion of the two viruses, relative to their proportion in the initial mixture, is determined. The logarithm of the ratio of the two competitors is plotted as a function of passage number. This yields a fitness vector, which is adjusted to an exponential equation. The antilogarithm of the vector slope quantitates the relative fitness between the reference and competitor virus (Holland et al., 1991; Duarte et al., 1992; Clarke et al., 1994; Escarmís et al., 1999; Domingo et al., 2001, 2019; Domingo, 2006; Quinones-Mateu and Arts, 2006; Sheldon et al., 2014) (Fig. 5.1). The two competing viruses can be distinguished by a phenotypic or a genotypic marker. A phenotypic marker can be any

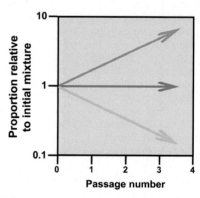

**FIGURE 5.1** Representation of fitness vectors. An equal mixture of the mutant and population to be tested and the reference virus is subjected to four serial passages in a host. The amount of the two competing viruses is determined at each passage. The logarithm of the proportion of the two viruses relative to the initial mixture is plotted as a function of passage number. In the upper, orange vector, the mutant or population displays a fitness advantage over the reference virus. In the horizontal, blue vector, both viruses have equal fitness. In the downward, green vector, the mutant virus displays a selective disadvantage relative to the reference. See text for studies where numerical quantifications of fitness can be found.

measurable trait that differs between the two viruses. Examples are resistance to a neutralizing monoclonal antibody or to an inhibitor so that the two viruses can be quantified by plating the virus at each passage in the presence and absence of the selective agent. Markers independent of an externally added selective agent that can be used for cytopathic viruses in cell culture are plaque morphology or size, provided the difference of these traits between the competing viruses is clear and stable; relying on plaque size has potential pitfalls derived from its dependence on environmental factors (presence of inhibitory polymers in the semi-solid medium, etc.), some of which may elude control.

The typical genotypic marker is a nucleotide sequence difference between the genome of the reference virus and the virus to be tested. The proportion of the two viruses can be estimated by the consensus nucleotide sequence of the population at each competition passage, by integrating the amounts of the nucleotides that distinguish the two viruses using the sequencing peak output, or through deep sequencing methods. Alternatively, specific primers can be designed to discriminate the two viral genomes by selective amplification on the basis of the nucleotide difference, using real-time RT-PCR for quantification (Yuste et al., 1999; Herrera et al., 2007; Wargo and Kurath, 2011; Sheldon et al., 2014). Fitness may be determined for an entire virus, or a viral gene or regulatory region inserted into an infectious clone, with a preselected genetic background (Centlivre et al., 2005; Domingo, 2006).

## 5.1.2 Power and limitations of fitness measurements

Comparison of fitness values is an important tool in viral genetics because fitness captures the overall capacity of a virus to replicate, independently of the step in the virus life cycle and the molecular mechanisms involved. The replicative capacity of a virus in vivo is medically relevant because it largely determines the viral load in infected individuals, and viral load can be a determinant of the progression of viral infections toward disease manifestations (Chapter 8). The fitness level may be influenced by any step in the virus life cycle, namely, entry into the host cells, uncoating of the viral genome, expression of viral proteins and genome replication (or transcription and replication), particle assembly, release from the cell, and stability of the free viral particle. Specific mutations can modify one or several processes that participate in the virus life cycle. Adaptation of Dengue virus to vertebrate versus invertebrate cells was linked to different types of noncoding RNAs that were derived from the 3′ untranslated region of the viral RNA generated as a consequence of point mutations (Filomatori et al., 2017). Elucidation of the steps responsible for a fitness variation is of great interest regarding the understanding of virus-cell interactions, and a prerequisite to designing experiments to define the molecular interactions behind fitness differences. For some purposes, however, the fitness value by itself may be sufficiently informative to assist in understanding virus behavior under a set of environmental conditions.

A number of limitations must be taken into consideration in performing and interpreting fitness measurements by growth-competition experiments in any host. A major limitation of the assays carried out in vivo stems from the fact that a fitness value depends on the environment in which fitness is measured, and an entire host organism represents a heterogeneous, highly compartmentalized environment. Compartmentalization is believed to have played an important role in the origin of complex life forms on earth (Chapter 1), and it has been maintained as an essential attribute of all differentiated organisms. For this reason, an array of environmental conditions is unavoidably presented to all viruses during an infection process. A virus may display high relative fitness in one cell type and low fitness in another cell type. A cell-dependent fitness difference was used to demonstrate that low fitness is an important determinant of viral extinction through the transition into error catastrophe: the same clonal population of foot-and-mouth disease virus (FMDV) could be easily extinguished in a cell line in which FMDV fitness was low, but not in a cell line in which FMDV fitness was high (Pariente et al., 2001) (Chapter 9).

The environment-dependent nature of fitness means that a ranking of fitness values established in cell culture may not reflect fitness differences in the natural hosts of the virus. This limitation of cell culture or *ex vivo* measurements to mirror fitness differences in vivo has been amply recognized, particularly in several studies that have addressed human immunodeficiency virus type 1 (HIV-1) fitness in vivo [reviewed in (Quinones-Mateu and Arts, 2006; Martinez-Picado and Martinez, 2008; Clementi and Lazzarin, 2010; Domingo et al., 2019)]. For similar reasons, the relative fitness of HIV-1 isolates may yield a different ranking when fitness is measured in peripheral blood mononuclear cells (PBMCs) or in established cell lines. Effort has been put in understanding HIV-1 fitness cost as a barrier to escape from antiretroviral agents (Harrigan et al., 1998; Weber et al., 2007; Martinez-Picado and Martinez, 2008; Quinones-Mateu et al., 2008), with the development of computational models proposed to assist in treatment planning (Gopalakrishnan et al., 2014) (fitness cost of drug resistance is discussed in Chapter 8).

The fundamental limitations of fitness values in vivo due to environmental heterogeneity are not the only concern regarding fitness determinations. Even in cell culture experiments, several key variables may also affect fitness values. If cells are infected at low multiplicity of infection (MOI), initially each of the two viruses involved in the competition for fitness measurement will replicate in separate cells. However, as the infection proceeds, if the yield of the two viruses is sufficiently high and the infection is not interrupted, a second round of high multiplicity infection will occur in the same virus passage. The consequence of this infection design is that there will be two waves of replication: a first wave without intracellular competition between the reference and competitor viruses, and a second wave of intracellular competition between the two viruses in need of exploiting cell resources (ribosomes and tRNA for viral protein expression, structures where replication complexes and replication factories are located, cellular proteins needed for viral RNA replication, nucleotide substrates and other intracellular metabolites, etc.). Some of these resources may not be limiting or display a preference for any of the two competing viruses, but it cannot be excluded that other resources may favor replication of some viral mutants (or genome subpopulations) over others. When this happens, an advantage of one viral population over another may be manifested with more intensity when both replicate in the same cell than when they replicate in separate cells.

Viruses can adapt to the MOI (Chapter 3), thus rendering essential the control of MOI values in comparative growth-competition assays. The situation can become even more complex in the case in which the yield of one virus is higher than the other as a result of the first infection wave, and if the extent of intracellular competition in the second infection wave affects the relative replication capacity of the two viruses. Furthermore, intracellular competition may involve complementation or interference between the mutant clouds originated by the two viruses (Novella, 2003; García-Arriaza et al., 2004; Wilke et al., 2004; Domingo, 2006; Perales et al., 2010; Shirogane et al., 2019) (see Chapter 3).

There are alternative assay designs to minimize replicative phases at different MOI. Viral vectors and subgenomic replicon constructs can be designed to limit replication to a single cycle, avoiding subsequent, second wave cell infections [as examples, see (Mansky, 1998; Scholle et al., 2004)]. For natural viral isolates or laboratory viruses or clones that should be studied unchanged, one possibility is to carry out parallel, independent infections with each virus separately. The virus titer at a defined time (or at multiple times) postinfection can be used as a surrogate fitness value, avoiding intracellular competition between the viruses under comparison. This procedure was used to compare the fitness of FMDV clones subjected to plaque transfers since comparative experiments showed that the individual viral yield led to fitness values comparable to those obtained by growth-competition experiments [compare procedures in (Escarmís et al., 1999, 2002)]. An alternative possibility is to restrict each of the serial infections with the competing viruses to a single round of high or low MOI infection, in the latter case by limiting the duration of infection to prevent the high MOI phase. This and other alternatives pose technical difficulties that have been previously discussed (Wilke et al., 2004; Domingo, 2006; Wargo and Kurath, 2012).

The above considerations render essential for the growth-competition assays to define as precisely as possible the experimental conditions (MOI, number of cells and their passage history, cell density in the case of cell suspensions or confluence status in the case of cell monolayers, the time postinfection at which the virus is collected at each passage, etc.), and to use the same conditions for comparative fitness measurements. Furthermore, since unavoidable mutation (and in many cases also recombination) events occur stochastically in the course of the competitions, it is necessary to carry out triplicate fitness determinations in parallel, and advisable to limit the number of passages to four. In the case of fitness determinations in vivo, well-characterized host organisms should be used; when possible, inbred, sex-, and age-matched pathogen-screened animals kept under controlled conditions, etc. These precautions should minimize confounding effects resulting from random genetic changes and environmental differences in the course of the competitions (Box 5.1).

### 5.1.3 Dissection of fitness determinants

The usual aim of fitness determinations is to quantify relative overall multiplication capacities of variant viruses. Fitness assays may also be

---

**BOX 5.1**

### Main points about viral fitness

- It measures the replication capacity of a virus relative to a reference isolate, mutant or population of the same virus.
- It is important because it captures an overall performance of the virus, and it can be used to plan virological and biochemical tests to identify the basis of fitness increase or decrease.

- A fitness value is relative to an environment. It is not a universal value for a viral mutant or population.
- Fitness differences in vivo are useful to compare the replicative or pathogenic potential of a virus. Fitness in vivo is a combined outcome of multiple influences due to the heterogeneous environment provided by differentiated organisms.

designed to test differences in the efficacy of specific steps in the virus life cycle (virus entry by determining the kinetics of internalization of two mutant viruses, synthesis of viral proteins, the release of virus from cells, etc.). An example of fitness measurements on the basis of single-step competitions (for entry, replication, and release of virus from cells) was the elucidation that an increased thermal stability of the viral particle harboring RNA genomes with internal deletions contributed to the selective advantage of a segmented genome version of FMDV over its corresponding nonsegmented form (Ojosnegros et al., 2011). Segmentation took place when a biological clone of FMDV was subjected to 260 serial infections in baby hamster kidney (BHK-21) cells at high MOI (see Section 2.11 and Fig. 2.8 in Chapter 2 for a detailed account of this laboratory experiment). The increased stability of the bipartite virus version was probably associated with a lower packaging density of the genomes with deletions inside the geometrically constrained capsid. The fitness decrease inflicted by thermal instability of the particles with full-length genomes coincided with the fitness decrease determined by the competition between the two viral forms (Ojosnegros et al., 2011).

For viral quasispecies, the term environment regarding its influence on fitness has to be taken in its broadest sense, with multiple participating factors. Physical parameters, such as temperature and composition of the culture medium, as well as biological factors contributed by the cells used for the infection, are all part of the environment. If two competing viruses display different thermal stability, because they differ in amino acids at the capsid or surface proteins, temperature will affect fitness, and the effect will be critically dependent on the time that the assembled viral particles virus spend in the extracellular culture medium relative to the time in the intracellular environment, prior to exit from the cell. Time in the extracellular environment must be included as an additional step of the replication cycle that can affect a fitness value.

Although not widely appreciated, quasispecies theory teaches us that from the point of view of any replicating entity, the mutant spectrum itself must be considered as part of the environment (Stadler, 2016). That is, the same initial viral genome replicating in two identical cells will soon face different environments because the chances of two mutant spectra evolving with the same genome distributions are very low. Mutant spectrum composition, subjected to indeterminacies derived from the random occurrence of mutation and recombination, may modulate the behavior of the entire quasispecies (Chapter 3). Furthermore, a viral population can modify its fitness without alteration of the consensus sequence (González-López et al., 2005; Grande-Pérez et al., 2005; Sánchez-Campos et al., 2008), again suggesting that important information may be lost if viral populations are characterized solely on the basis of consensus sequences (see also Chapter 9).

## 5.2 The challenge of fitness *in vivo*

The limitations of fitness determinations are accentuated in the assays performed in vivo because of the multiple steps and changing environments involved in the progress of a viral infection in an entire organism, be it an animal or a plant. A relevant factor is the effect of chance in the form of genetic drift that in the course of a fitness assay may modify the composition of the viral population relative to the composition it would have if selection were the only evolutionary force (Section 3.2 in Chapter 3). Experiments with FMDV by F. Sobrino and colleagues constituted the first direct experimental measurements of fitness in vivo, and documented fitness variations among components of the mutant spectrum of a virus replicating in an animal. FMDV variants isolated during an infection of a pig displayed different capacities to compete with a reference clone in coinfections carried out in pigs (Carrillo et al., 1998). FMDV variants found at low frequency in the viral quasispecies in vivo were subjected to competition with the parental, standard isolate using pigs as the host. Different proportions of the two competitors were found in different animal lesions (aphthae), thus documenting for the first time the fitness variations among components of FMDV quasispecies in vivo. The results provided evidence that each individual aphtha harbors virus with a different mutant spectrum composition, thus accentuating the stochastic component of FMDV transmission. Replacements of subsets of variants by others have been reported that suggest preferential colonization of specific organs [(Sanz-Ramos et al., 2008) among other examples]. Coxsackievirus with deletions in the 5′-terminal region of the genome can persist in the pancreas of mice, where they displace the corresponding parental genomes (Tracy et al., 2015).

Fitness assays in cell culture may encourage extending the determinations in vivo. This is the case with hepatitis C virus (HCV). The availability of a cell culture system for HCV (Lindenbach et al., 2005; Wakita et al., 2005; Zhong et al., 2005) opened the way to experimental studies on HCV evolution in cell culture. For the first time, it has been possible to obtain HCV from the same evolutionary lineages displaying 2- to 3-fold fitness differences (Perales et al., 2013; Sheldon et al., 2014; Moreno et al., 2017). High HCV fitness was associated with significant resistance to multiple anti-HCV inhibitors that target a viral or cellular function, including lethal mutagens (Chapter 9), and inhibitors that display a high barrier to resistance (Sheldon et al., 2014; Gallego et al., 2016, 2018). It must be emphasized that the fitness effect on HCV resistance to antiviral agents is not due to the reaching of standard resistance mutations due to enhanced viral replication. Fitness-associated resistance is not due to the established drug resistance mutations characterized in patients who fail therapy. No such mutations are either detected in the initial high fitness populations or become dominant when the virus has overcome the resistance imposed by the drug. Different models have been proposed for fitness-associated drug resistance, and they are discussed in Chapters 8 for standard nonmutagenic inhibitors, and in Chapter 9 for lethal mutagens.

The likely possibility that fitness may also affect treatment response in vivo renders extremely relevant to find means to measure (or at least estimate) fitness of HCV replicating in infected patients. Unless methods to efficiently grow natural HCV isolates ex vivo are found, it would be desirable to identify in vivo parameters that can be used as surrogate fitness values, and attempts to reach this goal are now in progress, including computations based on sequencing data (Lorenzo-Redondo et al., 2014; Seifert and Beerenwinkel, 2016). Thus, model studies have suggested valuable relevance for HCV fitness regarding antiviral treatments, but the challenge is to find parameters that portray HCV fitness in infected patients, and that could be applied to other pathogenic viruses.

Despite obvious difficulties in their completion and interpretation, fitness determinations have been (and continue being) of great value in the studies of viral population dynamics in cell culture, ex vivo and in vivo. Concepts relevant to medical research and to clinical practice, such as the connection between fitness and viral load, fitness costs of drug resistance mutations, identification of compensatory mutations for fitness gain, are now amenable to quantification thanks to fitness assays (Quinones-Mateu and Arts, 2006; Martinez-Picado and Martinez, 2008; Iyidogan and Anderson, 2014; Domingo et al., 2019) (Box 5.1) (Chapter 8).

## 5.3 Fitness landscapes

The historical origins and theoretical developments that led to the concepts behind fitness landscapes have been reviewed (Schuster, 2016). They are a pictorial display in the form

POSITION IN SEQUENCE SPACE
ENVIRONMENTAL CHANGE

**FIGURE 5.2**   Putative and schematic fitness landscapes for viruses. In both panels, high and low fitness values are represented as mountain peaks and valleys, respectively. Spheres and thin arrows indicate possible fitness variations (increases by upward movements, and decreases by downward movements) of a virus. Note the presence of a plateau at the center of the upper landscape, where movements do not entail a fitness cost. The two parameters that here we consider that can mediate a fitness change are the position in sequence space of the virus (due to genetic modifications) or an environmental change (assuming a fixed position in sequence space). The two variables are indicated with a double head, thick arrow at the bottom of the scheme. Examples of the effect of these two parameters on fitness levels, and evidence that the second, highly rugged landscape is a more faithful representation of fitness variation in viruses are presented in the text. Note that despite ruggedness, the same fitness value is shared by many points in the abscissa (discontinuous horizontal line in bottom panel).

of mountains and valleys that represent high fitness values at the top of mountains and low fitness values down the valleys (Fig. 5.2). Such representations were used first by S. Wright (Wright, 1931, 1988), and they have been extensively applied to represent transitions of viral populations from a high fitness value (high replicative and survival capacity; location at the top of a mountain or a plateau) to a low fitness value (a point down in a valley). In the representations in two dimensions, fitness values are plotted in ordinate, but it is not always clear what is plotted in abscissa. Here, we will consider that the variable in the abscissa is either the genomic nucleotide sequence (for a given environment) or the environment (for a given genomic sequence) (Fig. 5.2). The genomic sequence may signify the consensus sequence of a viral population or, more accurately, the position of a mutant cloud in sequence space. According

to this representation, a virus that acquires a mutation that leads to an amino acid that directly or indirectly affects the active site of the polymerase will most probably drive the virus from a fitness peak down toward a valley. In a specific case involving the FMDV polymerase, several substitutions in the enzyme that were selected by passaging the virus in the presence of the mutagenic purine nucleoside analog ribavirin produced a fitness decrease. In a related experiment, the mutant cloud that was produced by treating a clonal population of FMDV with the mutagenic pyrimidine base analog 5-fluorouracil displayed a sixfold reduction in fitness relative to a nonmutagenized control. In addition, the cloud exerted an interfering activity on the replication of standard FMDV RNA (González-López et al., 2004, 2005). The mutagenic treatment did not result in any variation of the consensus sequence of the population.

Therefore, in this case, the element responsible for the downhill movement in the fitness landscape was an altered cloud that kept its consensus sequence invariant.

The alternative plot is to represent in the abscissa an environmental change applied to a sequence or cloud at a given position in sequence space. Again, continuing with FMDV as a model, a clonal population displayed high fitness in BHK-21 cells, but low fitness in Chinese hamster ovary (CHO) cells (we mentioned this difference in Section 5.1.2 in connection with fitness-dependent extinction). In this case, the downward movement in the landscape refers to the same sequence (a clonal population with limited cloud amplitude) transported to a different cell type (from BHK-21 to Chinese hamster ovary cells). The downhill movement may also be triggered by an intracellular environmental change within the same cellular type. If a given genomic sequence aptly replicating inside a cell is suddenly surrounded by an inhibitor of viral replication, its replicative fitness will decrease because of the presence of the inhibitor, and the virus will move from a fitness peak down to a valley. If the virus acquires an inhibitor resistance mutation, its fitness in the presence of the inhibitor will increase, and the virus will move from a valley to a peak (Coffin, 1995; Quinones-Mateu and Arts, 2006) (Fig. 5.2). Fitness is relative to an environment, and the presence or absence of a drug represents a distinct and important change in the environment. The presence of an inhibitor (or any other perturbing factor) modifies the fitness landscape. Thus, in analyzing evolutionary transitions, the experimentalist must be aware that both position in sequence space and environment can affect the fitness value of a virus. Both factors (and perhaps even additional influences) may act conjointly rather than independently. It is noteworthy that environment-driven fitness changes can occur without modifications of the consensus sequence. This may come about by at least two different mechanisms: (i) an effect

of the mutant spectrum composition *per se*, not translated in the dominance of alternative genomes (Lorenzo-Redondo et al., 2011) (Chapter 3), or (ii) transitions of a mutant spectrum without the selection of specific genomes. This is the case of the "melting" of information during the transition into error catastrophe promoted by mutagenic agents (Chapter 9). More detailed knowledge of the shape of high fitness peaks and fitness valleys during evolutionary transitions, and how the tie between cloud composition and environmental change is manifested, will have to wait for multiple fitness measurements and comparisons of many virus-host systems, a field, which is only in its infancy.

### 5.3.1 Justification of ruggedness in fitness landscapes for viruses

Very simple and unrealistic fitness landscapes are frequently used to depict fitness. A single peak landscape is often assumed for mathematical tractability in theoretical studies, implying a constant environment. In this case, fitness variation is considered a function of the genomic nucleotide sequence. Of the two types of fitness landscapes for viruses depicted in Fig. 5.2, the second rugged landscape appears to be the one that best describes evolving viruses. The main argument is that fitness variations are the norm as a result of displacements in sequence space or environmental modifications, despite acknowledging that interactions among mutations (rescuing effect of compensatory mutations also termed *positive epistasis*) may buffer or eliminate the negative effects of individual mutations. The way a fitness plateau is depicted (for example toward the middle of the first landscape in Fig. 5.2) suggests that there is little or no fitness variation despite a movement in sequence space or environmental modification. There are multiple points in the neighboring sequence space, where the virus has similar fitness. It does not exclude the presence of narrow valleys or pits due to some single mutations. We favor the view that flat landscapes (or robust viral populations and clones; see Section 5.7) is not the norm for viruses, and that rugged landscapes are the ones that best represent reality while evolution is at play.

A different matter is that well-settled products of some optimization process might have taken advantage of epistatic interactions to reach some point of fitness stability that facilitates the continued presence of such well-adapted genomes. What we witness when focusing on the process before reaching stability points is that some mutational pathways evolve toward dominance while others do not. The latter is the raw material for potential alternative or future dominance. The beauty—but at the same time annoying complexity—is that what is fleeting a given time can be semi-permanent another time, and vice versa. These considerations justify that we are not only concerned with genomes that work in comfortable regions of fitness landscapes, but also with those that have a fleeting existence. A fitness change due to point mutations (that move to the vicinity of sequence space) is not incompatible with distant regions in sequence space yielding identical fitness in a given environment (discontinuous horizontal line in the bottom panel of Fig. 5.2). Conversely, according to this proposal, the same genomic sequence or the same cloud) may display the same fitness in different environments. The possibilities outlined here are relevant to the mechanisms of viral disease emergence (Chapter 7) and are worth being investigated.

The nascent evolutionary events are those in which fitness landscapes manifest their ruggedness. There is much evidence for preferring rugged fitness landscapes for viruses, and it recapitulates several features of virus evolution portrayed in other chapters of this book. Briefly, the main experimental arguments are as follows: (i) many point mutations are deleterious when performance is measured under a defined environment.

It means that just moving one position in sequence space often changes the replicative and survival capacity of a virus. Moving several positions should be even more deleterious. Again, positive epistasis may compensate deleterious effects of individual mutations, but reaching them may imply considerable searching in sequence space; the focus is in the process. (ii) Interfering genomes that are responsible for the loss of viral infectivity can cease to be interfering just by one additional mutation. (iii) Variants from the same virus isolated from different organs of an infected host often display different fitness. (iv) Finally, as emphasized in the previous sections of this chapter, fitness is extremely environment-dependent. The reader will find many literature references in several chapters of the book that support the above statements. From current experimental evidence, rugged landscapes are the ones most suitable to depict virus evolution, although simple landscapes are needed for progress in theoretical models.

## 5.4 Population factors on fitness variations: collective fitness and perturbations by environmental heterogeneity

Fitness of a viral population is influenced by the average replicative capacity of the individuals that compose the population, modified by modulatory effects of mutant spectra (Chapter 3). Some features of population dynamics quite reproducibly may have an effect on fitness: serial infections involving passages with large numbers of cells and viruses tend to result in fitness gain while repeated bottleneck events lead to average fitness decreases (Fig. 5.3) (Chao, 1990; Duarte et al., 1992; Novella et al., 1995; Escarmís et al., 1996, 1999; Yuste et al., 2000, 2005), a process that has been documented with FMDV in swine (Carrillo et al., 2007). In the latter study, 20 pig to pig contact transfers of FMDV isolates O/Taiwan/97 resulted in reduction of virus titer, interruption

**FIGURE 5.3**  The effect of passage conditions on fitness variation. A virus subjected to large, unrestricted population passages (large arrow) tends to increase its fitness due to quasispecies optimization in the considered environment. In contrast, a viral population subjected to repeated bottleneck events (plaque-to-plaque transfers in the laboratory, discontinuous arrow) tends to decrease its fitness through the accumulation of mutations, reflected in a modification of the consensus sequence (bottom horizontal lines). Experimental evidence in support of this scheme is given in the text and in Chapter 6.

of transmissibility, and establishment of a carrier state in swine, a demonstration of how the population regime of a virus can influence its biological behavior.

In the course of large population passages, the Darwinian principles of genetic variation, competition, and selection (Chapter 3) will generously operate on the genome clouds to progressively enrich the population in those subpopulations capable of increased infectious progeny production. In contrast, repeated bottleneck events acting on highly mutable entities result unavoidably in the accumulation of mutations reflected in a modification of the consensus sequence. Since, on average, mutations tend to be deleterious rather than beneficial, the result of mutation accumulation will be fitness decrease. Several experimental evolution studies that have led to this conclusion are further described in Chapter 6. Here, we deal with the main point regarding fitness; a virus can move

from high to low fitness values in a rugged fitness landscape (compare Figs. 5.2 and 5.3) depending on the passage regimen. Evidence of these passage regimen-dependent fitness transitions in vivo is limited due to technical difficulties, such as the control of viral population size in vivo and the likely and unavoidable existence of stages with large and low population sizes during the infection.

Fitness dynamics provides an interpretation of Haldane's dilemma (Haldane, 1937) when the concept is applied to viruses. Haldane calculated that, if an individual has a lower than optimal genotype at $n$ *loci*, and such deviation of optimality is reflected in a $(1 - S_i)$ decrease in fitness, the resulting fitness will be $(1 - S_1) (1 - S_2) \dots (1 - S_n)$ times as the optimal. Since many *loci* are polymorphic and selection acts on many *loci*, average fitness values for individuals must be very low, which seems to be contrary to observations with natural populations. A numerical example (Williams, 1992) illustrates the problem. If 1000 *loci* have an average fitness loss of 1%, the average individual will display a fitness of $10^{-5}$ relative to the fitness of the best possible genotype. This will result in a thoroughly inadequate gene pool if a good proportion of fitness components interact multiplicatively. For asexual populations subjected to quasispecies dynamics, mutant distributions are ranked according to fitness (Eigen and Schuster, 1979) (Chapter 3). Most individuals in the distributions have suboptimal fitness (below that of the master genome) due to similar arguments put forward by Haldane. Yet, many individuals that die (go extinct) leave their place to other, temporarily, more fit individuals that, despite being suboptimal, have an acceptable mutational load for survival, and give rise to new distributions in a continuous reiteration of the process. That is, new individuals will again generate progeny that will be suboptimal at many *loci*, reinitiating the competition-displacement cycle. A "superfit" virus would not survive for a long time because it would deplete the hosts needed to sustain its replication. Viruses and humans base their existence in coping with limitations. Haldane's dilemma in general genetics is largely obsolete, but its consideration for virus evolution leads to the realization that the continuous replenishment of nascent viral subpopulations and the relative nature of fitness values, aided by accelerated evolutionary events, explains a continuous biological competence of highly variable viruses.

The significant conclusions on the factors that promote increases or decreases of viral fitness (recapitulated in Fig. 5.3) have been established with viral populations and clones in the laboratory or with natural isolates ex vivo. Many interesting phenomena and principles of general genetics have been experimentally tested with viruses, based on fitness variations following highly biased population regimes or competitions among viruses. Muller's ratchet, Red Queen hypothesis, or the competitive exclusion principle are examples. Since the experiments

that probed such fundamental concepts in genetics fall into the area of experimental evolution, they are described in Chapter 6.

## 5.5 Quasispecies memory and fitness recovery

Quasispecies memory means the presence in the mutant spectrum of a specific subpopulation of genomes by virtue of its having been dominant at an early evolutionary phase of the same virus lineage (Ruiz-Jarabo et al., 2000). The basis of this class of molecular memory is fitness gain during the process of viral replication in the presence of a selective agent. This is the reason to explain it here once the concepts of selection and fitness have been covered. Memory was first described using two markers of FMDV: resistance to neutralization by a monoclonal antibody —a phenotypic marker related to a set of specific mutations at a viral gene that encodes a capsid protein— and an elongated internal oligoadenylate tract in the FMDV genome, generated as a result of repeated plaque-to-plaque transfers (Chapter 6). The experiments intended to show the presence of quasispecies memory were inspired in immunological memory, and more generally in the behavior of the immune system viewed as a complex adaptive system (Frank, 1996; Farber et al., 2016). The course of the observations on quasispecies memory was as follows: monoclonal antibody-escape mutants were selected by multiplying FMDV in the presence of the antibody, and they became dominant in the population, as expected. When the mutants were passaged in the absence of antibody, true revertants arose and became dominant in the population, which is expected if the escaped mutants had endured some fitness cost. The key finding was that the frequency of escape mutants that remained in the population after revertants had taken over, increased from a basal level that ranged from $4 \times 10^{-3}$ to $5 \times 10^{-4}$ to a memory level of $1 \times 10^{-2}$ to $5 \times 10^{-1}$, and the difference was statistically significant (Ruiz-Jarabo et al., 2000) (Fig. 5.4).

A similar observation was made with the genetic modification that consisted of the elongation of four adenylate residues located between the two functional AUG protein synthesis initiation codons in the FMDV genome. The number of adenylates increased by two and up to 23 residues in individual clones that had been subjected up to 30 serial plaque transfers, with considerable intraclone heterogeneity (Escarmís et al., 1996). This oligoadenylate elongation has never been seen in natural isolates of FMDV, and the likely reason why it repeatedly occurred in independent FMDV clones is that the viral polymerase has an inherent tendency to slippage or misalignment mutagenesis when copying homopolymeric template residues (Chapter 2). Usually, negative selection eliminates such altered genomes because they display alterations in protein expression as the molecular basis of fitness decrease (Escarmís et al., 1996). Negative selection is attenuated during plaque transfers of the virus, probably, because of a diminished intracellular competition among variants, imposed by movement constraints of viral particles as they exit cells (Chapter 6). Therefore, when clones with the elongated oligoadenylate tract were subjected to large population passages, they were overgrown by genomes with the four standard adenylates, as the population recovered fitness (compare with the contrast between fitness gain and fitness loss summarized in Fig. 5.3). Here again, the frequency of genomes with the elongated oligonucleotide remained at memory levels, even following 50 large population passages. Clones with a higher than the standard number of adenylate residues were found in 11%—17% of the molecular and biological clones analyzed. In contrast, none out of 40 molecular clones from the control population devoid of memory deviated from the standard sequence (Ruiz-Jarabo et al., 2000; Arias et al., 2001), and again the difference was

statistically significant (Fig. 2.4B in Section 2.3 of Chapter 2 depicts FMDV genomes with the internal oligoadenylate elongation, considered to be a consequence of misalignment mutagenesis).

In the process of being selected, those genomes harboring the memory markers increased their replicative fitness so that when they went back to the status of minority genome subpopulations, they reached a higher frequency dictated by their fitness level (remember that mutant spectrum components are ranked according to the fitness as explained in Chapter 3) (Fig. 5.4). Once established, the memory genomes can gain replicative fitness in parallel with the dominant components of the quasispecies, in agreement with the Red Queen hypothesis (Arias et al., 2004). Consistent with the proposed mechanism of memory implementation, it was shown that memory levels were dependent on the relative fitness of the parental populations that underwent a selection of memory markers. A 7.6-fold higher fitness resulted in 30- to 100-fold higher memory level (Ruiz-Jarabo et al., 2002). The frequency of memory genomes decreased with the number of serial passages in the course of FMDV replication in a process akin to "diffusion" in sequence space (see Section 3.7 in Chapter 3). The decay was strikingly reproducible in parallel viral lineages (Ruiz-Jarabo et al., 2003). This reproducible behavior is one of several findings that denote the potential of viral quasispecies to display a deterministic behavior, as discussed in Section 3.6.2 of Chapter 3. Memory can influence evolution under many types of environmental modifications. Capsid variants of hepatitis A virus (HAV) were selected in response to an artificial shut-off of host cell protein synthesis. When virus readapted to the initial environmental conditions, some genomes were rapidly replaced by others that had probably remained at memory levels (Costafreda et al., 2014).

The demonstration of quasispecies memory was extended to HIV-1 in vivo by C. Briones and colleagues who distinguished two classes

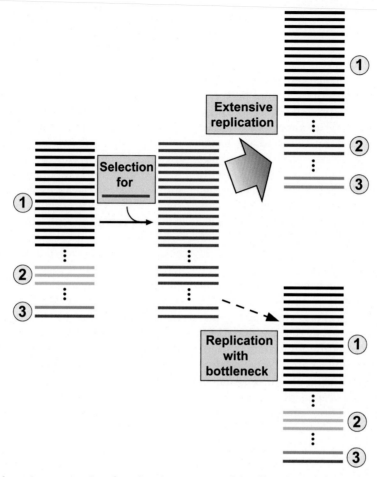

**FIGURE 5.4**  A schematic representation of quasispecies memory and the effect of a population bottleneck on memory. For clarity, the different scattered symbols on lines used to represent mutations in other figures of this book, have been omitted. Instead, frequency levels of genome subpopulations are depicted by horizontal lines with different colors. Level indicated by circled number one includes the most frequent genomes in the quasispecies (black horizontal lines in the distribution drawn on the left). A second and third frequency level is indicated with numbers 2 and 3, and green, blue and red genomes. When a low-frequency red genome is selected, it dominates the population (middle distribution). When replicated in the absence of the selective constraint (extensive replication, big arrow), revertants become dominant and occupy frequency level 1 (black lines in the upper right quasispecies). Because in the process of selective amplification the red genomes have attained higher fitness than the parental red genomes in the initial distribution, they occupy level 2 rather than level 1 (compare first distribution on the left with the upper distribution on the right). Level 2 is termed as *memory level*. When replication is allowed, and then a bottleneck intervenes, the result is the dominance of revertants, together with the exclusion of low-frequency genomes, leading to memory loss. A distribution similar to the original one will be produced (discontinuous arrow and bottom right distribution). The red genome is now present again at its basal frequency, which depends on the mutational pressure operating in the system. See text for experimental justification of this scheme.

of memory. One is the replicative memory that we have just examined in the case of FMDV and HAV. The second class of memory is termed *reservoir, cellular,* or *anatomical memory,* which is due to proviral DNA integration in the target cells. Once integrated, the viral DNA replicates as part of the cellular chromosome, which becomes a reservoir of ancestral sequences. The reservoir memory was identified by a phylogenetic analysis of consensus sequences obtained at different times after the initiation of retroviral therapy in HIV-1-infected patients, while replicative memory was evidenced by an extensive clonal analysis of HIV-1 samples from patients at different times of treatment, using multidrug resistance markers [detailed in (Briones et al., 2003) and reviewed in (Briones and Domingo, 2008)]. The two classes of memory were supported both by experimental findings and a theoretical model that considered both replicative memory and integration of ancient sequences as proviral DNA in the cellular genome (Briones et al., 2003; Briones and Bastolla, 2005). These studies on memory have permitted an interpretation of the reemergence of ancestral sequences

during virus evolution, and several observations are consistent with virus behavior being influenced by memory genomes (Borrow et al., 1997; Wyatt et al., 1998; Karlsson et al., 1999; Briones et al., 2000; Lau et al., 2000; García-Lerma et al., 2001; Imamichi et al., 2001; Kijak et al., 2002; Charpentier et al., 2004). Box 5.2 summarizes the main features of quasispecies memory and its implications. After being established with viruses, the memory concept was extended to *Drosophila melanogaster*, thus reinforcing its applicability to various types of biological systems (Teotonio et al., 2009).

## 5.5.1 Implications of quasispecies memory: harbinger mutations

Fitness-dependent memory is a collective property of the entire quasispecies and not of its individual components, as shown by memory loss when the population was subjected to a bottleneck event (Ruiz-Jarabo et al., 2000; Briones and Domingo, 2008) (Fig. 5.4 and Box 5.2). Memory has a number of theoretical and practical implications that extend to medical

---

**BOX 5.2**

### Main features of quasispecies memory

- The presence of memory genomes in a viral lineage is an expected consequence of quasispecies dynamics, and of the fitness gain of a viral population while subjected to a selective amplification.
- Memory levels are dependent on the relative fitness of the viral genomes destined to become memory genomes.
- Upon replication, memory genomes gain fitness in parallel with the entire population that harbors them.

- Memory genomes can modify the response of viral quasispecies to selective constraints directed to traits embodied in the memory genomes.
- Population bottlenecks erase memory because bottlenecks exclude low-frequency genomes, and memory genomes are present at low frequency.

practice. On the theoretical side, a way of expressing the collective behavior of a viral quasispecies is that its response to a selective constraint may depend on whether the population had previously experienced the same constraint, or whether the population had gone through a bottleneck or not. Memory can facilitate the response of a collectivity of genomes to selective constraints that occur intermittently in the life cycle of an evolutionary lineage. Examination of the types of selective constraints that have been identified as operating in nature suggests that many of them are not exceptional or unique, but rather, they are repeated occasionally or even regularly (physiological responses dependent on circadian rhythms, cycles of temperature variations, exposures to the same species of predators, etc.). In the case of viruses, if a response to recurrent constraints involves selection of genome subpopulations, once the selective force wanes, the storage of the responsive subpopulation at memory levels guarantees preparedness to confront the next identical or related constraint. The preparedness effect is, however, abolished whenever a bottleneck event excludes the memory genomes from the evolving population (Fig. 5.4 and Box 5.2). Since bottlenecks are increasingly recognized as abundant in the life cycles of viruses, it is uncertain how frequently memory of the type described here can indeed have an influence in virus behavior. It may have a major effect during prolonged chronic infections involving large viral loads, also known to be an important determinant of treatment response and disease progression (Chapter 8).

A practical application of quasispecies memory may be found in treatment planning and treatment modifications for chronic human infections, for example, for HIV-1-, HBV-, or HCV-infected patients. When an antiviral treatment is interrupted because the virus in the patient is dominated by mutants resistant to the inhibitor (or inhibitors) in use, inhibitor-sensitive subpopulations will replenish the mutant spectrum if the resistance mutations inflicted a fitness cost. Since the resistant mutants increased their fitness in the course of selection, they will not return to basal levels, but they will occupy a memory level. When time elapses, if no population bottlenecks intervene and no alternative treatments are available for the patient, when the previous inhibitor is administered again, the virus may find a way to resist the new round of treatment because of the presence of the relevant resistant genomes at memory levels. Many, but not all, studies have indicated that the presence in mutant spectra of low-frequency resistance mutations may jeopardize treatment efficacy. In cases in which minority mutations may confer partial resistance to inhibitors, it is important to avoid a treatment that includes again the same drugs used in the previous treatment period (Briones and Domingo, 2008). This point is discussed again in Chapter 8 in relation to the relevance of quasispecies dynamics in disease control.

Some mutations that are maintained at memory levels can be a predictor of later dominance in the same evolutionary lineage. Such mutations are termed *harbinger mutations*. They were reported during a clonal analysis of FMDV during a process of fitness gain (Arias et al., 2001). The presence of harbinger mutations is likely to become more evident as successive samples of an evolving viral population are examined through deep sequencing methodology since virus evolution consists of successive replacements of some viral genome subpopulations by others. Thus, quasispecies dynamics can have as one of its consequences the presence of a molecular memory of past selection events, and memory levels are decisively influenced by fitness values.

## 5.6 The relationship between fitness and virulence

Virulence of a virus (or of any parasite) is the ability to induce pathology in a host. Viruses generally encode one or several virulence determinants either in regulatory regions or in viral proteins. Pathological manifestations can be due to the direct action of viral gene products on cells of specific tissues or organs, or indirectly through the host response to the invading virus that can trigger immunopathological processes. Increasing evidence suggests that during their epidemic spread most viruses may undergo genetic modifications that can either enhance or attenuate their virulence. For example, the Ebola virus that caused the African outbreak of 2014–15, termed *strain Guinea-2014* (with cases of contagion outside Africa) was derived from strain Zaire-1976 (Gire et al., 2014). While mortality of the Zaire-1976 strain was around 90%, the estimated mortality of the Guinea-2014 strain was about 50% (although caution should be exercised concerning the precision of these estimates in areas with poor health infrastructure and presence of confounding factors). The decrease of mortality probably conferred the virus an epidemiological fitness advantage (see Section 5.9) by favoring virus spread among the susceptible human population. While the previous Ebola outbreaks in the 20th century caused infection in hundreds of individuals, the Guinea-2014 outbreaks infected tens of thousands of individuals, representing a higher risk of disease expansion and reemergence in distant areas. For virtually any virus there have been studies that document that genetic change, sometimes a minimal change can modify virus virulence.

Virulence, like fitness, is relative to a biological environment. The response to an infection is influenced by the host immune response, and probably, other mainly unknown influences. About 99% of poliovirus (PV) infections are asymptomatic, and this was also true at the peak of the poliomyelitis epidemics in the middle of the twentieth century (Koike and Nomoto, 2010). Yet, in a small proportion of infected individuals invasion of the central nervous system and neurological disease is produced, likely influenced by viral and host determinants. Since the time that infectious clones became available to express altered viruses, it was shown that one or a few mutations engineered in the clones could alter virulence and that excessive attenuation rendered putative vaccines ineffective. This is one of the reasons why very few engineered vaccines have replaced the classical inactivated and attenuated antiviral vaccines. Fitness must be adjusted to attain the required degree of attenuation of a live virus; this is a challenge for vaccines based on viral genome modifications.

Regarding the influence of the immune system, a debilitated immune response may favor viral infections for obvious reasons: the arms race between the virus and the host is tilted in favor of the virus. Furthermore, immunocompromised individuals (because of inherited or acquired immunodeficiency) may shed large amounts of PV, and they are termed *supershedders* or *superspreaders*; they may contribute to the spread of PV, and may constitute a factor of disease maintenance or reemergence (Martin et al., 2000; Martin, 2006; Dunn et al., 2015). Related studies had previously been conducted with an immunodeficient child persistently infected by influenza virus (Rocha et al., 1991) (Section 7.7.1 in Chapter 7). An excellent model that related quasispecies (intra-host) dynamics of poliovirus and vaccine-derived virus shed by an immunodeficient patient was formulated by O. M. Kew, V.I. Agol and colleagues (Gavrilin et al., 2000), in line with the need of understanding the short-term versus long-term evolutionary rate paradox (Geoghegan and Holmes, 2018) (the higher intra-host than inter-host evolutionary rates discussed in Chapter 7). In the events described in these in vivo studies, fitness variations as viruses encounter new environments play an essential role.

Attenuation of a virus or any pathogen to produce a live-attenuated vaccine, using the classical multipassage in an alternate host (cell culture or animal), can be viewed as a process of accumulation of mutations that alters relative fitness values. Indeed, such extensive passage renders the virus fit to replicate in the alternative environment. When this adaptation results in an adequate degree of fitness loss in the authentic host, the virus may no longer manifest disease potential. The derivation of an attenuated virus by serial passage of the pathogenic counterpart in an alternative host is an example of a virus increasing its fitness in a nonnatural environment concomitantly with a decrease in fitness and virulence for the authentic host. The serial passages of rabies virus in rabbits, a process termed *lapinization* (from the French word "lapin," rabbit) carried out by L. Pasteur is an example of the preparation of a vaccine for human use. Tissue culture-adapted PV can be highly attenuated for humans and monkeys, rendering a live vaccine virus, which is highly fit in cell culture (at least more fit than the initial virulent isolate), but unable to replicate in the host cell subsets associated with causing pathology. This does not mean that the tissue culture-adapted PV is not capable of replicating in *any* host cells. In fact, it must replicate at least to the extent of evoking a protective immune response if the virus has value as a live vaccine. Thus, fitness is not only host-dependent but organ-, tissue-, and cell population-dependent. The decrease in disease potential will often be associated with a decrease of viral fitness in those cells relevant to disease manifestations, or those cells needed to reach the ones involved in the disease. Studies are needed to quantify organ-dependent variation in viral fitness.

Some models of the evolution of virulence propose that viruses that would naturally cause harm and death in their main natural hosts would be selected against because depletion of the host population would prevent long-term survival of the virus. Several lines of evidence support the occurrence of virulent viral strains mainly as a result of the selection of variant viruses from essentially avirulent populations. Cardiovirulent coxsackievirus B (CVB) may arise commonly in infected neonates with an immature immune system. In this specific environment, some CVB mutants may gain access to the heart, where they may adapt to replicate efficiently, causing myocarditis (Tracy et al., 2006). This view has a conceptual parallelism with the mechanisms that lead to the emergence of a new human viral disease caused by a virus from some natural animal reservoir in which the virus is harmless or nearly so. Out of many variants generated in replicating viral quasispecies and that can potentially become established in a new host species, only the chance contact with the new host in which a new quasispecies can be selected, can lead to a viral disease emergence (compare the proposed mechanism for CVB cardiovirulence with suggested mechanisms of viral disease emergence in Chapter 7). The scenario of an efficient replication of a viral subpopulation in a specific organ, tissue, or cell subset as being associated with disease is supported in the case of CVB by the finding of $10^8$ $TCID_{50}$ (tissue culture infectious dose needed to kill 50% of cells) per gram of pediatric heart tissue obtained at death (Chapman et al., 2008).

The observations just summarized raise the issue of the relationship between viral virulence and fitness, as defined in Section 5.1, that is, the relative capacity of a virus to produce progeny in a given environment. Models of virulence have been surrounded by controversy, perhaps because of the difficulties of capturing the complex set of evolutionary and biochemical ingredients that result in what we term virulence. In particular, the connection between viral fitness and virulence has been a much-debated matter, involving experimentalists and theoreticians alike. Some models assume a direct relationship between fitness and capacity to produce disease (Lenski and May 1994; Poulin and Combes, 1999; Brown et al., 2006). Experimental studies with viruses and other types of pathogens have provided evidence either in favor or against such direct relationship between fitness and virulence. Among other examples, the diabetogenic phenotype of CVB correlates with replication efficiency (Kanno et al., 2006). In a comparative study of two divergent strains of the whispovirus white spot syndrome virus, virulence—measured by cumulative mortality rates of its shrimp host—correlated with replicative fitness in vivo (Marks et al., 2005). A direct correlation between the level of RNA replication and cytopathology was also observed in engineered alphavirus replicons (Frolov et al., 1999). In other cases, a direct connection between fitness and virulence has been assumed, but not proven experimentally. A greater replicative fitness of historical versus current HIV-1 isolates was taken as evidence of attenuation of HIV-1 as the AIDS pandemic advanced (Arien et al., 2005).

Other studies favor the view that fitness and virulence do not necessarily correlate. Despite similar replication of R5-tropic and X4-tropic clones of HIV-1 in mitogen-activated T cells, X4 tropic clones were transferred more efficiently than R5-tropic clones to CD4 (+) T cells. The increased transfer could contribute to the competitive advantage of X4 viruses in AIDS patients (Arien et al., 2006). In a related finding in a different biological setting, nucleopolyhedrovirus that was transmitted early to its host, the moth *Lymantriadispar*, was more virulent than virus transmitted late, despite the latter being more productive because it could use more host tissue for replication (Cooper et al., 2002). Lack of correlation between fitness and virulence was also observed with some plant viruses (Escriu et al., 2003) and in some fungal infections (Nielsen et al., 2005).

An observation with FMDV has unveiled one of the possible molecular mechanisms of a lack of correlation between fitness and virulence. A biological clone of FMDV acquired high fitness and virulence, measured in this case as the cell-killing capacity of populations that had been subjected to large population passages in BHK-21 cells. This was the expected result of competitive optimization of the replicating quasispecies. Subsequent plaque-to-plaque transfers of the clone (serial bottleneck events; Fig. 5.5 and

**FIGURE 5.5** Evolution of fitness and virulence of an FMDV clone subjected to multiple plaque-to-plaque transfers. Fitness was measured by growth-competition assays as described in Section 5.1.1, and virulence as the capacity of the virus to kill the host BHK-21 cells. Fitness determinants were scattered throughout the viral genome and decreased considerably as a result of the transfers. In contrast, virulence determinants were concentrated on more specific genomic regions and decreased to a lower extent than fitness as a result of plaque transfers. The drawing is a summary of a study by (Herrera et al., 2007). See text for experimental justification and additional references on the relationship between fitness and virulence.

Chapter 6) resulted in profound fitness loss, but only a minimal decrease of virulence (Herrera et al., 2007). Comparison of the behavior of constructed chimeric viruses mapped the fitness-decreasing mutations throughout the genome, while virulence determinants were concentrated on some genomic regions. Thus, virulence was more robust than fitness to the effects of mutations that accumulate upon serial plaque transfers (Escarmis et al., 1996, 2002). It must be emphasized that in this study, FMDV fitness and virulence was measured in the same biological environment provided by BHK-21 cells. Thus, depending on the passage regimen, viral fitness and virulence can follow different evolutionary trajectories (Herrera et al., 2007) (Fig. 5.5).

A distribution of fitness-enhancing mutations throughout the viral genome has also been observed in HCV passaged in human hepatoma cells in culture (Perales et al., 2013; Moreno et al., 2017). Incidentally, the observation that fitness determinants are scattered throughout a viral genome is expected from the fact underlined in Chapters 2 and 4 that most mutations in viral genomes are not neutral. A minimal change can deoptimize a virus, as reflected in a fitness decrease in the considered environment. The results of the link between fitness and virulence carried out with FMDV also imply that in theoretical models of virulence it is not justified to equate virulence with replicative capacity, again an example of an experimental finding that should be entered into theoretical studies to render them more realistic. What determines virulence in vivo is far more complex than what determines cell killing in cell culture. More work is needed to evaluate whether the fitness-virulence relationships that distill from cell culture models may be valid in the in vivo environments.

## 5.7  Fitness landscapes for survival: the advantage of the flattest

The first fitness landscape depicted in Fig. 5.2 includes a flat plateau with a virus whose movement does not entail any fitness variation. If we take a position in sequence space as the variable in the horizontal axis, what this part of the figure means is that a virus can accept mutations without its fitness being affected. The term robustness has been used to denote the invariance of a phenotypic trait despite the occurrence of mutations or, in more general terms, the preservation of the phenotype in the face of perturbations (de Visser et al., 2003; Elena, 2012). *In silico* results with digital organisms documented a selective advantage of the organism that live in a flat fitness surface, an event that, in terms of fitness landscapes, has been popularized with

the terms "the advantage of the flattest" or "survival of the flattest" (Wilke et al., 2001).

Despite evidence summarized in Section 5.3.1 that a rugged version of fitness landscapes is more realistic for viruses (bottom landscape in Fig. 5.2) one can envision areas of sequence space in which a virus is more resistant than average to the effect of mutations, or that a virus type has evolved to be more resistant to mutations. Several possible mechanisms of virus robustness have been proposed (Elena et al., 2006). Possibilities include evolution toward a genomic base composition or a codon usage that does not change the frequency of misincorporations but limits the proportion of amino acid substitutions prone to functional alterations, among many other possibilities one could imagine. The main problem is that there is a deficit of experimental descriptions of robustness with a precise supportive molecular mechanism. Coxsackievirus B3 (CVB3) exhibited higher vulnerability than PV to lethal mutagenesis by ribavirin that could not be attributed to increased incorporation of ribavirin-triphosphate (RTP) (Graci et al., 2012). Furthermore, CVB3 manifested increased sensitivity to the random introduction of a pyrimidine analog to the viral RNA, with the major effect being at the level of the encoded proteins rather than at the level of an RNA structure, with a lower tolerance of nonsynonymous mutations. How differences in robustness might relate to the lifestyle of viruses and their ability to respond to constraints is largely unknown.

High fidelity mutants due to a faithful polymerase cannot be considered a robustness mechanism because what such viruses do is to introduce a lower number of mutations, not that phenotypic traits are unaltered despite the acquisition of mutations. Some experimental observations are consistent with some virus mutants being more robust than others regarding mutation tolerance. Early evidence that a standard VSV virus may accept more mutations than one of its neutral derivatives will be presented in Chapter 6 in a discussion of the multiple contributions made by J.J. Holland and his colleagues on the translation to virology of concepts of population genetics, using VSV as the model system. Viral robustness has also been suggested as a possible mechanism to decrease the efficacy of lethal mutagenesis treatments (Sanjuan et al., 2007), with a seeming agreement with the results of Graci and colleagues on reduced mutational robustness of CVB3 (Graci et al., 2012). However, a distinction is in place here. One issue is that a virus naturally occupies a position of higher robustness than another virus (in the previous example, PV vs. CVB3), and a different issue is that within the time frame of a transition toward extinction— prompted by a virus-specific mutagenic agent— a viral population can migrate toward (and remain in) a position of sequence space, where robustness is significantly enhanced. This second possibility is, in my view, extremely unlikely, and it was not sustained by experiments on lethal mutagenesis of lymphocytic choriomeningitis virus (LCMV) (Martin et al., 2008). Furthermore, a theoretical study by C. Wilke and colleagues (O'Dea et al., 2009) did not support the concept that enhanced mutagenesis can displace viruses toward regions of increased robustness [reviewed by (Elena, 2012)]. Both the experiments and the theoretical model suggest that the prediction of a robustness-mediated failure of lethal mutagenesis was probably stretching too much the "robustness of robustness" so to speak; this is discussed in Chapter 9 in connection with lethal mutagenesis as an antiviral approach, and its mechanistic differences with the action of standard nonmutagenic inhibitors used in antiviral pharmacology.

Although difficult to prove, there might be instances in nature that a transient high fitness peak can be reached by a virus that has as a cost a high sensitivity to additional mutations, exactly the opposite to advantage of the flattest that we could call "disadvantage of the sharp." It is tempting to speculate that an example of such an occurrence was a virulent influenza virus (IV) whose hemagglutinin gene acquired a fragment of ribosomal RNA by nonhomologous recombination (Khatchikian et al., 1989). No such mosaic influenza viruses have been perpetuated in nature, suggesting that such constructs when they occur, may drive the virus toward extinction. They may become less tolerant of additional genetic modifications.

## 5.8 Fitness and function

Molecular virology has developed very important tools to dissect viral functions by engineering infectious molecular clones with preselected modifications, followed by an examination of their biological consequences. Reverse genetics was first developed by C. Weissmann with the first site-directed mutagenesis design using in vitro replication of bacteriophage Qβ (see Chapter 3). A number of derived technologies (i.e., oligonucleotide-mediated site-directed mutagenesis of DNA) have been applied to the majority of RNA and DNA viruses. Here fitness comes as a relevant parameter concerning the difference between a virus having the attribute of "viable" as such versus "viable" for evolutionary success. Many constructs that are currently investigated with viruses (readily accessible to literature searches) yield infectious viruses that would have only a minimal chance to survive in an evolutionarily competitive scenario. What an infectious construct tells us is that basic functional performance of a virus is operating, and this is extremely informative of the molecular requirements for a virus to minimally express its replicative program. We should be aware, however, that an infectious construct may be too debilitated to have prospects of its being established as a competitive biological element in nature or even in cell culture or in ex vivo experiments. In support of this assertion is the fact that replication of many such constructs leads to variant forms that display fitness increases relative to the parental construct, with an accumulation of multiple mutations in their genomes. Such fitness optimization through the accumulation of mutations is observed with entire viruses and subgenomic replicons. The advantage conferred by a mutant spectrum over its individual components may be one of the reasons for the poor performance of specific constructs. Therefore, what is functional biochemically or in a single round of infection may be nonfunctional from the point of view of long-term survival. The dynamics of genetic plasticity of viruses applies equally to natural viral genomes than laboratory constructs because error-prone replication remains.

### 5.8.1 Very low fitness versus lethality

Screenings of the effects of randomly introduced mutations in viral genomes have suggested that a large proportion (as much as 40%) of mutations are highly detrimental or lethal (Sanjuan et al., 2004; Fernandez et al., 2007). The penetration into the mutant spectrum composition achieved by deep sequencing is providing evidence of mutations that are present at very low frequency but that are not detection artifacts. Their presence raises the question of whether mutations that were considered to be lethal based on the absence of detection of progeny were, in reality, highly deleterious but not lethal. This is not a trivial distinction because what we are learning from quasispecies dynamics as described by deep sequencing is that genomes that can replicate only minimally may nevertheless acquire additional mutations that rescue them toward an increase of frequency. The need to reconsider the lethal character of mutations applies to other microbial entities (Jee et al., 2016). The possible biological relevance of very low-frequency mutations in viral and cellular systems has not been explored.

## 5.9 Epidemiological fitness

Epidemiological fitness is a semi-quantitative concept supported by the prevalence of some viral genomic sequences over others in a given geographical area, based on diagnostic surveys, and the spread of a particular variant virus into neighbor territories. A high epidemiological fitness may be due to a combination of parameters, such as the average rate of virus replication in different infected host individuals, the immune status of the host population, virus transmissibility, virus particle stability in the environment, etc. that altogether confer a high reproductive ratio ($R_0$, number of infected contacts per infected host; Chapter 7). For example, an FMDV of serotype O termed *O/PanAsia* was first isolated in India in 1990, and in the following years the virus displaced other FMDV strains as it spread to other continents (Knowles and Samuel, 2003). At that time, FMDV O/PanAsia displayed high epidemiologic fitness, for reasons that are not understood. Several sublineages of a successor virus termed O/PanAsia-II expanded in subsequent decades (Mahy and Belsham, 2017).

In 2009, a new H1N1 IV that had originated in swine spread into the human population. It partially displaced other H1N1 and H3N1 viruses that were circulating at the time, although in subsequent years the new virus coexisted with other, previously dominant influenza virus genotypes. In this case, the high epidemiologic fitness displayed by the new H1N1 virus when it irrupted into the human population was likely associated with having acquired a new genomic composition, including segments that encode antigenic determinants. A plausible model is that the new segment constellation conferred the virus high replicative capacity in a human population that had a partially protective immune response addressed to other antigenic forms of IV (see Chapter 7 on antigenic shift and drift as evolutionary events).

The relative immunogenicity of different viral lineages may affect dominance in the field. This was the case of Dengue virus type 1 (DENV-1) lineage 6. Higher epidemiological fitness than lineage 1 was associated with diminished immunogenicity of lineage 6 that led to higher viremia and consequent transmissibility (Pinheiro et al., 2018).

In summary, the fitness of a virus (or any pathogen) in vivo is determined basically by the complete genome of the pathogen and its expression products in interaction with the physical and biological environments provided by the host cell or organism.

## 5.10 Overview and concluding remarks

Fitness is an important parameter to describe the survival potential of a virus (be it a clone, population, mutant, or natural isolate) when confronted with related competitors in a given environment. Fitness, as a measure of relative replication capacity, can influence the progress of viral infections in vivo and, in consequence, the disease manifestations in the infected organisms. Application of deep sequencing should provide new prospects for fitness quantification and, equally important, the fine molecular details of the dominance of some variant subsets in competition with many others.

The argument has been made that fitness landscapes, a classic representation of fitness values and variations in the form of mountains and valleys, should be very rugged for viruses as they replicate in their natural environments. One of the reasons is the frequent deleterious effects of individual mutations, evidenced by many of the examples included in different chapters of this book. There is a general debate in biology on the relationship between fitness and virulence, and it is unlikely that this debate is solved solely by studies with viruses. There are several elements that complicate attempts at clarification, the least trivial being that the very exact meaning of both fitness and virulence is itself object of debate.

One of the consequences of the fitness gain that accompanies a selection event is the presence of memory genomes in viral quasispecies. Many of the implications of memory as part of quasispecies dynamics, and also for the planning at antiviral treatments, have not been fully appreciated.

Deep sequencing is unveiling the presence of many low-frequency mutations in viral populations. This introduces uncertainty regarding the meaning of lethality of mutations. It is not easy to distinguish mutations that evoke lethality in the genomes harboring them (that is, produce genomes with fitness zero) versus mutations that lead to low (but not zero) fitness genomes. The relevance of the reservoirs of low-frequency mutations remains largely unexplored, and it may depend on the nature of the selective forces to be confronted by the population.

Even more complex than the standard fitness concept for a virus in a defined biological environment are attempts to define in quantitative terms fitness in the field, the so-called *epidemiological fitness*. Again, multiple factors participate in conferring a virus an epidemiological advantage, and several such factors are probably unrelated to the fitness parameters we determine in the laboratory. This is a key point in dealing with viruses as populations that will again be treated in Chapter 7 in relation to the emergence of viral disease agents (see Summary Box).

## Summary Box

- Fitness is a measure of the relative efficacy of two viruses to produce progeny in a given environment.
- Fitness landscapes are a pictorial representation of the variation of fitness values as a function of the genomic sequence or the environment. They serve to visualize fitness gains, fitness losses, the advantage of the flattest, and the disadvantage of the sharpest.
- Molecular evidence suggests that fitness landscapes for viruses are extremely rugged.
- Fitness of a viral population may vary without any modification of the consensus sequence.

- A consequence of fitness increase during the selection of viral subpopulations is the establishment of memory genomes in the population. Memory levels are occupied by those genomes that were dominant at an earlier evolutionary phase of the same lineage. The presence of molecular memory in viral quasispecies has several theoretical and medical implications.
- Mutations previously defined as lethal may just evoke low fitness in the genomes harboring them.
- Epidemiological fitness is a semiquantitative parameter that reflects the capacity of a virus to become dominant in the field.

## References

Arias, A., Lázaro, E., Escarmís, C., Domingo, E., 2001. Molecular intermediates of fitness gain of an RNA virus: characterization of a mutant spectrum by biological and molecular cloning. J. Gen. Virol. 82, 1049–1060.

Arias, A., Ruiz-Jarabo, C.M., Escarmis, C., Domingo, E., 2004. Fitness increase of memory genomes in a viral quasispecies. J. Mol. Biol. 339, 405–412.

Arien, K.K., Troyer, R.M., Gali, Y., Colebunders, R.L., Arts, E.J., et al., 2005. Replicative fitness of historical and recent HIV-1 isolates suggests HIV-1 attenuation over time. AIDS 19, 1555–1564.

Arien, K.K., Gali, Y., El-Abdellati, A., Heyndrickx, L., Janssens, W., et al., 2006. Replicative fitness of CCR5-using and CXCR4-using human immunodeficiency virus type 1 biological clones. Virology 347, 65–74.

Ayala, F.J., 1965. Relative fitness of populations of *Drosophila serrata* and drosophila birchii. Genetics 51, 527–544.

Ayala, F.J., 1969. Evolution of fitness. V. Rate of evolution of irradiated populations of Drosophila. Proc. Natl. Acad. Sci. U.S.A. 63, 790–793.

Borrow, P., Lewicki, H., Wei, X., Horwitz, M.S., Peffer, N., et al., 1997. Antiviral pressure exerted by HIV-1-specific cytotoxic T lymphocytes (CTLs) during primary infection demonstrated by rapid selection of CTL escape virus. Nat. Med. 3, 205–211.

Briones, C., Bastolla, U., 2005. Protein evolution in viral quasispecies under selective pressure: a thermodynamic and phylogenetic analysis. Gene 347, 237–246.

Briones, C., Domingo, E., 2008. Minority report: hidden memory genomes in HIV-1 quasispecies and possible clinical implications. AIDS Rev. 10, 93–109.

Briones, C., Mas, A., Gomez-Mariano, G., Altisent, C., Menendez-Arias, L., et al., 2000. Dynamics of dominance of a dipeptide insertion in reverse transcriptase of HIV-1 from patients subjected to prolonged therapy. Virus Res. 66, 13–26.

Briones, C., Domingo, E., Molina-París, C., 2003. Memory in retroviral quasispecies: experimental evidence and theoretical model for human immunodeficiency virus. J. Mol. Biol. 331, 213–229.

Brown, N.F., Wickham, M.E., Coombes, B.K., Finlay, B.B., 2006. Crossing the line: selection and evolution of virulence traits. PLoS Pathog. 2, e42.

Carrillo, C., Borca, M., Moore, D.M., Morgan, D.O., Sobrino, F., 1998. *In vivo* analysis of the stability and fitness of variants recovered from foot-and-mouth disease virus quasispecies. J. Gen. Virol. 79, 1699–1706.

Carrillo, C., Lu, Z., Borca, M.V., Vagnozzi, A., Kutish, G.F., et al., 2007. Genetic and phenotypic variation of foot-and-mouth disease virus during serial passages in a natural host. J. Virol. 81, 11341–11351.

Centlivre, M., Sommer, P., Michel, M., Ho Tsong Fang, R., Gofflo, S., et al., 2005. HIV-1 clade promoters strongly influence spatial and temporal dynamics of viral replication in vivo. J. Clin. Investig. 115, 348–358.

Chao, L., 1990. Fitness of RNA virus decreased by Muller's ratchet. Nature 348, 454–455.

Chapman, N.M., Kim, K.S., Drescher, K.M., Oka, K., Tracy, S., 2008. 5′ terminal deletions in the genome of a coxsackievirus B2 strain occurred naturally in human heart. Virology 375, 480−491.

Charpentier, C., Dwyer, D.E., Mammano, F., Lecossier, D., Clavel, F., et al., 2004. Role of minority populations of human immunodeficiency virus type 1 in the evolution of viral resistance to protease inhibitors. J. Virol. 78, 4234−4247.

Clarke, D.K., Duarte, E.A., Elena, S.F., Moya, A., Domingo, E., et al., 1994. The red queen reigns in the kingdom of RNA viruses. Proc. Natl. Acad. Sci. U.S.A. 91, 4821−4824.

Clementi, M., Lazzarin, A., 2010. Human immunodeficiency virus type 1 fitness and tropism: concept, quantification, and clinical relevance. Clin. Microbiol. Infect. 16, 1532−1538.

Coffin, J.M., 1995. HIV population dynamics in vivo: implications for genetic variation, pathogenesis, and therapy. Science 267, 483−489.

Cooper, V.S., Reiskind, M.H., Miller, J.A., Shelton, K.A., Walther, B.A., et al., 2002. Timing of transmission and the evolution of virulence of an insect virus. Proc. Biol. Sci. 269, 1161−1165.

Costafreda, M.I., Perez-Rodriguez, F.J., D'Andrea, L., Guix, S., Ribes, E., et al., 2014. Hepatitis A virus adaptation to cellular shutoff is driven by dynamic adjustments of codon usage and results in the selection of populations with altered capsids. J. Virol. 88, 5029−5041.

de Visser, J.A., Hermisson, J., Wagner, G.P., Ancel Meyers, L., Bagheri-Chaichian, H., et al., 2003. Perspective: evolution and detection of genetic robustness. Evolution 57, 1959−1972.

Dobzhansky, T., Ayala, F.J., Stebbins, G.L., Valentine, J.W., 1977. Evolution. Freeman and Company, San Francisco.

Domingo, E., 2006. Quasispecies: concepts and implications for virology. Curr. Top. Microbiol. Immunol. 299, 83−140.

Domingo, E., Holland, J.J., 1997. RNA virus mutations and fitness for survival. Annu. Rev. Microbiol. 51, 151−178.

Domingo, E., Biebricher, C., Eigen, M., Holland, J.J., 2001. Quasispecies and RNA Virus Evolution: Principles and Consequences. Landes Bioscience, Austin.

Domingo, E., de Avila, A.I., Gallego, I., Sheldon, J., Perales, C., 2019. Viral fitness: history and relevance for viral pathogenesis and antiviral interventions (Pathog. Dis. 77, ftz021.

Duarte, E., Clarke, D., Moya, A., Domingo, E., Holland, J., 1992. Rapid fitness losses in mammalian RNA virus clones due to Muller's ratchet. Proc. Natl. Acad. Sci. U.S.A. 89, 6015−6019.

Dunn, G., Klapsa, D., Wilton, T., Stone, L., Minor, P.D., et al., 2015. Twenty-eight years of poliovirus replication in an immunodeficient individual: impact on the global polio eradication initiative. PLoS Pathog. 11, e1005114.

Eigen, M., Schuster, P., 1979. The Hypercycle. A Principle of Natural Self-Organization. Springer, Berlin.

Elena, S.F., 2012. RNA virus genetic robustness: possible causes and some consequences. Curr. Opin. Virol. 2, 525−530.

Elena, S.F., Carrasco, P., Daros, J.A., Sanjuan, R., 2006. Mechanisms of genetic robustness in RNA viruses. EMBO Rep. 7, 168−173.

Escarmís, C., Davila, M., Charpentier, N., Bracho, A., Moya, A., et al., 1996. Genetic lesions associated with Muller's ratchet in an RNA virus. J. Mol. Biol. 264, 255−267.

Escarmís, C., Dávila, M., Domingo, E., 1999. Multiple molecular pathways for fitness recovery of an RNA virus debilitated by operation of Muller's ratchet. J. Mol. Biol. 285, 495−505.

Escarmís, C., Gómez-Mariano, G., Dávila, M., Lázaro, E., Domingo, E., 2002. Resistance to extinction of low fitness virus subjected to plaque-to- plaque transfers: diversification by mutation clustering. J. Mol. Biol. 315, 647−661.

Escriu, F., Fraile, A., Garcia-Arenal, F., 2003. The evolution of virulence in a plant virus. Evolution 57, 755−765.

Farber, D.L., Netea, M.G., Radbruch, A., Rajewsky, K., Zinkernagel, R.M., 2016. Immunological memory: lessons from the past and a look to the future. Nat. Rev. Immunol. 16, 124−128.

Fernandez, G., Clotet, B., Martínez, M.A., 2007. Fitness landscape of human immunodeficiency virus type 1 protease quasispecies. J. Virol. 81, 2485−2496.

Filomatori, C.V., Carballeda, J.M., Villordo, S.M., Aguirre, S., et al., 2017. Dengue virus genomic variation associated with mosquito adaptation defines the pattern of viral non-coding RNAs and fitness in human cells. PLoS Pathog. 13, e1006265.

Fisher, R.A., 1930. The Genetical Theory of Natural Selection. Oxford University Press, Oxford.

Frank, S.A., 1996. The design of natural and artificial adaptive systems. In: Rose, M.R., Lauder, G.V. (Eds.), Adaptation. Academic Press, San Diego, pp. 451−505.

Frolov, I., Agapov, E., Hoffman Jr., T.A., Pragai, B.M., Lippa, M., et al., 1999. Selection of RNA replicons capable of persistent noncytopathic replication in mammalian cells. J. Virol. 73, 3854−3865.

Gallego, I., Sheldon, J., Moreno, E., Gregori, J., Quer, J., et al., 2016. Barrier-independent, fitness-associated differences in sofosbuvir efficacy against hepatitis C virus. Antimicrob. Agents Chemother. 60, 3786−3793.

Gallego, I., Gregori, J., Soria, M.E., García-Crespo, C., García-Alvarez, M., et al., 2018. Resistance of high fitness hepatitis C virus to lethal mutagenesis. Virology 523, 100−109.

García-Arriaza, J., Manrubia, S.C., Toja, M., Domingo, E., Escarmís, C., 2004. Evolutionary transition toward defective RNAs that are infectious by complementation. J. Virol. 78, 11678−11685.

García-Lerma, J.G., Nidtha, S., Blumoff, K., Weinstock, H., Heneine, W., 2001. Increased ability for selection of zidovudine resistance in a distinct class of wild-type HIV-1 from drug-naive persons. Proc. Natl. Acad. Sci. U.S.A. 98, 13907–13912.

Gavrilin, G.V., Cherkasova, E.A., Lipskaya, G.Y., Kew, O.M., Agol, V.I., 2000. Evolution of circulating wild poliovirus and of vaccine-derived poliovirus in an immunodeficient patient. A unifying model. J. Virol. 74, 7381–7390.

Geoghegan, J.L., Holmes, E.C., 2018. Evolutionary virology at 40. Genetics 210, 1151–1162.

Gire, S.K., Goba, A., Andersen, K.G., Sealfon, R.S., Park, D.J., et al., 2014. Genomic surveillance elucidates Ebola virus origin and transmission during the 2014 outbreak. Science 345, 1369–1372.

González-López, C., Arias, A., Pariente, N., Gómez-Mariano, G., Domingo, E., 2004. Preextinction viral RNA can interfere with infectivity. J. Virol. 78, 3319–3324.

González-López, C., Gómez-Mariano, G., Escarmís, C., Domingo, E., 2005. Invariant aphthovirus consensus nucleotide sequence in the transition to error catastrophe. Infect. Genet. Evol. 5, 366–374.

Gopalakrishnan, S., Montazeri, H., Menz, S., Beerenwinkel, N., Huisinga, W., 2014. Estimating HIV-1 fitness characteristics from cross-sectional genotype data. PLoS Comput. Biol. 10, e1003886.

Graci, J.D., Gnadig, N.F., Galarraga, J.E., Castro, C., Vignuzzi, M., et al., 2012. Mutational robustness of an RNA virus influences sensitivity to lethal mutagenesis. J. Virol. 86, 2869–2873.

Grande-Pérez, A., Gomez-Mariano, G., Lowenstein, P.R., Domingo, E., 2005. Mutagenesis-induced large fitness variations with an invariant arenavirus consensus genomic nucleotide sequence. J. Virol. 79, 10451–10459.

Haldane, J.B.S., 1937. The effect of variation on fitness. Am. Nat. 71, 337–349.

Harrigan, P.R., Bloor, S., Larder, B.A., 1998. Relative replicative fitness of zidovudine-resistant human immunodeficiency virus type 1 isolates in vitro. J. Virol. 72, 3773–3778.

Herrera, M., García-Arriaza, J., Pariente, N., Escarmís, C., Domingo, E., 2007. Molecular basis for a lack of correlation between viral fitness and cell killing capacity. PLoS Pathog. 3, e53.

Holland, J.J., de la Torre, J.C., Clarke, D.K., Duarte, E., 1991. Quantitation of relative fitness and great adaptability of clonal populations of RNA viruses. J. Virol. 65, 2960–2967.

Imamichi, H., Crandall, K.A., Natarajan, V., Jiang, M.K., Dewar, R.L., et al., 2001. Human immunodeficiency virus type 1 quasi species that rebound after discontinuation of highly active antiretroviral therapy are similar to the viral quasi species present before initiation of therapy. J. Infect. Dis. 183, 36–50.

Iyidogan, P., Anderson, K.S., 2014. Current perspectives on HIV-1 antiretroviral drug resistance. Viruses 6, 4095–4139.

Jee, J., Rasouly, A., Shamovsky, I., Akivis, Y., Steinman, S.R., et al., 2016. Rates and mechanisms of bacterial mutagenesis from maximum-depth sequencing. Nature 534, 693–696.

Kanno, T., Kim, K., Kono, K., Drescher, K.M., Chapman, N.M., et al., 2006. Group B coxsackievirus diabetogenic phenotype correlates with replication efficiency. J. Virol. 80, 5637–5643.

Karlsson, A.C., Gaines, H., Sallberg, M., Lindback, S., Sonnerborg, A., 1999. Reappearance of founder virus sequence in human immunodeficiency virus type 1-infected patients. J. Virol. 73, 6191–6196.

Khatchikian, D., Orlich, M., Rott, R., 1989. Increased viral pathogenicity after insertion of a 28S ribosomal RNA sequence into the haemagglutinin gene of an influenza virus. Nature 340, 156–157.

Kijak, G.H., Simon, V., Balfe, P., Vanderhoeven, J., Pampuro, S.E., et al., 2002. Origin of human immunodeficiency virus type 1 quasispecies emerging after antiretroviral treatment interruption in patients with therapeutic failure. J. Virol. 76, 7000–7009.

Knowles, N.J., Samuel, A.R., 2003. Molecular epidemiology of foot-and-mouth disease virus. Virus Res. 91, 65–80.

Koike, S., Nomoto, A., 2010. Poliomyelitis. In: Ehrenfeld, E., Domingo, E., Roos, R.P. (Eds.), The Picornaviruses. ASM Press, Washington DC, pp. 339–352.

Lau, D.T., Khokhar, M.F., Doo, E., Ghany, M.G., Herion, D., et al., 2000. Long-term therapy of chronic hepatitis B with lamivudine. Hepatology 32, 828–834.

Lenski, R.E., May, R.M., 1994. The evolution of virulence in parasites and pathogens: reconciliation between two competing hypotheses. J. Theor. Biol. 169, 253–265.

Lindenbach, B.D., Evans, M.J., Syder, A.J., Wolk, B., Tellinghuisen, T.L., et al., 2005. Complete replication of hepatitis C virus in cell culture. Science 309, 623–626.

Lorenzo-Redondo, R., Borderia, A.V., Lopez-Galindez, C., 2011. Dynamics of in vitro fitness recovery of HIV-1. J. Virol. 85, 1861–1870.

Lorenzo-Redondo, R., Delgado, S., Morán, F., Lopez-Galindez, C., 2014. Realistic three dimensional fitness landscapes generated by self organizing maps for the analysis of experimental HIV-1 evolution. PLoS One 9, e88579.

Mahy, B.W.J., Belsham, G.J., 2017. Overview of foot-and-mouth disease and its impact as a re-emergent viral infection. In: Sobrino, F., Domingo, E. (Eds.), Foot-and-mouth Disease Virus. Current Research and Emerging Trends. Caister Academic Press, Norfolk, UK, pp. 417–426.

Mansky, L.M., 1998. Retrovirus mutation rates and their role in genetic variation. J. Gen. Virol. 79 (Pt 6), 1337–1345.

Maree, A.F., Keulen, W., Boucher, C.A., De Boer, R.J., 2000. Estimating relative fitness in viral competition experiments. J. Virol. 74, 11067–11072.

Marks, H., van Duijse, J.J., Zuidema, D., van Hulten, M.C., Vlak, J.M., 2005. Fitness and virulence of an ancestral white spot syndrome virus isolate from shrimp. Virus Res. 110, 9–20.

Martin, J., 2006. Vaccine-derived poliovirus from long term excretors and the end game of polio eradication. Biologicals 34, 117–122.

Martin, J., Dunn, G., Hull, R., Patel, V., Minor, P.D., 2000. Evolution of the Sabin strain of type 3 poliovirus in an immunodeficient patient during the entire 637-day period of virus excretion. J. Virol. 74, 3001–3010.

Martin, V., Grande-Perez, A., Domingo, E., 2008. No evidence of selection for mutational robustness during lethal mutagenesis of lymphocytic choriomeningitis virus. Virology 378, 185–192.

Martinez-Picado, J., Martinez, M.A., 2008. HIV-1 reverse transcriptase inhibitor resistance mutations and fitness: a view from the clinic and ex vivo. Virus Res. 134, 104–123.

Moreno, E., Gallego, I., Gregori, J., Lucía-Sanz, A., Soria, M.E., et al., 2017. Internal disequilibria and phenotypic diversification during replication of hepatitis C virus in a noncoevolving cellular environment. J. Virol. 91, e02505–e02516.

Nielsen, C., Keena, M., Hajek, A.E., 2005. Virulence and fitness of the fungal pathogen *Entomophaga maimaiga* in its host *Lymantria dispar*, for pathogen and host strains originating from Asia, Europe, and North America. J. Invertebr. Pathol. 89, 232–242.

Novella, I.S., 2003. Contributions of vesicular stomatitis virus to the understanding of RNA virus evolution. Curr. Opin. Microbiol. 6, 399–405.

Novella, I.S., Duarte, E.A., Elena, S.F., Moya, A., Domingo, E., et al., 1995. Exponential increases of RNA virus fitness during large population transmissions. Proc. Natl. Acad. Sci. U.S.A. 92, 5841–5844.

O'Dea, E.B., Keller, T.E., Wilke, C.O., 2009. Does mutational robustness inhibit extinction by lethal mutagenesis in viral populations? PLoS Comput. Biol. 6, e1000811.

Ojosnegros, S., Garcia-Arriaza, J., Escarmís, C., Manrubia, S.C., Perales, C., et al., 2011. Viral genome segmentation can result from a trade-off between genetic content and particle stability. PLoS Genet. 7, e1001344.

Pariente, N., Sierra, S., Lowenstein, P.R., Domingo, E., 2001. Efficient virus extinction by combinations of a mutagen and antiviral inhibitors. J. Virol. 75, 9723–9730.

Perales, C., Lorenzo-Redondo, R., López-Galíndez, C., Martínez, M.A., Domingo, E., 2010. Mutant spectra in virus behavior. Future Virol. 5, 679–698.

Perales, C., Beach, N.M., Gallego, I., Soria, M.E., Quer, J., et al., 2013. Response of hepatitis C virus to long-term passage in the presence of alpha interferon: multiple mutations and a common phenotype. J. Virol. 87, 7593–7607.

Pinheiro, T.M., Mota, M.T.O., Watanabe, A.S.A., Biselli-Perico, J.M., Drumond, B.P., et al., 2018. Viral immunogenicity determines epidemiological fitness in a cohort of DENV-1 infection in Brazil. PLoS Neglected Tropical Diseases 12, c0006525.

Poulin, R., Combes, C., 1999. The concept of virulence: interpretations and implications. Parasitol. Today 15, 474–475.

Quinones-Mateu, M.E., Arts, E.J., 2006. Virus fitness: concept, quantification, and application to HIV population dynamics. Curr. Top. Microbiol. Immunol. 299, 83–140.

Quinones-Mateu, M.E., Moore-Dudley, D.M., Jegede, O., Weber, J., Arts, E.J., 2008. Viral drug resistance and fitness. Adv. Pharmacol. 56, 257–296.

Rocha, E., Cox, N.J., Black, R.A., Harmon, M.W., Harrison, C.J., et al., 1991. Antigenic and genetic variation in influenza A (H1N1) virus isolates recovered from a persistently infected immunodeficient child. J. Virol. 65, 2340–2350.

Ruiz-Jarabo, C.M., Arias, A., Baranowski, E., Escarmís, C., Domingo, E., 2000. Memory in viral quasispecies. J. Virol. 74, 3543–3547.

Ruiz-Jarabo, C.M., Arias, A., Molina-París, C., Briones, C., Baranowski, E., et al., 2002. Duration and fitness dependence of quasispecies memory. J. Mol. Biol. 315, 285–296.

Ruiz-Jarabo, C.M., Miller, E., Gomez-Mariano, G., Domingo, E., 2003. Synchronous loss of quasispecies memory in parallel viral lineages: a deterministic feature of viral quasispecies. J. Mol. Biol. 333, 553–563.

Sánchez-Campos, S., Dominguez-Huerta, G., Díaz-Martínez, L., Tomás, D.M., Navas-Castillo, J., et al., 2018. Differential shape of geminivirus mutant spectra across cultivated and wild hosts with invariant viral consensus sequences. Front. Plant Sci. 9, 932.

Sanjuan, R., Moya, A., Elena, S.F., 2004. The distribution of fitness effects caused by single nucleotide substitutions in an RNA virus. Proc. Natl. Acad. Sci. U.S.A. 101, 8396–8401.

Sanjuan, R., Cuevas, J.M., Furio, V., Holmes, E.C., Moya, A., 2007. Selection for robustness in mutagenized RNA viruses. PLoS Genet. 3, e93.

Sanz-Ramos, M., Diaz-San Segundo, F., Escarmís, C., Domingo, E., Sevilla, N., 2008. Hidden virulence determinants in a viral quasispecies in vivo. J. Virol. 82, 10465–10476.

Scholle, F., Girard, Y.A., Zhao, Q., Higgs, S., Mason, P.W., 2004. Trans-packaged West Nile virus-like particles: infectious properties in vitro and in infected mosquito vectors. J. Virol. 78, 11605–11614.

Schuster, P., 2016. Quasispecies on fitness landscapes. Curr. Top. Microbiol. Immunol. 392, 61–120.

Seifert, D., Beerenwinkel, N., 2016. Estimating fitness of viral quasispecies from next generation sequencing data. Curr. Top. Microbiol. Immunol. 392, 181–200.

Sheldon, J., Beach, N., Moreno, E., Gallego, I., Piñeiro, D., et al., 2014. Increased replicative fitness can lead to decreased drug sensitivity of hepatitis C virus. J. Virol. 88, 12098–12111.

Shirogane, Y., Watanabe, S., Yanagi, T., 2019. Cooperation between different variants: a unique potential for virus evolution. Virus Res. 264, 68–73.

Spiess, E.B., 1977. Genes in Populations. John Wiley & Sons, New York.

Stadler, P.F., 2016. Evolution of RNA-based networks. Curr. Top. Microbiol. Immunol. 392, 219–229.

Teotonio, H., Chelo, I.M., Bradic, M., Rose, M.R., Long, A.D., 2009. Experimental evolution reveals natural selection on standing genetic variation. Nat. Genet. 41, 251–257.

Tracy, S., Chapman, N.M., Drescher, K.M., Kono, K., Tapprich, W., 2006. Evolution of virulence in picornaviruses. Curr. Top. Microbiol. Immunol. 299, 193–209.

Tracy, S., Smithee, S., Alhazmi, A., Chapman, N., 2015. Coxsackievirus can persist in murine pancreas by deletion of 5′ terminal genomic sequences. J. Med. Virol. 87, 240–247.

Wakita, T., Pietschmann, T., Kato, T., Date, T., Miyamoto, M., et al., 2005. Production of infectious hepatitis C virus in tissue culture from a cloned viral genome. Nat. Med. 11, 791–796.

Wargo, A.R., Kurath, G., 2011. In vivo fitness associated with high virulence in a vertebrate virus is a complex trait regulated by host entry, replication, and shedding. J. Virol. 85, 3959–3967.

Wargo, A.R., Kurath, G., 2012. Viral fitness: definitions, measurement, and current insights. Curr. Opin. Virol. 2, 538–545.

Weber, J., Henry, K.R., Arts, E.J., Quinones-Mateu, M.E., 2007. Viral fitness: relation to drug resistance mutations and mechanisms involved: nucleoside reverse transcriptase inhibitor mutations. Curr. Opin. HIV AIDS 2, 81–87.

Wilke, C.O., Wang, J.L., Ofria, C., Lenski, R.E., Adami, C., 2001. Evolution of digital organisms at high mutation rates leads to survival of the flattest. Nature 412, 331–333.

Wilke, C.O., Reissig, D.D., Novella, I.S., 2004. Replication at periodically changing multiplicity of infection promotes stable coexistence of competing viral populations. Evolution 58, 900–905.

Williams, G.C., 1992. Natural Selection. Domains, Levels and Challenges. Oxford University Press, Oxford, New York.

Wright, S., 1931. Evolution in mendelian populations. Genetics 16, 97–159.

Wright, S., 1988. Surfaces of selective value revisited. Am. Nat. 131, 115–123.

Wyatt, C.A., Andrus, L., Brotman, B., Huang, F., Lee, D.H., et al., 1998. Immunity in chimpanzees chronically infected with hepatitis C virus: role of minor quasispecies in reinfection. J. Virol. 72, 1725–1730.

Yuste, E., Sanchez-Palomino, S., Casado, C., Domingo, E., Lopez-Galindez, C., 1999. Drastic fitness loss in human immunodeficiency virus type 1 upon serial bottleneck events. J. Virol. 73, 2745–2751.

Yuste, E., López-Galíndez, C., Domingo, E., 2000. Unusual distribution of mutations associated with serial bottleneck passages of human immunodeficiency virus type 1. J. Virol. 74, 9546–9552.

Yuste, E., Bordería, A.V., Domingo, E., López-Galíndez, C., 2005. Few mutations in the 5′ leader region mediate fitness recovery of debilitated human immunodeficiency type 1 viruses. J. Virol. 79, 5421–5427.

Zhong, J., Gastaminza, P., Cheng, G., Kapadia, S., Kato, T., et al., 2005. Robust hepatitis C virus infection in vitro. Proc. Natl. Acad. Sci. U.S.A. 102, 9294–9299.

# Virus population dynamics examined with experimental model systems

## Abbreviations

**BHK** Baby Hamster Kidney cells
**DI** Defective Interfering
**FMDV** Foot-and-Mouth Disease Virus
**HIV-1** Human Immunodeficiency Virus type 1
**HCV** Hepatitis C Virus
**IFN** Interferon
**IRES** Internal Ribosome Entry Site
**IV** Influenza Virus
**miRNA** micro-RNA
**MAb** Monoclonal Antibody
**MOI** Multiplicity of Infection
**NCCR** Noncoding Control Region
**PBMCs** Peripheral Blood Mononuclear Cells
**PV** Poliovirus
**VSV** Vesicular Stomatitis Virus

## 6.1 Value of experimental evolution

There are good reasons to suspect that we are not aware of many of the influences that guide the evolution of viruses in nature and that only a few of the influences that we can identify can be properly quantified. A rapid review of the major concepts expressed in previous chapters (the chance origin of mutations, complex selective forces, and random events that alter population compositions in an almost incessant manner, potential fidelity modifications of viral polymerases, rapid diversification of viruses even within an individual host, interactions of cooperation among variants within an infecting population, fitness variations due to changes in genomic sequences or in the environment, etc.) forces us to realize that the interpretation of how and why viral populations evolve in nature must be based on largely indirect evidence and a considerable number of assumptions. An understanding of the basic principles that preside over the generation of viral diversity and how some viral forms help or replace others should be based not only on indirect evidence but also on the design of experiments in which a limited number of variables can be examined. Ideally, we need to investigate the operation of one variable at a time in an evolutionary outcome, and then try to extend the simplified approach to integrate and interpret the effect of multiple variables acting conjointly.

Some of the concepts explained in previous chapters can indeed be translated into one defined variable. If we ignore intrinsic population heterogeneity, we can compare the effect of viral population size by examining the outcome of infection of a cell culture or an animal with 0, 1, 10, $10^2$, or $10^3$ infectious units of a virus. We can fix the amount of virus to $10^2$ and examine the influence of prior treatment with 0, 1, 10, $10^2$, or $10^3$ doses of any compound that might modify the virus, the cell, or the virus-cell interaction with the aim of investigating the virus life cycle or finding a new antiviral target. Parallel replicas may help in limiting the effects of differences in virus population composition. We may also consider physico-chemical variables, such as temperature or ionic composition of the cell culture medium. A diagram of some experimental possibilities with viruses that can grow in cell culture is depicted in Fig. 6.1. For some virus-cell systems, persistent infections are readily established because the virus displays limited cytopathology. In other cases, persistence is only established with cells that survive from being killed by the virus (Section 6.4). Remarkable insights have been obtained during the last 4 decades by subjecting clonal, complex, natural, or reconstructed viral populations to defined constraints in the laboratory, inspired by the types of constraints that we suspect may operate in vivo. This is one of the major

objectives of experimental evolution in virology. The studies included in this chapter serve the dual purpose of identifying the experimental origin of some of the conclusions drawn in other chapters, and of providing additional complementary information on the basic concepts of virus evolution.

## 6.2 Experimental systems in cell culture and *in vivo*

For viruses that grow in cell culture, the experimental designs may involve serial passages of cytolytic viruses or persistently infected cells under different infection conditions (i.e., by modifying virus or cell population numbers) or environmental alterations (absence or presence of drugs or antibodies) (Fig. 6.1). The passage of a cytolytic virus involves the infection of fresh (uninfected) cells at each passage, implying that

FIGURE 6.1    Scheme of possible laboratory experiments with a virus that can grow in cell culture. A virus sample from an infected host can be used to infect a cell culture (upper branch A), and the progeny can be passaged serially in cytolytic or persistent infections under different selective pressures. The virus can be diluted and plated to characterize biological clones from the population (middle branch B). The virus may also be adapted to an alternative host by serially passaging the progeny virus produced in the new host (branch C, at the bottom). These and additional possible designs form the basis of experimental evolution, and several examples are discussed in the text, with literature references.

no cell evolution can take place. In contrast, the passage of persistently infected cells involves successive rounds of cell multiplication with the resident virus replicating in the cells. The number of duplications that the cells undergo can be estimated from the number of cells seeded on the plate and their number when they reach confluence or are taken for the next passage. In this type of system, both the cells and the resident virus may evolve (Section 6.4). Increasing the number of passages can enhance coevolutionary differences of cells and resident virus relative to their parental counterparts (their `phenotypic profile at the beginning of the experiment). When bottlenecks are introduced (meaning a reduction of the number of cells per passage), they will affect both viruses and cells, with consequences for the evolutionary outcome.

During cytolytic infections in which only the virus can evolve, population size in experimental designs has two components: the multiplicity of infection (MOI, or the number of infectious particles added per cell) and the total amount of virus used in each infection. In the case of plaque-to-plaque transfers, first performed by L. Chao using bacteriophage φ6 (Chao, 1990), the population size is reduced to one infectious unit in each transfer, representing an extreme passage regime that has been highly informative of the profound molecular alterations associated with fitness decrease undergone by the components of mutant spectra (Section 6.5). The viral population size has yet another relevant influence that can be explained with a numerical example. In standard virus passages (not plaque-to-plaque transfers), infection of $10^3$ cells with $10^4$ infectious particles yields the same MOI than infection of $10^6$ cells with $10^7$ particles (in both cases the MOI is 10 infectious particles per cell). However, the two infections may lead to different outcomes because the number of genomes that initiate infection is 1000-fold larger in the second case, and the virus is genetically heterogeneous. Specifically, any relevant mutant present at a frequency of $10^{-4}$ in the viral

populations will have a high probability of exclusion from participation in the subsequent infection round if the latter is initiated with $10^3$ infectious particles. In contrast, the mutant will be included in infections started with $10^7$ particles. [The same concept applies to the loss of memory genomes in viral quasispecies due to intervening bottlenecks that exclude subsets of minority and memory genomes for the next infection rounds (Section 5.5 in Chapter 5; see also "more is different" applied to viral infections, discussed in Section 3.9 of Chapter 3).]

The presence and maintenance of a specific mutant in a viral population depends on the virus population size and, therefore, it is an "extrinsic" property of a viral population, as opposed to the "intrinsic" properties that are independent of the population size (Domingo and Perales, 2012) ("intrinsic" vs. "extrinsic" properties of mutant spectra were discussed in Chapter 3 in connection with biological complexity as applied to viruses). A natural isolate of a virus may also be adapted to some alternative hosts in vivo depending on the barriers to be confronted by the virus in the potential new host (Fig. 6.1 and Chapter 4). The basic designs depicted in Fig. 6.1 can be extended and modified including complex scenarios, such as environmental heterogeneity (presence of multiple cell types on the same dish, or artificial migration of virus among different cell lines, or between cell lines and host organism, etc.), including the alternation between mammalian and insect cells in the study of arbovirus evolution (Coffey and Vignuzzi, 2011; Coffey et al., 2008; Novella et al., 1999a, 2007; Calisher and Higgs, 2018; Ciota, 2019; among other investigations and reviews). In a study with vesicular stomatitis virus (VSV) populations passaged in different cell lines, virus adaptation was cell-specific in the absence of cell flow, and fitness in all environments decreased with migration rate (Cuevas et al., 2003). VSV populations often consist of clonal mixtures that differ in interferon (IFN) induction or susceptibility (Marcus et al., 1998), and the variant interplay could modulate IFN production (Marcus, 1982). IFN-stimulating VSV affected the fitness of neighboring viruses, with possible implications for virus escape to innate immunity (Domingo-Calap et al., 2019). The scope of possibilities of designed experiments to learn about virus evolution is truly remarkable. However, we have to be aware that viruses change continuously in the course of experiments, and that validation of proposed mechanisms for natural infections is largely pendent.

## 6.2.1 "To culture is to disturb"

The transfer of a viral isolate into an alternative host (either a cell culture, an explant, or an intact organism) implies a perturbation regarding the representation of the parental quasispecies in the new host for two reasons: (i) the involvement of a bottleneck event whose intensity depends on the amount of virus in the biological sample relative to the total amount in the infected parental host, and (ii) the change of biological environment from the donor into the recipient host (Fig. 6.2). Elaborating (i) further, a virus sample contains only a subset of the genomes present in the infected host, and the relative fitness of genome subsets may not represent the fitness of the entire population, if it were feasible to introduce the entire population into the recipient host. Both chance and selection events will modify the genome composition that will enter subsequent rounds of multiplication (Fig. 6.2). This important point was first noted by A. Meyerhans, S. Wain-Hobson, and their colleagues in a comparison of the *tat* gene from sequential human immunodeficiency virus type 1 (HIV-1) isolates, and peripheral blood mononuclear cells (PBMCs) infected with the natural isolates. The study unveiled the difficulties of defining HIV-1 infections in molecular terms, and the authors coined the following sentence that became popular in virus evolution: "to culture is to disturb" (Meyerhans et al., 1989).

It gives a very pertinent image of what is hidden behind quasispecies dynamics. In the 3 decades elapsed since that statement, its value and underlying mechanisms have become increasingly obvious.

Significant genetic and phenotypic differences between natural isolates and their cell culture-adapted counterparts have been observed in several viruses, including DNA viruses. Polyomaviruses are widespread in humans, but they are rarely pathogenic except for immunocompromised individuals. One of the regions of the polyomavirus genome, the noncoding control region (or NCCR), is quite conserved among natural, usually nonpathogenic isolates. When these viruses are adapted to cell culture, the NCCR undergoes rearrangements involving deletions and duplications. Rearrangements are observed in variants that cause pathology, and they can be readily grown in cell culture. An interesting possibility is that NCCR variants display high replication rates associated with pathogenic potential (Gosert et al., 2010). The connection between replication rate, viral load, and disease progression is discussed in Chapter 8. Genome rearrangements were also identified in regions with repeated sequences in cell culture-adapted cytomegaloviruses (Murphy and Shenk, 2008). These observations with DNA viruses, as well as their relevance for viral persistence and the possible implication of micro-RNA (miRNA) expression have been reviewed (see Imperiale and Jiang, 2015, and references therein).

## 6.2.2 Experimental evolution in vivo

The field of experimental evolution includes designed experiments with viruses in their natural or alternative potential hosts. The objective is to probe concepts of viral evolution and pathogenesis, which are interconnected. There is extensive literature with plant and animal viruses that have contributed to the major concepts discussed in this book. Because the studies are very numerous, here, we will discuss

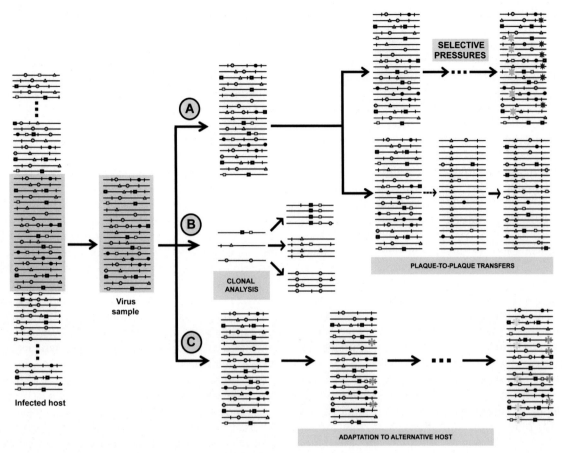

**FIGURE 6.2** The implication of mutant spectra and population size in experimental designs. This scheme is parallel to that shown in Fig. 6.1, except that infected objects have been replaced by mutant spectra in those same objects. From left to right: the infected host has a huge mutant spectrum whose depiction would occupy millions of columns as the one drawn on the left. The virus sample taken from the infected host includes only a subset of genomes (even if they amount to hundred-thousands), here shadowed in gray. Upper branch A: When this genomic subset is passaged in cells under selective pressure, the mutant spectrum will be modified, and genomes with specific mutations (green and red asterisks) will increase in dominance. If the same subset is subjected to plaque-to-plaque transfers, mutations accumulate beginning with a founder genome that will hitchhike two initial mutations (red triangle and horizontal line). In the middle branch B the three biological clones will have different initial mutations, and heterogeneity will increase upon expansion of the population. In the bottom branch C, adaptation to a new host will imply increasing dominance of new mutations (yellow and blue asterisks). Many examples of genetic variation due to the types of effects shown here in a diagrammatic form are discussed in the text and in other chapters of the book.

some selected examples, to again underline general concepts.

Studies that rose considerable controversy relate to the elucidation of amino acid substitutions in avian influenza virus (IV) that mediated pathogenicity and human-to-human transmission potential of the virus (reviews in Bouvier and Lowen, 2010; Schrauwen et al., 2014; Imperiale et al., 2018; Lipsitch, 2018). The ferret is a useful animal model for IV research since it produces respiratory symptoms similar to those in humans. The use of ferret as the animal model showed that amino acid substitutions in the receptor-binding domain of the hemagglutinin and the polymerase PB2 were important for the transmission of an avian IV among ferrets. In one of the experiments to unveil critical determinants for human-to-human transmission, an avian H9N2 IV was adapted to replication in mammals by serially passaging the avian IV in ferrets. Considerable IV heterogeneity that was diagnostic of active population dynamics in the animals was recorded. The ferret-adapted virus was transmitted efficiently among ferrets, indicating that, not surprisingly, an avian IV can adapt to be transmissible among humans (Sorrell et al., 2009). Likewise, several different mutations, including those that affect receptor preference, are needed for effective airborne transmission of an avian H5N1 IV (Herfst et al., 2012).

These and other experiments to unveil transmissibility of avian IV in humans open the possibility of screening of avian IVs in search for mutations that may approximate avian viruses to represent a zoonotic threat for humans. The controversy surrounding the need to carry out these types of experiments ("gain-of-function" experiments) has, in my view, two major components: the uncertainties derived from specifying and publicizing mutations that may render avian IV a biological weapon, and the danger of laboratory escape of a modified IV that can be highly pathogenic and transmissible among humans, causing a devastating disease, in the shadow of the 1918 influenza pandemic. With regard to the first concern, unfortunately, there are many biological weapons available without the need for new ones, and it is very likely that pathogen-enhancing mutations in general (and for IV in particular) do not have a universal value and may change even for closely related isolates of a virus (even, we may speculate, in the course of preparing new stocks for ill-intentioned purposes). The threat of IV lies in the intrinsic variation potential of this multi-host pathogen (Ma et al., 2019). Concerning the second concern, indeed, high containment facilities and strict protocols must be used for experiments on directed changes of virus virulence and transmissibility.

Foot-and-mouth disease virus (FMDV) naturally displays a broad host range and has contributed examples of biological modifications as a result of designed laboratory experiments in cell culture and in vivo. In cell culture, multiple passages of an FMDV clone in the standard BHK-21 cell line resulted in a dramatic expansion of host cell tropism (Ruiz-Jarabo et al., 2004) whose possible significance in vivo is unknown. Regarding experimental FMDV evolution in vivo, as discussed in Chapter 4, a molecular analysis of the adaptation of a swine FMDV to the guinea pig documented the influence of substitutions in nonstructural viral proteins on the host range (Section 4.4.1). An earlier analysis involved a collaborative study between Madrid, Rio de Janeiro, and Buenos Aires teams. It consisted in the genetic and antigenic analysis of sequential FMDV samples extracted from cattle during a persistent infection established experimentally with a clonal virus population derived from a cattle FMDV isolate (Gebauer et al., 1988). The reason to bring this study here is not only because it is an early, interesting, and informative example of experimental evolution in vivo, but also because its results were published at the time when the extensive genetic diversity of HIV-1 was being discovered. Many virologists regarded variation of HIV-1 as unusual, even unique, perhaps only paralleled by that of IV, with the concept of antigenic shift and drift of IV well established at the time (Chapter 7). H. Temin took the analysis of variation of FMDV reported by F. Gebauer and colleagues to emphasize to his retrovirology colleagues that HIV-1 was not "unique but merely different" (Temin, 1989). In the study with FMDV, the virus was recovered from the esophageal-pharyngeal area, the site of FMDV persistence in ruminants, and the virus was examined for up to 539 days postinfection. Despite the infection originating from a biological clone, the sequential samples displayed genetic heterogeneity and dominance of viral subpopulations. Moreover, the persistent virus evolved at rates as high as $0.9 \times 10^{-2}$ to $7.4 \times 10^{-2}$ substitutions per nucleotide and year (s/nt/y), which is as high or even higher than the rate calculated for HIV-1 [$10^{-2}$ to $3.7 \times 10^{-4}$ s/nt/y according to several studies (Hahn et al., 1986; Korber et al., 2000; Shankarappa et al., 1999)]. The mutations were certainly not neutral since several of them affected the reactivity of the virus with antibodies. This controlled experiment showed that FMDV underwent extensive genetic and antigenic variation during persistence in cattle. Thus, H. Temin could emphasize that HIV-1 is not unique concerning variation potential. If a virus is confronted with a focused selective pressure in vivo—as is the case of FMDV in the pharyngeal region where an active local mucosal immune response is triggered—it can reach remarkable rates of evolution that are $10^6$- to $10^7$-fold higher than average values for cellular genes (Holland et al., 1982) (Chapter 7).

In agreement with the model study by F. Gebauer and colleagues with FMDV, persistent viral infections in hosts that display an active immune response (albeit insufficient to clear the virus), may constitute a source of antigenic variants that may occasionally be transmitted to new hosts, or may remain essentially confined to the persistently infected individual (Vosloo and Thomson, 2017). In the latter case, successive waves of variants may be selected to prevent virus clearance by the immune system thus contributing to persistence (Clements et al., 1988; Narayan et al., 1981; Pawlotsky, 2006; Richman et al., 2003; Sponseller et al., 2007, among other examples). Given the potential coevolution of antigenic sites and receptor-recognition domains (Section 4.5 in Chapter 4), persistent infections in animals are a potential threat for the zoonotic emergence of human pathogens, accentuated by the possibility of recombination between the persistent and a related virus from an external source.

## 6.3 Viral dynamics in controlled environments: alterations of viral subpopulations

Experimental studies that revealed a sustained heterogeneity in replicating viral populations in nature and model infections have spanned 5 decades and have been based on widely different technologies: the very early RNA T1 oligonucleotide fingerprint used to analyze biological clones of Qβ RNA (Domingo et al., 1978) and other RNA viruses [(Trent et al., 1990) and other articles in the same volume], the CirSeq design for deep sequencing applied to poliovirus (PV) populations (Acevedo et al., 2014), or nanopore sequencing, with many analyses of mutant spectrum dynamics in between [(Andino and Domingo, 2015; Braun et al., 2018), among other studies and summaries].

The initial findings with bacteriophage Qβ illustrate how difficult it is for a virus to reach population equilibrium. The phage was multiplied for many years in its host *Escherichia coli* since its isolation from Kyoto feces in 1961 by I. Watanabe and colleagues (Miyake et al., 1967). It was passaged in the laboratories of I. Watanabe, S. Spiegelman (in Urbana), and C. Weissmann (in New York, 1965—67 and in Zürich, 1967—74). There is no record that the virus had been biologically cloned during these multiple passages in *E. coli*. In 1974, a clone termed A.S. (from A. Shapira who was working with M. Billeter in Zürich) was isolated, and its T1 oligonucleotide fingerprint was compared with that of a reconstructed stock of the uncloned Qβ population. The comparison revealed that the uncloned stock was heterogeneous since one oligonucleotide that was quantified as present in full molar amount in clone A.S. was found in submolar amounts in the uncloned stock, and had been replaced by a mutant oligonucleotide [explained in (Domingo et al., 1978)]. The resolved oligonucleotides that represented about 10% of the genome had been sequenced and mapped by

M. Billeter, although the results were published years later (Billeter, 1978). These very early comparisons reinforced a suspicion that was frequently commented in the discussions held in Zürich: that bacteriophage Qβ, and other RNA bacteriophages, were probably highly heterogeneous [see Weissmann et al. (1973) for statements about potential heterogeneity at the time that nucleotide sequencing was slow and cumbersome]. Earlier indirect evidence of genetic instability of RNA viruses (abundance of temperature-sensitive mutants in virus stocks, frequent reversion of phenotypic markers, etc.) was reviewed (Domingo and Holland, 1988). The observations with bacteriophage Qβ prompted the discovery of viral quasispecies (as described in Chapter 3), but the reason to bring them here is to emphasize that even after extensive passage without cloning, no population equilibrium with one dominant genome type was reached in the Qβ population.

From the current knowledge of quasispecies dynamics, we can interpret that one of the reasons why no equilibrium with a defined consensus sequence was produced lies in the multiple possibilities of exploration of sequence space, open to any virus during replication as a result of the stochastic generation of mutations. Nonequilibrium might have been favored by the mutual effects (positive or negative, sometimes termed epistatic effects) among mutations in the same genome, and interactions among different genomes, or sampling effects during passages, among other factors. It is worth (albeit not easy) picturing that a tendency toward equilibrium (meaning a trend toward a steady distribution of mutant forms) has to be based on the number and types of mutants available to the viral population at a given time. When new mutants are generated stochastically, the pathway toward equilibrium may change. A very large population will generate a siginificant number of alternatives toward equilibrium to choose from, as compared to a small population. This is why deterministic features of quasispecies are more likely to be observed with large viral populations (Section 3.6.2 in Chapter 3).

No equilibrium can be assumed (and much less so the absence of mutations!) even for a virus with a long history of multiplication in the same environment (Box 6.1). Recent results with hepatitis C virus (HCV) have reinforced this conclusion (Moreno et al., 2017). Furthermore, environments which might be regarded as "constant" may nevertheless be highly patchy due to cell-to-cell variations even under controlled bacterial or animal cell culture conditions.

The initial FMDV passage experiments in cell culture that provided evidence of quasispecies dynamics for an animal virus were carried out with a biological clone of a swine isolate (Sobrino et al., 1983). The viral genome displayed increases and decreases in the molar proportion of T1 oligonucleotides, with heterogeneity levels estimated in an average of two to eight mutations per genome in the mutant spectrum, as compared with the corresponding consensus sequence. Passage resulted in the adaptation of the virus to the cultured cells, as documented by an increase of infectious progeny production. In parallel with the FMDV work, the experimental studies by J.J. Holland and colleagues with VSV contributed significantly to the understanding of viral quasispecies, starting with pioneer experiments on the generation of defective interfering (DI) particles and their competition with the standard virus (Holland et al., 1979, 1982). The studies carried out in J.J. Holland's laboratory in San Diego are summarized in several chapters of this book (Novella, 2003).

Application of standard molecular cloning and sequencing, and deep sequencing to the analysis of viral populations during experimental infections have supported the view that viral populations are composed of many genome subpopulations, and that their evolution is best described as the replacement of some viral subpopulations by others (Acevedo et al., 2014; Baccam et al., 2003; Sobrino et al., 1983). Because conditions are far from population equilibrium, such replacements occur even in the absence of an externally applied selective pressure. A selective pressure acts as a guiding force to decant viral subpopulations in favor of those

BOX 6.1

## Some Teachings of Experimental Virus Evolution

- Mutant viruses are continuously arising in viral populations. Even prolonged passage in the same environment does not mean that a population equilibrium has been attained. An invariant consensus sequence does not imply the absence of mutations. It means that the genomes are mutating continuously to yield the same consensus.

- Virus evolution consists in the replacement of some viral subpopulation by others, due to random events or in response to selective constraints.

- The reorganization of viral subpopulations is profoundly altered by bottleneck events.
- The model studies predict that the effect of a bottleneck depends on its size and the fitness of the parental population.
- Work conducted under the guidance of J.J. Holland has permitted testing experimentally with viruses several theoretical proposals, hypotheses, and principles of general genetics. They include:
- Muller's ratchet
- Competitive Exclusion principle
- Red Queen hypothesis

---

that best respond to the constraint, often after many transient, abortive attempts, as also observed in natural infections (Cale et al., 2011; Fischer et al., 2010; Kortenhoeven et al., 2015; Tsibris et al., 2009) (Fig. 6.3). In other terms, different areas of sequence space are dynamically occupied prior to the occurrence of selective constraints as well as in response to selective demands (Chapters 3 and 7). Without an environmental change, the mutant spectrum explorations have an essential component of randomness, whereas, when a selective force is applied, the cloud is guided toward regions that can best confront the environmental demand.

## 6.4 Persistent infections in cell culture: virus-cell coevolution

Persistent infections in vivo may result from failure of immune surveillance systems to clear a virus, from infection of cells that exert functions related to the immune response, or from other mechanisms that limit viral population numbers and cell killing. Persistent infections with or without pathology may involve the integration of genetic material of the virus into host cells or maintenance of the virus replicating by its standard mechanisms, but with modulation of viral population numbers. Persistent infections are abundant in most biological phyla that have been examined [for reviews of mechanisms and biological effects of viral persistence for the host cells see (Ahmed and Chen, 1999; Nash et al., 2015; Nathanson and Gonzalez-Scarano, 2007; Oldstone, 2006; Roossinck, 2014; Fan et al., 2018)]. A feature of persistent infections is that the population numbers of infectious particles often remain limited and constrained to a specific environmental domain where viruses are enclosed (liver, kidney, brain, etc.). Under these circumstances, the $r$ strategy typical of overt virus infections (success based on rapid reproduction to cope with different environments) is transformed into one closer to the $K$

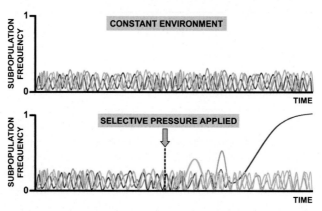

**FIGURE 6.3** Diagram of dynamic genome subpopulations in the course of virus replication. Four different colors have been chosen to depict fluctuations of four genomic classes. In a real population, thousands of genomes may be involved in each infected cell. When a selective pressure is applied, multiple transient selection events occur (green and blue outstanding waves) only to be finally displaced by a winning subset of genomes here represented in red. The time frame in the abscissa is left without units on purpose because the diagram can represent events in a few infected cells during minutes or in a persistently infected organism during years. See text for examples and references.

strategy (limited progeny adapted to a specific environment) which is typical of large animals (compare with Section 4.1 in Chapter 4).

Cell culture systems have provided excellent tools to address the mechanisms of persistence of DNA and RNA viruses, including herpesviruses, polyomaviruses, parvoviruses, picornaviruses, reoviruses, flaviviruses, and retroviruses [(Herrera et al., 2008; Imperiale and Jiang, 2015; Olivares et al., 2013), among other examples]. Here, we will analyze some of these systems that have contributed new insights into the interaction of cells and viruses in a controlled environment, and in particular, virus-host coevolution during persistence. That is, a mutual influence is exerted on the evolution of two interacting biological entities (discussed also in Section 4.5 of Chapter 4). Takemoto and Habel reported the first evidence of virus-cell coevolution with the picornavirus Coxsackievirus A9 (Takemoto and Habel, 1959). R. Ahmed, B.N. Fields, and colleagues documented the role of cells and coevolution of reovirus and L cells during persistent infection in cell culture (Ahmed et al., 1981). These early studies were followed by experiments with the lymphotropic minute

virus of mice (Ron and Tal, 1985, 1986), reovirus (Chiarini et al., 1983), polyomavirus (Delli Bovi et al., 1984), FMDV (de la Torre et al., 1988), and HCV (Zhong et al., 2006) (several systems are summarized in Table 6.1).

Persistent infectious of FMDV were established in BHK-21 and IBRS-2 cells by growing the cells that survived a cytolytic infection (the frequency of survivors was $10^{-3}$ to $10^{-4}$). During persistence, the resident virus changed genetically and phenotypically: the heterogeneity of the resident viral genomes was estimated in $5 \times 10^{-4}$ substitutions per nucleotide, and the virus displayed temperature sensitivity and small plaque morphology (de la Torre et al., 1985, 1988). During FMDV persistence in BHK-21 cells, the virus became increasingly virulent for the parental BHK-21 cells, and in turn, the cells became progressively more resistant to the parental FMDV (de la Torre et al., 1985, 1988). The resistance of the cells to the parental virus was not due to an impairment of virus attachment, penetration, or uncoating, but to some intracellular block as evidenced by virological and cell fusion experiments (de la Torre et al., 1988, 1989a).

**TABLE 6.1** Examples of virus-cell coevolution in cell culture.

| Virus cell | Main findings | References |
|---|---|---|
| Coxsackievirus A9 HeLa | Carrier cells showed increased resistance to the virus. Virus underwent antigenic variation, plaque size reduction, increased virulence for HeLa cells and decreased virulence for suckling mice | Takemoto and Habel (1959) |
| Reovirus L | Mutant cells selected during persistence displayed partial resistance to reinfection with the parental virus, and readily reestablished persistence. Mutant virus selected during persistence overgrew the parental virus | Ahmed et al. (1981) |
| Bovine rotavirus AU-BEK | Persistence was dependent on the presence of fetal calf serum. Cells evolved to be highly resistant to the virus | Chiarini et al. (1983) |
| Polyomavirus (Py) Friend erythroleukemia (FL) | Viral genomes integrate at high temperature (large T-inactive). Viral genomes rescued at permissive temperature. Coevolution with selection of virus-free Py-resistant cells or cells shedding Py variants | Delli Bovi et al. (1984) |
| Lymphotropic minute virus of mice L | Host range virus mutants and virus-resistant cells were selected. Reconstruction of the persistently infected culture required mutant virus and mutant cells. Resistance to infection was due to an intracellular block that affected the synthesis of viral single-stranded DNA | Ron and Tal (1985) |
| Herpes simplex virus type 1 Human lymphoblastoid CEM | Cured cells were partially resistant to parental virus. Virus isolated during persistence displayed increased virulence for the parental CEM cells | Cummings and Rinaldo (1989) |
| Reovirus L | Mutation in cells and virus affect an early step of the virus replication cycle. Amino acid substitutions, which alter virus sigma 1 protein oligomerization, mediate the infection of virus-resistant L cells. The capsid is a determinant of reovirus adaptation | Dermody et al. (1993) and Wilson et al. (1996) |
| Poliovirus HEp-2c | Persistently infected cells cured of the resident poliovirus (PV) displayed selective permissivity to the parental PV (Mahoney strain) but partial resistance to several PV mutants. The restriction was mapped at an early phase (adsorption or uncoating) of the infection | Calvez et al. (1995) |
| Mouse hepatitis DBT | Murine astrocytoma (DBT) cells resistant to mouse hepatitis virus (MHV) were selected; resistance included diminished expression of the MHV receptor. MHVs with increased avidity for the receptor —and that recognized additional receptors—were selected | Chen and Baric (1996) |
| Rotavirus MA104 | Virus rescued from persistently infected cells produced higher viral yield than the parental virus in persistently infected cells that were cured of the virus. Mutations in virus and cells affected virus entry | Mrukowicz et al. (1998) |
| Hepatitis C virus Human hepatoma cells | Virus variants with accelerated replication kinetics and higher peak titers were selected during persistence. The altered phenotype was associated with a substitution in the envelope E2 protein. Cells with decreased permissivity to the virus due to a block of virus entry, RNA replication, or both were selected | Zhong et al. (2006) |

The coevolutionary process involved both viral and cellular heterogeneity. J.C. de la Torre and colleagues analyzed a total of 248 stable BHK-21 cell lineages established from individual cells isolated by limiting dilution from the persistently infected cultures (de la Torre et al., 1989a,b). They distinguished 6 cell categories based on morphology, duplication rate, and resistance to the FMDV that initiated persistence. The study indicated that the coexistence of the cell with the virus led to the selection of cells with genetic modifications that resulted in altered phenotypes. In particular, increased cell transformation (evidenced by rapid cell duplication, loss of contact inhibition, and growth in semisolid agar medium) correlated with resistance to FMDV, although a possible mechanism that could link the two phenotypic traits was not investigated. Contrary to expectations, it was variation of the BHK-21 cells and not of the virus the critical event for the establishment of the persistent infection with FMDV (Martin Hernandez et al., 1994). Mutual influences that lead to both cell and virus modifications provide long-term stability to the system without either the cells or the virus eliminating its partner (except for documented "crises of cytopathology" during which high virus titers are reached, suggesting a transient imbalance in favor of the virus) (Herrera et al., 2008). The generality of these proposals is supported by observations with several virus-host systems (Table 6.1). The cell culture models can help in understanding coevolutionary events between viruses and their host organisms, with implications for emerging and re-emerging viral infections (Fedeli et al., 2018). At the time when most studies on virus-mediated cellular modifications were performed, there was no awareness that epigenetic modifications in the cells as a result of viral infection may have also played a role in modifying the intracellular environment. This is a facet open to the study of persistent viral infections, with new therapeutic possibilities (Moos et al., 2017).

## 6.4.1 Back again 4000 million years: contingency in evolution

In an excursion toward the question of contingency in evolution, M. Herrera and colleagues, 20 years later, again repeated the entire process of establishment and maintenance of FMDV persistence, starting from frozen virus and cell stocks (Herrera et al., 2008). The main conclusion with two new persistent BHK-21-FMDV lineages established in parallel was that the virus-cell coevolution displayed features very similar to those observed in the first persistent lineage, although there were differences in some of the mutations in FMDV. However, two amino acid substitutions which affected residues around a pore located at the capsid 5-fold axis were selected in the three persistent lineages. One of the substitutions (D9 → A in capsid protein VP3) was later selected in other FMDV serotypes that having originated in acute infections in animals, were passaged in evolved BHK-21 cells that displayed partial resistance to the virus (de la Torre et al., 1988; Díez et al., 1990; Escarmís et al., 1998). This substitution belongs to the category of "joker" substitutions in the sense of providing a selective advantage to the virus in a variety of sequence contexts and environments (Section 4.9 in Chapter 4).

The two newly established BHK-21-FMDV persistent lineages displayed strikingly parallel features in the variation of progeny production and imbalances in the ratio of positive versus negative-strand viral RNA, that were found at the same passage number (Herrera et al., 2008). This parallel behavior reflects a deterministic feature of virus evolution that has been encountered with other virus-host systems (see Section 6.7.1 and the general concept of deterministic vs. stochastic quasispecies in

Section 3.6.2 of Chapter 3). Interestingly, a deterministic behavior should be generally favored by large viral population sizes, but in this case, a drastic decrease in population size preceded its manifestation. This unexpected association was interpreted as deterministic behavior being due to strict selective demands that guide the virus toward one (or very few) biological solutions to respond to the constraint (Herrera et al., 2008). The reproducibility of the main features of FMDV persistence in independent establishment events was suggested to mean that when a biological system is highly constrained, there may be limited room for contingency. Thus, we can now consider the question: If we could rewind the tape of evolution and play it again, would it turn out to be similar or different from what we know? According to S.J. Gould, if we could rewind the tape to the remote past, evolution would turn out entirely different (Gould, 1989). The answer proposed by M. Herrera and colleagues was that the tape would turn out equally, differently or "in-between" depending on the constraints operating at the time we start to retape. If the tape was situated in the RNA world around $3.6 \times 10^9$ to $4.2 \times 10^9$ years ago or perhaps prior to the Cambrian explosion, about $5 \times 10^8$ years ago (Conway Morris, 1998; Gould, 1989) there would be ample room for divergence due to equally valid alternative solutions for evolution of biological organization and complexity. In contrast, once the constraints inherent to functional cells and viruses are in place, the system might be forced to enter a route conductive to a seemingly deterministic behavior (see Chapter 1 for the time frame in which the increasingly chemical and precellular complexity evolved on Earth). The fact that RNA viruses can accumulate mutations at about one million faster rate than most differentiated organisms (that is, that they represent exceedingly accelerated versions of biological evolution) may underlie

unpredictable versus reproducible evolution (Gutierrez et al., 2019), decanted by factors that are not well understood.

It is tempting to make a connection between virus-cell coevolution in cell culture and the models of virus origins that contemplated vesicle-wrapped primitive cellular and viral entities (Section 1.5.5 in Chapter 1). What the experiments of viral persistence in cell culture show is that as part of coevolutionary mechanisms, a cell may diminish the expression of surface proteins that act as receptors for the coevolving virus or even alter fundamental properties such as the rate of cellular multiplication. Cells persistently infected with PV express mutated forms of the PV receptor CD155 (Gosselin et al., 2003; Pavio et al., 2000), and persistently infected BHK-21 cells increased their degree of cellular transformation (de la Torre et al., 1988, 1989b). [For review of picornavirus persistence, see Colbère-Garapin and Lipton (2010).] It has been proposed that in the primitive biosphere early vesicles evolved to limit their invasion by virus-like elements, and they did so by building protective walls that culminated in primitive receptor-dependent virus entry into increasingly autonomous precellular entities. The insights provided by current cell-virus coevolution models indicate that such kind of mutual influences occur with naked or enveloped viruses without the need for the virus to integrate its genome (or part of it) into the genetic material of the coevolving cell (Colbère-Garapin and Lipton, 2010; de la Torre et al., 1988). During persistent infectious in cell culture, virus titers change in ways partially dictated by their own quasispecies dynamics and coevolving carrier cells.

An area of experimental evolution that puts its emphasis on the virus rather than the host cells consists in subjecting the virus to repeated bottlenecks that are experimentally realized through plaque-to-plaque transfers (Figs. 6.1 and 6.2). In this design, the cell obviously plays the essential role of hosting viral replication, but it cannot evolve since fresh (uninfected) cells are used for each virus transfer. A lot has been learned from this class of experiments.

## 6.5 Teachings from plaque-to-plaque transfers

Bottlenecks increase the stochastic component of evolutionary events and may modify the course of selection (Chapter 3). In the present section, we review evidence that population bottlenecks may have additional and even more profound influences in the composition of viral quasispecies, by permitting hidden minority genomes to surface in populations. This important line of investigation of viral genetics was initiated by L. Chao while working with the tripartite double-stranded RNA bacteriophage φ6 (Chao, 1990). He demonstrated an average fitness decrease in φ6 clones subjected to serial plaque transfers in its host bacterium *Pseudomonas phaseolicola*. The results constituted the first experimental support for the operation of Muller's ratchet, a concept from theoretical biology explained in Section 6.5.1. However, the molecular basis of fitness decrease was not investigated in this first study.

The experimental design of L. Chao was extended by J.J. Holland and his colleagues to the animal virus VSV growing in mammalian cells (Duarte et al., 1992). The results agreed with those of L. Chao and documented variable, but in some cases, severe fitness drops by subjecting VSV to only 20 serial plaque-to-plaque transfers. A relevant observation was the large difference in the extent of fitness decrease among biological VSV clones (those derived from a common parental population) subjected to the same passage regime, which is by itself a reflection of the remarkable heterogeneity within a VSV population. A second highly significant observation was that VSV clones isolated from a population with a long history of passage in BHK-21

cells displayed a more pronounced fitness decrease when subjected to plaque-to-plaque transfers in the alternative hosts HeLa or MDCK cells than in BHK-21 cells to which the virus was better adapted (Duarte et al., 1992). This difference probably reflects the fact that when the virus replicated in a less adequate environment provided by the unfamiliar host cell, a larger proportion of genetic variants occupied the most frequent class of genomes. The genomes forced to be more represented are those more likely to be picked at random from individual plaques to enter the following transfer with a higher initial mutational load. The result is in agreement with models that predict that fluctuating selection contributes to the accumulation of deleterious mutations in selfing populations (Ho and Agrawal, 2018). J.J. Holland and colleagues extended these studies with VSV to show fitness decreases at different plating temperatures and contrasted the fitness loss associated with plaque transfers with fitness gain upon large population passages in the same host cells. Debilitated clones rapidly regained fitness when they were passaged as large populations (Clarke et al., 1993). A study of alternation between passage regimes evidenced that the fitness increase that occurs during two successive large population passages was not sufficient to overcome the decrease produced by a single bottleneck passage (Duarte et al., 1993).

The model studies of J.J. Holland, I.S. Novella, and colleagues with VSV represented considerable progress in the understanding of virus evolution, and some of the information provided by the results is not (even today!) sufficiently considered in interpreting the consequences of bottlenecks and population expansions in vivo (Chapter 5). A pertinent study was the quantification of the effect of the bottleneck size (how many infectious particles participate in the bottleneck passage) on viral fitness. This effect depends on the fitness of the population subjected to the bottleneck (Novella et al., 1995) (Fig. 6.4). When the starting population has

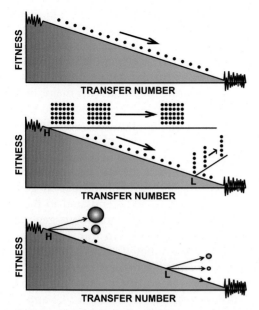

**FIGURE 6.4** The effect of bottleneck size on fitness is dependent on the fitness of the population subjected to the bottleneck. At the top, plaque-to-plaque transfers (bottleneck size 1, black dots) lead to fitness decrease dependent on the number of transfers (downward arrow). The middle diagram shows that when the starting population has high fitness (point H in the graph), a bottleneck size of 30 (pool of virus from 30 plaques) is necessary just to maintain the fitness value (horizontal arrow). In contrast, when the starting point has low fitness (point L in the graph), a bottleneck size of five is sufficient to increase fitness (upward arrow). In the bottom diagram, bottleneck sizes are depicted as spheres of different size to schematically summarize the bottleneck size requirements to maintain or increase fitness depending on the initial fitness (H, high; L, low). The concept expressed in this diagram is based on studies with VSV by I.S. Novella, J.J. Holland, and colleagues, as described in the text. The fluctuations in fitness values at very high and very low-fitness values have been observed experimentally, and their interpretation and implications are also discussed in the text.

low fitness (point L in the middle and bottom schemes of Fig. 6.4), few particles per transfer are sufficient for the passaged population to gain fitness. In contrast, when the initial population has high fitness (point H in the middle and bottom schemes of Fig. 6.4), a large number of particles is needed just to maintain fitness. An even larger number is necessary to increase

fitness above the initial level of point H. These observations acquire additional biological significance for fitness evolution of viral populations in the event of bloc transmissions of multiple viral particles into the same cell [(Altan-Bonnet et al., 2019) and several articles in the same special issue of Virus Research].

If these observations on fitness dependence of fitness evolution were operative in the patchy environments provided by nature (still to be proven), they could contribute to maintaining equilibrium between population numbers of viruses and their hosts. Large fitness increases of viruses would be prevented since whenever a bottleneck of any size is reached, it will act to limit fitness to an extent which is commensurate with the fitness value already attained by the relevant population (see also Section 6.6). This modulating effect is credible in view of the evidence of the frequent occurrence of bottlenecks in nature (compare Fig. 6.4 with the evidence of bottlenecks during the arbovirus life cycle described in Section 4.10 of Chapter 4). It is tempting to consider that the equilibrium between viral and host populations numbers that had to exist from ancestral times (Chapter 1) is not only due to interaction and escape strategies between viruses and host immune systems (Chapter 4), but also due to in-built self-regulatory mechanisms inherent to viral genome dynamics that limit excessive replication capacity in the face of a far more inflexible cellular world. Bottleneck size may be an evolvable trait that has contributed to some balance between virus and host population numbers.

Repeated bottlenecks not only decrease fitness as summarized in previous paragraphs but, in addition, they may have long-term detrimental effects on virus adaptability. Even when a bottlenecked VSV population had the same fitness than a virus maintained under large population passages, it displayed lower adaptability (Novella, 2004). This result means that it is not fitness per se that determines the adaptive potential of a virus, but it is the part of sequence space that it occupies that does. Different areas of sequence space may provide a mutant cloud with comparable fitness, but with different capacity to respond to perturbations (see Section 6.7.1 in this chapter and Section 3.7 in Chapter 3). A similar and rather counterintuitive conclusion is reached when comparing mutation frequencies (or other complexity measurements) of viral populations: the same high mutation frequency may either permit viral survival or drive the virus toward extinction. It is not the mutation frequency value per se that counts, but the context in which the mutations occur (the experimental values that justify such an assertion are discussed in Chapter 9).

Once more, we come to the conclusion of the importance of viral population numbers in fitness variations and in guiding evolutionary episodes. It is obvious that some nuances must be introduced into the rather simplified view that bottlenecks lead to fitness loss (Chapter 5). There is no absolute population size value that guarantees the avoidance of fitness decrease. The bottleneck size that leads to decrease, maintenance, or increase of fitness is dependent on the initial fitness and on the position of the virus in sequence space. It is quite clear that in this type of studies, experimental evolution has gone far ahead than field observations, partly due to the difficulties in following fitness evolution and in controlling bottleneck sizes in vivo. With these clarifications, we are now in a position to begin to approach the general question (or dilemma) between asexual and sexual modes of reproduction, as a precedent to approach clonality versus nonclonality in biological evolution to be addressed in the closing Chapter 10.

## 6.5.1 Muller's ratchet and the advantage of sex

Fitness loss due to bottleneck events (with all its participating parameters discussed in the previous section) provides experimental support to a theoretical proposal made by H.J. Muller, known as Muller's ratchet (Maynard-Smith, 1976; Muller, 1964; Nowak and Schuster, 1989). The proposal is that small populations of asexual organisms that display high mutation rates will tend to incorporate mutations (the majority deleterious) in an irreversible, ratchet-type mechanism unless recombination can yield the initial type of genome devoid of mutations (often termed the zero mutation class of genomes). Fitness decrease due to the accumulation of mutations proposed by H.J. Muller in general terms is expected to be accentuated in the case of viral populations with continuous generation of new mutant genomes. In mutant spectra of viruses, the least mutated class of genomes will be the one displaying highest replicative fitness and will correspond to the master sequence that dominates the population (Chapter 3). With the plaque-to-plaque transfer design, there is a probability that in each plating (a step or click in the ratchet) the least mutated class of genomes is lost. Therefore, the viral population is forced to regenerate a distribution from which again the least mutated class will be lost in the next ratchet click. The system is doomed to rapid deterioration and extinction unless mechanisms operate to restore fitter genomes.

The operation of Muller's ratchet appears as rather general in viruses since, in addition to bacteriophage φ6 and the animal virus VSV (studies summarized in Section 6.5), other viral systems have documented fitness loss associated with serial bottlenecks (de la Iglesia and Elena, 2007; Escarmís et al., 1996; Jaramillo et al., 2013; Yuste et al., 1999). The studies by C. Escarmís and colleagues with FMDV contributed decisively to define the molecular basis of fitness loss in an RNA virus, and to unveil unusual genetic lesions (with phenotypic consequences) that hide in mutant spectra (detailed in the coming Section 6.5.2). Muller's ratchet operates not only in viruses but also in bacteria, protozoa, plants, fish, as well as in chromosomal and mitochondrial DNA (Allen et al., 2009; Andersson and Hughes, 1996; Bell, 1988; Coates, 1992; Engelstadter, 2008; Leslie and Vrijenhoek, 1980; Loewe, 2006; Moran, 1996; Zhang et al., 2019).

The concept that sex conferred an advantage to organisms by providing new gene combinations for adaptation dates back to A. Weismann (Bell, 1982; Weismann, 1889—1892). Avoidance of the detrimental effects of Muller's ratchet is believed to have been one of the driving forces to introduce genetic recombination and sex in the reproduction and evolution of living systems (Agrawal, 2006; Barton and Charlesworth, 1998; Maynard Smith and Szathmáry, 1999; Maynard-Smith, 1976). In the words of G. Bell: "Sex acts as an editor which detects serious copying errors and enables the genetic message to be transmitted without contamination" (Bell, 1988). It is worth comparing Bell's statement with the molecular evidence discussed in Chapters 2 and 3 that copying errors during virus replication are not corrected because of the absence of proofreading exonucleases (or related) activities in RNA viruses. Lack of correcting activities is remarkable from the point of view of an evolution rich in lateral gene transfers because proofreading-repair activities are available not only from DNA polymerases but also from the coronavirus polymerases (and likely other large RNA genomes to be discovered). From the perspective of long-term evolution, the biological benefits of high mutation rates for many viruses find support in the absence of incorporation into their genomes of error correction domains available in a biosphere prone to genomic exchanges. This relates to the issue of clonal evolution in viruses, an important feature of long-term virus evolution (Chapter 10). Now we return to plaque-to-plaque transfers because there is more to learn from them.

## 6.5.2 Molecular basis of fitness decrease: deep fluctuations, massive extinctions, and rare survivors

The genomic RNA of 19 biological clones of FMDV that had undergone fitness decrease as a consequence of 30 serial plaque-to-plaque transfers was sequenced and the sequences compared with those of the parental clones, and to the parental populations subjected to large population passages. Many mutations that had never been detected in natural isolates or laboratory clones and populations of FMDV were found. A total of 69 mutations affecting the internal ribosome entry site (IRES), L-protease, P1- and polymerase-coding regions were identified (Escarmís et al., 1996). Notably, nine out of 19 clones showed an extension of an internal tract of four adenylate residues that precedes the second functional AUG. Depending on the clone, the extension was of one, or several residues and up to 23 additional adenylates in one of the clones. This elongation affected translation, was one of the determinants of fitness decrease, and reverted when the clones were subjected to large population passages. It was proposed that the elongation that constitutes a hot spot for variation was prompted by slippage mutagenesis of the polymerase. This genetic lesion in the FMDV genome was given as an example of mutational instruction in Section 2.3 of Chapter 2, and it served as one of the genetic markers for quasispecies memory when it reverted upon large population passages of the altered clones (Section 5.5 in Chapter 5). The molecular instability associated with this oligoadenylate elongation is also suggested by the fact that genomes with different numbers of adenylate residues coexisted in the same viral plaque.

Another rare lesion was a point deletion located between the two functional AUG protein synthesis initiation codons that led to a predicted stop codon for the proteins synthesized from the first AUG. A significant finding is that of the total number of mutations that had accumulated at the end of the 30 transfers, one-third of them were already present in the first transfer, although a rather steady accumulation of mutations was observed when the clones were subjected to many additional transfers. A possible limitation in the accumulation of mutations will be examined again in connection with limited tolerance to mutations, and the concept of contingent neutrality (Section 6.7.1).

Additional (up to 409) plaque-to-plaque transfers unveiled new rare genetic lesions, unusual phenotypes, and remarkable resistance to extinction (Escarmís et al., 2002, 2008). The most salient phenotypic change was that after more than 100 transfers, FMDV became noncytolytic, was unable to form visible plaques, and it could readily establish a persistent, noncytocidal infection that normally would be established only from the few cells that survived a cytolytic infection (see Section 6.4). Thus, strikingly, a fundamental property of the virus-host interaction, such as the capacity to kill cells [a marker of virus virulence (see Section 5.6 in Chapter 5)!], was altered as a result of repeated bottleneck passages. In fact, if infectivity is judged by the capacity to produce plaques, these multiply transferred clones underwent decreases of specific infectivity (the ratio between the amount of infectivity and that of viral RNA) of at least 140-fold relative to their corresponding parental biological clone. Such a reduction is enormous, and it should serve to compare the decrease of specific infectivity that accompanies virus extinction by lethal mutagenesis. (Decreases of specific infectivity are mentioned in Section 4.3 of Chapter 4 as one of the consequences of suboptimal codon usage, and are revisited in Chapter 9 as a diagnostic parameter of lethal mutagenesis). A proof that pursuing molecular analysis of clones subjected to many bottleneck transfers may provide new information was documented by the discovery of an amino acid substitution in capsid protein VP1 that produced virus thermosensitivity, and exerted an effect at a distance in the processing of the FMDV polyprotein (Escarmís et al., 2009), unveiling a new feature of picornavirus protein processing (Martínez-Salas and Ryan, 2010). Unusual FMDVs isolated after plaque transfers are summarized in Box 6.2.

Several mechanisms have been proposed to explain why so many unusual, often unique,

detrimental mutations can be rescued in clones subjected to many plaque transfers. One model emphasizes that negative selection is attenuated during plaque transfers in the sense that no truly competitive optimization of the mutant spectrum is allowed, except for intracellular competition in each individual cell. The competitive mixing of genomes is limited in the cell monolayer whose individual cells become infected by the wave of progeny virus that had its initial focus in the cell hit by the virus in the corresponding transfer. Even accepting that intraplaque MOI might be high, no competition among many of the newly arising genomes can take place in any of the cells. Therefore, the displacement of unfit genomes by fitter ones is limited. A necessity to explain the frequency and types of mutations observed is that the virus must be subjected to high mutational pressure. Although mutation rates for RNA viruses are discussed in Chapter 2, it must be stated here that many of the studies that allowed the quantification of high mutation rates came from experimental evolution designs. The fact that one of the FMDV clones at transfer 409 differed from its parental clone in 122 mutations implies a mutation frequency of $1.5 \times 10^{-2}$ s/nt (Escarmís et al., 2008). Remarkably when FMDV is subjected to mutagenesis, a ten-fold lower frequency can drive the virus to extinction (Chapter 9). The observed mutability during plaque transfers renders perfectly understandable multiple reversion events scored in picornavirus genomes, that were met with skepticism by some (de la Torre et al., 1992; Domingo et al., 2010).

Another, not mutually exclusive, mechanism behind rare mutations is that many extinction events take place in the course of the transfers and that the low-fitness survivors acquire compensatory mutations that allow plaque formation unless a noncytolytic phenotype that still allows intracellular RNA replication to take place is produced.

C. Carrillo, D.L. Rock, and colleagues applied an in vivo protocol of 20 serial swine-to-swine

BOX 6.2

## Unusual Genotypes and Phenotypes in Foot-and-Mouth Disease Virus Subjected to Plaque-to-Plaque Transfers

- Mutations never found in other populations of the same virus subjected to other passage regimes, including a rare one nucleotide deletion.

- Noncytocidal mutants that can establish a persistent infection in cell culture without intervening cell killing.
- Amino acid substitutions in the capsid that can affect thermal stability and polyprotein processing.

contact transmission that resembles the plaque-to-plaque transfer design (Carrillo et al., 2007) (see also Section 5.4 in Chapter 5). Interestingly, profound phenotypic changes occurred in the virus after several transfers, including reduction of virulence and establishment of a carrier state in pigs, previously thought to be typical of ruminants. Several mutations accumulated in the viral genome, suggestive of the operation of Muller's ratchet in vivo. This study validated the FMDV cell culture model as a source of information on the consequences of bottleneck events. The swine study also proved the feasibility of serial bottleneck passages in animals that may be highly informative of potential viral alterations associated with transmission events.

## 6.6 Limits to fitness gain and loss

A key issue of theoretical and experimental interest in the studies of fitness evolution is whether fitness of a virus population can grow indefinitely or it has a limit, reaching a plateau value. Limitations in the capacity to occupy sequence space suggest that there must be a limit to fitness gain and that, in the case of viruses this limit may be imposed by the viral population size. Some experimental studies and theoretical predictions support that increases or decreases of fitness reach a plateau, as the result of mutational effects and the ratio of beneficial to deleterious mutations (average mutations will increase the fitness of a very low-fitness population but decrease the fitness of a high-fitness population) (Silander et al., 2007). However, the situation may not be so simple. I.S. Novella and colleagues demonstrated that exponential increase of VSV fitness has a limit but that when the limit is reached, stochastic fitness fluctuations occur (Novella et al., 1999b). Variations are represented by the zig-zag lines at the upper left side of the fitness diagrams shown in Fig. 6.4. Since, the results of I.S. Novella and colleagues suggest that VSV reached a fitness value whose further increase could not be guaranteed by the replicating population size attained. An area of sequence space is reached where the newly arising mutations are not steadily incorporated, but rather they produce unpredictable fitness jumps.

A similar fluctuation of fitness values was observed in the case of FMDV subjected to many plaque-to-plaque transfers (Lázaro et al., 2003). Variations are represented by the zig-zag lines at the lower right side of the fitness diagrams shown in Fig. 6.4. An example of

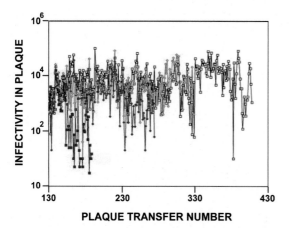

**PLAQUE TRANSFER NUMBER**

**FIGURE 6.5** The fluctuation pattern of fitness values (infectivity in plaque) of FMDV subclones subjected to a maximum of 400 serial plaque-to-plaque transfers. Results for four different subclones (color-coded) for passage 130–400 are shown. Only one of the subclones (identified in black color) survived to produce plaques to the end of the transfer series. The other clones (coded red, blue, and green) became noncytopathic at different passages. *Modified from Escarmís, C., Lazaro, E., Arias, A., Domingo, E., 2008. Repeated bottleneck transfers can lead to non-cytocidal forms of a cytopathic virus: implications for viral extinction. J. Mol. Biol. 376, 367–379, with permission of the authors.*

four FMDV subclones is presented in Fig. 6.5. The fluctuating pattern at low-fitness values followed a Weibull statistical distribution (Weibull, 1951), which was taken to mean that complex virus-cell interactions contribute to the level of progeny in each individual plaque. The results reinforce the concept of extreme resistance of viruses to extinction at the population level despite many extinctions at the individual level. The reason is that when fitness is very low, the probability of stochastic occurrence of beneficial mutations can rescue subpopulations from their fate toward extinction (Escarmís et al., 2002).

Fitness instability might underlie the transition toward genome segmentation underwent by FMDV upon extended high MOI passage in cells, an experiment intended to see if FMDV reached a fitness plateau (García-Arriaza et al., 2004). In the course of hundreds of passages, the virus

diversified in two distinct subpopulations that exhibited a competition-colonization dynamics previously shown by D. Tilman to operate in classical ecological systems (Tilman, 1994). The viruses diversified into colonizers, which were efficient in killing cells, and competitors that modulated cell killing. Thus, internal quasispecies interactions can regulate virus virulence (Ojosnegros et al., 2010). This balance was maintained until around passage 260 at which the viral genome became segmented: the monopartite genome was outcompeted by the two segments that complemented each other to replicate and kill cells (García-Arriaza et al., 2004, 2006; Moreno et al., 2014; Ojosnegros et al., 2011). This transition was discussed in Section 2.11 of Chapter 2 in connection with a mutation-driven genome transition involving RNA recombination. The transition toward segmentation might have been favored by fitness instability (the zig-zag lines depicted at the high fitness regions in Fig. 6.4) when fitness increase reached its population size-dependent limits (Novella et al., 1999b); the observations are in line with models that predict stochastic evolutionary outcomes associated with mutational load (Zhao et al., 2019).

## 6.7 Competitive exclusion principle and Red Queen hypothesis

J.J. Holland and colleagues examined the fate of two competing VSV clones that displayed approximately equal fitness. The two clones coexisted for several generations until one of the populations rapidly outgrew the other (Clarke et al., 1994). This observation is schematically depicted in Fig. 6.6, and agrees with the competitive exclusion principle of population genetics (Gause, 1932, 1971). This principle states that one competing species will always outcompete the other provided no niche differentiation exists between them. In a related formulation, the resource-based competition theory asserts that two

**FIGURE 6.6** Schematic representation of the Competitive Exclusion principle and the Red Queen hypothesis applied to RNA viruses. At the top, two populations of fitness equal to control I (initial population), were serially passaged in cell culture. After many passages in which both populations coexisted at about the same frequency, suddenly one of the populations (A, yellow arrow, or B, green arrow) displaced the other. This behavior is in agreement with the competitive exclusion principle of population genetics. Below, both populations A (winner) and B (looser) gained fitness relative to the control I. This observation agrees with the Red Queen hypothesis. See text for implications and references.

with two VSV quasispecies. The genetic and phenotypic constraints of *Drosophila* within the time frame of the experiment contrasts with the much more voluble and dynamic nature of two competing VSV mutant clouds. Indeed, the eventual exclusion of one VSV quasispecies by the other is expected as a consequence of the random, probably infrequent occurrence of advantageous mutations in a genome of one of the populations and not in the other. Competitive exclusion has been proposed as a mechanism of elimination of one hepatic virus by infection with another hepatic virus reported in the clinic (Amaku et al., 2013).

During the competition process, both the winners and the losers gained fitness at similar rates, in agreement with the Red Queen hypothesis: "No species can ever win and new adversaries grinningly replace the losers" (Van Valen, 1973). Among the competing viral quasispecies in which the mutant spectrum can modulate the behavior of the ensemble, infrequently arising, superior mutants are likely to perturb the population in such a way as to exclude or maintain at low levels all other mutants present in the competing quasispecies. Until such exclusion occurs, most members of the mutant spectrum gain replicative fitness, including those present at "memory" levels (Section 5.5 in Chapter 5). As expressed by the Red Queen in Lewis Carrol's *Through the Looking Glass*: "It takes all the running you can do to stay in the same place." It appears as if the mutational background associated with some competitive optimization of quasispecies affected both competing populations in a similar way, while the stochastic occurrence of a saliently beneficial mutation disrupted coexistence. The results also imply that the frequency of mutations that are advantageous enough to upset the coexistence of the two populations is low. This is expected from the fact that detrimental and lethal mutations are generally far more frequent than advantageous mutations (unless the populations were poorly adapted to the environment) (Fig. 6.6).

consumers that share a single, limiting resource cannot stably coexist in a spatially and temporally homogeneous habitat. It is assumed that there are no other intervening ecological factors. In testing the competitive exclusion principle experimentally, the precise definition of "limiting resource" is essential. In a series of classical experiments, F.J. Ayala showed that two species of *Drosophila* could coexist for many generations in competition for limited resources (Ayala, 1971). The genetic and phenotypic inflexibility of a differentiated organism such as *Drosophila* may explain its very different behavior as compared

## 6.7.1 Contingent neutrality in virus

The behavior of competing VSV populations was not always unpredictable. In parallel competitions of a wild-type VSV and a surrogate marked subclone of equal fitness, predictable nonlinear behavior of the two populations was characterized (Quer et al., 1996). After nearly constant times, the two viruses competing in a constant cell culture environment followed different trajectories. There was a reproducible tendency of the wild type to gain fitness at a higher rate than the surrogate mutant. Thus, despite the stochastic occurrence of mutations, a nearly deterministic evolutionary behavior was observed (Tsimring et al., 1996). A number of environmental perturbations (presence of DIs, increased temperature during viral replication, or limited, enhanced mutagenesis) led to an accelerated dominance of the wild type over the mutant (Quer et al., 2001). Comparison of consensus nucleotide sequences of the entire VSV genomes showed that the mutant genome had acquired a number of mutations with respect to the wild type, and suggested that neutrality of the mutant relative to the wild type was maintained provided the environment was not perturbed. The behavior of the mutant relative to the wild type was described as being of "contingent neutrality." The presence of mutations rendered the virus less robust to accept additional mutations. In terms of fitness landscapes, the results can be interpreted as if the wild type lies on a relatively flat (or less rugged) fitness surface that preserves its replicative efficacy despite the occurrence of mutations. In contrast, the mutant lies on a sharper fitness peak prone to fitness decrease upon mutation or environmental stress (Sections 5.3 and 5.7 in Chapter 5).

Thus, the behavior of the competing viral subpopulations, a frequent occurrence in quasispecies dynamics may be affected by the mutational load in the competing genomes to the point of accelerating a selective advantage by virtue of limited tolerance to acquire new mutations.

## 6.8 Studies with reconstructed quasispecies

Experimental evolution offers the possibility—that has so far been exploited only minimally—of reconstructing complex quasispecies swarms with specific mutants or viral subpopulations, to examine their behavior during replication under different environmental conditions. These types of experiments in which population evolution can now be followed by deep sequencing methodologies would actually represent an important contribution to the study of complexity, a field of science in need of experimental approaches.

In an early study, an FMDV quasispecies was reconstructed with 19 antigenic variants of the virus, each identified by an amino acid substitution at the major antigenic site that conferred resistance to a monoclonal antibody (MAb). The mutants were added to the mutant spectrum of a biological clone of FMDV, at a concentration typical of the antibody-resistant mutants found in FMDV. The reconstructed quasispecies was allowed to replicate in the absence and presence of the MAb, and the resulting populations analyzed. In the populations passaged in the presence of the antibody, but not in the control population, 10 out of the 19 mutant introduced became dominant, indicating the selection of a mutant cloud that shared the required phenotype (Perales et al., 2005). In a subsequent study, the reconstruction was carried out using matched pairs of distinguishable MAb-escape mutants of the same antigenic site. Each mutant of a pair differed from the other in 11- or 33-fold in fitness. The analysis of the populations subjected to antibody selection revealed the dominance of the corresponding high-fitness mutants. Thus, relative viral fitness can influence significantly the mutant repertoire selected by neutralizing antibody (Martin et al., 2006). Such fitness effects are likely to underlie the response to any selective constraint and should be considered in the interpretation of the effect of selective forces acting on complex viral populations.

## 6.9 Quasispecies dynamics in cell culture and in vivo

Some authors have argued that quasispecies dynamics is valid for viruses replicating in cell culture, but that its basic principles are not adequate to understand virus behavior in vivo. A reexamination of the recent evidence derived from the application of deep sequencing to in vivo systems renders the suggestion of a fundamental difference between cell culture and in vivo quasispecies dynamics untenable. One can argue that in vivo systems offer highly complex, compartmentalized environment, which delays even further than a controlled cell culture system a possible approach toward population equilibrium. What we have learned, however, from the several model experimental studies summarized in this chapter, both using cell culture systems and animals, is that there are no fundamental differences in the quasispecies behavior of viruses in cell culture and in vivo. Perhaps a significant example is provided by the accumulation of mutations in the FMDV genomes and the profound phenotypic changes observed both in cell culture plaque-to-plaque transfers and in serial bottleneck transmission of the virus in swine (Section 6.5.2). Evidence in support of the value of quasispecies to interpret viral population dynamics in vivo is increasing as the new deep sequencing tools are applied to evolving viral populations.

## 6.10 Overview and concluding remarks

The possibilities of experimental evolution to gain new insights into the mechanisms of virus evolution are enormous, and they remain largely unexploited. In this chapter, we have summarized cell culture and in vivo designs that have revealed fundamental features of virus evolution, including some that have provided experimental confirmation of some concepts of population biology.

Coevolution of cells and viruses appears to be quite common during persistent infections in cell culture. Several reproducible traits of virus and cell variation are shared by widely different viral pathogens and cell types. It is tempting to speculate that coevolutionary interactions reflect the inheritance of old-time relationships between primitive forms of viruses and cells.

Experimental evolution opens the possibility to examine the effects of extreme population regimes in fitness evolution, notably the result of massive infections versus the most profound bottleneck restriction: serial infections limited to one infectious particle per passage. Teachings of plaque-to-plaque transfers include the evidence of highly unusual mutations and phenotypes that contradict textbook types of viral properties. In particular, the recognized cytopathic nature of FMDV is turned into noncytopathic in clones rescued from the low-frequency levels of viral quasispecies. In a similar note, repeated bottleneck passages of FMDV in swine produced viruses that established a carrier state in swine, a type of virus-host interaction previously thought to occur only in ruminants.

Muller's ratchet, the competitive exclusion principle, Red Queen hypothesis, and contingent neutrality are concepts that have been established in experimental designs based on competitive, large population passages of marked mutant mixtures. In natural infections and coinfections, however, it may not be possible to distinguish whether competition is established among well-differentiated quasispecies (distinct cloud categories) or a continuum of variant viruses that span a large portion of sequence space including overlapping regions of different density. Experiments using reconstructed quasispecies with different cloud shape face a promising research time in which the mechanism by which some variants overgrow others, or interact with others to disturb or complement function can be elucidated through deep sequencing analyses. These studies can be framed under the concept of viral swarms acting as units of selection. This is a remarkable property that is probably shared by the microbial world and to a far less extent by differentiated organisms. The potential implications for the

diagnosis of viral disease and considerations for treatment options are very evident, and they will be discussed in the chapters that folow (see Summary Box).

---

## Summary Box

- Experimental evolution permits the establishment of many fundamental concepts of virus evolution that are facilitating the interpretation of observations in the far more complex natural scenarios.
- Designs intended to reproduce extreme passage regimes have unveiled many features of mutant spectra that are hidden to standard analyses based exclusively on consensus sequences. In particular, multiple plaque-to-plaque transfers have revealed that extremely unusual mutations can be found in low-fitness viruses that result in extreme phenotypes. Such unusual viral subpopulations replicate thanks to the presence of compensatory mutations that rescue a few genomes out of a great majority that are extinguished.
- The experimental studies have revealed the fitness dependence of fitness variation, a field of research still to be applied in vivo. Fitness increases and decreases have a limit imposed

either by insufficient population size or extremely low replicative fitness. In such extreme scenarios, unexpected evolutionary transitions may be favored.
- Several concepts of population genetics have found experimental support in work with viruses, notably Muller's ratchet, competitive exclusion principle, and the Red Queen hypothesis.
- The dynamics of competition between viral populations has defined the concept of contingent neutrality in virus evolution.
- The capacity to reconstruct quasispecies with selected types of mutants opens new research avenues to understand viral population dynamics under controlled conditions.
- It will be extremely interesting to reexamine the population dynamics in the different designs described in this chapter with the new tool of deep sequencing.

---

# References

Acevedo, A., Brodsky, L., Andino, R., 2014. Mutational and fitness landscapes of an RNA virus revealed through population sequencing. Nature 505, 686–690.

Agrawal, A.F., 2006. Evolution of sex: why do organisms shuffle their genotypes? Curr. Biol. 16, R696–R704.

Ahmed, R., Chen, S.Y. (Eds.), 1999. Persistent Viral Infections. John Wiley and Sons, New York.

Ahmed, R., Canning, W.M., Kauffman, R.S., Sharpe, A.H., Hallum, J.V., et al., 1981. Role of the host cell in persistent viral infection: coevolution of L cells and reovirus during persistent infection. Cell 25, 325–332.

Allen, J.M., Light, J.E., Perotti, M.A., Braig, H.R., Reed, D.L., 2009. Mutational meltdown in primary endosymbionts: selection limits Muller's ratchet. PLoS One 4, e4969.

Altan-Bonnet, N., Perales, C., Domingo, E., 2019. Extracelular vesicles: vehicles of en bloc viral transmission. Virus Res. 265, 143–149.

Amaku, M., Coutinho, F.A.B., Chaib, E., Massad, E., 2013. The impact of hepatitis A virus infection on hepatitis C virus infection: a competitive exclusion hypothesis. Bull. Math. Biol. 75, 82–93.

Andersson, D.I., Hughes, D., 1996. Muller's ratchet decreases fitness of a DNA-based microbe. Proc. Natl. Acad. Sci. U.S.A. 93, 906–907.

Andino, R., Domingo, E., 2015. Viral quasispecies. Virology 479–480, 46–51.

Ayala, F.J., 1971. Competition between species: frequency dependence. Science 171, 820–824.

Baccam, P., Thompson, R.J., Li, Y., Sparks, W.O., Belshan, M., et al., 2003. Subpopulations of equine infectious anemia virus Rev coexist *in vivo* and differ in phenotype. J. Virol. 77, 12122–12131.

Barton, N.H., Charlesworth, B., 1998. Why sex and recombination? Science 281, 1986–1990.

Bell, G., 1982. The Master Piece of Nature: The Evolution and Genetics of Sexuality. Croom-Helm, Ltd., London.

Bell, G., 1988. Sex and Death in Protozoa: The History of an Obsession. Cambridge University Press, Cambridge.

Billeter, M.A., 1978. Sequence and location of large RNase T1 oligonucleotides in bacteriophage Qbeta RNA. J. Biol. Chem. 253, 8381–8389.

Bouvier, N.M., Lowen, A.C., 2010. Animal models for influenza virus pathogenesis and transmission. Virus 2, 1530–1563.

Braun, T., Borderia, A.V., Barbezange, C., Vignuzzi, M., Louzoun, Y., 2018. Long-term context-dependent genetic adaptation of the viral genetic cñloud. Bioinformatics 1–9. https://doi.org/10.1093/bioinformatics/bty891.

Cale, E.M., Hraber, P., Giorgi, E.E., Fischer, W., Bhattacharya, T., et al., 2011. Epitope-specific CD8+ T lymphocytes cross-recognize mutant simian immunodeficiency virus (SIV) sequences but fail to contain very early evolution and eventual fixation of epitope escape mutations during SIV infection. J. Virol. 85, 3746–3757.

Calisher, C.H., Higgs, S., 2018. The discovery of arthropod-specific viruses in hematophagous arthropods: an open door to understanding the mechanisms of arbovirus and arthropod evolution? Annu. Rev. Entomol. 63, 87–103.

Calvez, V., Pelletier, I., Couderc, T., Pavio-Guedo, N., Blondel, B., et al., 1995. Cell clones cured of persistent poliovirus infection display selective permissivity to the wild-type poliovirus strain Mahoney and partial resistance to the attenuated Sabin 1 strain and Mahoney mutants. Virology 212, 309–322.

Carrillo, C., Lu, Z., Borca, M.V., Vagnozzi, A., Kutish, G.F., et al., 2007. Genetic and phenotypic variation of foot-and-mouth disease virus during serial passages in a natural host. J. Virol. 81, 11341–11351.

Chao, L., 1990. Fitness of RNA virus decreased by Muller's ratchet. Nature 348, 454–455.

Chen, W., Baric, R.S., 1996. Molecular anatomy of mouse hepatitis virus persistence: coevolution of increased host cell resistance and virus virulence. J. Virol. 70, 3947–3960.

Chiarini, A., Arista, S., Giammanco, A., Sinatra, A., 1983. Rotavirus persistence in cell cultures: selection of resistant cells in the presence of fetal calf serum. J. Gen. Virol. 64, 1101–1110.

Ciota, A.T., 2019. The role of coinfection and swarm dynamics in arbovirus transmission. Virus Res. 265, 88–93.

Clarke, D.K., Duarte, E.A., Moya, A., Elena, S.F., Domingo, E., et al., 1993. Genetic bottlenecks and population passages cause profound fitness differences in RNA viruses. J. Virol. 67, 222–228.

Clarke, D.K., Duarte, E.A., Elena, S.F., Moya, A., Domingo, E., et al., 1994. The red queen reigns in the kingdom of RNA viruses. Proc. Natl. Acad. Sci. U.S.A. 91, 4821–4824.

Clements, J.E., Gdovin, S.L., Montelaro, R.C., Narayan, O., 1988. Antigenic variation in lentiviral diseases. Annu. Rev. Immunol. 6, 139–159.

Coates, D.J., 1992. Genetic consequences of a bottleneck and partial genetic structure in the triggerplant *Stylidium coroniforme* (Stylidiaceae). Heredity 69, 512–520.

Coffey, L.L., Vignuzzi, M., 2011. Host alternation of chikungunya virus increases fitness while restricting population diversity and adaptability to novel selective pressures. J. Virol. 85, 1025–1035.

Coffey, L.L., Vasilakis, N., Brault, A.C., Powers, A.M., Tripet, F., et al., 2008. Arbovirus evolution *in vivo* is constrained by host alternation. Proc. Natl. Acad. Sci. U.S.A. 105, 6970–6975.

Colbère-Garapin, F., Lipton, H.L., 2010. Persistent infections. In: Ehrenfeld, E., Domingo, E., Roos, R.P. (Eds.), The Picornaviruses. ASM Press, Washington, DC, pp. 321–338.

Conway Morris, S., 1998. The Crucible of Creation. The Burgess Shale and the Rise of Animals. Oxford University Press, Oxford.

Cuevas, J.M., Moya, A., Elena, S.F., 2003. Evolution of RNA virus in spatially structured heterogeneous environments. J. Evol. Biol. 16, 456–466.

Cummings, P.J., Rinaldo Jr., C.R., 1989. Coevolution of virulent virus and resistant cells as a mechanism of persistence of herpes simplex virus type 1 in a human T lymphoblastoid cell line. J. Gen. Virol. 70 (Pt 1), 97–106.

de la Iglesia, F., Elena, S.F., 2007. Fitness declines in tobacco etch virus upon serial bottleneck transfers. J. Virol. 81, 4941–4947.

de la Torre, J.C., Davila, M., Sobrino, F., Ortin, J., Domingo, E., 1985. Establishment of cell lines persistently infected with foot-and-mouth disease virus. Virology 145, 24–35.

de la Torre, J.C., Martínez-Salas, E., Diez, J., Villaverde, A., Gebauer, F., et al., 1988. Coevolution of cells and viruses in a persistent infection of foot-and-mouth disease virus in cell culture. J. Virol. 62, 2050–2058.

de la Torre, J.C., de la Luna, S., Diez, J., Domingo, E., 1989a. Resistance to foot-and-mouth disease virus mediated by trans-acting cellular products. J. Virol. 63, 2385–2387.

de la Torre, J.C., Martínez-Salas, E., Díez, J., Domingo, E., 1989b. Extensive cell heterogeneity during persistent infection with foot-and-mouth disease virus. J. Virol. 63, 59–63.

de la Torre, J.C., Giachetti, C., Semler, B.L., Holland, J.J., 1992. High frequency of single-base transitions and extreme frequency of precise multiple-base reversion mutations in poliovirus. Proc. Natl. Acad. Sci. U.S.A. 89, 2531–2535.

Delli Bovi, P., De Simone, V., Giordano, R., Amati, P., 1984. Polyomavirus growth and persistence in Friend erythroleukemic cells. J. Virol. 49, 566–571.

Dermody, T.S., Nibert, M.L., Wetzel, J.D., Tong, X., Fields, B.N., 1993. Cells and viruses with mutations affecting viral entry are selected during persistent infections of L cells with mammalian reoviruses. J. Virol. 67, 2055–2063.

Díez, J., Dávila, M., Escarmís, C., Mateu, M.G., Dominguez, J., et al., 1990. Unique amino acid substitutions in the capsid proteins of foot-and-mouth disease virus from a persistent infection in cell culture. J. Virol. 64, 5519–5528.

Domingo, E., Holland, J.J., 1988. High error rates population equilibrium, and evolution of RNA replication systems. In: Domingo, E., Holland, J.J., Ahlquist, P. (Eds.), RNA Genetics. CRC Press, Boca Raton, FL, pp. 3–36.

Domingo, E., Perales, C., 2012. From quasispecies theory to viral quasispecies: how complexity has permeated virology. Math. Model. Nat. Phenom. 7, 32–49.

Domingo, E., Sabo, D., Taniguchi, T., Weissmann, C., 1978. Nucleotide sequence heterogeneity of an RNA phage population. Cell 13, 735–744.

Domingo, E., Perales, C., Agudo, R., Arias, A., Ferrer-Orta, C., et al., 2010. Mutation, quasispecies and lethal mutagenesis. In: Ehrenfeld, E., Domingo, E., Roos, R.P. (Eds.), The Picornaviruses. ASM Press, Washington, DC, pp. 197–211.

Domingo-Calap, P., Segredo-Otero, E., Durán-Moreno, M., Sanjuán, R., 2019. Social evolution of innate immune evasion of a virus. Nat. Microbiol. 4, 1006–1013.

Duarte, E., Clarke, D., Moya, A., Domingo, E., Holland, J., 1992. Rapid fitness losses in mammalian RNA virus clones due to Muller's ratchet. Proc. Natl. Acad. Sci. U.S.A. 89, 6015–6019.

Duarte, E.A., Clarke, D.K., Moya, A., Elena, S.F., Domingo, E., et al., 1993. Many-trillion fold amplification of single RNA virus particles fails to overcome the Muller's ratchet effect. J. Virol. 67, 3620–3623.

Engelstadter, J., 2008. Muller's ratchet and the degeneration of Y chromosomes: a simulation study. Genetics 180, 957–967.

Escarmís, C., Dávila, M., Charpentier, N., Bracho, A., Moya, A., et al., 1996. Genetic lesions associated with Muller's ratchet in an RNA virus. J. Mol. Biol. 264, 255–267.

Escarmís, C., Carrillo, E.C., Ferrer, M., Arriaza, J.F., Lopez, N., et al., 1998. Rapid selection in modified BHK-21 cells of a foot-and-mouth disease virus variant showing alterations in cell tropism. J. Virol. 72, 10171–10179.

Escarmís, C., Gómez-Mariano, G., Dávila, M., Lázaro, E., Domingo, E., 2002. Resistance to extinction of low fitness virus subjected to plaque-to-plaque transfers: diversification by mutation clustering. J. Mol. Biol. 315, 647–661.

Escarmís, C., Lazaro, E., Arias, A., Domingo, E., 2008. Repeated bottleneck transfers can lead to non-cytocidal forms of a cytopathic virus: implications for viral extinction. J. Mol. Biol. 376, 367–379.

Escarmís, C., Perales, C., Domingo, E., 2009. Biological effect of Muller's Ratchet: distant capsid site can affect picornavirus protein processing. J. Virol. 83, 6748–6756.

Fan, Y., Sanyal, S., Bruzzone, R., 2018. Breaking bad: how viruses subvert the cell cycle. Front. Cell Infect. Microbiol. 8, 396.

Fedeli, C., Moreno, H., Kunz, S., 2018. Novel insights into cell entry of emerging human pathogenic arenaviruses. J. Mol. Biol. 430, 1839–1852.

Fischer, W., Ganusov, V.V., Giorgi, E.E., Hraber, P.T., Keele, B.F., et al., 2010. Transmission of single HIV-1 genomes and dynamics of early immune escape revealed by ultra-deep sequencing. PLoS One 5, e12303.

García-Arriaza, J., Manrubia, S.C., Toja, M., Domingo, E., Escarmís, C., 2004. Evolutionary transition toward defective RNAs that are infectious by complementation. J. Virol. 78, 11678–11685.

García-Arriaza, J., Ojosnegros, S., Dávila, M., Domingo, E., Escarmis, C., 2006. Dynamics of mutation and recombination in a replicating population of complementing, defective viral genomes. J. Mol. Biol. 360, 558–572.

Gause, G.F., 1932. Experimental studies on the struggle for existence. J. Exp. Biol. 9, 389–402.

Gause, G.F., 1971. The Struggle for Existence. Dover, New York.

Gebauer, F., de la Torre, J.C., Gomes, I., Mateu, M.G., Barahona, H., et al., 1988. Rapid selection of genetic and antigenic variants of foot-and-mouth disease virus during persistence in cattle. J. Virol. 62, 2041–2049.

Gosert, R., Kardas, P., Major, E.O., Hirsch, H.H., 2010. Rearranged JC virus noncoding control regions found in progressive multifocal leukoencephalopathy patient samples increase virus early gene expression and replication rate. J. Virol. 84, 10448–10456.

Gosselin, A.S., Simonin, Y., Guivel-Benhassine, F., Rincheval, V., Vayssiere, J.L., et al., 2003. Poliovirus-induced apoptosis is reduced in cells expressing a mutant CD155 selected during persistent poliovirus infection in neuroblastoma cells. J. Virol. 77, 790–798.

Gould, S.J., 1989. Wonderful Life: The Burgess Shale and the Nature of History. W.W. Norton, New York.

Gutierrez, B., Escalera-Zamudio, M., Pybus, O.G., 2019. Parallel molecular evolution and adaptation in viruses. Curr. Opin. Virol. 34, 90–96.

Hahn, B.H., Shaw, G.M., Taylor, M.E., Redfield, R.R., Markham, P.D., et al., 1986. Genetic variation in HTLV-III/LAV over time in patients with AIDS or at risk for AIDS. Science 232, 1548–1553.

Herfst, S., Schrauwen, E.J., Linster, M., Chutinimitkul, S., de Wit, E., et al., 2012. Airborne transmission of influenza A/H5N1 virus between ferrets. Science 336, 1534–1541.

Herrera, M., Grande-Perez, A., Perales, C., Domingo, E., 2008. Persistence of foot-and-mouth disease virus in cell culture revisited: implications for contingency in evolution. J. Gen. Virol. 89, 232–244.

Ho, E.K.H., Agrawal, A.F., 2018. Mutation accumulation in selfing populations under fluctuating selection. Evolution 72, 1759–1772.

Holland, J.J., Grabau, E.A., Jones, C.L., Semler, B.L., 1979. Evolution of multiple genome mutations during long-term persistent infection by vesicular stomatitis virus. Cell 16, 495–504.

Holland, J.J., Spindler, K., Horodyski, F., Grabau, E., Nichol, S., et al., 1982. Rapid evolution of RNA genomes. Science 215, 1577–1585.

Imperiale, M.J., Jiang, M., 2015. What DNA viral genomic rearrangements tell us about persistence. J. Virol. 89, 1948–1950.

Imperiale, M.J., Howard, D., Casadevall, A., 2018. The silver lining in gain-of-function experiments with pathogens of pandemic potential. Methods Mol. Biol. 1836, 575–587.

Jaramillo, N., Domingo, E., Munoz-Egea, M.C., Tabares, E., Gadea, I., 2013. Evidence of Muller's ratchet in herpes simplex virus type 1. J. Gen. Virol. 94, 366–375.

Korber, B., Muldoon, M., Theiler, J., Gao, F., Gupta, R., et al., 2000. Timing the ancestor of the HIV-1 pandemic strains. Science 288, 1789–1796.

Kortenhoeven, C., Joubert, F., Bastos, A., Abolnik, C., 2015. Virus genome dynamics under different propagation pressures: reconstruction of whole genome haplotypes of West Nile viruses from NGS data. BMC Genomics 16, 118.

Lázaro, E., Escarmis, C., Perez-Mercader, J., Manrubia, S.C., Domingo, E., 2003. Resistance of virus to extinction on bottleneck passages: study of a decaying and fluctuating pattern of fitness loss. Proc. Natl. Acad. Sci. U.S.A. 100, 10830–10835.

Leslie, J.F., Vrijenhoek, R.C., 1980. Consideration of Muller's ratchet mechanism through studies of genetic linkage and genomic compatibilities in clonally reproducing Poeciliopsis. Evolution 34, 1105–1115.

Lipsitch, M., 2018. Why do exceptionally dangerous gain-of-function experiments in influenza? Methods Mol. Biol. 1836, 589–608.

Loewe, L., 2006. Quantifying the genomic decay paradox due to Muller's ratchet in human mitochondrial DNA. Genet. Res. 87, 133–159.

Ma, M.-J., Yang, Y., Fang, L.-Q., 2019. Highly pathogenic avian H7N9 influenza viruses: recent challenges. Trends Microbiol. 27, 93–95.

Marcus, P.I., 1982. Interferon induction by viruses IX. Antagonistic activities of virus particles modulate interferon production. J. Interferon Res. 2, 511–518.

Marcus, P.I., Rodriguez, L.L., Sekellick, M.J., 1998. Interferon induction as a quasispecies marker of vesicular stomatitis virus. J. Virol. 72, 542–549.

Martin Hernandez, A.M., Carrillo, E.C., Sevilla, N., Domingo, E., 1994. Rapid cell variation can determine the establishment of a persistent viral infection. Proc. Natl. Acad. Sci. U.S.A. 91, 3705–3709.

Martin, V., Perales, C., Davila, M., Domingo, E., 2006. Viral fitness can influence the repertoire of virus variants selected by antibodies. J. Mol. Biol. 362, 44–54.

Martínez-Salas, E., Ryan, M.D., 2010. Translation and protein processing. In: Ehrenfeld, E., Domingo, E., Roos, R.P. (Eds.), The Picornaviruses. ASM Press, Washington, DC, pp. 141–164.

Maynard Smith, J., Szathmáry, E., 1999. The Origins of Life: From the Birth of Life to the Origins of Language. Oxford University Press, Oxford.

Maynard-Smith, J., 1976. The Evolution of Sex. Cambridge University Press, Cambridge.

Meyerhans, A., Cheynier, R., Albert, J., Seth, M., Kwok, S., et al., 1989. Temporal fluctuations in HIV quasispecies in vivo are not reflected by sequential HIV isolations. Cell 58, 901–910.

Miyake, T., Shiba, T., Watanabe, I., 1967. Grouping of RNA phages by a Milipore filtration method. Jpn. J. Microbiol. 11, 203–211.

Moos, W.H., Pinkert, C.A., Irwin, M.H., Faller, D.V., Kodukula, K., 2017. Epigenetic treatment of persistent viral infections. Drug Dev. Res. 78, 24–36.

Moran, N.A., 1996. Accelerated evolution and Muller's rachet in endosymbiotic bacteria. Proc. Natl. Acad. Sci. U.S.A. 93, 2873–2878.

Moreno, E., Ojosnegros, S., Garcia-Arriaza, J., Escarmís, C., Domingo, E., et al., 2014. Exploration of sequence space as the basis of viral RNA genome segmentation. Proc. Natl. Acad. Sci. U.S.A. 111, 6678–6683.

Moreno, E., Gallego, I., Gregori, J., Lucía-Sanz, A., Soria, M.E., et al., 2017. Internal disequilibria and phenotypic diversification during replication of hepatitis C virus in a noncoevolving cellular environment. J. Virol. 91, e02505–e02516.

Mrukowicz, J.Z., Wetzel, J.D., Goral, M.I., Fogo, A.B., Wright, P.F., et al., 1998. Viruses and cells with mutations affecting viral entry are selected during persistent rotavirus infections of MA104 cells. J. Virol. 72, 3088–3097.

Muller, H.J., 1964. The relation of recombination to mutational advance. Mutat. Res. 1, 2–9.

Murphy, E., Shenk, T., 2008. Human cytomegalovirus genome. Curr. Top. Microbiol. Immunol. 325, 1–19.

Narayan, O., Clements, J.E., Griffin, D.E., Wolinsky, J.S., 1981. Neutralizing antibody spectrum determines the antigenic profiles of emerging mutants of visna virus. Infect. Immun. 32, 1045–1050.

Nash, A.A., Dalziel, R.G., Fitzgerald, J.R., 2015. Mims' Pathogenesis of Infectious Disease. Elsevier Science & Technology, Edinburgh.

Nathanson, N., Gonzalez-Scarano, F., 2007. Viral persistence. In: Nathanson, N. (Ed.), Viral Pathogenesis and Immunity. Academic Press, San Diego, pp. 130–145.

Novella, I.S., 2003. Contributions of vesicular stomatitis virus to the understanding of RNA virus evolution. Curr. Opin. Microbiol. 6, 399–405.

Novella, I.S., 2004. Negative effect of genetic bottlenecks on the adaptability of vesicular stomatitis virus. J. Mol. Biol. 336, 61–67.

Novella, I.S., Elena, S.F., Moya, A., Domingo, E., Holland, J.J., 1995. Size of genetic bottlenecks leading to virus fitness loss is determined by mean initial population fitness. J. Virol. 69, 2869–2872.

Novella, I.S., Hershey, C.L., Escarmis, C., Domingo, E., Holland, J.J., 1999a. Lack of evolutionary stasis during alternating replication of an arbovirus in insect and mammalian cells. J. Mol. Biol. 287, 459–465.

Novella, I.S., Quer, J., Domingo, E., Holland, J.J., 1999b. Exponential fitness gains of RNA virus populations are limited by bottleneck effects. J. Virol. 73, 1668–1671.

Novella, I.S., Ebendick-Corpus, B.E., Zarate, S., Miller, E.L., 2007. Emergence of mammalian cell-adapted vesicular stomatitis virus from persistent infections of insect vector cells. J. Virol. 81, 6664–6668.

Nowak, M.A., Schuster, P., 1989. Error thresholds of replication in finite populations mutation frequencies and the onset of Muller's ratchet. J. Theor. Biol. 137, 375–395.

Ojosnegros, S., Beerenwinkel, N., Antal, T., Nowak, M.A., Escarmis, C., et al., 2010. Competition-colonization dynamics in an RNA virus. Proc. Natl. Acad. Sci. U.S.A. 107, 2108–2112.

Ojosnegros, S., Garcia-Arriaza, J., Escarmis, C., Manrubia, S.C., Perales, C., et al., 2011. Viral genome segmentation can result from a trade-off between genetic content and particle stability. PLoS Genet. 7, e1001344.

Oldstone, M.B., 2006. Viral persistence: parameters, mechanisms and future predictions. Virology 344, 111–118.

Olivares, I., Sanchez-Jimenez, C., Vieira, C.R., Toledano, V., Gutierrez-Rivas, M., et al., 2013. Evidence of ongoing replication in a human immunodeficiency virus type 1 persistently infected cell line. J. Gen. Virol. 94, 944–954.

Pavio, N., Couderc, T., Girard, S., Sgro, J.Y., Blondel, B., et al., 2000. Expression of mutated poliovirus receptors in human neuroblastoma cells persistently infected with poliovirus. Virology 274, 331–342.

Pawlotsky, J.M., 2006. Hepatitis C virus population dynamics during infection. Curr. Top. Microbiol. Immunol. 299, 261–284.

Perales, C., Martin, V., Ruiz-Jarabo, C.M., Domingo, E., 2005. Monitoring sequence space as a test for the target of selection in viruses. J. Mol. Biol. 345, 451–459.

Quer, J., Huerta, R., Novella, I.S., Tsimring, L., Domingo, E., et al., 1996. Reproducible nonlinear population dynamics and critical points during replicative competitions of RNA virus quasispecies. J. Mol. Biol. 264, 465–471.

Quer, J., Hershey, C.L., Domingo, E., Holland, J.J., Novella, I.S., 2001. Contingent neutrality in competing viral populations. J. Virol. 75, 7315–7320.

Richman, D.D., Wrin, T., Little, S.J., Petropoulos, C.J., 2003. Rapid evolution of the neutralizing antibody response to HIV type 1 infection. Proc. Natl. Acad. Sci. U.S.A. 100, 4144–4149.

Ron, D., Tal, J., 1985. Coevolution of cells and virus as a mechanism for the persistence of lymphotropic minute virus of mice in L-cells. J. Virol. 55, 424–430.

Ron, D., Tal, J., 1986. Spontaneous curing of a minute virus of mice carrier state by selection of cells with an intracellular block of viral replication. J. Virol. 58, 26–30.

Roossinck, M.J., 2014. Metagenomics of plant and fungal viruses reveals an abundance of persistent lifestyles. Front. Microbiol. 5, 767.

Ruiz-Jarabo, C.M., Pariente, N., Baranowski, E., Dávila, M., Gómez-Mariano, G., et al., 2004. Expansion of host-cell tropism of foot-and-mouth disease virus despite replication in a constant environment. J. Gen. Virol. 85, 2289–2297.

Schrauwen, E.J., de Graaf, M., Herfst, S., Rimmelzwaan, G.F., Osterhaus, A.D., et al., 2014. Determinants of virulence of influenza A virus. Eur. J. Clin. Microbiol. Infect. Dis. 33, 479–490.

Shankarappa, R., Margolick, J.B., Gange, S.J., Rodrigo, A.G., Upchurch, D., et al., 1999. Consistent viral evolutionary changes associated with the progression of human immunodeficiency virus type 1 infection. J. Virol. 73, 10489–10502.

Silander, O.K., Tenaillon, O., Chao, L., 2007. Understanding the evolutionary fate of finite populations: the dynamics of mutational effects. PLoS Biol. 5, e94.

Sobrino, F., Dávila, M., Ortín, J., Domingo, E., 1983. Multiple genetic variants arise in the course of replication of foot-and-mouth disease virus in cell culture. Virology 128, 310–318.

Sorrell, E.M., Wan, H., Araya, Y., Song, H., Perez, D.R., 2009. Minimal molecular constraints for respiratory droplet transmission of an avian-human H9N2 influenza A virus. Proc. Natl. Acad. Sci. U.S.A. 106, 7565–7570.

Sponseller, B.A., Sparks, W.O., Wannemuehler, Y., Li, Y., Antons, A.K., et al., 2007. Immune selection of equine infectious anemia virus env variants during the long-term inapparent stage of disease. Virology 363, 156–165.

Takemoto, K.K., Habel, K., 1959. Virus-cell relationship in a carrier culture of HeLa cells and Coxsackie A9 virus. Virology 7, 28–44.

Temin, H.M., 1989. Is HIV unique or merely different? J. Acquir. Immune Defic. Syndr. 2, 1–9.

Tilman, D., 1994. Competition and biodiversity in spatially structured habitats. Ecology 75, 2–16.

Trent, D.W., Manske, C.L., Fox, G.F., Chu, M.C., Kliks, S.C., et al., 1990. The molecular evolution of Dengue viruses. Genetic variation and microevolution. In: Kurstak, E., Marusyk, R.G., Murphy, F.A., Van Regenmortel, M.H.V. (Eds.). Plenum Publishing Corporation, New York, pp. 293–315.

Tsibris, A.M., Korber, B., Arnaout, R., Russ, C., Lo, C.C., et al., 2009. Quantitative deep sequencing reveals dynamic HIV-1 escape and large population shifts during CCR5 antagonist therapy *in vivo*. PLoS One 4, e5683.

Tsimring, L.S., Levine, H., Kessler, D.A., 1996. RNA virus evolution via a fitness-space model. Phys. Rev. Lett. 76, 4440–4443.

Van Valen, L., 1973. A new evolutionary law. Evol. Theory 1, 1–30.

Vosloo, W., Thomson, G.R., 2017. Natural habitats in which foot-and-mouth disease viruses are maintained. In: Sobrino, F., Domingo, E. (Eds.), Foot-and-mouth Disease Virus. Current Research and Emerging Trends. Caister Academic Press, Norfolk, UK, pp. 179–210.

Weibull, W., 1951. A statistical distribution function of wide applicability. ASME J. Appl. Mech. 18, 293–297.

Weismann, A., 1891-1892. Essays upon Heredity and Kindred Subjects, vols. I and II. Clarendon Press, Oxford.

Weissmann, C., Billeter, M.A., Goodman, H.M., Hindley, J., Weber, H., 1973. Structure and function of phage RNA. Annu. Rev. Biochem. 42, 303–328.

Wilson, G.J., Wetzel, J.D., Puryear, W., Bassel-Duby, R., Dermody, T.S., 1996. Persistent reovirus infections of L cells select mutations in viral attachment protein sigma 1 that alter oligomer stability. J. Virol. 70, 6598–6606.

Yuste, E., Sánchez-Palomino, S., Casado, C., Domingo, E., López-Galíndez, C., 1999. Drastic fitness loss in human immunodeficiency virus type 1 upon serial bottleneck events. J. Virol. 73, 2745–2751.

Zhang, Y., Li, Y., Li, T., Shen, X., Zhu, T., et al., 2019. Genetic load and potential mutational meltdown in cancer cell populations. Mol. Biol. Evol. 36, 541–552.

Zhao, L., Abbasi, A.B., Illingworth, C.J.R., 2019. Mutational load causes stochastic evolutionary outcomes in acute RNA viral infections. Virus Evol. 5 vez 008.

Zhong, J., Gastaminza, P., Chung, J., Stamataki, Z., Isogawa, M., et al., 2006. Persistent hepatitis C virus infection in vitro: coevolution of virus and host. J. Virol. 80, 11082–11093.

# CHAPTER
# 7

# Long-term virus evolution in nature

## Abbreviations

**BCE** Before Christian Era
**CAT** Colonization-Adaptation Trade-off
**cccDNA** Covalently Closed Circular DNA
**CTL** Cytotoxic T Lymphocyte
**CRF** Circulating Recombinant Forms
**ELISA** Enzyme-Linked Immunosorbent Assay
**EMCV** Encephalomyocarditis Virus
**ET** Evolutionary Trace
**FAO** Food and Agriculture Organization of the United Nations
**FMD** Foot-and-Mouth Disease
**FMDV** Foot-and-Mouth Disease Virus
**HAV** Hepatitis A Virus
**HBV** Hepatitis B Virus
**HCV** Hepatitis C Virus
**HIV-1** Human Immunodeficiency Virus Type 1
**HLA** Human Leukocyte Antigen
**HTLV-1** Human T-Cell Lymphotropic Virus Type 1
**HTLV-2** Human T-Cell Lymphotropic Virus Type 2
**HRV** Human Rhinovirus
**ICTV** International Committee on Taxonomy of Viruses
**IV** Influenza Virus
**MARM** Monoclonal Antibody-Resistant Mutant
**ML** maximum likelihood
**MV** Measles Virus
**PAM** Percent Accepted Mutation
**PIR** Protein Information Resource
**PV** Poliovirus
**RV** Rabies Virus
**SARS** Severe Acute Respiratory Syndrome
**SIVcpz** Chimpanzee Simian Immunodeficiency Virus
**s/nt/y** Substitutions Per Nucleotide and Year
**URF** Unique Recombinant Form

**URL** Uniform Resource Locator
**WEEV** Western Equine Encephalitis Virus
**WNV** West Nile Virus

## 7.1 Introduction to the spread of viruses. Outbreaks, epidemics, and pandemics

Intrahost virus replication and evolution are the first steps in the process of virus diversification that continue with successive virus transmission events that are a condition for long-term survival in nature. Viruses are perpetuated as a consequence of many rounds of persistent or acute infections, with possible extracellular stages in which genomes remain basically invariant. Despite lacking direct evidence, we presume that multitudes of successive transmissions have allowed viruses to survive at least for thousands of years, probably undergoing continuous genetic change. Picornavirologists are familiar with an Egyptian stela dated 1550−1333 BCE. (18th Egyptian dynasty) that portrays the image of a man with an atrophic leg probably a consequence of infection with poliovirus (PV) or a related virus (Eggers, 2002). In this chapter, we deviate from the focus on how viral population numbers affect short-term survival and evolution, and we turn to features of viruses as they infect successive hosts to permit virus perpetuation in nature.

*Virus as Populations*
https://doi.org/10.1016/B978-0-12-816331-3.00007-6

Viruses can be transmitted vertically or horizontally. Vertical transmission occurs from parental organisms to their offspring, and it includes infection through the germline in animals and plants, from the mother to the embryo during fetal development, and also postnatal transmission to the newborn via blood, milk, or contact (Mims, 1981; Nash et al., 2015). In horizontal transmission, a virus spreads from infected individuals to susceptible recipients. We are most familiar with this type of transmission. It frequently gives rise to disease outbreaks (infection episodes localized in space and time that affect a few individuals), epidemics (that affect an ample geographical area and are often extended in time), and pandemics (that affect most areas of our planet), typically the periodic influenza pandemics.

All transmission modes have probably contributed to the maintenance of viruses in our biosphere. Persistent infections are likely to have played a major role when the number of individual humans or animals living in close contact was limited throughout the pre-agricultural era, earlier than 10,000 years ago. A favorable climate change during the Holocene (the geological epoch that began at the end of the Pleistocene, around 12,000 years before the present; compare with Chapter 1) was probably an important driver toward large-scale domestication of plants and animals, 10,000–7000 years BCE. From the behavior of current viruses, there might have always been a dynamics of virus change for adaptability within individual hosts, and transmissions among animals or plants. The probability of transmission increased as host population numbers rose with agricultural practices and urban life in the last several thousand years. Intensive agriculture must have contributed to accelerated sequence space exploration by viruses with consequences for the emergence of viral disease (Section 7.7). Probably, there has been a continuous dynamics of viral emergences, reemergences, and extinctions with patterns that may be parallel to those

observed in present-day viruses. Virology has existed as an organized scientific discipline with the possibility to isolate, store, and study viruses only for about one century. The challenge to reconstruct the events that might have led to viruses similar to the ones we isolate today was addressed in Chapter 1, with a critical first question being if viruses originated 4000 million years ago, or "only" 2000 million years ago (diagrams in Figs. 1.3 and 1.4, and Section 1.5 in Chapter 1). In this chapter, we are more modest in our aspirations, and we will analyze, with the tools of genomics, what happens when viruses evolve for months or years in what we call inter-host virus evolution.

Unfortunately, viruses have not left a fossil record (at least one that we can uncover with the available tools), since according to current paleontology, nitrogen- and phosphorus-rich molecules are unlikely to be protected in fossils older than 1 million years. At most, hundred to thousand years-old skeletal remains or frozen bodies that contain viruses have been analyzed, and sequences retrieved [(Muhlemann et al., 2018) and references therein]. Fortunately, there is a different historical record of ancient viral sequences in the DNA of differentiated organisms, in the form of integrated virus-like genetic elements. The research area that consists in rescuing ancestral viral sequences is termed paleo-virology, and it is providing information on viruses that circulated thousands of years or even millions of years ago in the case of viral sequences integrated into cellular DNA (Aswad and Katzourakis, 2012).

The presence of recognizable viral genomic sequences in present-day cellular DNA (Tomonaga et al., 2019) suggests a history of long-term interaction between viruses and cells, in support of some models of virus origins that propose long coevolution between precellular and cellular entities with virus-like elements (Chapter 1). Despite the current capacity to amplify tiny amounts of viral nucleic acids for nucleotide sequence determinations, proposals

7.2 Reproductive ratio as a predictor of epidemic potential. Indeterminacies in transmission events

227

on how long-term viral evolution might proceed have to be based mainly on the comparison of viral genomes and the structure of viral proteins from modern representatives of different virus groups. First, we should understand the basic concepts related to virus transmission, keeping in mind viral population numbers and the complexity of viral populations.

## 7.2 Reproductive ratio as a predictor of epidemic potential. Indeterminacies in transmission events

The basic reproductive ratio or basic reproduction number ($R_0$) is the average number of infected contacts per infected individual. At a population level, a value of $R_0$ larger than one means that a virus will continue its propagation among susceptible hosts if no environmental changes or external influences intervene. An $R_0$ value lower than one means that the virus is doomed to extinction at the epidemiological level under those specific circumstances. The basic models of infection dynamics were developed by R.M. Anderson, R.M. May, and M.A. Nowak, with inclusion of the following key parameters: rate $k$ at which uninfected hosts enter the population of susceptible individuals ($x$), their normal death rate ($u$) (so that the equilibrium abundance of uninfected hosts is $k/u$), number of infected hosts ($y$), mortality due to infection ($v$) (so that $1/u + v$ is the average lifetime of an infected host), a rate constant ($\beta$) that characterizes parasite infectivity (so that $\beta x$ is the rate of new infections and $\beta xy$ is the rate at which infected hosts transmit the virus to uninfected hosts). These parameters are schematically indicated in Fig. 7.1 and they provide a theoretical value for $R_0$ (Anderson and May 1991; Nowak and May 2000; Nowak, 2006; Woolhouse, 2017).

$R_0$ values are not a universal constant for viruses because, as discussed in Chapters 3 and 4, virus variation may affect viral fitness and

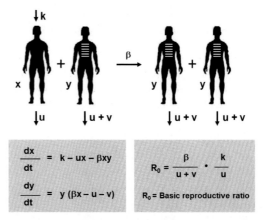

$$\frac{dx}{dt} = k - ux - \beta xy$$

$$\frac{dy}{dt} = y(\beta x - u - v)$$

$$R_0 = \frac{\beta}{u + v} \cdot \frac{k}{u}$$

$R_0$ = Basic reproductive ratio

FIGURE 7.1    A schematic representation of the main parameters of viral dynamics that enter the equations that predict the rate of variation of uninfected and infected (internal horizontal lines in the human figure) individuals (shaded box on the left) and the $R_0$ value (shaded box on the right). The meaning of parameters and literature references are given in the text.

viral load in infected individuals, and the latter, in turn, may influence the amount of virus that surfaces in a host to permit transmission (Delamater et al., 2019). Despite uncertainties, consistent $R_0$ values have been estimated for different viral pathogens based on field observations. Values of $R_0$ for human immunodeficiency virus type 1 (HIV-1) and severe acute respiratory syndrome (SARS) coronavirus range from two to five; for PV the range is 5−7, and for Ebola virus is 1.5−2.5. For the measles virus (MV), which is one of the most contagious viruses described to date, the $R_0$ reaches 12−18 (Heffernan et al., 2005; Althaus, 2014). Most isolates of the SARS coronavirus that circulated months after the emergence of this human pathogen had modest $R_0$ values, and this is consistent with SARS not having reached the pandemic proportions that were feared immediately following its emergence. In contrast, MV is highly transmissible, thus explaining frequent outbreaks as soon as a sizable population stops vaccinating its infants. This is an important problem, fueled by antivaccination campaigns without scientific basis. Since some of the parameters that enter the basic equations of viral dynamics depend on the nucleotide sequence of the viral genome, mutations may alter $R_0$ values, allowing some virus variants to overtake those that were previously circulating in the population (Fig. 7.2). Viral replication, fitness, load, transmissibility, and virulence are all interconnected factors that contribute to virus persistence in its broader sense of virus being perpetuated in nature. These parameters can affect both disease progression in an infected individual and transmissibility at the epidemiological level.

The difference between the number of infectious particles that participate in transmission and the total number of virus in an infected, donor organism provides a first picture of the indeterminacies involved in viral transmissions. The larger the population size and genetic heterogeneity of the virus in an infected individual, the higher will be the likelihood that independent transmission events have different outcomes. Individual susceptible hosts will receive subsets of related but nonidentical genomes. In a bright article that emphasized

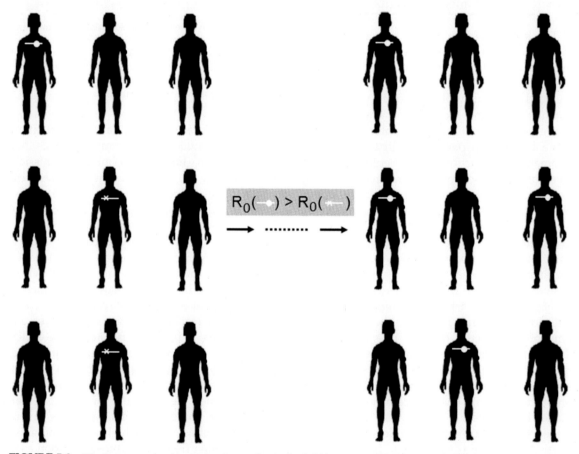

FIGURE 7.2  Displacement of a virus variant by another in the field by virtue of the latter displaying a higher $R_0$ value. The competing viruses are depicted as horizontal lines with a distinctive symbol. Differences in $R_0$ recapitulate part of the determinants of epidemiological fitness (Section 5.9 in Chapter 5). Concepts of competition among clones or populations within infected host organisms or cell cultures, treated in previous chapters, can be extended at the epidemiological level, with the appropriate choice of the key parameters. References are given in the text.

the molecular evidence and medical implications of quasispecies in viruses, J.J. Holland and colleagues wrote the following statement: "Therefore, the acute effects and subtle chronic effects of infections will differ not only because we all vary genetically, physiologically, and immunologically, but also because we all experience a different array of quasispecies challenges. These facts are easily overlooked by clinicians and scientists because disease syndromes are often grossly similar for each type of virus, and because it would appear to make no difference in a practical sense. However, for the person who develops Guillain-Barré syndrome following a common cold, or for the individual who remains healthy despite many years of HIV-1 infection, for example, it may make all the difference in the world" (Holland et al., 1992). The ever increasing number of identified human genes whose allelic forms influence viral infections provides strong support for the predictions of Holland et al. Indeterminacies in the process of virus spread can be viewed as an extension of the diversification due to bottleneck events in the case of virus transmission, as visualized in Figs. 6.1 and 6.2 in Chapter 6, when dealing with the limitations of the virus samples retrieved from an infected host as the starting material for experimental evolution approaches.

## 7.3 Rates of virus evolution in nature

Despite a necessarily approximate and imprecise knowledge of how many and which types of genomes participate in successive horizontal and vertical transmissions, we can obtain an overall estimate of the rate at which viruses evolve in nature. This is commonly done by comparing consensus genomic nucleotide sequences of viruses isolated at different times during an outbreak, epidemic, or pandemic.

The rate of evolution (also termed as the rate of fixation or rate of accumulation of mutations) is generally expressed as substitutions per nucleotide and year (s/nt/y). The term fixation is not the most adequate when dealing with virus evolution given that the term refers to a consensus that, in addition to being an average of the real sequences, has a fleeting dominance. However, the term is frequently used in the literature of general genetics and virus evolution. The rate of evolution is calculated from genetic distances between consensus viral genomic sequences of successive viral samples from a single persistently or acutely infected host, or from different host individuals infected at different times. Rates of evolution are only indirectly related to mutation rates and mutation frequencies that do not include a time factor in them. There have been several comparisons of rates of evolution for viruses that document the differences between RNA and DNA viruses (Jenkins et al., 2002; Hanada et al., 2004; Domingo, 2007). As was the case for mutant spectra (Section 3.3 in Chapter 3), some DNA viruses attain rates of evolution in nature that are very similar to those typical of RNA viruses (Duffy and Holmes, 2008). A few comparative values are given in Table 7.1.

Herpes simplex virus constitutes an example of a complex DNA virus for which, despite uncertainties (Firth et al., 2010) a calculated rate of evolution was $10^{-8}$ s/nt/y (Sakaoka et al., 1994), which is actually closer to the rate estimated for

TABLE 7.1 Some representative rates of virus evolution in nature.

| Virus or organism | Range of values |
| --- | --- |
| RNA viruses (riboviruses) | $10^{-2}$ to $10^{-5}$ |
| Retroviruses | $10^{-1}$ to $10^{-5}$ |
| Single-stranded DNA viruses | $10^{-3}$ to $10^{-4}$ |
| Double-stranded DNA viruses | $10^{-7}$ to $10^{-8}$ |
| Cellular genes of host organisms | $10^{-8}$ to $10^{-9}$ |

Values are expressed as substitutions per nucleotide and year. The range of values is based on several studies. Values depend on the virus or organism under study, the genomic region analyzed, and several factors discussed in the text.

cellular genes than for most viruses. However, its mutation frequencies, measured by independent procedures, are in the range of $7 \times 10^{-3}$ to $1 \times 10^{-5}$ (see also Section 7.4.2). The latter values may result from the selective agent targeting a replicating herpes simplex virus that has produced multiple variants, while the overall slow rate of evolution may be influenced by periods of latency. Slow evolution is expected for retroviruses such as human T-cell lymphotropic virus types 1 and 2 (HTLV-1 and HTLV-2) whose life cycles are dominated by the integrated provirus stage, with the viruses following the clonal expansion of their host cells (Melamed et al., 2014; Kulkarni and Bangham, 2018). Some single-stranded DNA viruses display rates of evolution typical of the rapidly evolving RNA viruses (Table 7.1).

Different genes of the same virus set may show different rates of evolution (i.e., the polymerase and other nonstructural proteins may evolve more slowly than structural proteins). Thus, a rate of evolution is far from being a universal feature of a virus. A comparison of rates of synonymous substitutions (under the assumption that synonymous substitutions do not affect protein function; see Chapter 2 for limitations of considering synonymous mutations as neutral) for several RNA viruses, yielded a range of evolutionary rates of $6 \times 10^{-2}$ to $1 \times 10^{-7}$ synonymous substitutions per synonymous site per year (Hanada et al., 2004). The values were recalculated from primary phylogenetic data using maximum likelihood (ML) (Section 7.6), under the assumption of the molecular clock, and inference of the ancestral nucleotide sequences at the tree nodes. The five orders of magnitude variation were attributed mainly to the degree of virus replication rather than to differences in error rate. We will deal with the molecular clock hypothesis (constant rate of accumulation of mutations) in Section 7.3.3, but the major features of virus evolution studied in previous chapters (mainly those typical of mutant swarm-forming RNA and DNA viruses)

should make us skeptical of similar evolution rates in different biological contexts. Rate variations were documented with HIV-1 subpopulations in different compartments of the human brain (Salemi et al., 2005). The data did not fit a "global" molecular clock for the virus in the brain, and "local" clocks showed that meninges and temporal lobe HIV-1 subpopulations evolved 30 and 100 times faster, respectively, than other HIV-1 populations in the brain. It is believed that these differences were due to random drift of viral sequences rather than selection for some genome types. An additional complication is that even restricting virus isolations to the same biological material in a standard epidemiological setting, several measurements indicated discontinuities in evolutionary rates. The discontinuities had at least two origins: the nonlinear effect of time, and some unique features of evolution occurring inside an infected host. These points are examined next.

## 7.3.1 Influence of the time of sampling

It has been known for decades that the calculated rates of virus evolution in nature depend on the genomic region analyzed and the time interval between the isolation of the viral samples on which the calculations are based. Noncumulative sequence changes in the hemagglutinin of influenza virus (IV) type C were found in an early study by Buonagurio et al. (1985). The authors proposed a cocirculation of variants that belonged to different evolutionary lineages. If multiple evolutionary pathways coexist in a given geographical area, and they establish a network of lineages that evolve with time, variations of calculated rates of evolution are expected, and they may distort the actual rate of evolution of individual lineages.

A second early observation was made during an episode of foot-and-mouth disease (FMD) in Spain. Estimates of the rate of evolution of the virus ranged from $<4 \times 10^{-4}$ to $4 \times 10^{-2}$ s/nt/y, depending on the genomic region analyzed, and the time period between isolations (Sobrino et al., 1986). Cocirculation of multiple heterogeneous foot-and-mouth disease virus (FMDV) samples ("evolving quasispecies") was proposed. The result to be emphasized here is that the calculated rates of evolution were extremely high (higher than $10^{-2}$ s/nt/y) if the two FMDVs compared were isolated at close time points, while lower values were calculated when the viruses were sampled from different animals at distant time points.

The dependence of the calculated rate of evolution during the epidemic spread of the virus on the time interval between virus isolations for sequence determination is expected for viruses that need not be transmitted by direct contact between an infected and a susceptible host. Some viruses remain infectious in the environment for prolonged time periods, until they reach a susceptible host in which to initiate replication rounds. This is the case of viruses transmitted by the fecal-oral route, such as enteroviruses. FMDV can adhere and remain infectious on many objects (fomites), including dust particles, food products with neutral pH, or insects that can transport the virus mechanically. Infectious FMDV can traverse long distances (many kilometers) on dust particles, people, trains, and the like. Even if some infectivity is lost, a few infectious particles are sufficient to infect an animal (Sellers, 1971, 1981). There are some classic examples of long-distance transport of FMDV, a virus subjected to close scrutiny due to its economic impact. One is the spread of SAT1 and A22 FMDV during the 1960s in Turkey along the railway line from the cattle-raising region of Lake Van to slaughterhouses in Istanbul [this and other examples are described in (Brooksby, 1981)]. Computer models have been developed to explain and predict possible airborne FMDV transmission in different geographical areas,

the origin of epidemics, and the effectiveness of contingency plans (Sorensen et al., 2000; Tidesley et al., 2017). [As an anecdote, in my experience as a member of the Research Group of the Standing Technical Committee for the Control of FMD of Food and Agriculture Organization of the United Nations (FAO) in the 1980s, FMD outbreaks in any country always came from somewhere else]. A time-dependent bias in evolutionary rates for viruses has been amply documented (Duchene et al., 2014; Aiewsakun and Katzourakis, 2016), including when rates are compared between present-day isolates and those that circulated thousands of years ago (Muhlemann et al., 2018).

For viruses that can remain infectious outside their hosts, and that do not need donor-recipient host contacts to perpetuate transmission chains, the time between isolations will influence the calculated rate of evolution based on genomic nucleotide sequences. The reason is that during the extracellular stages, the virus will not undergo genetic change, at least to the extent of variation during intracellular replication (possible mutations due to chemical damage in viral genomes is indicated in Section 2.2. of Chapter 2). The effect of nonreplicative time intervals in the rate of evolution is illustrated in Fig. 7.3.

Some complications should be considered in the interpretation of the analyses depicted in Fig. 7.3: (i) the consensus sequences determined to characterize the virus shed by each animal represent a simplification of the real genome composition of the virus. (ii) Individual animals vary in physiological and immunological status, and, obviously, they are not in line waiting to be infected; they move, gather around water and food sources, some are isolated, others in close contact with their peer, and so on. (iii) In this case, virus transport is assumed to be mechanical (on dust particles carried by wind, aerosols, insects, etc.) without additional viral replication during transport. However, subpopulations of

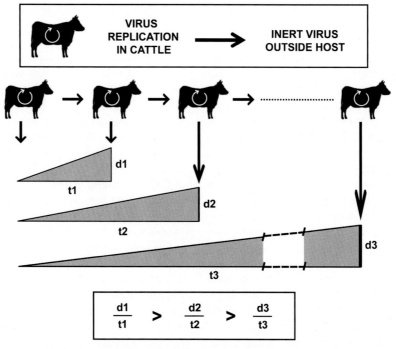

**FIGURE 7.3**  Inverse correlation between the time between viral isolations for consensus sequence determination, and the calculated rate of evolution. Animals sustain the replication (inside curved arrow) of a virus that will be transmitted to a susceptible animal. The time that the virus spends outside an animal (absence of replication) is depicted by a horizontal arrow. The time of virus isolation is given by $t1$, $t2$, and $t3$. The number of nucleotide differences in the virus isolates relative to the sequence of the initial reference virus (from the animal on the left) is given by $d1$, $d2$, and $d3$ (vertical arrows). Because of the increasing periods of stasis (addition of horizontal arrows), calculated rates of evolution given by the $d/t$ ratio will be higher the shorter the time interval between isolations. See text for additional models of time effects of evolutionary rates, and references.

the most environment-resistant particles, or particles that adhere best to the transporter object, may bias the composition of the virus that will reach an animal to pursue replication. Such events, occurring for 10 to 100 rounds of host infections, render the appalling virus diversity described in Chapter 1 a bit less appalling. Since several additional environmental circumstances are changeable and unpredictable, it is unlikely that rates of viral evolution in nature can remain invariant on the basis of some internal principle of constant mutation occurrence (as if the accumulation of mutations was as monotonous as radioactive decay!).

## 7.3.2 Interhost versus intrahost rate of evolution

Additional observations against constant mutational input with time have been made with HIV-1 and human and avian hepatitis B virus (HBV). The main finding is that interhost rates of evolution are lower than intrahost rates, even under a comparable set of epidemiological parameters. Several proposals have been made to account for this difference. A.J. Leslie and colleagues described cytotoxic T lymphocyte (CTL)-escape mutants of HIV-1 from infected patients. Some of the mutants reverted to the wild-type sequence after transmission to individuals negative for the human leukocyte antigen (HLA) alleles associated with long-term HIV-1 control (Leslie et al., 2004). Strong intrahost selective pressures and reversion of a part of the selected mutations upon transmission to a susceptible individual is one of the possible mechanisms behind diminished evolutionary rates when viruses from multiple host individuals are compared (Fig. 7.4, Box 7.1).

J.T. Herbeck, J.I. Mullins, and colleagues systematically observed lower nucleotide sequence divergence between HIV-1 isolates from different individuals sampled in primary infection than between isolates from individuals with advanced illness. HIV-1 regained some ancestral features when infecting a new host, again explaining a higher intrahost than the interhost evolutionary rate (Herbeck et al., 2006). In a study of HIV-1 transmission between several pairs of individuals over an 8-year period, A.D. Redd and colleagues reported that the viral populations found in the newly infected recipients were more closely related to ancestral sequences from the donor than to the sequences found in the donor near the time of transmission (Redd et al., 2012). Preferential transmission of ancestral sequences may also contribute to lower interhost than intrahost rates of evolution (Box 7.1).

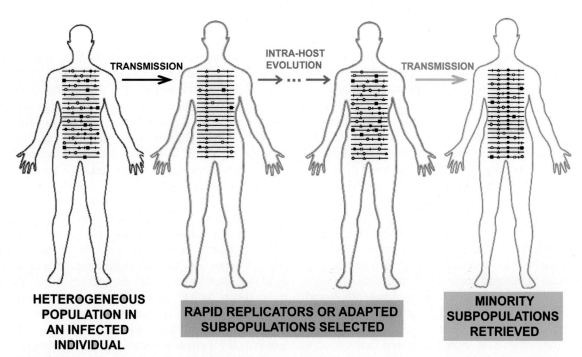

FIGURE 7.4  Scheme of a possible mechanism for faster intrahost than interhost virus evolution. Transmission events are represented by long arrows and intrahost evolution by short arrows (middle of the picture). The virus in the person on the left (black outline) has evolved to generate a complex mutant spectrum. However, only a subset of genomes are efficiently transmitted to the recipient person (brown outline). The virus in the recipient person evolves toward a complex mutant spectrum. Again, in this new mutant spectrum, only a minor set of genomes that resemble the ones in the first transmission is efficiently transmitted to the third person (green outline). The net result is that because at each transmission the genomes related to those that first entered the previous host have an advantage, rates of evolution will appear as slower than those within each host. Boxes at the bottom summarize the major event at each step. See text for additional related mechanisms and references.

---

BOX 7.1

## MODELS FOR NONLINEAR RATES OF EVOLUTION

- For viruses that remain infectious in the extracellular environment, stasis due to the absence of replication will result in rates of evolution inversely correlated with the time between the isolation of the compared viruses.
- Adapt and revert. Mutants that permit adaptation to a new host individual revert upon transmission.
- Preferential transmission of ancestral sequences. Despite diversification in any host,

ancestral sequences have a selective advantage in transmission. They may be retrieved from cellular memory (integrated provirus in HIV-1 or cccDNA in HBV).
- Colonization-adaptation trade-off. Sequential changes in the intensity of the host immune response favor dominance of some genome subpopulations over others. Upon transmission, ancestral minority subpopulations may become dominant.

---

K.A. Lythgoe and C. Fraser provided evidence that cycling of HIV-1 through long-lived memory CD4$^+$T cells is probably the main contributing factor to slower HIV-1 evolution at the epidemic level (Lythgoe and Fraser, 2012). Ancestral sequences of HIV-1 in infected individuals may arise by the activation of proviral sequences kept in the form of quasispecies memory. In this case, we refer to the type of molecular memory defined as a reservoir, anatomical, or cellular memory in Section 5.5 of Chapter 5. A related type of reservoir memory is found in HBV, in the form of covalently closed circular DNA (cccDNA) that persists in the nuclei of infected hepatocytes, and acts as a template for the synthesis of pregenomic RNA and viral mRNAs (Kay and Zoulim, 2007). In this case, a record of ancient sequences is registered in the cccDNA. It should be noted that memory levels are dependent on fitness values, as evidenced experimentally with FMDV and expected from the theoretical basis of memory implementation (Chapter 5). In consequence, the most abundant memory genomes established early in an infection might be those displaying the highest fitness early in infection,

and they might be better adapted to initiate infections than to sustain them (Fig. 7.4).

Additional mechanisms for the time dependence of evolutionary rates have been suggested for HBV. In a 17 years follow-up of several patients, HBV diversity increased during periods of active host immune response, and viral copy numbers decreased. When the immune response was weak, viral genome diversity decreased, and viral copy numbers increased; these periods are expected to be those of high transmissibility (Wang et al., 2010).

Endogenous hepadnaviruses are present in the genomes of several organisms. There is evidence that some of the integration events in avian hosts are at least 19 million years old. These integrated hepadnaviruses maintain about 75% nucleotide sequence identity with present-day hepadnaviruses, and the comparisons suggest that the long-term substitution rates are 10$^3$-fold lower than those for circulating avian HBVs (Gilbert and Feschotte, 2010). The permanence of viral genomic sequences in cellular DNA is a mechanism of evolutionary stasis, as it was emphasized in Chapter 3 with the comparison of the evolutionary rate of the retroviral v-*mos* gene and its cellular counterpart c-*mos* (Gojobori and Yokoyama, 1985), among other evidence. Considerable evolutionary stasis is also observed by comparing isolates of HTLV-1 and HTLV-2, whose replication displays a preference for maintaining its integration in cellular DNA (Melamed et al., 2014). For viruses that have a dual potential of error-prone replication and cellular DNA-like stasis, the permanence in cellular DNA may also contribute to reduced long-term evolutionary rates.

HBV quasispecies dynamics was examined in a virus that infected members of the same family that presumably acquired the virus through mother-to-infant transmission (Lin et al., 2015). Again, the intrahost evolutionary rate was higher than the interhost rate, and the latter decreased with the number of transmissions. The differences were mainly due to nonsynonymous substitutions at limited sites. These observations were interpreted as a rapid switch of HBV between colonization (invasion of new host) and adaptation (quasispecies optimization in the new host). The authors referred to the colonization-adaptation trade-off (CAT) model or alternations of virus facing an environment marked by a limited host immune response followed by a period of active immune response. In the former environment, viruses displaying rapid replication are selected, while in the latter environment, HBV escape mutants with lower productivity are selected. In each transmission, when the virus reaches a new host, the previously adapted subpopulations are overgrown by the rapidly replicating ones. Again, cccDNA can serve as a reservoir of ancient sequences.

In agreement with these proposals, rates of evolution measured in a single infected individual persistently infected with a continuously replicating virus tend to be higher than those observed with the same viruses isolated from different individuals (Morse, 1994; Domingo et al., 2001; Domingo, 2006). Slowly evolving viral genes may nevertheless undergo episodes of rapid evolution and, vice versa, a rapidly evolving gene may be transiently static. This should be considered in statistical approaches to evolution (Gaucher et al., 2002).

The viruses that colonize the human gut are extremely diverse and evolving at different rates. From all the evidence, the gut virome composition is unique to each individual human. It includes single-stranded DNA viruses that evolve at rates $>10^{-5}$ substitutions per nucleotide and day (an extremely high rate!); other genomes, particularly those of temperate bacteriophages, display lower evolutionary rates, partly due to their being replicated by high fidelity bacterial polymerases when the prophage is integrated into host DNA (Minot et al., 2013). Thus, both the viral populations that infect our tissues and organs and those that infect the microbes that colonize us (1%−3% of the mass of a healthy human is composed of microorganisms) are diverse, unique to each of us, and evolving at different rates.

The reader will find in the literature additional proposals to account for the contrast between rapid intrahost evolution versus relative long-term conservation when comparing ancestral sequences [(Ali and Melcher, 2019; Simmonds et al., 2019), among other studies; see also Section 7.3.1]. It would not come as a surprise if several factors acted conjointly to produce the observed rate discrepancies. There is a need to integrate the approaches and conclusions of short-term evolution centered on quasispecies dynamics, and of long-term evolution based on the phylodynamics methodology (Geoghegan and Holmes, 2018).

### 7.3.3 Rate discrepancies and the clock hypothesis

The molecular clock hypothesis dwindles as a conceptual framework because, from all evidence, virus evolution is far from being dictated by a steady accumulation of mutations in viral genomes. The major event that we have learned by comparing the genomic composition of individual viral populations is that evolution is guided by "replacement of subpopulations" rather than "accumulation of mutations in genomes." This conceptual change is as important for the understanding of long-term evolution of viruses as it is the consideration of the wild type as a cloud of mutants in the definition of a viral population, a major input from quasispecies theory (Chapter 3).

There are additional arguments against the operation of a molecular clock in virus evolution. According to the clock hypothesis, the rate of accumulation of mutations coincides with the rate at which mutations arise in infected individuals. This holds for neutral mutations, and as mentioned in several chapters of this book, very few mutations occurring in highly compact viral genomes are truly neutral (with no functional consequences in any environment; see Section 2.3 in Chapter 2). Even for neutral evolution, spatial asymmetries in populations are sufficient to perturb the molecular clock rate, as documented with a theoretical model of broad applicability (Allen et al., 2015). However, quasispecies and the operation of a clock are not totally irreconcilable. An epidemiological study with FMDV suggested that viral quasispecies could produce a transient molecular clock due to the periodic sampling of components of the mutant spectrum in transmission (Villaverde et al., 1991). In this case, both time differences between transmissions and spatial heterogeneities would blur the transiently observed regularity.

## 7.4 Long-term antigenic diversification of viruses

Viruses can change their antigenic properties gradually, in a process termed antigenic drift, or suddenly, in a process termed antigenic shift. The distinction between antigenic drift and shift was established with IV (Gething et al., 1980; Webster, 1999; Parrish and Kawaoka, 2005). The shift in IV is due to genome segment reassortment that incorporates new hemagglutinin or neuraminidase genes. In monopartite viruses, the difference between gradual and drastic antigenic change has also been established (Martínez et al., 1991b).

The antigenic diversification of one FMDV serotype was examined over a six-decade period by comparing amino acid sequences of the major antigenic sites of the virus isolated in three continents (Martínez et al., 1992). The evolution of the capsid genes was associated with linear accumulation of synonymous mutations, but not of amino acid substitutions. Remarkably, the antigenic variation over 6 decades was due to fluctuations among limited combinations of amino acid residues without net accumulation of amino acid substitutions over time (Fig. 7.5). This result suggests that constraints at the protein level may maintain a long-term virus identity at the antigenic level. In a related observation on the interhost evolution of HIV-1 mentioned in Section 7.3.2, HIV-1 recovered ancestral features when infecting a new host (Herbeck et al., 2006). Thus, multiple constraints in viruses may limit the rate and mode of long-term diversification, resulting in different numbers of circulating serotypes among viruses of the same family.

FIGURE 7.5 Evolution of a major antigenic site of FMDV over 4 decades. The sequence at the top is that of amino acid residues 129–151 of capsid protein VP1 of FMDV C2 Pando, isolated in Uruguay in 1944. Key amino acid positions did not diverge in a linear fashion and isolates over 4 decades displayed only two types of amino acid residues at each position. See text for possible mechanisms and references.

## 7.4.1 Widely different number of serotypes among genetically variable viruses

A puzzling question in evolutionary virology is that despite sharing high mutation rates, some viruses display extensive antigenic diversity in nature reflected in multiple serotypes, while other viruses maintain a relatively invariant antigenic structure, with only one serotype recorded. For the latter group of viruses, the same vaccine can maintain its efficacy over many decades; examples are rabies virus (RV) and MV, two RNA viruses that show remarkable genetic diversity in nature, and estimates of mutation rates and frequencies comparable to other RNA viruses. Antigenic constancy is a determinant of long-lasting immunity after infection or vaccination. MV infection produces lifelong immunity (probably as a result of several factors) while patients that have cleared the hepatitis C virus (HCV) can be reinfected by the same virus. Cases of patients infected with HCVs of different genotypes and subtypes are increasingly identified, as more refined diagnostic tests are utilized, and the virus diversifies in nature (Hedskog et al., 2019).

No correlation between virus structure (or its morphotype) and antigenic diversity has been found. Among structurally closely related viruses, differences in antigenic diversity are apparent. A dramatic case is that of the picornaviruses, since encephalomyocarditis virus (EMCV) or hepatitis A virus (HAV) have a single serotype, while human rhinoviruses (HRVs) have been divided into more than 100 serotypes. Other picornaviruses have intermediate numbers of serotypes: three in the case of PV and seven in the case of FMDV. Although it may seem that a diverse antigenic structure may predict a broad host range, this is actually not the case. HAV is highly specialized for the human host, while EMCV infects more than 30 species, including mammals, birds, and invertebrates (Knowles et al., 2010).

Several, not mutually exclusive models, have been proposed to account for differences in the antigenic stability (number of serotypes) among viruses:

- Differences in mutation rate, either the average value for the entire genome or the local mutation rate at the genomic sites that encode antigenic determinants.
- The presence of some dominant and invariant antigenic sites that evoke long-lasting antibodies in the infected hosts, and that obscure other antigenic sites that produce different antibodies that have a limited impact on the antigenic profile of the virus.
- Differences among the assays used for serotype classification. If a universal and standard procedure to classify virus isolates in different serotypes were applied, differences among viruses would be largely lost.
- Difference in the history of virus circulation. Ancient viruses that undergo many rounds of genome replication in each infected host have had an opportunity to diversify antigenically in a manner not possible with viruses that have a more limited history of circulation among susceptible hosts. According to this model, antigenic diversification of some viruses currently viewed as antigenically invariant will take place during the next hundreds of years if their circulation continues.
- Some viruses have antigenic sites that cannot vary because they are under severe constraints to accept amino acid substitutions. Antigenic variants may exist as low-fitness subpopulations, but their frequency is too low to modify the results of the diagnostic tests used for serological classifications.

Consideration of these possibilities requires examining some experimental data on virus antigenicity. First, as a conceptual precision, we assume that the number of serotypes is essentially determined by amino acid sequences located in the virus particle and that either directly or indirectly can affect the interaction of the virus with antibodies. Neutralizing and nonneutralizing antibodies may contribute to serological distinctions, depending on the assays performed for serotyping. Serum neutralization tests will identify differences in sensitivity to neutralization, while Enzyme-Linked Immuno-Sorbent Assay (ELISA) tests will capture reactivity by all raised antibodies.

Antibodies can be obtained from infected natural hosts, or from some laboratory animals which are not a natural host for the virus. An ensemble of amino acid residues forms an antigenic determinant which is usually composed of multiple epitopes [defined here as a unit of interaction with a monoclonal antibody (MAb)]. Epitopes can be either continuous (also termed linear) or discontinuous (also termed structured). Continuous epitopes are those whose primary amino acid sequence has the information to react with the cognate antibody. Discontinuous epitopes are those whose reactive residues come from distant positions of the same protein or residues of different proteins. Many overlapping epitopes can be found within the same antigenic site. Epitopes can include modified amino acid residues such as glycosylated amino acids. Reactivity of discontinuous epitopes with the cognate antibody is generally lost as a consequence of denaturation of the proteins that form the epitope.

With these introductory clarifications, we are now in a position to examine the different possibilities listed above to account for differences in the number of serotypes among genetically variable viruses.

There is no correlation between limited antigenic diversity and low average mutation rate. Mutation rates and frequencies for RNA viruses fall in the range of $10^{-5}$ to $10^{-3}$ substitutions per nucleotide (Chapter 2). However, mutation rates along a viral genome are not uniform, as evidenced by the occurrence of hot spots for variation. Influences such as nucleotide sequence

context or RNA structure may conceivably alter mutation rates. It was proposed that a predicted double-stranded RNA at the region encoding the major antigenic site of FMDV might increase the polymerase error rate locally and give rise to multiple amino acid substitutions (Weddell et al., 1985). While at some specific sites polymerases may be more error-prone than average, subsequent evidence for FMDV indicated that antigenic variation is due to amino acid substitutions at different antigenic sites and that even variation at the major site can be mediated by distant amino acids on the viral capsid (Rowlands et al., 1983; Geysen et al., 1984; Mateu et al., 1990; Feigelstock et al., 1996). Later molecular studies have not provided evidence that viruses may have a large number of serotypes because their polymerases are more error-prone when copying regions encoding amino acids that belong to antigenic sites. Therefore, the possibility that differences in mutation rates can determine a different number of circulating serotypes is unlikely.

Most viruses include multiple antigenic sites, and antibodies are raised against several surface proteins to produce an array of neutralizing and nonneutralizing antibody molecules. Taking picornaviruses again as an example, the number of antigenic domains (each composed of multiple epitopes) varies between one and four (Mateu, 1995, 2017; Fry and Stuart, 2010). There is no evidence that a restriction on the number of sites or epitopes or that the expression of a salient class of antibody molecule may explain a 100-fold difference in the number of serotypes among picornavirus genera. Thus, the second proposal is unlikely to be correct.

The difference among classification assays argument does not have an easy response. Indeed, there is no universal procedure used to classify viruses serologically, and therefore, strictly speaking, there is the possibility that a different number of serotypes could be obtained using alternative classification procedures.

FMDV is a pertinent example. Its seven serotypes are defined on the basis of a very stringent test that cannot be performed with human viruses for obvious reasons: the absence of cross-protection resulting from vaccination or infection with a given FMDV. Infection or vaccination with FMDV of one serotype does not confer protection against FMDV of a different serotype. In contrast, the subtype classification of FMDV was based on serological assays, such as cross-neutralization or complement fixation tests, usually using sera raised in guinea pigs. These assays allowed classification of FMDV in more than 65 serological subtypes. Subtyping was stopped when it was realized that using increasingly discriminatory assays such as reactivity with MAbs, virtually any new isolate could define a new subtype [(Mateu et al., 1988); see (Domingo et al., 1990; Sobrino and Domingo, 2004; Mateu, 2017) for review of serotype and subtype classification of FMDV]. Despite these considerations, it is unlikely that serological assays using in vitro tests would be responsible for a 100-fold difference between two human pathogens such as HRV and HAV. Thus, it does not seem justified to attribute antigenic constancy to an artifact derived from diagnostic procedures.

More extensive virus circulation will favor genetic and antigenic diversification, and a single serotype may evolve into multiple serotypes. What we describe today as the antigenic profile of virus groups is a snapshot of an evolving process. Genotype differentiation is actually being witnessed during the expansion of HCV pandemics, partly due to a true genetic diversification of the virus as it circulated over the last decades, and partly due to increasing capacity of virus surveillance, and molecular and phylogenetic tools for genome analysis. The reader can find an illustration of this point by comparing the expanded phylogenetic HCV tree from six to seven genotypes and the subtype ramifications, published by P. Simmonds and colleagues in 1993 and 2014 [compare (Simmonds et al., 1993) and (Smith et al., 2014)], and continuing new subtype identification (Hedskog et al., 2019). Although it cannot be excluded that time might tend to equalize the number of serotypes among viruses, current evidence does not justify blaming differences in the extent of virus circulation to settle this issue.

We come to constraints at antigenic sites that limit the number of accepted amino acid substitutions as a model for antigenic invariance. It is the preferred model of molecular virologists. The initial concept was proposed by M.G. Rossmann, in his canyon hypothesis (Rossmann, 1989), based on studies with HRV14. A canyon in the virus preserves the receptor-binding site inside while permitting amino acid substitutions that affect antigenicity, without consequences for receptor recognition. A physical and functional separation between receptor and antibody binding allows extensive antigenic variation. Clearly, in many viruses, there is an overlap between antigenic and receptor recognition sites (Section 4.5 in Chapter 4), that could limit antigenic variation. A difference in constraints is also supported by a structural comparison carried out by J.M. Casasnovas and his colleagues of the interaction of PV and HRV16 with their respective cellular receptors that revealed a receptor-binding site more accessible in PV than in HVR16, rendering the latter suited to escape antibody neutralization (Xing et al., 2000). This would render HRV a picornavirus prone to antigenic variation, as indeed found in nature. Thus, constraints imposed by the requirement to interact with the cellular receptor may explain the limited capacity for antigenic diversification, and perhaps with the contribution of other influences, the puzzle of widely different antigenic types despite similarly high genome mutability. Additional structural and functional studies with viruses of different families are necessary to substantiate this proposal.

## 7.4.2 Similar frequencies of monoclonal antibody-escape mutants in viruses differing in antigenic diversity

Fitness cost is expected to limit antibody escape (Louie et al., 2018) but this may be reflected in mutant stability but not in the frequency of MAb-escape mutants [monoclonal antibody-resistant mutant (MARM) frequencies] in laboratory experiments. Comparison of MARM frequencies of different viruses shows comparable values independently of the number of circulating serotypes of the virus (Table 7.2). In particular, the cardiovirus Mengo virus (one serotype) displays similar MAR frequencies than HRV (100 serotypes). In fact, none of the RNA and DNA viruses listed in Table 7.2 deviate from a broad range of MARM frequencies of $10^{-3}$ to $10^{-5}$, except for substitutions at some discontinuous epitopes of FMDV (Lea et al., 1994). In some studies, the stability of the selected escape mutants was tested after a few passages in cell culture, but in other studies, the lack of reversion of the antigenic change was not ascertained. Two

FMDV escape mutants showed a selective disadvantage over the parental wild-type virus (fitness decrease); upon continued replication, the mutants acquired fitness-enhancing mutations without reversion of the antigenic change (Martínez et al., 1991a).

Unless the escape mutations are selectively neutral, the expectation is that MARM frequencies may be an underestimate of the real rate at which the amino acid substitutions occur. Thus, it is possible that following selection by an antibody, some mutants may decrease in frequency due to a fitness cost, or that their level is maintained due to additional compensatory mutations acquired by the replicating genomes (Fig. 7.6). Viruses that are highly constrained for antigenic variation may be diagnosed through fitness decrease of MARM mutants despite their occurring at similar rates as those that affect unconstrained sites. This is a concept similar to the distinction between fitness and function that we made in Section 5.8 of Chapter 5. That is, the occurrence of an antigenic change

**TABLE 7.2**    Frequency of monoclonal antibody-resistant mutants (MARMs) for some viruses.

| Virus | Monoclonal antibody-resistant mutant (MARM) frequencies | References |
|---|---|---|
| Poliovirus | $10^{-4}$ to $10^{-5}$ | Emini et al. (1982) and Minor et al. (1983, 1986) |
| Mengovirus | $3 \times 10^{-3}$ to $5 \times 10^{-5}$ | Boege et al. (1991) |
| Foot-and-mouth disease virus | $10^{-4}$ to $10^{-5}$ (continuous epitopes) | Martínez et al. (1991a) |
|  | $10^{-4}$ to $10^{-7}$ (discontinuous epitopes) | Lea et al. (1994) |
| Rhinovirus | $10^{-4}$ to $10^{-5}$ | Sherry et al. (1986) |
| Hepatitis A virus | $3 \times 10^{-3}$ | Stapleton and Lemon (1987) |
| Vesicular stomatitis virus | $0.5 \times 10^{-4}$ to $1 \times 10^{-4}$ | Holland et al. (1990) |
| Rabies virus | $10^{-4}$ | Wiktor and Koprowski (1980) |
| Measles virus | $9 \times 10^{-5}$ | Schrag et al. (1999) |
| Sindbis virus | $10^{-3}$ to $10^{-5}$ | Stec et al. (1986) |
| Canine parvovirus | $10^{-3.4}$ to $10^{-5.4}$ | Smith and Inglis (1987) |
| Herpes simplex virus | $1 \times 10^{-5}$ | Smith and Inglis (1987) |

7.5 Comparing viral genomes. Sequence alignments and databases

241

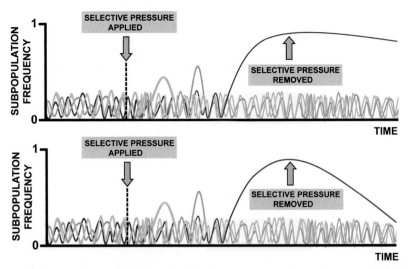

**FIGURE 7.6** Stability of selected mutant subpopulations when selective pressure is removed. This scheme is the same portrayed in Fig. 3.6 of Chapter 3, but shifted toward the right of the time scale. Five different colors have been chosen to depict fluctuations of four genomic classes. In a real population, thousands of genomes may be involved in each infected cell. In the present diagram, the red line represents genomes selected for their resistance to neutralizing monoclonal or polyclonal antibodies. Once the antibody pressure is removed, the mutant genomes may remain dominant either because the relevant substitutions do not affect viral fitness or because compensatory mutations have been acquired (top diagram). In contrast, upon removal of the antibody pressure, the proportion of selected genomes may fade with time due to a fitness cost and absence of compensatory mutations (bottom diagram). In the latter case, the antibody-resistant mutant will not contribute to the long-term antigenic diversification of the virus.

does not guarantee that the change will be perpetuated in nature and contribute to natural antigenic diversification. Again, fitness should be considered as a relevant parameter, and fitness effects on antigenic stability have been largely unexplored.

## 7.5 Comparing viral genomes. Sequence alignments and databases

The likely multiple origins of viruses, followed by extended events of interaction with evolving host organisms of all phyla have produced myriad viral particles that at the present time outnumber cells by a factor of 10 (Chapter 1). The way to put order into such diversity is to classify viruses as done periodically by the International Committee on Taxonomy of Viruses (ICTV) (http://www.ictvonline.org/).

Computational procedures developed to study phylogenetic relationships in evolutionary biology are routinely applied to virology to establish relationships among closely or distantly related viruses (Page and Holmes, 1998; Hall, 2001; Felsenstein, 2004; Salemi and Vandamme, 2004; Yang, 2006; Russell, 2014). No phylogenetic tree that connects the viruses that have been characterized to date can be derived in a reliable way, not even a tree for DNA or for RNA viruses. What we can do is to produce trees for related viruses that probably share a common ancestor. Many data banks are available for viruses to retrieve sequences for comparison with new isolates. Despite the fact that data banks are periodically updated, some are listed in Table 7.3, and can serve as the starting point to reach the desired uniform resource locator (URL) to implement a procedure for genome characterization. Prior to any

**TABLE 7.3** Information on nucleotide and amino acid sequence alignment programs, Data Banks, and phylogenetic procedures.

| Identification | URL (or reference) | Contents |
|---|---|---|
| EMBL Nucleotide Sequence Database | http://www.ebi.ac.uk/ena | All reported sequences. General database |
| GenBank, the NIH Genetic Sequence Database | http://www.ncbi.nlm.nih.gov/genbank | All reported sequences. General database. |
| DNA Data Bank of Japan | http://www.ddbj.nig.ac.jp | All reported sequences. General database. |
| UNIPROT | http://www.ebi.ac.uk/uniprot | Protein sequences. |
| Protein Data Bank (PDB) | http://www.rcsb.org/ | Protein structure data. |
| Virus Particle Explorer | http://viperdb.scripps.edu/ | Virus structures and structure-derived properties: capsid interactions, residue contributions to protein-protein interactions. Links to sequences and taxonomic information. |
| Viral Genomes Project | http://www.ncbi.nlm.nih.gov/genome/viruses | Complete or nearly complete viral genome sequences. Additional information. Includes Pairwise Seqeunce Comparisons (PASC) within viral families. |
| The Influenza Sequence Database | http://www.fludb.org/ | Sequences, tools for the analysis of hemagglutinin and neuraminidase sequences. |
| Picornavirus Sequence Database | http://www.viprbrc.org/brc/home.spg?decorator=picorna | Sequence and specific references for different picornavirus isolates. |
| Plant viruses | http://www.dpvweb.net | Description of Plant Viruses (DPV). Expanding data bank of viruses, viroids and satellites of plants, fungi and protozoa. |
| Potyvirus Database | http://www.danforthcenter.org/iltab/potyviridae/ | Taxonomy, references, and sequence databases of members of the *Potyviridae* family. |
| Calicivirus Sequence Database | http://www.viprbrc.org/brc/home.spg?decorator=calici | Sequences, information, and specific references for different calicivirus isolates. |
| HPV Sequence Database | http://pave.niaid.nih.gov/ | Human papilomavirus, sequences, analysis, and alignment tools. |
| HIV Sequences Database | http://www.hiv.lanl.gov<br>http://www.hiv.lanl.gov/content/sequence/RESDB/ | Sequences, drug resistance. Molecular immunology and vaccine trials. Analysis tolls. |
| HIV Drug Resistance Database | http://hivdb.stanford.edu/ | Sequence, correlations genotype-phenotype, and genotype-antiretroviral treatment. Sequence analysis tools. |
| Hepatitis C virus Database | http://hcv.lanl.gov/<br>http://www.hcvdb.org | Sequences and genome analysis tools. |
| Hepatitis virus Database | http://s2as02.genes.nic.ac.jp/ | Hepatitis B, C and E virus sequences. |

7.5 Comparing viral genomes. Sequence alignments and databases

243

**TABLE 7.3** Information on nucleotide and amino acid sequence alignment programs, Data Banks, and phylogenetic procedures.—cont'd

| Identification | URL (or reference) | Contents |
|---|---|---|
| Poxvirus Bioinformatics Research Center | http://www.poxvirus.org/ | Poxvirus genomes. |
| Viral Bioinformatics Resource Center | http://athena.bioc.uvic.ca/ | Large DNA viruses (Poxviruses, African Swine Fever Viruses, Iridoviruses, Baculoviruses). |
| Human Endogenous Retroviruses Database | http://herv.img.cas.cz | Human endogenous retroviruses, and genome analysis tools. |
| VIDA, Virus Database at University College London | http://www.biochem.ucl.ac.uk/bsm/virus_database/VIDA.html | Homologous protein families from herpes, pox, papilloma, corona and arteriviruses. |
| Subviral RNA Database | http://subviral.med.uottawa.ca/cgi-bin/home.cgi | Sequences and prediction of RNA secondary structures. |
| VirOligo Compilation Lab | http://viroligo.okstate.edu/ | Database of oligonucleotides used in virus detection and identification. Technical information and several links to original sequence information. |
| ViPR Virus Pathogen Resource | http://viprbrc.org/ | Sequences of multiple virus families. |
| Chromas | http://technelysium.com.au/?page_id=13 | Software for DNA sequencing. |
| FORMAT CONVERSION | http://hcv.lanl.gov/content/sequence/FORMAT_CONVERSION/form.html | Program that converts the sequence(s) to a different user-specified format. |
| BEAST, Tracer | http://beast.bio.ed.ac.uk/Tracer | Phylogenetic inferences. Bayesian methods. |
| MODELTEST | http://darwin.uvigo.es/software/modeltest.html | Determination of nucleotide substitution model. Phylogenetic derivations. |
| PHYLIP package | http://evolution.genetics.washington.edu/phylip.html | Programs for inferring phylogenies. |
| PAUP | http://paup.csit.fsu.edu/ | Software for inference of evolutionary trees. |
| BiBiServ | http://bibiserv.techfak.uni-bielefeld.de/splits | Bielefeld Bioinformatics Service. |
| Bowtie2 | http://bowtie-bio.sourceforge.net/bowtie2/index.shtml | It is a tool for aligning sequencing reads to long reference sequences. |
| SAM tools (Sequence Alignment/Map) | http://samtools.sourceforge.net/ | A generic format for storing large nucleotide sequence alignments. |
| Free Bayes | https://github.com/ekg/freebayes | FreeBayes is a Bayesian genetic variant detector designed to find small polymorphisms. |
| EMBOSS | http://www.ebi.ac.uk/Tools/emboss/ | Sequence analysis. |
| T-Coffee | http://www.tcoffee.org/ | Tools for Computing, Evaluating, and Manipulating Multiple Alignments. |

*(Continued)*

**TABLE 7.3**　Information on nucleotide and amino acid sequence alignment programs, Data Banks, and phylogenetic procedures.—cont'd

| Identification | URL (or reference) | Contents |
| --- | --- | --- |
| IGV, Integrative Genomics Viewer | http://www.broadinstitute.org/igv/ | The Integrative Genomics Viewer (IGV) is a high-performance visualization tool for interactive exploration of large, integrated genomic datasets. It supports a wide variety of data types, including array-based and next-generation sequence data, and genomic annotations. |
| CLC Genomics Workbench | https://www.qiagenbioinformatics.com/ | Analysis of deep sequencing data |
| RDP4, version 4.36 beta | P.D. Martin et al. Virus Evol. 1, vev003, 2015 | Detection of recombination |
| Dintor | https://dintor.eurac.edu/ | Tools for the analysis of genomic and proteomic data sets |
| Geneious Basic | https://www.geneious.com/ | Organization of biological data, including nucleotide and amino acid sequences |
| G-large—INS—1 in MAFFT | https://mafft.cbrc.jp/alignment/software/mpi.html | Large-scale sequence alignments |
| WIT | http://www.algorithm-skg.com/wit/home.html | Alignment of large number of reads obtained by deep sequencing |
| MEGA 7 | https://www.megasoftware.net/ | Extended MEGA version for the analysis of large data sets |
| Quasispecies Diversity | https://www.bioconductor.org/packages/release/bioc/html/QSutils.html | Set of utility functions for viral quasispecies analysis with deep sequencing data |
| Meta PGN | https://github.com/peng-ye/Meta PGN | Visualization of pangenome networks |

comparative study of nucleotide or amino acid sequences (not only to establish phylogenetic relationships, but also to calculate genetic distances, to identify regulatory regions, functional domains, and structural motifs, to design oligonucleotide primers for amplification, or other applications) it is essential to align sequences accurately, and some programs for sequence alignments are also given in Table 7.3, including processing of large data sets obtained by deep sequencing (Kearse et al., 2012; Kumar et al., 2016, 2019; Lai and Verma, 2017; Nakamura et al., 2018), and visualization of pangenome networks (Peng et al., 2018).

Databases differ in format and contents, which may include prediction of traits derived from sequence information (RNA secondary structures, antiviral drug sensitivity levels, assignments to homologous protein families, etc.). Some of them offer a link with the *web* page of the ICTV, thus providing background information to assign newly determined sequences to current taxonomic groups. A structure-based amino acid sequence alignment of protein homologs can be carried out based on three-dimensional structures of proteins. Such types of amino acid sequence alignments may help in the identification of relevant

7.6 Phylogenetic relationships among viruses. Evolutionary models

**245**

structural and functional motifs. Sequence variability among a set of aligned sequences can be quantitated by the number of variable sites, mean pairwise diversity, mutation frequency, and other estimators (i.e., the Watterson's estimator) (Page and Holmes, 1998; Mount, 2004; Salemi and Vandamme, 2004) (see also Chapter 3 for parameters used to quantify mutant spectrum complexity).

Relevant information on protein evolution can be derived from alignment of the protein sequence of related viruses (or of isolates from one virus, or for components of the same mutant spectrum) and analyzing the statistical acceptability of the divergent amino acids at each position (Feng and Doolittle, 1996) with consideration of amino acid location when a three-dimensional structure is known (Farheen et al., 2017). Statistical acceptability derives from the chemical nature and shape of the amino acid side chains, their structural context, and also the limitations that the genetic code imposes on amino acid replacements (Porto et al., 2005). The basic assumption is that the more conserved the amino acid sequences, and the more similar are the variant amino acids, it is more likely that the proteins are derived from a common ancestor. M. Dayhoff pioneered the early comparison of protein sequences establishing a protein information resource (PIR) in the middle of the 20th century. Tables named PAM (percent accepted mutation) were constructed, and several evolved versions such as BLOSUM matrices, based on the BLOCKS database, are used to compare protein sequences. The BLOSUM62 amino acid substitution matrix groups amino acids according to their chemical structure and provides a probability of occurrence of each amino acid replacement: zero, amino acid replacement expected by chance; positive number, replacement found more often than by chance; and negative number, replacement found less often than by chance.

## 7.6 Phylogenetic relationships among viruses. Evolutionary models

The URLs listed in Table 7.3 give access to computational analyses that allow sequence alignments and derivation of phylogenetic trees, which are extremely informative of middle- and long-term evolutionary change of viruses (Page and Holmes, 1998; Notredame et al., 2000; Mount, 2004; Salemi and Vandamme, 2004; Holmes, 2008, 2009). Application of phylogenetic methods to virus evolution requires careful consideration of the evolutionary models to be used, including probabilities of the different types of nucleotide and amino acid replacements, and the rates at which they may occur. Statistical methods (i.e., likelihood ratio tests) are available to select an adequate model for a given data set (Salemi and Vandamme, 2004). At the nucleotide sequence level, it is often assumed that when transitions are more frequent than transversions in a set of related sequences, no saturation of mutation took place. In contrast, when transversions are more frequent than transitions, saturation is presumed (Xia and Xie, 2001). Parameter $\alpha$ applied to amino acid sequence alignments (e.g., using the program AAml from PAML package, version 3.14) takes into account multiple amino acid replacements per site, as well as unequal substitution rates among sites (Yang et al., 2000). Parameter $\alpha$ can be calculated using the amino acid replacement matrix WAG available in the program MODELTEST (Posada and Crandall, 1998). Despite their obvious utility, it is unlikely that these statistical procedures which were developed on the assumption of successions of defined sequences (rather than mutant clouds) can capture the complexities underlying long-term evolution of viruses in nature.

Phylogenetic reconstructions based on nucleotide (and deduced amino acid) sequence alignments are generally possible with selected genes of relatively close viruses (i.e., that belong to the same family). The main methods used to derive evolutionary trees are: maximum parsimony, distance, ML, Bayesian methods of phylogenetic inference, and splits-tree analysis [reviewed in (Eigen, 1992; Page and Holmes, 1998; Mount, 2004; Salemi and Vandamme, 2004; Sullivan, 2005; Holmes, 2008; Beale et al., 2018; Geoghegan and Holmes, 2018)] (Table 7.3).

Maximum parsimony predicts the minimal mutation steps needed to produce the observed sequences from ancestor sequences. It is most suitable for closely related sequences. Often, all possible trees are examined before a consensus tree is produced, and, therefore, the method is time-consuming. Most programs based on maximum parsimony assume the operation of a molecular clock, with the limitations that were discussed in Section 7.3.

Distance methods are based on the calculation of genetic distances between any two sequences of a multiple sequence alignment. Large genetic distances require a correction for multiple mutational steps (i.e., Kimura 2-parameter distance). Most distance methods can handle large numbers of sequences, and results are relatively reliable even when a molecular clock does not operate. Commonly applied distance methods include neighbor-joining (NJ) (that does not assume a molecular clock and yields an unrooted tree), several variant versions of NJ, and the unweighted pair group method with arithmetic mean (UPGMA, a clustering method that assumes a molecular clock and produces a rooted tree). The software package TREECON was developed to derive NJ trees.

ML methods use probability calculations to derive a branching pattern from the mutations at different positions of the nucleic acids under study. They can estimate both distances and the most accurate mutational pathway between sequences. Generally, supercomputers are needed when many sequences are compared since all possible trees are examined. ML methods are included in several programs listed

in Table 7.3. Bayesian methods (based on conditional probabilities derived by Baye's rule) (Huelsenbeck et al., 2001; Ronquist and Huelsenbeck, 2003; Huelsenbeck and Dyer, 2004; Tonkin-Hill et al., 2019) have the advantage of increased speed of data processing, but they still require time to avoid incorrect inferences.

Splits-tree procedures are based on split-decomposition theory or statistical geometry, and they provide a geometrical representation of the distance relationships in sequence space (Eigen, 1992; Dopazo et al., 1993; Salemi and Vandamme, 2004) (Chapter 3). The procedure has been used to analyze rapidly evolving viral sequences (http://bibiserv.techfak.uni-bielefeld.de/splits/); methods that allow the inclusion of insertions and deletions have been adapted to the splits-tree program (Cheynier et al., 2001). Phylogenetic trees can be presented as rooted trees (with a reference out-group) and unrooted trees.

When possible it is advisable to apply different phylogenetic procedures to compare tree topologies. Resampling methods (i.e., bootstrapping, jackknifing, etc.) are used to assess the statistical reliability of the trees (Page and Holmes, 1998; Salemi et al., 1998; Mount, 2004; Salemi and Vandamme, 2004). A tree defines clades or lineages of a virus attending to groupings by relatedness. Different tree topologies can be obtained when analyzing different genes of the same virus set. Discordant phylogenetic positions of two different genes of the same virus are suggestive of recombination that should be evaluated statistically (Worobey, 2001; Salemi and Vandamme, 2004; Martin et al., 2005). Recombination is frequent, and in some viruses recombination is intimately linked to the replication mechanism (Chapters 2 and 10).

The more conserved genes (i.e., those encoding the polymerase or other nonstructural proteins) may permit the establishment of phylogenetic relationships among some distant virus groups. Examples are the clustering of a number of animal and plant RNA viruses as supergroups (Morse, 1994). Families of DNA-dependent DNA polymerases group some bacterial and bacteriophage DNA polymerases with some eukaryotic polymerases (Morse, 1994; Villarreal, 2005), in support of the active exchange of modules during coevolution of viruses and their hosts (Botstein, 1980, 1981; Zimmern, 1988). In contrast to conserved genes, variable genes (typically, capsid proteins and surface glycoproteins) serve to establish short-term evolutionary relationships within the same virus group, including the survey of virus variation during outbreaks, epidemics, and pandemics (Gorman et al., 1992; Martínez et al., 1992; Morse, 1994; Gavrilin et al., 2000).

Distantly related viruses, with no discernible nucleotide or amino acid sequence identity, can sometimes be grouped on the basis of the three-dimensional structures of viral proteins. The evolutionary trace (ET) clustering method combines phylogenetic partition of sequences with structural information (Chakravarty et al., 2005), and it may help in identifying functionally relevant domains shared by divergent isolates in particular highly variable capsid and surface viral proteins. ET can be applied to proteins and nucleic acids, and its clustering features may reveal conserved structures that are overlooked when all sequences are compared together. As explained in Chapter 1, the great diversity of amino acid sequences recorded among viral structural proteins (several URL links in Table 7.3) are actually reduced to a limited number of morphotypes at the structural level. In another approach, the probabilities of equivalence between pairs of residues in viral proteins are converted into evolutionary distances (Bamford et al., 2005; Ravantti et al., 2013). The structure-based classification has grouped the coat protein of icosahedral viruses in separate classes, each of which, interestingly, embraces different domains of life (Archaea, Bacteria, and Eukarya) (McTavish et al., 2017). A lineage of structurally related viruses includes tailed bacteriophages and the herpesviruses,

suggesting that parts of the genomes of complex viruses may have a very ancient origin. They might have belonged to viruses that infected primitive cells before the latter diverged into the domains of life that we identify in our biosphere (Bamford et al., 2005; Villarreal, 2005) (compare with models of virus origins in Chapter 1).

Viral clades may cluster with clades of their host species, suggesting either virus-host coadaptation or an extended parasite-host relationship, with limited possibilities of jumping the host barrier (Section 7.7). Hantaviruses and their rodent hosts (Plyusnin and Morzunov, 2001), lyssaviruses and bat species, spumaviruses and their primate hosts, and herpesviruses and their vertebrate hosts, are some among other examples of long-term host-virus coevolution (McGeoch and Davison, 1999; Woolhouse et al., 2002; Switzer et al., 2005; Voskarides et al., 2018). [See, however, a discussion on time scale discrepancies of coevolutionary rates (Sharp and Simmonds, 2011), and compare with section 7.3.3].

## 7.7 Extinction, survival, and emergence of viral pathogens. Back to the mutant clouds

The viral groups defined by phylogenetic methods may or may not occupy a defined geographical location. It will depend on whether viral vectors or infected individuals carry the virus over long distances or not. A defined phylogenetic group may include viruses that produce similar or different pathology. This is because the capacity of a virus to cause disease may depend on modest genetic change (i.e., one or a few amino acid substitutions) that does not alter its position in a phylogenetic tree. It is important to emphasize that, independently of the time frame considered, the tips of phylogenetic trees are a cloud of mutants, that genomes within the cloud are the origin of future diversification pathways, and that individual

**FIGURE 7.7** Diversification of HIV-1 from the time of introduction into the human population of retroviral simian ancestors SIVcpz from chimpanzees. Group M diversified into at least nine subtypes plus about 100 circulating recombinant forms (CRF), and multitudes of unique recombinant forms (URFs) that have not reached epidemiological relevance (box on the right). Genetic and antigenic diversifications are discussed in the text.

cloud components may differ in pathogenic potential. Fig. 7.7 summarizes the diversification of HIV-1, since it entered the human population. Once HIV-1 originated from multiple introductions of a chimpanzee simian immunodeficiency virus (SIVcpz), the four major HIV-1 groups M, O, N, and P were generated, and group M evolved into the multiple subtypes and recombinant forms that circulate at present. Many factors determine the pathogenic potential of any of the HIV-1 subtypes and the newly arising recombinant forms. HIV-1 is a notorious case of successful emergence of a new viral pathogen from a zoonotic reservoir of a related virus. HIV-1 populations are extremely complex and in continuous evolution (Hamelaar et al., 2019).

The relevance of the mutant cloud in determining viral fitness and survival was documented by comparing five isolates of West Nile virus (WNV) that had identical consensus sequences and differed in the mutant spectrum, as analyzed by dep sequencing (Kortenhoeven et al., 2015) (Fig. 7.8). The study concerned a WNV lineage 2 that circulated in Europe during the beginning of the 21st century. Environmental changes modified the haplotype composition while maintaining an invariant consensus sequence, an example of

"perturbation" manifested only at the level of the mutant spectrum (see Section 6 in Chapter 6).

There is evidence that some viruses that once produced human disease might be now extinct. One example is provided by a putative viral agent of Economo's disease (also termed lethargic encephalitis or epidemic encephalitis), a degenerative disease of the brain that produced a loss of neurons. The disease had an acute phase of variable duration and intensity, followed by a chronic phase, sometimes with a late onset of symptoms. The disease showed a seasonal character with a maximum incidence in late winter. The first cases were recorded in Eastern Europe in 1915, and the disease was first described by Baron C. Von Economo in Vienna in 1917. During 1920–23 the disease attained pandemic proportions, although the number of cases and mortality were limited. It was estimated that between 1917 and 1929, about 100,000 cases occurred in Germany and Great Britain, and then mysteriously, the number of cases decreased, and the disease disappeared (Ford, 1937). At the time it was suspected that a virus similar to IV or some picornavirus might have been the etiological agent of this disease, but no proof could be provided. Occasional cases of lethargic encephalitis are diagnosed, and there is evidence of a possible connection of the once epidemic disease agent with postencephalitic Parkinsonism (Vilensky et al., 2010; Bigman and Bobrin, 2018).

There are additional examples of disease incidence decline that may be the prelude of virus extinction in some geographical areas. The alphavirus western equine encephalitis virus (WEEV) was an important human and animal pathogen in the Americas during the first part of the 20th century. The number of isolations of WEEV both from vertebrate and insect species declined during the second half of the 20th century, and two main mechanisms have been proposed: (i) fitness decrease due to operation of Muller's ratchet (concept described in Section 6.5 of Chapter 6), with elimination of

7.7 Extinction, survival, and emergence of viral pathogens. Back to the mutant clouds

249

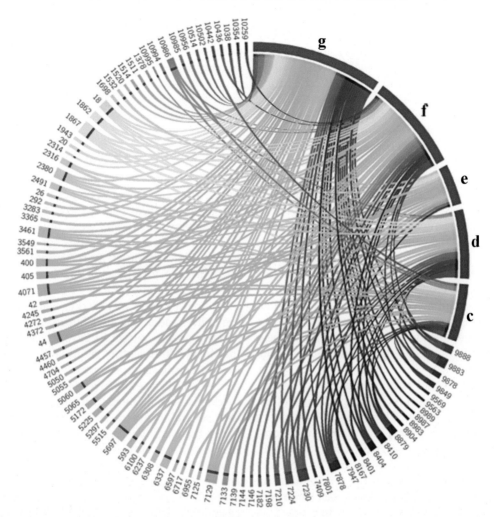

**FIGURE 7.8** Visualization of the complexity of mutant spectra (haplotype composition) of five isolates of WNV denoted by c, d, e, f, and g (magenta bars) that have identical consensus sequence. The color lines connect the genomes where the same single nucleotide change occurs. Similarly, color-coded ribbons indicate that the same mutation occurs in two genomes in the same position. Nucleotide positions are numbered next to the outer rim of the circle. *Figure reproduced from Kortenhoeven et al. (2015). BMC Genomics is an open-access journal, and the article can be reproduced under the terms of the Creative Commons Attribution. The figure has been reproduced with the permission of the authors.*

virulence-associated mutations by genetic drift, or (ii) ecological factors such as decline of population numbers of susceptible host species, combined with increased vaccine coverage in horses (Bergren et al., 2014). It is not clear which is the major mechanism behind the decline, and whether it represents a firm, irreversible trend,

or genetic and ecological factors may vary to produce a reemergence of this virus as a pathogen.

FMDV, the agent of the economically most important disease of cattle and other farm animals circulated until recently as seven different serotypes termed A, O, C, Asia 1, SAT1, SAT2, and SAT3, and each serotype as multiple sub-types and antigenic variants (review in Sobrino and Domingo, 2004). Interestingly, in the 1980s, the incidence of serotype C FMDV decreased to the point that at the beginning of the 21st century this FMDV serotypes was considered nearly extinct and its eradication feasible. It cannot be totally excluded, however, that type C FMDV is replicating in some persistently infected rumi-nant in some remote part of our planet and that the virus reemerges again. If not, its ecological niche has been occupied by FMDVs of other serotypes. This is one important issue behind virus eradication (smallpox in the late 1970s or rinderpest in 2011): the possibility that the niche left by an eradicated pathogen is occupied by a related pathogen. A.E. Gorbalenya, E. Wimmer, and colleagues examined the possible evolu-tionary origin of present-day PV, and that other picornaviruses might occupy the PV niche in the event of its eradication (Jiang et al., 2007). Their phylogenetic analysis suggests that PV could originate from a C-cluster coxsackie A virus through amino acid substitutions in the capsid that led to a change of receptor specificity (other cases are discussed in Chapter 4). They gener-ated chimeras of PV and its putative ancestors, and some of them were viable and pathogenic for transgenic mice expressing the PV receptor. The authors suggest that in a world without antiPV neutralizing antibodies, coxsackieviruses may mutate to generate a new PV-like agent.

Thus, despite virology being a very recent scientific discipline, there is ample evidence of the emergence of new viral pathogens, as well as some cases of extinction due to human inter-ventions, and possible extinctions by natural influences. Viruses may evolve with regard to the symptoms they inflict upon their hosts. An increase of severity of Dengue virus infection has been observed in some world areas, consist-ing of neurological manifestations in patients with dengue fever or dengue hemorrhagic fever (Cam et al., 2001), among other examples of human and veterinary viral diseases. The

dynamics of extinction of mutant viruses and their replacement by other forms is a continuous process, as the cycles of birth-death for any organism, but in a highly accelerated fashion.

We now turn to the pressing problem of the emergence and reemergence of viral disease.

### 7.7.1 Factors in viral emergence

New human viral pathogens emerge or ree-merge at a rate of about one per year, represent-ing an important concern for public health, alerting established and new organizations of the need of preparedness (Brechot et al., 2018). Emergence is defined as the appearance of a new pathogen for a host, while reemergence often refers to the reappearance of a viral pathogen, following a period of absence. Being a popular topic, the reader will find numerous books and reviews on the subject. It is worth emphasizing that in the 20th century many authors took the lead in emphasizing the problem of viral emer-gences and the need to investigate the underlying mechanisms, notably S.S. Morse and J. Lederberg [see several chapters of Morse (1993, 1994)]. Given the adaptive capacity of viruses, in partic-ular, the RNA viruses, the reader will certainly suspect that genetic variation of viruses must be one of the factors involved in viral emergences. Indeed, most of the high-impact new viral dis-eases recorded recently or historically are due to RNA viruses. A statement by J. Lederberg reflects our vulnerability in the face of the nearly unlim-ited potential of viruses to vary: "Abundant sour-ces of genetic variation exist for viruses to learn new tricks, not necessarily confined to what hap-pens routinely or even frequently" (Lederberg, 1993). The situation is even more complex because a genetic variation of viruses is only one of many ingredients that promote the intro-duction of new viral pathogens in the human population. A report issued by the U.S. Institute of Medicine in 2003, analyzed and documented 13 factors that, individually or in combination, participate in the emergence of microbial disease. They include a number of sociological, environmental, and ecological influences that act to promote the emergence and reemergence of viruses, bacteria, fungi, and protozoa (Smolinski et al., 2003) (Box 7.2).

Here, we will deal briefly with those factors of viral emergence related to the virus and host population numbers, in line with the focus of this book. Other aspects have been covered elsewhere (Antia et al., 2003; Haagmans et al., 2009; Wang and Crameri, 2014; Lipkin and Anthony, 2015; among others). The emergence of a viral disease can be regarded as a consequence of virus adaptation to a new environment, therefore, involving the concepts and mechanisms dissected in previous chapters. In particular, a relevant parameter is the variation of viral fitness in different environments (Domingo, 2010; Wargo and Kurath, 2011; Domingo et al., 2019).

Fitness can directly or indirectly impact any of the three steps involved in viral disease emergence or reemergence, which can be summarized as follows:

- Introduction of a virus into a new host species.
- Establishment (replication) of the virus in the new host.

- Dissemination of the virus among individuals of the new host species (transmissibility) to produce outbreaks, epidemics, or pandemics.

For the introduction and establishment steps, replicative fitness is critical while for the dissemination step, epidemiological fitness plays a major role (Chapter 5).

Two population numbers are relevant to the establishment step: the number of infectious viral particles shed by the infected donor host, and the number of potential new hosts that come into contact with the infected donor. We are now aware that even if two viral populations shed by an infected host have an identical number of infectious particles, not all mutant spectra might have the genomes subpopulations to permit the establishment in the human host (Fig. 7.9). There is a natural lottery regarding which quasispecies subpopulations will hit which host. In the words of J.J. Holland and his colleagues: "Although new RNA virus diseases of humans will continue to emerge at indeterminate intervals, the viruses themselves will not really be new, but rather mutated and rearranged to allow infection of new hosts, or to cause new disease patterns. It is important to remember that every quasispecies genome swarm in an infected individual is

---

BOX 7.2

## FACTORS IN THE EMERGENCE OF MICROBIAL DISEASE

- Microbial change and adaptation.
- Human susceptibility to infection: impaired host immunity and malnutrition.
- Climate and weather.
- Changing ecosystems: vector ecology; reservoir abundance; and distribution.
- Human demographics and behavior: population growth; aging; and urbanization.
- Economic development and land use.
- International travel and commerce.
- Technology and industry.

- Breakdown of public health measures.
- Poverty and social inequality.
- War and famine.
- Lack of political will.
- Intent to harm: bioterrorism and agroterrorism.

*Points summarized from Smolinski, M.S., Hamburg, M.A., Lederberg, J., 2003. Microbial Threats to Health, Emergence, Detection and Response. The National Academies Press, Washington, DC.*

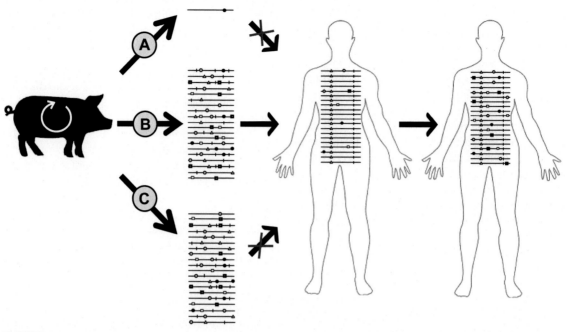

**FIGURE 7.9** Relevance of virus population size and mutant spectrum composition in the zoonotic transmission of a virus. Only a subset of the genomes that surface an infected pig may be able to establish an infection in humans. The scheme indicates that a single genome that reached a human was not adequate to establish infection (pathway A). When multiple genomes reached the human (pathways B and C), only those that included a subset that displayed a minimum fitness in humans were able to initiate an infection and expand in the new host (pathway B). For pathways A and C, events are as if the contact between donor and recipient host had not taken place (arrows with cross). See text for implications.

unique and 'new' in the sense that no identical population of RNA genomes has ever existed before and none such will ever exist again" (Holland et al., 1992).

The number of infectious particles shed by an infected host determines the probability of transmission to susceptible hosts (Section 7.2), and of producing an emergence in a new host species. Viral population numbers and the number of transmissible particles can be largely amplified in immunocompromised individuals, and have been termed super-spreaders; they can contribute large amounts of variant viruses to the transmission lottery (Rocha et al., 1991; Paunio et al., 1998; Gavrilin et al., 2000; Khetsuriani et al., 2003; Small et al., 2006; Odoom et al., 2008; Woolhouse, 2017). Concerning the recipient hosts, the higher the number of potentially susceptible hosts that

come into contact with an infected donor, the higher the probability of establishment of an emergent infection. It is likely that the advent of agricultural practices some 10,000 years ago, combined with increased contacts between humans and animals, inaugurated a time of new viral emergences. In the new scenario, viruses could shift from a persistent (low interhost transmission) mode into an acute (high interhost transmission) infection mode.

Not only population numbers are important, but the connections between the spatial habitats of a potential donor and recipient hosts are also highly relevant (Fig. 7.10). As correctly emphasized by S.S. Morse, changes in viral traffic may allow viruses to come near potential new hosts that had never been encountered before. Several sociological and ecological factors that

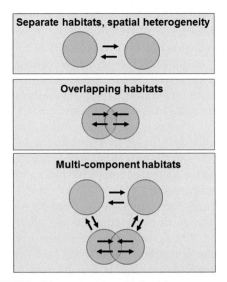

**FIGURE 7.10** Types of habitats that may limit or facilitate interaction between hosts that can potentially establish an emergent infection in a local habitat. In separate habitats, contacts are restricted while in overlapping habitats, contacts are facilitated. In most cases, habitats cannot be reduced to the standard extremes, and are multicomponent habitats with various degrees of complexity. See text for implications for viral emergences.

can impact directly or indirectly the accessibility to an infected donor play a role. A typical example that connects several of the points listed in Box 7.2 is provided by the increase of arbovirus vectors during a humid season due to the climate change because insect larvae can proliferate on water reservoirs. Increased travel may put humans infected with arboviruses in contact with the flourishing insect vector population. Climate change may modify the migration routes of some birds, again putting these potential vertebrate hosts in contact with infected animals and insect vectors.

Other points listed in Box 7.2 are worth commenting: close human-to-human contacts are favored by urbanization. In 1975, there were five megacities in the world (meaning cities with more than 10 million human population) while in 2018 the number was 33, with a projection of 43 for 2030. Humans in close contact are,

in addition, highly mobile. At present, it is possible to go around the world in about 36 h (if you choose the adequate airports) which represents a 1000-fold increase in spatial mobility of humans relative to the mobility in the year 1800. The 2014—15 Ebola epidemics in Africa was made worse by the breakdown of public health measures, poverty, and lack of political will of local and international agencies to put efforts in stopping transmission; Ebola outbreaks in Africa continue at the time of this writing. Underdeveloped countries are a reservoir of viral infections that represent a global threat due to several of the points listed in Box 7.2 [(Smolinski et al., 2003); several chapters of Singh (2014)].

Concerning the establishment and dissemination steps, the molecular mechanisms of quasispecies optimization in the new host environment apply. The underlying events are those presided by the extended Darwinian concepts of variation, competition, and selection, with the perturbations derived from stochastic effects (treated in different chapters of this book).

It is worth emphasizing parallelism between the steps involved in viral disease emergence and the steps that mediate entrance and establishment of any organism in a new habitat, which is a process well studied in ecology (van Driesche and van Driesche, 2000).

### 7.7.2 Complexity revisited

The meanings of complexity in virology were discussed in Section 3.9 of Chapter 3, one of them being the inability to explain a whole as the sum of its parts (Solé and Goodwin, 2000). R.V. Solé and B. Goodwin define the sciences of complexity as "the study of those systems in which there is no simple and predictable relationship between levels, between the properties of parts and of wholes." Several levels of complexity can be identified in the events that give rise to the emergence of a viral disease (Domingo, 2010; Sáiz et al., 2014). One level of complexity concerns the behavior of viral populations which is often determined by interactions among components of mutant spectra in a way that cannot be predicted by the individual components of the population, even if we knew them!

The second level of complexity that can have an impact on the emergence of viral disease stems from the environmental, sociological, and ecological variables that must converge for a virus from some animal reservoir to come into contact and successfully infect a new host, for example, a human. Despite close surveillance, the emergences of viral diseases are unpredictable. Experts anticipate new influenza pandemics to arise somewhere in Asia from the avian reservoirs of IV; however, in 2009, the new influenza pandemic originated in Central America. Paradoxically, despite a general agreement that surveillance of human and zoonotic virus reservoirs should be intensified using new molecular tools (i.e., mutant spectrum analyses to go beyond the consensus sequences), the reality is that what we have learned are the reasons why viral emergences are unpredictable. The "abundant sources of genetic variation" that was emphasized by J. Lederberg (see Section 7.7.1) should be extended to refer to "abundant sources of complexity in viral emergences." For the time being, we have to be ready to react once the emergence has already occurred. Research funding agencies have a lot to do in favoring "reactive" rather than "preventive" antiviral policies.

## 7.8 Overview and concluding remarks

Viruses have survived because they have undergone multiple rounds of vertical and horizontal transmission in their host organisms and because occasionally they have found new suitable hosts where to replicate. Among the many parameters involved, in this chapter, we have emphasized the relevance of virus and host population numbers for sustained transmissions and the long-term maintenance of viral entities. A point that is often either ignored or not sufficiently emphasized is that the quasispecies nature of viral populations introduces an element of uncertainty regarding which types of mutants are transmitted to new hosts. Despite being a complication that cannot be easily handled, it is a fact that should stimulate new approaches to the surveillance of virus transmission and the identification of the founder viruses in new infections.

A steady accumulation of mutations during the evolution of organisms was a well-accepted proposal that agreed with the neutral theory of molecular evolution developed last century. One suspects that this agreement resulted in a premature preference for a regular clock of steady incorporation of mutations to work in the case of viruses. The evidence is that there are multiple molecular mechanisms that render the operation of a molecular clock for viruses very unlikely, and perhaps fortuitous in some cases. Several possible mechanisms of variable evolutionary rates have been discussed, and further clarification is expected from entire genome sequencing applied to viruses during outbreaks and epidemics. Viruses are probably not the best biological systems to obtain experimental evidence in support of the clock hypothesis.

A puzzling and pendent issue in viral evolution is the interpretation of the widely different number of viral serotypes, despite viruses sharing comparably large mutation rates and frequencies. Different possibilities have been examined, and a slight preference for variable constraints acting on the amino acid residues that determine the antigenic properties of viruses has been expressed. Again, additional work is necessary to solve this interesting problem.

Procedures for sequence alignments and the establishment of phylogenetic relationships among related viruses have been summarized, with some indications to find useful URL sites for sequence alignment and processing of large data sets. The comparison of genomic sequences (and their encoded amino acids) of new viral isolates with those of the viruses characterized to date is important given the increasing number of new viruses discovered in all kinds of natural habitats.

The important problem of viral emergences and reemergences has been treated with emphasis on the concept of complexity. There are multiple interacting influences that converge to produce the emergence or reemergence of a viral pathogen, one of them being the heterogeneity of viral populations at the genetic and phenotypic level. Despite considerable methodological progress, we are still in the realm of uncertainty regarding the prediction of when and where a new viral pathogen will emerge (see Summary Box).

---

### Summary Box

- Long-term evolution of viruses is the result of a history of virus transmission among hosts. Basic principles of transmission dynamics must take into consideration sampling effects and the inherent heterogeneity of viral populations.
- Rates of evolution of viruses in nature are extremely high as compared to the estimated rate for their host organisms. Contrary to some tenets of neutral evolution, rates of viral evolution are not constant with time. In particular, several mechanisms explain why intrahost virus evolution is faster than interhost evolution.
- Several procedures for sequence alignments and derivation of phylogenetic trees allow a partial description of virus diversification in nature.
- Antigenic diversification of viruses is subjected to constraints that differ among viruses. Some viruses have a single serotype while others have 100 serotypes. Several possible mechanisms may contribute to this difference.
- The emergence and reemergence of new viral pathogens is a multifactorial event with a clear influence of host and virus population numbers. Several levels of complexity participate in the emergence of a new pathogen, rendering the event highly unpredictable.

## References

Aiewsakun, P., Katzourakis, A., 2016. Time-dependent rate phenomenon in viruses. J. Virol. 90, 7184–7195.

Ali, A., Melcher, U., 2019. Modeling of mutational events in the evolution of viruses. Viruses 11, E418.

Allen, B., Sample, C., Dementieva, Y., Medeiros, R.C., Paoletti, C., et al., 2015. The molecular clock of neutral evolution can be accelerated or slowed by asymmetric spatial structure. PLoS Comput. Biol. 11, e1004108.

Althaus, C.L., 2014. Estimating the reproduction number of Ebola virus (EBOV) during the 2014 outbreak in West Africa. PLoS Currents Outbreaks. https://doi.org/10.1371/currents.outbreaks.91afb5e0f279e7129e7056095255b288.

Anderson, R.M., May, R.M., 1991. Infectious Diseases of Humans. Oxford University Press, Oxford.

Antia, R., Regoes, R.R., Koella, J.C., Bergstrom, C.T., 2003. The role of evolution in the emergence of infectious diseases. Nature 426, 658–661.

Aswad, A., Katzourakis, A., 2012. Paleovirology and virally derived immunity. Trends Ecol. Evol. 27, 627–636.

Bamford, D.H., Grimes, J.M., Stuart, D.I., 2005. What does structure tell us about virus evolution? Curr. Opin. Struct. Biol. 15, 655–663.

Beale, G., Dellicour, S., Suchard, M.A., Lemey, P., Vrancken, B., 2018. Recent advances in computational phylodynamics. Curr. Opin. Virol. 31, 24–32.

Bergren, N.A., Auguste, A.J., Forrester, N.L., Negi, S.S., Braun, W.A., et al., 2014. Western equine encephalitis virus: evolutionary analysis of a declining alphavirus based on complete genome sequences. J. Virol. 88, 9260–9267.

Bigman, D.Y., Bobrin, B.D., 2018. Von Economo's disease and postencephalitic parkinsonism responsive to carbidopa and levodopa. Neuropsychiatric Dis. Treat. 14, 927–931.

Boege, U., Kobasa, D., Onodera, S., Parks, G.D., Palmenberg, A.C., et al., 1991. Characterization of Mengo virus neutralization epitopes. Virology 181, 1–13.

Botstein, D., 1980. A theory of modular evolution for bacteriophages. Ann. N. Y. Acad. Sci. 354, 484–491.

Botstein, D., 1981. A modular theory of virus evolution. In: Fields, B.N., Jaenisch, R., Fox, C.F. (Eds.), Animal Virus Genetics. Academic Press, New York, pp. 363–384.

Brechot, C., Bryant, J., Endtz, H., Griffin, D.E., Lewin, S.R., et al., 2018. International meeting of the global virus network. Antivir. Res. 163, 140–148.

Brooksby, J.B., 1981. Surveillance and control of virus diseases: Europe, Middle East and Indian sub-continent. In: Gibss, E.P.J. (Ed.), Virus Diseases of Food Animals, International Perspectives, vol. I. Academic Press Inc., London, pp. 69–78.

Buonagurio, D.A., Nakada, S., Desselberger, U., Krystal, M., Palese, P., 1985. Noncumulative sequence changes in the hemagglutinin genes of influenza C virus isolates. Virology 146, 221–232.

Cam, B.V., Fonsmark, L., Hue, N.B., Phuong, N.T., Poulsen, A., et al., 2001. Prospective case-control study of encephalopathy in children with dengue hemorrhagic fever. Am. J. Trop. Med. Hyg. 65, 848–851.

Chakravarty, S., Hutson, A.M., Estes, M.K., Prasad, B.V., 2005. Evolutionary trace residues in noroviruses: importance in receptor binding, antigenicity, virion assembly, and strain diversity. J. Virol. 79, 554–568.

Cheynier, R., Kils-Hutten, L., Meyerhans, A., Wain-Hobson, S., 2001. Insertion/deletion frequencies match those of point mutations in the hypervariable regions of the simian immunodeficiency virus surface envelope gene. J. Gen. Virol. 82, 1613–1619.

Delamater, P.L., Street, E.J., Leslie, T.F., Yang, Y.T., Jacobsen, K.H., 2019. Complexity of the basic reproduction number ($R_0$). Emerg. Infect. Dis. 25, 1–4.

Domingo, E., 2006. Quasispecies: concepts and implications for virology. Curr. Top. Microbiol. Immunol. 299. Springer, Berlin.

Domingo, E., 2007. Virus evolution. In: Knipe, D.M., Howley, P.M. (Eds.), Fields Virology, fifth ed. Lippincott Williams & Wilkins, Philadelphia, pp. 389–421.

Domingo, E., 2010. Mechanisms of viral emergence. Vet. Res. 41, 38.

Domingo, E., Mateu, M.G., Martínez, M.A., Dopazo, J., Moya, A., et al., 1990. Genetic variability and antigenic diversity of foot-and-mouth disease virus. In: Kurkstak, E., Marusyk, R.G., Murphy, S.A., Van-Regenmortel, M.H.V. (Eds.), Applied Virology Research. Plenum Publishing Co., New York, pp. 233–266.

Domingo, E., Biebricher, C., Eigen, M., Holland, J.J., 2001. Quasispecies and RNA Virus Evolution: Principles and Consequences. Landes Bioscience, Austin.

Domingo, E., de Avila, A.I., Gallego, I., Sheldon, J., Perales, C., 2019. Viral fitness: history and relevance for viral pathogenesis and antiviral interventions. Pathogens and Disease 77 ftz021.

Dopazo, J., Dress, A., von Haeseler, A., 1993. Split decomposition: a technique to analyze viral evolution. Proc. Natl. Acad. Sci. U.S.A. 90, 10320–10324.

Duchene, S., Holmes, E.C., ho, S.Y.W., 2014. Analyses of evolutionary dynamics in viruses are hindered by a time-dependent bias in rate estimates. Proc. Boil. Sci. 281, 20140732.

Duffy, S., Holmes, E.C., 2008. Phylogenetic evidence for rapid rates of molecular evolution in the single-stranded DNA begomovirus tomato yellow leaf curl virus. J. Virol. 82, 957–965.

Eggers, H.J., 2002. History of poliomyelitis and poliomyelitis research. In: Semler, B.L., Wimmer, E. (Eds.), Molecular Biology of Picornaviruses. ASM Press, Washington, DC, pp. 3–14.

Eigen, M., 1992. Steps towards Life. Oxford University Press, Oxford.

Emini, E.A., Jameson, B.A., Lewis, A.J., Larsen, G.R., Wimmer, E., 1982. Poliovirus neutralization epitopes: analysis and localization with neutralizing monoclonal antibodies. J. Virol. 43, 997–1005.

Farheen, N., Sen, N., Nair, S., Tan, K.P., Madhusudhan, M.S., 2017. Depth dependent amino acid substitution matrices and their use in predicting deleterious mutations. Prog. Biophys. Mol. Biol. 128, 14–23.

Feigelstock, D.A., Mateu, M.G., Valero, M.L., Andreu, D., Domingo, E., et al., 1996. Emerging foot-and-mouth disease virus variants with antigenically critical amino acid substitutions predicted by model studies using reference viruses. Vaccine 14, 97–102.

Felsenstein, J., 2004. Inferring Phylogenies. Sinauer Associates, Sunderland, MA.

Firth, C., Kitchen, A., Shapiro, B., Suchard, M.A., Holmes, E.C., et al., 2010. Using time-structured data to estimate evolutionary rates of double-stranded DNA viruses. Mol. Biol. Evol. 27, 2038–2051.

Feng, D.F., Doolittle, R.F., 1996. Progressive alignment of amino acid sequences and construction of phylogenetic trees from them. Methods Enzymol. 266, 368–382.

Ford, F.R., 1937. Diseases of the Nervous System in Infancy, Childhood and Adolescence. Charles C. Thomas, Springfield, IL.

Fry, E.E., Stuart, D., 2010. Virion structure. In: Eherenfeld, E., Domingo, E., Roos, R.P. (Eds.), The Picornaviruses. ASM Press, Washington, DC, pp. 59–71.

Gaucher, E.A., Gu, X., Miyamoto, M.M., Benner, S.A., 2002. Predicting functional divergence in protein evolution by site-specific rate shifts. Trends Biochem. Sci. 27, 315–321.

Gavrilin, G.V., Cherkasova, E.A., Lipskaya, G.Y., Kew, O.M., Agol, V.I., 2000. Evolution of circulating wild poliovirus and of vaccine-derived poliovirus in an immunodeficient patient: a unifying model. J. Virol. 74, 7381–7390.

Geoghegan, J.L., Holmes, E.C., 2018. Evolutionary virology at 40. Genetics 210, 1151–1162.

Gething, M.J., Bye, J., Skehel, J., Waterfield, M., 1980. Cloning and DNA sequence of double-stranded copies of haemagglutinin genes from H2 and H3 strains elucidates antigenic shift and drift in human influenza virus. Nature 287, 301–306.

Geysen, H.M., Meloen, R.H., Barteling, S.J., 1984. Use of peptide synthesis to probe viral antigens for epitopes to a resolution of a single amino acid. Proc. Natl. Acad. Sci. U.S.A. 81, 3998–4002.

Gilbert, C., Feschotte, C., 2010. Genomic fossils calibrate the long-term evolution of hepadnaviruses. PLoS Biol. 8, e1000459.

Gojobori, T., Yokoyama, S., 1985. Rates of evolution of the retroviral oncogene of Moloney murine sarcoma virus and of its cellular homologues. Proc. Natl. Acad. Sci. U.S.A. 82, 4198–4201.

Gorman, O.T., Bean, W.J., Webster, R.G., 1992. Evolutionary processes in influenza viruses: divergence, rapid evolution, and stasis. Curr. Top. Immunol. 176, 75–97.

Haagmans, B.L., Andeweg, A.C., Osterhaus, A.D., 2009. The application of genomics to emerging zoonotic viral diseases. PLoS Pathog. 5, e1000557.

Hall, B.G., 2001. Phylogenetic Trees Made Easy: A How-To Manual for Molecular Biologists. Sinauer Association Inc., Sunderland, MA.

Hamelaar, J., Elangovan, R., Yun, J., Dickson-Tetteh, L., Fleminger, I., et al., 2019. Global and regional molecular epidemiology of HIV-1, 1990-2015: a systematic review, global survey, and trend analysis. Lancet Infect. Dis. 19, 143–155.

Hanada, K., Suzuki, Y., Gojobori, T., 2004. A large variation in the rates of synonymous substitution for RNA viruses and its relationship to a diversity of viral infection and transmission modes. Mol. Biol. Evol. 21, 1074–1080.

Hedskog, C., Parhy, B., Chang, S., Zeuzem, S., Moreno, C., et al., 2019. Identification of 19 novel hepatitis C virus subtypes—Further expanding HCV classification. Open Forum Infect. Dis. 6, ofz076.

Heffernan, J.M., Smith, R.J., Wahl, L.M., 2005. Perspectives on the basic reproductive ratio. J. R. S. Interface R. Soc. 2, 281–293.

Herbeck, J.T., Nickle, D.C., Learn, G.H., Gottlieb, G.S., Curlin, M.E., et al., 2006. Human immunodeficiency virus type 1 env evolves toward ancestral states upon transmission to a new host. J. Virol. 80, 1637–1644.

Holland, J.J., Domingo, E., de la Torre, J.C., Steinhauer, D.A., 1990. Mutation frequencies at defined single codon sites in vesicular stomatitis virus and poliovirus can be increased only slightly by chemical mutagenesis. J. Virol. 64, 3960–3962.

Holland, J.J., de La Torre, J.C., Steinhauer, D.A., 1992. RNA virus populations as quasispecies. Curr. Top. Microbiol. Immunol. 176, 1–20.

Holmes, E.C., 2008. Comparative studies of RNA virus evolution. In: Domingo, E., Parrish, C.R., Holland, J.J. (Eds.), Origin and Evolution of Viruses, second ed. Elsevier, Oxford, pp. 119–134.

Holmes, E.C., 2009. The evolution and emergence of RNA viruses. Oxford Series in Ecology and Evolution. Oxford University Press, New York.

Huelsenbeck, J.P., Dyer, K.A., 2004. Bayesian estimation of positively selected sites. J. Mol. Evol. 58, 661–672.

Huelsenbeck, J.P., Ronquist, F., Nielsen, R., Bollback, J.P., 2001. Bayesian inference of phylogeny and its impact on evolutionary biology. Science 294, 2310–2314.

Jenkins, G.M., Rambaut, A., Pybus, O.G., Holmes, E.C., 2002. Rates of molecular evolution in RNA viruses: a quantitative phylogenetic analysis. J. Mol. Evol. 54, 156–165.

Jiang, P., Faase, J.A., Toyoda, H., Paul, A., Wimmer, E., et al., 2007. Evidence for emergence of diverse polioviruses from C-cluster coxsackie A viruses and implications for global poliovirus eradication. Proc. Natl. Acad. Sci. U.S.A. 104, 9457–9462.

Kay, A., Zoulim, F., 2007. Hepatitis B virus genetic variability and evolution. Virus Res. 127, 164–176.

Kearse, M., Moir, R., Wilson, A., Stones-Havas, S., Cheung, M., et al., 2012. Geneius basic: an integrated and extendable desktop software platform for the organization and analysis of sequencing data. Bioinformatics 28, 1647–1649.

Khetsuriani, N., Prevots, D.R., Quick, L., Elder, M.E., Pallansch, M., et al., 2003. Persistence of vaccine-derived polioviruses among immunodeficient persons with vaccine-associated paralytic poliomyelitis. J. Infect. Dis. 188, 1845–1852.

Knowles, N.J., Hovi, T., King, A.M.Q., Stanway, G., 2010. Overview of taxonomy. In: Ehrenfeld, E., Domingo, E., Roos, R.P. (Eds.), The Picornaviruses. ASM Press, Washington, DC, pp. 19–32.

Kortenhoeven, C., Joubert, F., Bastos, A., Abolnik, C., 2015. Virus genome dynamics under different propagation pressures: reconstruction of whole genome haplotypes of west Nile viruses from NGS data. BMC Genomics 16, 118.

Kulkarni, A., Bangham, C.R.M., 2018. HTLV-1: regulating the balance between proviral latency and reactivation. Front. Microbiol. 9, 449.

Kumar, S., Stecher, G., Tamura, K., 2016. MEGA 7: molecular evolutionary genetics analysis version 7.0 for bigger datasets. Mol. Biol. Evol. 33, 1870–1874.

Kumar, S., Agarwal, S., Ranvijay, 2019. Fast and memory efficient approach for mapping NGS reads to a reference genome. J. Bioinform. Comput. Biol. 17, 1950008.

Lai, D., Verma, M., 2017. Large-scale sequence comparison. Methods Mol. Biol. 1525, 191–224.

Lea, S., Hernández, J., Blakemore, W., Brocchi, E., Curry, S., et al., 1994. The structure and antigenicity of a type C foot-and-mouth disease virus. Structure 2, 123–139.

Lederberg, J., 1993. Viruses and humankind: intracellular symbiosis and evolutionary competition. In: Morse, S.S. (Ed.), Emerging Viruses. Oxford University Press, Oxford, pp. 3–9.

Leslie, A.J., Pfafferott, K.J., Chetty, P., Draenert, R., Addo, M.M., et al., 2004. HIV evolution: CTL escape mutation and reversion after transmission. Nat. Med. 10, 282–289.

Lin, Y.Y., Liu, C., Chien, W.H., Wu, L.L., Tao, Y., et al., 2015. New insights into the evolutionary rate of hepatitis B virus at different biological scales. J. Virol. 89, 3512–3522.

Lipkin, W.I., Anthony, S.J., 2015. Virus hunting. Virology 479–480C, 194–199.

Louie, R.H.Y., Kaczorowski, K.J., Barton, J.P., Chakraborty, A.K., McKay, M.R., 2018. Fitness landscape of the human immunodeficiency virus envelope protein that is targeted by antibodies. Proc. Natl. Acad. Sci. U.S.A. 115, E564–E573.

Lythgoe, K.A., Fraser, C., 2012. New insights into the evolutionary rate of HIV-1 at the within-host and epidemiological levels. Proc. Biol. Sci. 279, 3367–3375.

Martin, D.P., Williamson, C., Posada, D., 2005. RDP2: recombination detection and analysis from sequence alignments. Bioinformatics 21, 260–262.

Martínez, M.A., Carrillo, C., Gonzalez-Candelas, F., Moya, A., Domingo, E., et al., 1991a. Fitness alteration of foot-and-mouth disease virus mutants: measurement of adaptability of viral quasispecies. J. Virol. 65, 3954–3957.

Martínez, M.A., Hernández, J., Piccone, M.E., Palma, E.L., Domingo, E., et al., 1991b. Two mechanisms of antigenic diversification of foot-and-mouth disease virus. Virology 184, 695–706.

Martínez, M.A., Dopazo, J., Hernández, J., Mateu, M.G., Sobrino, F., et al., 1992. Evolution of the capsid protein genes of foot-and-mouth disease virus: antigenic variation without accumulation of amino acid substitutions over six decades. J. Virol. 66, 3557–3565.

Mateu, M.G., 1995. Antibody recognition of picornaviruses and escape from neutralization: a structural view. Virus Res. 38, 1–24.

Mateu, M.G., 2017. The foot-and-mouth disease virion: structure and function. In: Sobrino, F., Domingo, E. (Eds.), Foot-and-Mouth Disease Virus. Current Research and Emerging Trends. Caister Academic Press, Norfolk, UK, pp. 61–105.

Mateu, M.G., Da Silva, J.L., Rocha, E., De Brum, D.L., Alonso, A., et al., 1988. Extensive antigenic heterogeneity of foot-and-mouth disease virus of serotype C. Virology 167, 113–124.

Mateu, M.G., Martínez, M.A., Capucci, L., Andreu, D., Giralt, E., et al., 1990. A single amino acid substitution affects multiple overlapping epitopes in the major antigenic site of foot-and-mouth disease virus of serotype C. J. Gen. Virol. 71, 629–637.

McGeoch, D.J., Davison, A.J., 1999. The molecular evolutionary history of the herpesviruses. In: Domingo, E., Webster, R., Holland, J. (Eds.), Origin and Evolution of Viruses. Academic Press, San Diego, pp. 441–465.

McTavish, E.J., Drew, B.T., Redelings, B., Cranston, K.A., 2017. How and why to build a unified tree of life. Bioassays 39. https://doi.org/10.1002/bies.201700114.

Melamed, A., Witkover, A.D., Laydon, D.J., Brown, R., Ladell, K., et al., 2014. Clonality of HTLV-2 in natural infection. PLoS Pathog. 10, e1004006.

Mims, C.A., 1981. Vertical transmission of viruses. Microbiol. Rev. 45, 267–286.

Minor, P.D., Schild, G.C., Bootman, J., Evans, D.M., Ferguson, M., et al., 1983. Location and primary structure of a major antigenic site for poliovirus neutralization. Nature 301, 674–679.

Minor, P.D., Ferguson, M., Evans, D.M., Almond, J.W., Icenogle, J.P., 1986. Antigenic structure of polioviruses of serotypes 1, 2 and 3. J. Gen. Virol. 67, 1283–1291.

Minot, S., Bryson, A., Chehoud, C., Wu, G.D., Lewis, J.D., et al., 2013. Rapid evolution of the human gut virome. Proc. Natl. Acad. Sci. U.S.A. 110, 12450–12455.

Morse, S.S., 1993. Emerging Viruses. Oxford University Press, Oxford.

Morse, S.S., 1994. The Evolutionary Biology of Viruses. Raven Press, New York.

Mount, D.W., 2004. Bioinformatics. Sequence and Genome Analysis. Cold Spring Harbor Laboratory Press, Cold Spring Harbor, NY.

Muhlemann, B., Margaryan, A., Damgaard, P. de B., Allentoft, M.E., Vinner, L., et al., 2018. Ancient human parvovirus B19 in Eurasia reveals its long-term association with humans. Proc. Natl. Acad. Sci. U.S.A. 115, 7557–7562.

Nakamura, T., Yamada, K.D., Tomii, K., Katoh, K., 2018. Parallelization of MAFFT for large-scale multiple sequence alignments. Bioinformatics 34, 2490–2492.

Nash, A.A., Dalziel, R.G., Fitzgerald, J.R., 2015. Mims' Pathogenesis of Infectious Disease. Elsevier Science & Technology, Edinburgh.

Notredame, C., Higgins, D.G., Heringa, J., 2000. T-Coffee: a novel method for fast and accurate multiple sequence alignment. J. Mol. Biol. 302, 205–217.

Nowak, M.A., 2006. Evolutionary Dynamics. The Belknap Press of Harvard University Press, Cambridge, Massachusetts and London, England.

Nowak, M.A., May, R.M., 2000. Virus Dynamics. Mathematical Principles of Immunology and Virology. Oxford University Press Inc., New York.

Odoom, J.K., Yunus, Z., Dunn, G., Minor, P.D., Martin, J., 2008. Changes in population dynamics during long-term evolution of sabin type 1 poliovirus in an immunodeficient patient. J. Virol. 82, 9179–9190.

Page, R.D.M., Holmes, E.C., 1998. Molecular Evolution. A Phylogenetic Approach. Blackwell Science Ltd., Oxford.

Parrish, C.R., Kawaoka, Y., 2005. The origins of new pandemic viruses: the acquisition of new host ranges by canine parvovirus and influenza A viruses. Annu. Rev. Microbiol. 59, 553–586.

Paunio, M., Peltola, H., Valle, M., Davidkin, I., Virtanen, M., et al., 1998. Explosive school-based measles outbreak: intense exposure may have resulted in high risk, even among revaccinees. Am. J. Epidemiol. 148, 1103–1110.

Peng, Y., Tang, S., Wang, D., Zhong, H., Jia, H., et al., 2018. Meta PGN: a pipeline for construction and graphical visualization of annotated pangenome networks. GigaScience 7. https://doi.org/10.1093/gigascience/giy121.

Plyusnin, A., Morzunov, S.P., 2001. Virus evolution and genetic diversity of hantaviruses and their rodent hosts. Curr. Top. Microbiol. Immunol. 256, 47–75.

Porto, M., Roman, H.E., Vendruscolo, M., Bastolla, U., 2005. Prediction of site-specific amino acid distributions and limits of divergent evolutionary changes in protein sequences. Mol. Biol. Evol. 22, 630–638.

Posada, D., Crandall, K.A., 1998. Modeltest: testing the model of DNA substitution. Bioinformatics 14, 817–818.

Ravantti, J., Bamford, D., Stuart, D.I., 2013. Automatic comparison and classification of protein structures. J. Struct. Biol. 183, 47–56.

Redd, A.D., Collinson-Streng, A.N., Chatziandreou, N., Mullis, C.E., Laeyendecker, O., et al., 2012. Previously transmitted HIV-1 strains are preferentially selected during subsequent sexual transmissions. J. Infect. Dis. 206, 1433–1442.

Rocha, E., Cox, N.J., Black, R.A., Harmon, M.W., Harrison, C.J., et al., 1991. Antigenic and genetic variation in influenza A (H1N1) virus isolates recovered from a persistently infected immunodeficient child. J. Virol. 65, 2340–2350.

Ronquist, F., Huelsenbeck, J.P., 2003. MrBayes 3: Bayesian phylogenetic inference under mixed models. Bioinformatics 19, 1572–1574.

Rossmann, M.G., 1989. The canyon hypothesis. Hiding the host cell receptor attachment site on a viral surface from immune surveillance. J. Biol. Chem. 264, 14587–14590.

Rowlands, D.J., Clarke, B.E., Carroll, A.R., Brown, F., Nicholson, B.H., et al., 1983. Chemical basis of antigenic variation in foot-and-mouth disease virus. Nature 306, 694–697.

Russell, D.J., 2014. GramAlign: fast alignment driven by grammar-based phylogeny. Methods Mol. Biol. 1079, 171–189.

Sáiz, J.C., Sobrino, F., Sevilla, N., Martin, V., Perales, C., et al., 2014. Molecular and evolutionary mechanisms of viral emergence. In: Singh, S.K. (Ed.), Viral Infections and Global Change. John Wiley & Sons Inc., Hoboken, NJ, pp. 297–325.

Sakaoka, H., Kurita, K., Iida, Y., Takada, S., Umene, K., et al., 1994. Quantitative analysis of genomic polymorphism of herpes simplex virus type 1 strains from six countries: studies of molecular evolution and molecular epidemiology of the virus. J. Gen. Virol. 75 (Pt 3), 513–527.

Salemi, M., Vandamme, A.M., 2004. The phylogeny handbook. A Practical Approach to DNA and Protein Phylogeny. Cambridge University Press, Cambridge.

Salemi, M., Vandamme, A.M., Gradozzi, C., Van Laethem, K., Cattaneo, E., et al., 1998. Evolutionary rate and genetic heterogeneity of human T-cell lymphotropic virus type II (HTLV-II) using isolates from European injecting drug users. J. Mol. Evol. 46, 602–611.

Salemi, M., Lamers, S.L., Yu, S., de Oliveira, T., Fitch, W.M., et al., 2005. Phylodynamic analysis of human immunodeficiency virus type 1 in distinct brain compartments provides a model for the neuropathogenesis of AIDS. J. Virol. 79, 11343–11352.

Schrag, S.J., Rota, P.A., Bellini, W.J., 1999. Spontaneous mutation rate of measles virus: direct estimation based on mutations conferring monoclonal antibody resistance. J. Virol. 73, 51–54.

Sellers, R.F., 1971. Quantitative aspects of the spread of foot-and-mouth disease. Vet. Bull. 41, 431−439.

Sellers, R.F., 1981. Factors affecting the geographical distribution and spread of virus diseases of food animals. In: Gibbs, E.P.J. (Ed.), Virus Diseases of Food Animals, International Perspectives, vol. I. Academis Press Inc, London, pp. 19−29.

Sharp, P.M., Simmonds, P., 2011. Evaluating the evidence of virus/host co-evolution. Curr Opin Virol 1, 436−441.

Sherry, B., Mosser, A.G., Colonno, R.J., Rueckert, R.R., 1986. Use of monoclonal antibodies to identify four neutralization immunogens on a common cold picornavirus, human rhinovirus 14. J. Virol. 57, 246−257.

Simmonds, P., Holmes, E.C., Cha, T.A., Chan, S.W., McOmish, F., et al., 1993. Classification of hepatitis C virus into six major genotypes and a series of subtypes by phylogenetic analysis of the NS-5 region. J. Gen. Virol. 74 (Pt 11), 2391−2399.

Simmonds, P., Aiewsakun, P., Katzourakis, A., 2019. Prisoners of war-host adaptation and its constraints on virus evolution. Nat. Rev. Microbiol. 17, 321−330.

Singh, S.K., 2014. Viral Infections and Global Change. John Wiley & Sons Inc., Hoboken, NJ.

Small, M., Tse, C.K., Walker, D.M., 2006. Super-spreaders and the rate of transmission of the SARS virus. Physica D 215, 146−158.

Smith, D.B., Inglis, S.C., 1987. The mutation rate and variability of eukaryotic viruses: an analytical review. J. Gen. Virol. 68, 2729−2740.

Smith, D.B., Bukh, J., Kuiken, C., Muerhoff, A.S., Rice, C.M., et al., 2014. Expanded classification of hepatitis C virus into 7 genotypes and 67 subtypes: updated criteria and genotype assignment web resource. Hepatology 59, 318−327.

Smolinski, M.S., Hamburg, M.A., Lederberg, J., 2003. Microbial Threats to Health, Emergence, Detection and Response. The National Academies Press, Washington, DC.

Sobrino, F., Domingo, E., 2004. Foot-and-Mouth Disease: Current Perspectives. Horizon Bioscience, Wymondham, England.

Sobrino, F., Palma, E.L., Beck, E., Dávila, M., de la Torre, J.C., et al., 1986. Fixation of mutations in the viral genome during an outbreak of foot-and-mouth disease: heterogeneity and rate variations. Gene 50, 149−159.

Solé, R., Goodwin, B., 2000. Signs of Life. How Complexity Pervades Biology. Basic Books, New York.

Sorensen, J.H., Mackay, D.K., Jensen, C.O., Donaldson, A.I., 2000. An integrated model to predict the atmospheric spread of foot-and-mouth disease virus. Epidemiol. Infect. 124, 577−590.

Stapleton, J.T., Lemon, S.M., 1987. Neutralization escape mutants define a dominant immunogenic neutralization site on hepatitis A virus. J. Virol. 61, 491−498.

Stec, D.S., Waddell, A., Schmaljohn, C.S., Cole, G.A., Schmaljohn, A.L., 1986. Antibody-selected variation and reversion in Sindbis virus neutralization epitopes. J. Virol. 57, 715−720.

Sullivan, J., 2005. Maximum-likelihood methods for phylogeny estimation. Methods Enzymol. 395, 757−779.

Switzer, W.M., Salemi, M., Shanmugam, V., Gao, F., Cong, M.E., et al., 2005. Ancient co-speciation of simian foamy viruses and primates. Nature 434, 376−380.

Tidesley, M.J., Probert, W.J.M., Woolhouse, M.E.J., 2017. Mathematical models of the epidemiology and control of foot-and-mouth disease. In: Sobrino, F., Domingo, E. (Eds.), Foot-and-mouth Disease Virus. Current Research and Emerging Trends. Caister Academic Press, Norfolk, UK, pp. 385−408.

Tomonaga, K., Suzuki, N., Berkhout, B., 2019. Integration of viral sequences into eukaryotic host genomes: legacy of ancient infections. Virus Res. 262, 1 [Preface of a special issue of Virus Research on integration of viral sequences in host genomes].

Tonkin-Hill, G., Lees, J.A., Bentley, S.D., Frost, S.D.W., Corander, J., 2019. Fast hierarchical Bayesian analysis of population structure. Nucleic Acids Res. 47, 5539−5549.

van Driesche, J., van Driesche, R., 2000. Nature Out of Place: Biological Invasions in the Global Age. Island Press, Washington, DC.

Vilensky, J.A., Gilman, S., McCall, S., 2010. Does the historical literature on encephalitis lethargica support a simple (direct) relationship with postencephalitic Parkinsonism? Mov. Disord. 25, 1124−1130.

Villarreal, L.P., 2005. Viruses and the Evolution of Life. ASM Press, Washington, DC.

Villaverde, A., Martínez, M.A., Sobrino, F., Dopazo, J., Moya, A., et al., 1991. Fixation of mutations at the VP1 gene of foot-and-mouth disease virus. Can quasispecies define a transient molecular clock? Gene 103, 147−153.

Voskarides, K., Christaki, E., Nikolopoulos, G.K., 2018. Influenza virus-host coevolution. A predator-prey relationship? Front. Immunol. 9, 2017. https://doi.org/10.3389/fimmu.2018.02017.

Wang, L.F., Crameri, G., 2014. Emerging zoonotic viral diseases. Rev. Sci. Tech. (Int. Off. Epiz.) 33, 569−581.

Wang, H.Y., Chien, M.H., Huang, H.P., Chang, H.C., Wu, C.C., et al., 2010. Distinct hepatitis B virus dynamics in the immunotolerant and early immunoclearance phases. J. Virol. 84, 3454−3463.

Wargo, A.R., Kurath, G., 2011. In vivo fitness associated with high virulence in a vertebrate virus is a complex trait regulated by host entry, replication, and shedding. J. Virol. 85, 3959−3967.

Webster, R.G., 1999. Antigenic variation in influenza viruses. In: Domingo, E., Webster, R.G., Holland, J.J. (Eds.), Origin and Evolution of Viruses. Academic Press, San Diego, pp. 377−390.

Weddell, G.N., Yansura, D.G., Dowbenko, D.J., Hoatlin, M.E., Grubman, M.J., et al., 1985. Sequence variation in the gene for the immunogenic capsid protein VP1 of foot-and-mouth disease virus type A. Proc. Natl. Acad. Sci. U.S.A. 82, 2618—2622.

Wiktor, T.J., Koprowski, H., 1980. Antigenic variants of rabies virus. J. Exp. Med. 152, 99—112.

Woolhouse, M.E., 2017. Quantifying transmission. Microbiol. Spectr. 5 https://doi.org/10.1128/microbiolspec.MTBP-0005-2016.

Woolhouse, M.E., Webster, J.P., Domingo, E., Charlesworth, B., Levin, B.R., 2002. Biological and biomedical implications of the co-evolution of pathogens and their hosts. Nat. Genet. 32, 569—577.

Worobey, M., 2001. A novel approach to detecting and measuring recombination: new insights into evolution in viruses, bacteria, and mitochondria. Mol. Biol. Evol. 18, 1425—1434.

Xia, X., Xie, Z., 2001. DAMBE: software package for data analysis in molecular biology and evolution. J. Hered. 92, 371—373.

Xing, L., Tjarnlund, K., Lindqvist, B., Kaplan, G.G., Feigelstock, D., et al., 2000. Distinct cellular receptor interactions in poliovirus and rhinoviruses. EMBO J. 19, 1207—1216.

Yang, Z., 2006. Computational Molecular Evolution. Oxford University Press, Oxford.

Yang, Z., Nielsen, R., Goldman, N., Pedersen, A.M., 2000. Codon-substitution models for heterogeneous selection pressure at amino acid sites. Genetics 155, 431—449.

Zimmern, D., 1988. Evolution of RNA viruses. In: Domingo, E., Holland, J.J., Ahlquist, P. (Eds.), RNA Genetics. CRC Press Inc., FL, pp. 211—240.

# Quasispecies dynamics in disease prevention and control

## Abbreviations

**Ab** Antibody
**AIDS** Acquired Immuno Deficiency Syndrome
**$CC_{50}$** Cytotoxic Concentration 50
**CTL** Cytotoxic T Cell
**DAA** direct-acting antiviral
**FMDV** Foot-and-Mouth Disease Virus
**FU** 5-Fluorouracil
**HAV** Hepatitis A Virus
**HBV** Hepatitis B Virus
**HCV** Hepatitis C Virus
**HIV-1** Human Immunodeficiency Virus Type 1
**$IC_{50}$** Inhibitory Concentration 50
**IFN-α** Interferon-alpha
**IV** Influenza Virus
**MAb** Monoclonal Antibody
**MOI** Multiplicity of Infection
**NNRTIs** Nonnucleoside Reverse Transcriptase Inhibitors
**NRTIs** Nucleoside/nucleotide Reverse Transcriptase Inhibitors
**RT** Reverse Transcriptase
**TI** Therapeutic Index

## 8.1 Medical interventions as selective constraints

Medical interventions have dramatically increased over the last century, and in the case of infectious diseases, the discovery and development of antibiotics and antiviral agents have represented a very powerful external selective constraint imposed upon replicating microbes. Hundreds of antiviral agents have been developed since the second half of the 20th century, and viruses can generally find evolutionary pathways to continue replication in their presence. The same is true of antibiotics and replication of bacteria. The belief that bacterial diseases were on their way toward extinction was quite widespread in the middle of the 20th century. Sir M. Burnet wrote in his 1966 textbook: "And since bacterial infections are, with unimportant exceptions, amenable to treatment with one or other of the new drugs, our real problems are likely to be concerned with virus diseases" (Burnet, 1966). In 1932, a prominent Spanish medical doctor, G. Marañón, declared: "In the year 2000 cancer will be a historical disease. Infections will be almost entirely absent as a cause of mortality." (On a personal note, when I joined the University of California, Irvine in 1969, to work as postdoctoral student with R.C. Warner, I attended some biology courses in which the teachers expressed to students that infectious diseases would disappear in a few decades as a consequence of the use of antibiotics and antiviral agents). Predictions in science tend to fail.

The optimistic view was not unanimous. A. Fleming, the discoverer of penicillin, recognized the adaptive capacity of bacteria and suggested that bacteria would inevitably find ways of resisting the damage to them caused by antimicrobial drugs [quoted from the document "New antimicrobial drugs" from the European Academies Science Advisory Council, November 2014 (www.easac.eu); see also Chapter 10]. Furthermore, there was early evidence of selection of *Mycobacterium tuberculosis* mutants resistant to streptomycin (Mitchison, 1950). Antibiotic resistance in bacteria has similarities and differences with antiviral resistance in viruses, and they are compared in Chapter 10.

We are now very aware that one of the major problems in antiviral therapy is the nearly systematic selection of drug-resistant virus mutants, which is often associated with treatment failure. Other external influences, such as vaccination or immunotherapy, particularly using monoclonal antibodies, can also evoke the selection of viral subpopulations capable of replicating in the presence of those components inherent to an immune response. Thus, selective constraints intended to limit RNA virus replication meet with the broad and dynamic repertoire of variants ingrained in quasispecies dynamics. Two space-time levels of the effects of drugs or vaccines are distinguished in coming sections: (i) short-term consequences for the individual in the form of treatment or vaccination failure and (ii) long-term consequences at the population level in the field, or vaccine-driven evolution of the antigenic properties of viruses.

There are other medical interventions that may alter virus survival. Individuals who are immunocompromised as a consequence of treatment after organ transplantation or those subjected to anticancer chemotherapy become particularly vulnerable to viral infections. Enhanced viral replication can favor pathological manifestations in the affected individual as well as the spread of a large number of viruses into the environment, with consequences for the emergence and reemergence of viral disease (Section 7.7 in Chapter 7).

## 8.2 Different manifestations of virus evolution in the prevention and treatment of viral disease

Viral diseases are an important burden for human health and agriculture (Bloom and Lambert, 2003). Virus evolution, through the basic mechanisms exposed in previous chapters, can influence the two major strategies to combat viral infections: prevention by vaccination and treatment by antiviral inhibitors. For the design of new antiviral vaccines, a critical issue is the diversity displayed in the field by the virus to be controlled. The natural evolution of the virus may result in the circulation of one major antigenic type or the cocirculation of multiple antigenic forms. The vaccine composition (independently of the type of vaccine; see Section 8.3.1) must match the antigenic composition of the virus to be controlled. Hepatitis A virus (HAV) circulates as a single serotype, while foot-and-mouth disease virus (FMDV) circulates as seven serotypes and diverse subtypes, and the antigenic types are unevenly distributed in different geographical locations. A monovalent vaccine made of the prevailing antigenic type of HAV should be sufficient to confer protection, while a multivalent vaccine composed of several types or subtypes is required to confer protection against FMDV, and the antigenic composition of the vaccine should match the circulating viruses in each geographical region. This is why antiFMD vaccines of different compositions are used in different world areas at a given time, and vaccine composition must be periodically updated to maintain its efficacy. Thus, one effect of virus evolution relevant to vaccine design derives from the necessity to prepare a vaccine that mirrors the antigenic composition of the virus to be controlled. In the case of live-attenuated antiviral vaccines, the evolution of the vaccine virus while it replicates in the vaccinee is a risk factor to produce virulent derivatives.

The invasion of a susceptible host by a virus and the ensuing viral replication can be regarded as a step-wise process during which the virus must adapt to a series of selective pressures presented by the host, notably the immune response. The outcome can be either viral clearance (elimination of the infection) or virus survival and progression toward an acute or a persistent infection. Administration of antiviral agents is an additional selective constraint that limits viral replication. Evolutionary mechanisms may either succeed in the selection of mutants resistant to the antiviral agent that will permit the infection to continue or fail in sustaining the infection, resulting in the clearing of the virus from the organism.

Treatment planning, one of the aims of the new antiviral pharmacological interventions, based on information of viral genomic sequences present in each infected patient, has parallels with vaccine composition design. For vaccines, the information comes from the analyses of antigenic composition of circulating viruses, and for antiviral agents, the information comes from the mutant spectrum composition of the virus to be controlled in the infected patient.

## 8.3 Antiviral vaccines and the adaptive potential of viruses

World-wide vaccination campaigns made possible the eradication of human smallpox [with the official declaration by the World Health Organization (WHO) in 1980] and animal rinderpest [with the official declaration by the World Organization for Animal Health, Office International des Epizooties (OIE) in 2011]. The number of new cases has decreased as a result of vaccination programs against several viral diseases, including measles or hepatitis B (Bloom and Lambert, 2003), and substantial progress has been made toward the eradication of poliomyelitis (Chumakov and Kew, 2010; Himman, 2017). Only the deliberate decision not to vaccinate (for religious reasons or misinformation

campaigns) or lack of vaccine accessibility (for socioeconomic circumstances) jeopardizes vaccine efficacy. These facts demonstrate that at least some viral diseases can be controlled on a global basis by vaccination, an unprecedented achievement of human and animal health.

Despite the huge economic investment, however, there are important viral diseases such as acquired immunodeficiency syndrome (AIDS), hepatitis C, or viral hemorrhagic fevers for which no effective vaccines are available. For some diseases such as human influenza or animal FMD, vaccines are accessible, but they require periodic updating to approximate the antigenic composition of the vaccine to that of the circulating virus (Section 8.2). In the case of influenza virus (IV), a major change in antigenic composition can occur through antigenic shift, in which the virus acquires new hemagglutinin and neuraminidase genes by genome segment reassortment (Section 7.4 in Chapter 7), with the first evidence obtained by G. Laver as early as 1971 (for the early history of influenza, its causative virus, and vaccine designs, see Beveridge, 1977; Kilbourne, 1987). Antigenic variation of viruses, whatever the mechanism might be, can affect vaccine efficacy and in some cases, the extreme rapid intra and interhost evolution of a virus may render an effective vaccine unfeasible at least with the current tools of vaccinology. CD8$^+$ T cell responses may act soon after infection and promote the selection of escape mutants (Bull et al., 2015). The difficulties for the control of virus disease derived from the adaptive potential of viruses (Domingo, 1989; Domingo and Holland, 1992; Bailey et al., 2004; Hamelaar et al., 2019) require the judicious application of existing tools and innovative approaches that are still in their infancy.

### 8.3.1 Some requirements for the design of vaccines to control highly variable viruses

A first basic requisite for the preparation of a vaccine against a viral agent is the understanding of the immune response evoked by the virus when it infects the organism to be protected (activation of B and T lymphocytes for antibody production, cellular responses, and generation of memory cells) and correlates of protection (Bloom and Lambert, 2003; Van Regenmortel, 2012; Hagan et al., 2015; Cunningham et al., 2016; Rolland, 2019). Despite social pressure to rapidly obtain a vaccine, for each virus-host system, well-designed experiments are necessary to try to establish the determinants of protection, which is not a simple issue. The discussions in coming paragraphs are focused on the relevance of virus evolution in vaccine efficacy, irrespective of the type of protection afforded by the vaccine. What we term "protection" may mean the total absence of replication of the infecting virus (termed "sterilizing" immunity) or absence of disease manifestations despite infection and virus replication. As a general initial statement, which is widely accepted by vaccinologists, a vaccine is likely to be effective when it evokes an immune response that is similar to the response elicited by the authentic viral pathogen when it produces disease successfully overcome by the infected organism (Evans and Kaslow, 1997; Bloom and Lambert, 2003). We refer to this as the basic principle of vaccinology. When infection by an antigenically constant virus produces lifelong immunity (i.e., measles virus infection), a vaccine is likely to evoke long-lasting protection. In contrast, if a patient cured of a virus can be reinfected by the same (or a closely related) virus (i.e., hepatitis C virus infection) a vaccine—at least one prepared by standard methodology—is unlikely to evoke protection.

Some points to be considered in the design of antiviral vaccines are listed in Box 8.1. They are intended to minimize the selection of vaccine-escape mutants and favor the success of vaccination campaigns. Some of the recommendations deserve further comment. First, a basic knowledge of virus evolutionary dynamics and how it affects virus antigenic stability (or lack of) is essential. The fact that a methodology is available (i.e., vectors that can express large amounts of

---

## BOX 8.1

## VACCINE DESIGNS AND VACCINATION STRATEGIES FOR ANTIGENICALLY VARIABLE VIRUSES

- Prior to the planning of a vaccine strategy, it is essential to review what is known about genetic and antigenic variation of the virus to be controlled (whether it is a DNA or RNA virus displaying high- or low-fidelity replication, antigenic diversity in the field, location of B- and T-cell epitopes, etc.).
- Carry out research to understand the correlates of protection.
- In keeping with the basic principle of vaccinology, from the point of view of inducing a protective response the preferred order of vaccine types is as follows: live

attenuated > whole virus inactivated = empty viral particles > multiple immunogenic viral proteins > a single immunogenic viral protein > mixtures of synthetic peptides, dendrimeric scaffolds, peptide arrays > a single synthetic peptide.

- International vaccination programs should be carried out as quickly as possible.
- Programs to update the antigenic composition of vaccines should be implemented.

*Based on Domingo and Holland (1992).*

antigens displaying good immunogenicity) does not guarantee vaccine efficacy, and even less if correlates of protection are not understood. The order of efficacy of different vaccine designs proposed in Box 8.1 is justified both by the basic principle of vaccinology and by the mechanisms of selection of antibody (Ab)- and cytotoxic T-cell (CTL)-escape mutants by viruses. Single amino acid substitutions at B- and T-cell epitopes in viral proteins are often sufficient to elude neutralization by the corresponding cognate-specific antibody or to escape recognition by a clonal CTL population. For many viruses, the frequency of monoclonal antibody-escape mutants has been measured in $10^{-4}$ to $10^{-6}$, even in clonal populations obtained under controlled laboratory conditions and that have undergone a limited number of replication rounds (Section 7.4.2 in Chapter 7). Generation of immune-escape variants can result in lack of vaccine efficacy, contribute to viral persistence (Pircher et al., 1990; Weidt et al., 1995; Ciurea et al., 2000, 2001; Richman et al., 2003; Pawlotsky, 2006), and provoke vaccination-induced virus evolution (Section 8.3.2). In human immunodeficiency virus type 1 (HIV-1), antibody-escape variants are incessantly being produced in vivo to the point that virus replication continues despite the antibody response (Richman et al., 2003; Bailey et al., 2004).

The frequency of selection of mutants that can escape a number ($n$) of components in which we could hypothetically separate a global immune response is far lower than the frequency of escape to a single ($a$, $b$, $c$, etc.) of the $i$ components of the response. Making a simple mathematical abstraction that is applicable also to antiviral-escape mutants (Section 8.4), the frequency of mutants that escape $n$ components of an immune response is the product of frequencies of escape to each individual component [$10^{-a} \times 10^{-b} \times 10^{-c} \times \ldots 10^{-i} = 10^{-(a + b + c + \ldots i)}$]. This is an oversimplification because it is not realistic to dissect the selective impact of a complex immune response into discrete components. A virus generally includes multiple antigenic sites and each of them is often composed of several overlapping or nonoverlapping epitopes; in addition, a virus has several T-cell epitopes in different structural and nonstructural proteins, and each epitope displays a different degree of relative dominance. However, the above abstraction reflects the advantage of stimulating the host immune system with a sufficiently broad array of B- and T-cell epitopes to prevent selection of vaccine-escape mutants due to a high genetic and phenotypic barrier (compare with the barrier to drug resistance described in Section 8.4.2). Therefore, selection of vaccine-escape viral mutants is more likely with synthetic peptidic vaccines, than with whole virus-attenuated or inactivated vaccines because the latter present a broad epitope repertoire to the immune system. Selection of escape-mutants by peptidic vaccines that evoked partial protection of cattle was documented with FMDV (Taboga et al., 1997; Tami et al., 2003). The arguments in favor of multiepitopic presentation are also endorsed by a notorious scarcity of licensed peptidic vaccines for viral diseases despite horrendous economic investments (orders of magnitude greater than investments in quasispecies research!). Use of a complex, multiepitopic vaccine, however, need not prevent long-term selection of antigenic virus variants as a result of vaccine usage, an important still largely underexplored topic discussed in Section 8.3.2.

Experimental evolution with FMDV has opened the way to a new generation of antiviral vaccines that share features of attenuated and inactivated vaccines. This new design is based on the conversion of the monopartite FMDV genome into a segmented genome version that dominated the population after extensive high MOI passages (Section 2.12 in Chapter 2 and Section 6.6 in Chapter 6). For the segmented virus version to be able to produce progeny, the two genome classes (which are encapsidated into separate particles) must reach the same cell. Therefore, administration of the segmented virus preparation should lead to an immune response akin to that evoked by inactivated viral particles, followed by a self-limiting infection. The vaccine was tested successfully in mice and swine, the authentic host of the parental virus (Rodriguez-Calvo et al., 2010). Potential concerns with this type of vaccine are that the standard (monopartite) genome can be reconstructed by recombination in the early stages of replication in the animal. A safety level is included in that particular FMD vaccine because of multiple mutations that accumulated during the transition toward genome segmentation, which deviated the genome sequence from the one present in the original swine isolate (Moreno et al., 2014). Exploration of evolutionary mechanisms that result in altered forms of viruses may open new possibilities for vaccine design.

New prospects for attenuated vaccines have been opened with the engineering of viruses with suboptimal replication fidelity or deoptimized codon or codon pair usage (Coleman et al., 2008; Vignuzzi et al., 2008; Cheng et al., 2015). Altered polymerase copying fidelity often leads to virus attenuation (Gnadig et al., 2012; Graham et al., 2012; Korbouk et al., 2014; Rozen-Gagnon et al., 2014; Van Slyke et al., 2015). A critical issue with live-attenuated vaccines based on deviation from the standard mutation rate is the stability of the attenuation trait. Not only true revertants or other site revertants of the polymerase may arise and displace the vaccine virus, but other viral proteins may affect nucleotide incorporation (Smith et al., 2013, 2015; Stapleford et al., 2015; Agudo et al., 2016).

## 8.3.2 Vaccination-induced evolution

When a virus circulates in a population where vaccinated and unvaccinated host individuals coexist, and the vaccine does not induce sterilizing immunity, viruses with an altered antigenic profile might be selected. The larger the overall effective population size of the circulating virus, and the longer the virus is allowed to replicate in such a scenario, the higher the probability of incorporation of compensatory mutations that yield high-fitness antigenic variants.

These events in the case of vaccines used in veterinary medicine are particularly significant because they may alter the cell tropism and host range of viruses, thus increasing the possibilities of their zoonotic transmission into humans (Schat and Baranowski, 2007). Evidence of vaccination-induced DNA and RNA virus evolution is increasing, and it has been documented with bovine respiratory syncytial virus, bovine herpesvirus-1, Marek's disease virus, porcine circovirus 2, and classical swine fever virus, among others [(Valarcher et al., 2000; Muylkens et al., 2006; Ji et al., 2014; Kekarainen et al., 2014; Constans et al., 2015; Yoo et al., 2018); reviews in Gandon et al., 2003; Schat and Baranowski, 2007)]. The timing of dominance of CTL-escape mutants of the simian immunodeficiency virus (SIV) was influenced by vaccination, and the process could be analyzed by penetration into the mutant spectra of the relevant viral populations (Loh et al., 2008).

For human viruses, evidence of vaccine-escape mutants has been obtained for hepatitis A and B viruses. Vaccination-associated escape mutants of HAV with substitutions around the immunodominant site of the virus were identified in a cohort of HIV-1, HAV doubly infected individuals (Perez-Sautu et al., 2011). The study suggested that an incomplete vaccination schedule, combined with the HIV-1-produced immunosuppression might have contributed to high-HAV loads, thus facilitating the generation and dominance of antigenic variants. In Taiwan, the prevalence of mutants at a major antigenic determinant of the surface antigen of hepatitis B virus (HBV) tripled in 1 decade, and it has been suggested that this increase of prevalence might be due to the ample vaccination coverage in the region (Hsu et al., 1999). Other studies also suggest the circulation of HBV mutants associated with vaccine escape and diagnosis failures (Di Lello et al., 2019).

Vaccines can rarely afford protection to all vaccinated individuals due to many factors that include variations in vaccine receptivity factors due to polymorphisms in genes involved in the adaptive immune response, immunosuppression of the vaccine recipient, the insufficient time between vaccination and exposure to the viral pathogen, and antigenic differences between the vaccine strain and circulating viruses. In addition, for massive vaccinations in veterinary medicine, damage to the vaccine (during transport, storage, etc.) and improper administration are additional problems. Vaccination may occasionally promote the selection not only of antigenic variants but also host cell tropism, host range, or virulent variants (Swayne and Kapczynski, 2008; Kirkwood, 2010; Read et al., 2015; Rolland, 2019). It is not known to what extent the widespread use of vaccination can contribute to antigenic variation relative to other factors (persistence of antibodies from previous infections, genetic drift due to genetic bottlenecks, etc.). However, our current understanding of virus dynamics should encourage investigations on the genetic and antigenic modifications of breakthrough viruses that arise from vaccinated individuals as compared with changes in viruses from unvaccinated host populations.

Reversion of live-attenuated vaccine viruses into virulent forms is a cause of disease derived from the evolutionary potential of viruses. In the case of attenuated Sabin poliovirus vaccine, the rate of vaccine-associated poliomyelitis among those vaccinated for the first time was one per 500,000 to 750,000 vaccinees, and the rate of those receiving the second vaccine dose was about one in 12 million (reviewed in Rowlands and Minor, 2010). Attenuated antiFMD vaccines were used in some countries during the second half of the 20th century, but a reversion to virulence forced the halting of the vaccination programs.

Vaccine-escape mutants may arise due to ineffective vaccines, and concomitant factors, such as immunosuppression. The escape mutants may remain confined to the unsuccessfully vaccinated host or may spread to other susceptible individuals, and attain different degrees of epidemiological relevance. Escape mutants may be direct mutants of the infecting virus or may originate by recombination between the infecting virus and other coinfecting related viruses, as observed with poliovirus and bovine herpesvirus-1 [Kew et al., 2002; Thiry et al., 2006; among other studies]. Reiteration of vaccine selection and fitness increase processes over many generations of vaccinees (be it humans or animals) may result in accelerated virus evolution. Since systematic use of vaccines for humans and animals in intensive production units is relatively recent in terms of evolutionary time (less than 100 years, and in some cases even only a few decades), it is still premature to evaluate whether vaccination is a significant factor in promoting long-term virus evolution.

## 8.4 Resistance to antiviral inhibitors

The first description of virus resistant to an antiviral inhibitor was by J. Barrera-Oro, H.J. Eggers, I. Tamm, and colleagues working with enteroviruses and guanidine hydrochloride and 2-(alpha-hydroxybenzyl)-benzimidazole as inhibitors (Eggers and Tamm, 1961; Melnick et al., 1961). These early results that suggested that antiviral-resistant mutants could be readily selected have been amply confirmed with many viruses and inhibitors in cell culture and in vivo. Indeed, the selection of viral mutants resistant to antiviral agents is an extremely frequent occurrence that has been known for decades, although it became widely recognized in the course of development and clinical use of antiretroviral agents to treat HIV-1 infections and AIDS.

The description of drug-escape mutants has been based on three main groups of observations:

- Detection of antiviral-resistant mutants in patients during treatment. When a reverse genetics system is available, the suspected mutation should be introduced in an infectious clone and resistance ascertained and quantified in cell culture or in vitro enzyme assays.
- Selection of resistant mutants in cell culture, by subjecting the viruses to passages in the presence of inhibitors. The viral population size is an important variable in this type of experiment (Section 8.4.1).
- Calculation of the frequency of resistant mutants by plating a virus in the absence and presence of the antiviral agent similar to the assays to calculate the frequency of monoclonal antibody (MAb)-resistant mutants (described in Chapter 7, Section 7.4.2).

In the three groups of observations, the frequency at which a specific escape mutant is found depends on a number of barriers to resistance (Section 8.4.2).

Traditionally, the fact that a drug can select virus-resistant mutants is regarded as a proof of the selectivity of the drug, as opposed to unspecific or toxic effects on the host cell that indirectly impair virus replication (Herrmann and Herrmann, 1977; Golan and Tashjian, 2011). Selection of viral mutants resistant to antiviral inhibitors is a major problem for the control of viral disease for two main reasons: (i) because it often results in virus breakthrough (increase of viral load) resulting in treatment failure and (ii) because resistant virus variants may become epidemiologically relevant, with the consequent decrease of inhibitor efficacy at the population level (Domingo and Holland, 1992; Huang et al., 2019).

Increasing numbers of antiviral agents have been developed based on the three-dimensional structure of viral proteins and their complexes with natural and synthetic ligands, in efforts

that have engaged academic institutions and pharmaceutical companies. Others have been incorporated as a result of drug repositioning, that is, the discovery of antiviral activity of compounds licensed for other medical purposes, often with the help of computational tools (Ab Ghani et al., 2019). Antiviral agents may target viral or cellular proteins involved in any step of the virus life cycle. They may interact with virions and inhibit an early step of infection, such as the attachment to the host cell, penetration into the cell, or uncoating to liberate the genetic material of the virus inside the cell. Other agents interfere with the synthesis of viral nucleic acids or viral protein processing, particle assembly, or virus release from cells.

Selection of resistant mutants has been described for virtually any chemical type of antiviral agent directed to any step of the infectious cycle of DNA or RNA viruses, including important pathogens, such as herpesviruses, picornaviruses, IV, HBV, and hepatitis C virus (HCV). Several reviews and articles have covered the theoretical basis of drug resistance, and consequences for treatment management [as examples see (Domingo et al., 2001b) and previous versions in Progress in Drug Research (Richman, 1994, 1996; Ribeiro and Bonhoeffer, 2000; Domingo et al., 2001a, 2012; Menendez-Arias, 2013; Perales, 2018; Mokaya et al., 2018; Nitta et al., 2019; Pawlotsky, 2019), and the articles in the Current Opinion of Virology volume edited by L. Menendez-Arias and D. Richman (Menendez-Arias and Richman, 2014)]. Therefore, the general mechanisms that confer adaptability to viruses are very effective in finding drug-escape pathways through molecular mechanisms that are summarized in Section 8.5.

## 8.4.1 Replicative load and antiviral resistance

Considering the implications of quasispecies dynamics explained in previous chapters, the following statement will be obvious to the reader: "If a single mutation is able to confer resistance to an antiviral agent, and the mutation does not cause a significant selective disadvantage to the virus (fitness decrease) in the considered environment, a drug-resistant virus mutant will be present in most, if not all, virus populations" (Domingo, 1989). If a virus replicates in such a way that a population size of $10^4$ can never be achieved in a single population, it is extremely unlikely that any drug-resistance mutation (or any mutation associated with a phenotypic change) that is generated at a frequency of $10^{-4}$ or lower will be propagated from that viral population (Perales et al., 2011).

Selection of escape mutants depends on the replicative load, and the concentration of inhibitor attained at the sites of virus replication. Consider different cell or tissue compartments in which an antiviral inhibitor reaches different concentrations (exerts different intensity of selection) (Fig. 8.1). In each compartment, there are multiple replication complexes. A mutation conferring resistance to the inhibitor will occur at the same rate in each of them, assuming that the mutation rate is independent of the presence of the inhibitor. However, after its occurrence, the proportion of viral RNAs harboring the mutation will decrease depending on the inhibitor concentration. The time at which the effect of the inhibitor will be manifested depends on the inhibitor target. In the example of Fig. 8.1 we assume that the concentration of inhibitor-resistant mutants will decrease in the replication complexes, reaching resistant mutant frequencies of $10^{-3}$, $10^{-4}$, and $10^{-5}$ in compartments 1, 2, and 3, respectively. The frequency of inhibitor-resistant mutants in the entire cell, tissue, or organism at that time will be given by the weighted average of mutation frequencies at the individual compartments. In the case of a virus-producing viremia, assuming no bottleneck effects or differential selection for other

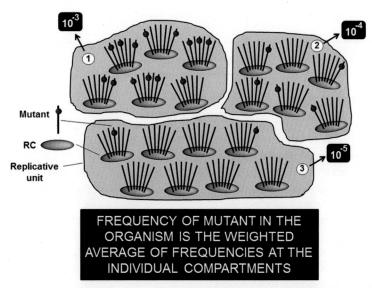

FREQUENCY OF MUTANT IN THE ORGANISM IS THE WEIGHTED AVERAGE OF FREQUENCIES AT THE INDIVIDUAL COMPARTMENTS

**FIGURE 8.1** Frequency of a drug-resistant mutant in different compartments (subcellular site, cell, tissue, or organ) of the same organism. Three compartments labeled 1, 2, and three are drawn. Replication complexes are depicted as ellipses and replicating genomes as lines. A replicative unit is defined here as a set of replication complexes. The inhibitor-resistant mutant is represented by a red circle in a genome. Compartments 1, 2, and 3 reach increasing concentration of the inhibitor, rendering inhibitor-resistant mutant frequencies of $10^{-3}$, $10^{-4}$, and $10^{-5}$, respectively. See text for the difference between occurrence and presence of the resistant mutant, and implications of compartmentalization.

traits, the frequency of resistant mutants calculated for the virus in blood should reflect the average frequency in all compartments that supply virus to blood. Low inhibitor concentration in a compartment will favor the selection of the resistant mutant that can either be archived as an adaptive reservoir or penetrate other compartments, depending on the sequence of events of virus spread.

If two or more independent mutations can confer resistance to an inhibitor, the probability of occurrence of an inhibitor-resistant mutant is equal to the sum of probabilities of occurrence of each mutation. For multiple mutations, the probability will be the sum of probabilities of the different mutations, a frequent case in viruses since they often display several evolutionary pathways to drug resistance. The probability of finding a viral genome resistant to two or more inhibitors directed to different targets is given by the product of probabilities of resistance to each of the individual inhibitors. The basic probability considerations regarding the frequency of occurrence of

inhibitor-resistant viral mutants are summarized in Box 8.2. When two or more mutations occur in the same genome, they may be subjected to epistatic effects, meaning either increase (positive epistasis) or decrease (negative epistasis) of viral fitness (see Section 2.3 of Chapter 2 for the concept of epistasis).

The diversity of chemical structures of the antiviral compounds that can select for escape mutants is illustrated in Figs. 8.2 and 8.3 with the formulae of some antiviral agents in current or historical use. They include relatively simple organic molecules, nucleoside analogs, and complex heterocyclic compounds with a variety of residues ($CH_3-$, $C=O$, $NH$, $NH_2$, $F$, and $Cl$) that may contribute to interactions with viral proteins or alter the electronic structure of neighbor bonds thus modifying the interaction behavior of some atoms. For all of them, resistant viral mutants have been identified, despite barriers imposed upon the virus to reach a drug-resistance phenotype.

---

## BOX 8.2

## PROBABILITY OF SELECTION OF INHIBITOR-ESCAPE MUTANTS

- If there are two or more different mutations that produce the same inhibitor-resistance phenotype, and once one of the mutations is present additional mutations are no longer necessary to produce the phenotype, the probability of achieving the phenotypic change is equal to the sum of probabilities of finding each mutation individually.

- If two or more independent mutations must happen to produce resistance to an inhibitor, the probability of occurrence of the necessary mutations is equal to the product of

probabilities of occurrence of each mutation individually.

- If a virus is inhibited by an inhibitor combination, and the mutations that confer resistance to each inhibitor are independent (no cross-resistance is involved), the probability of a combination-resistant mutant to arise is equal to the product of probabilities of resistance to the individual mutations.

These probability calculations are applicable to other mutation-dependent virus variations.

**FIGURE 8.2** Some inhibitors of picornaviruses and influenza virus. The inhibitors are (1) Dichloroflavan (4′,6-dichloroflavan). (2) Disoxaril, 5-[7-[4-(4,5 dihydro-2-oxazolyl) phenoxyl] heptyl]-3-methyl-isoxazole (WIN 51,711). (3) WIN 52,084. (4) Arildone, 4-[6-(2-chloro-4-methoxyphenoxy) hexyl]-3,5-heptanedione. (5) Enviroxime, anti-6-[(hydroxyimino)-phenyl]-1-[(−methylethyl) sulfonylimidazol-2-amine]. (6) Amantadine, (1-amino-adamantane). (7) Rimantadine, (α-methyl-1-adamantane methylamine). (8) Oseltamivir (trade name Tamiflu), ethyl (3R, 4R, 5S)-5-amino-4-acetamido-3-(pentan-3-yloxy)-cyclohex-1-ene-1-carboxylate. (9) Zanamivir (trade name Relenza), (2R, 3R, 4S)-4-guanidino-3-(prop-1-en-2-ylamino)-2-((1R, 2R)-1,2,3-trihydroxypropyl)-3,4-dihydro-2H-pyran-6-carboxylic acid.

## 8.4.2 Barriers to drug resistance

The impediments for a virus to attain resistance to an inhibitor are divided into genetic, phenotypic, and mutant swarm (population) barriers to resistance (Box 8.3).

The genetic barrier to resistance to a specific inhibitor is not a universal value for a virus group, since it may be affected by genetic differences among natural viral isolates. The diversification of HCV into genotypes 1a and 1b influenced the genetic barrier to resistance to the NS3/4A protease inhibitor telaprevir [formula (20) in Fig. 8.3]. One of the amino acid substitutions that confer resistance to telaprevir is R155K in NS3. In genotype 1a, the triplet encoding R155 is AGA; therefore, a single nucleotide transition G → A can yield the triplet AAA, which encodes K. In genotype 2b, the triplet encoding R-155 is CGA; therefore, two nucleotide changes (transversion C → A and transition G → A) are required to reach AAA, the triplet encoding K. Reaching the alternative AAG codon for K would require the same or a larger number of mutations (see Section 4.3.1 in Chapter 4 for another example of how the synonymous codon usage can influence an

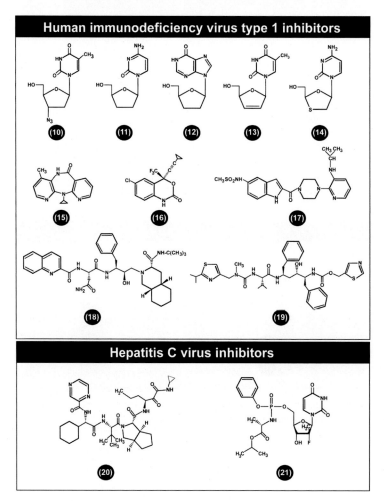

**FIGURE 8.3**    Some inhibitors of human immunodeficiency virus type 1 (antiretroviral agents) and hepatitis C virus. The inhibitors are (10) Zidovudine (AZT), 1-[(2R, 4S, 5S)-4-Azido-5(hydroxymethyl)oxolan-2-yl]-5-methylpyrimidine-2,4-dione. (11) Zalcitabine (ddC), 4-amino-1-((2R, 5S)-5-(hydroxymethyl)tetrahydrofuran-2-yl)pyrimidin-2(1H)-one. (12) Didanosine (ddI), 9-((2R, 5S)-5-(hydroxymethyl) tetrahydrofuran-2-yl)-3H-purin-6(9H)-one. (13) Stavudine (d4T), 1-[(2R, 5S)-5-(hydroxymethyl)-2,5-dihydrofuran-2-yl]-5-mehyl-1,2,3,4-tetrahydropyrimidine-2,4-dione. (14) Lamivudine (3 TC), 4-amino-1-[(2R, 5S)-2-(hydroxymethyl)-1,3-oxathiolan-5-yl]-1,2-dihydropyrimidin-2-one. (15) Nevirapine, 11-cyclopropyl-4-methyl-5, 11-dihydro-6H-dipyrido[3,2-b:2′,3′-e][1,4] diazepin-6-one. (16) Efavirenz, (4S)-6-chloro-4-(2-cyclopropylethynyl)-4-(trifluoromethyl)-2,4-dyhidro-1H,3,1-benzoxazin-2-one. (17) Delavirdine, N-[2-({4-[3-(propan-2-ylamino) pyridin-2-yl] piperazin-1-yl} carbonyl)-1H-indol-5-yl]methanesulfonamide. (18) Saquinavir, (2S)−N-[(2S,3R)-4-[(3S)-3-(tert-butylcarbamoyl)-decahydroisoquinolin-2-yl]-3-hydroxy-1-phenylbutan-2-yl]-2-(quinolin-2-ylformamido)butanediamide. (19) Ritonavir, 1,3-thiazol-5-ylmethyl N-[(2S,3S, 5S)-3-hydroxy-5-[(2S)-3-methyl-2-{[methyl({[2-(propan-2-yl)-1,3-thiazol-4-yl] methyl})carbamoyl]amino}butanamido]-1,6-diphenylhexan-2-yl]carbamate. (20) Telaprevir, (1S,3aR,6aS)-2-[(2S)-2-[[(2S)-2-cyclohexyl-2-(pyrazine-2-carbonylamino) acetyl]amino]-3,3-dimethylbutanoyl]-N-[(3S)-1-1(cyclopropylamino)-1,2-dioxohexan-3-yl]-3,3a,4,5,6,6a-hexahydro-1H-cyclopenta [c]pyrrole-1-carboxamide.    (21) Sofosbuvir, isopropyl(2S)-2[[[(2R,3R,4R,5R)-5-(2,4-dioxopyrimidin-1-yl)-4-fluoro-3-hydroxy-4-methyl-tetrahydrofuran-2-yl]methoxy-phenoxy-phosphoryl]amino]propanoate. Additional drugs currently used in antiviral therapy can be found in the references quoted in the text and in Chapter 9.

---

BOX 8.3

## BARRIERS TO DRUG RESISTANCE IN VIRUSES

- Genetic barrier: Number and types of mutations needed to acquire the resistance trait.
- Phenotypic barrier: Fitness cost imposed by the resistance mutations. The cost may be due to effects at the RNA level, protein level, or both. A high fitness cost may result in reversion of the relevant mutation in the absence of the drug or incorporation of

compensatory mutations that increase viral fitness.
- Mutant swarm barrier: Suppressive effect of mutant spectra may impede the dominance of resistant mutants.
- The combined effect of the different barrier classes determines the ease of dominance of drug-resistant mutants in populations subjected to the selective pressure of one or multiple inhibitors.

---

evolutionary outcome). The requirement of transitions versus transversions to reach the resistant phenotype will affect the genetic barrier to resistance. Most viral polymerases tend to produce transition mutations more readily than transversions presumably because in the course of RNA elongation it is easier to misincorporate a purine by another purine than by a pyrimidine, and the same for pyrimidine misincorporations (Chapter 2). Mutation preference is one of several factors that determine the frequency of drug-escape mutants. Thus, evolution may diversify viruses to display different genetic barriers to the same drugs. Since in many cases, several independent mutations may confer resistance to the same drug to complicate matters even more, it also has to be considered that the genetic barrier to one inhibitor may be affected by the presence of other inhibitors (Beerenwinkel et al., 2005).

The phenotypic barrier to drug resistance is equivalent to the fitness cost inflicted upon the virus by the mutations and corresponding amino acid substitution(s) required for resistance [Fitness cost is treated in Chapter 4 (Section 4.6) and in Chapter 7 (Section 7.4.2) in connection

with the frequency of monoclonal antibody- or cytotoxic T-cell-escape mutants in viral populations]. When a drug-resistance mutation inflicts a high fitness cost, a likely result is a reversion of the mutation when the virus replicates in the absence of the drug. An alternative outcome is that compensatory mutations are introduced in the genome so that viral fitness increases while maintaining the inhibitor-resistance mutation. The two outcomes are not mutually exclusive and may contribute to the multiple, transient selection pathways observed by the application of deep sequencing to monitor the response of a viral population to specific selective force (Tsibris et al., 2009; Fischer et al., 2010; Cale et al., 2011; Kortenhoeven et al., 2015) (see Section 6.3 in Chapter 6). A high fitness cost may prevent or delay the selection of escape mutants. Sofosbuvir [formula (21) in Fig. 8.3] is a very effective NS5B (viral polymerase) inhibitor of HCV. Amino acid substitution S282T in NS5B has been associated with sofosbuvir resistance, and the substitution has been detected in patients and in some natural isolates of HCV. In one of several clinical studies on sofosbuvir efficacy, the mutant spectrum composition of HCV genotype 2b in an infected patient treated with the drug was followed by deep sequencing of the virus at baseline (prior to initiation of treatment), in the course of treatment, and posttreatment. The frequency of S282T was 0.05% at baseline, indicating preexistence of resistance mutations despite no exposure of the virus to the drug (Section 8.6). Two days after initiation of sofosbuvir treatment, the level of S282T decreased to 0.03%, and viral breakthrough was detected 4 weeks later when 99.8% of the viral population included S282T. During the posttreatment period, genomes with the wild-type S282 amino acid regained dominance that was attributed to the true reversion of mutant genomes rather than the outgrowth of baseline wild-type genomes (Hedskog et al., 2015). This result suggests a high phenotypic barrier for sofosbuvir, and that HCV has mechanisms to overcome the barrier. The complexities of virus-host interactions render the elucidation of the pathways exploited by a virus to overcome the phenotypic barrier to a drug a highly empirical endeavor. The hope is that a combination of inhibitors that display a high barrier to resistance may impede escape and drive the virus to extinction (Chapter 9).

The mutant swarm barrier to resistance is a consequence of the interfering interactions that operate within quasispecies, and that are described in Chapter 3 (Section 3.8). It is a particular case of interference that can delay or impede the increase of frequency of a resistance mutation (Crowder and Kirkegaard, 2005; Kirkegaard et al., 2016). The possible contribution of mutant swarms to facilitate or prevent the dominance of drug-resistant mutants in infected patients is still largely unexplored.

It is difficult to anticipate how the three types of barrier listed in Box 8.3 may result in a level of drug resistance for a particular virus, in a particular host individual, in a specific target organ, at a given time. Additional influences are drug pharmacokinetics, drug penetration into different cells, tissues, and organs where the virus replicates (see Fig. 8.1), and prior history of virus replication in the infected host. It is not surprising that the study of drug resistance in viruses remains fundamentally descriptive.

### 8.4.3 Drug efficacy, mutant frequencies, and selection of escape mutants

When independent amino acid substitutions can lead to resistance to the same drug, alternative evolutionary pathways may be followed depending on tRNA abundances, mutational preferences, and relative nucleotide substrate concentrations at the virus replication sites. If resistance requires two or more amino acid substitutions, the genetic barrier will be correspondingly increased (Section 8.4.1 and Box 8.2).

Quantification of barriers to resistance in experiments in cell culture requires a prior characterization of the drug to be tested when acting on the cell culture-adapted virus as it infects a specific cell line. The two basic parameters to be determined are the toxicity of the drug for the host cell, and its capacity to inhibit the production of infectious virus. Toxicity is quantified by the concentration of drug that kills a given percentage (generally 50%, but sometimes another value) of cells under the conditions used in the infection. It is expressed as the cytotoxic concentration 50 ($CC_{50}$), as depicted in Fig. 8.4. Toxicity may depend on the cell concentration, the extent of confluence in a cell monolayer, and the metabolic state of the cell (resting vs. actively dividing). The capacity of inhibition is quantified by the concentration of inhibitor that reduces the infectious progeny production by a given percentage (generally 50%, but sometimes another value) under the defined conditions of the infection, including a multiplicity of infection (MOI). It is expressed as the inhibitory concentration 50 ($IC_{50}$), as depicted in Fig. 8.4. The therapeutic index (TI) is given by the quotient $CC_{50}/IC_{50}$, and although generally used for in vivo experiments of drug efficacy testing, it can also be applied to cell culture measurements.

**Cell toxicity and inhibitory activity of antiviral agents**

FIGURE 8.4 Schematic representation of two experiments to determine the concentration of an inhibitor needed to kill 50% of cells in culture ($CC_{50}$ value, left) and the concentration of inhibitor that reduces the viral production to 50% ($IC_{50}$ value, right). The therapeutic index is the quotient between $CC_{50}$ and $IC_{50}$ (box at the bottom). Similar tests can be performed with tissue explants or animals, under controlled environmental conditions. See text for pharmacological implications.

The three parameters, $CC_{50}$, $IC_{50}$, and TI, are not universal for a virus and a drug since they may be influenced by the composition of the viral population and environmental factors, as repeatedly expressed for other features of viruses in the present book. As a guide, TI values of 100 or more suggest the excellent performance of an antiviral agent; values higher than ten are acceptable, but values lower than ten predict limited efficacy. The quantitative effects of a drug may vary when analyzing a single round of infection versus multiple rounds in serial passages, or when comparing in vivo versus cell culture experiments. $CC_{50}$ and $IC_{50}$ values serve as a guide to decide the range of the drug concentration to be used in serial passage experiments to evaluate the possible selection of inhibitor-resistant mutants and to estimate the genetic barrier.

The possibility to overcome a genetic barrier depends on the virus population size. For viruses that replicate in cell culture, it is possible to estimate the minimal viral population size

needed to select a drug-resistant mutant which is generally positively correlated with the genetic barrier (Fig. 8.5). In the hypothetical example of the figure, a viral population is composed of inhibitor-sensitive viruses (blue spheres), and a low level of inhibitor-resistant viruses (red spheres). The proportion of inhibitor-resistant viruses is given by the mutational pressure (e.g., at a frequency of $10^{-4}$, which is increased in the picture for clarity). Passage of a small amount of virus (e.g., $10^{2}$ infectious virus in the small circle at the upper part of the figure) will exclude the mutant virus (red spheres) that will be maintained at the basal level dictated by mutational pressure in the course of passages (limited to two in the figure for simplicity). Selection of escape mutants is precluded by the limited population size at each transfer. In contrast, if the population size used for the successive infections is sufficiently large ($>10^{4}$, larger circles at the bottom that surround both sensitive and resistant viruses), the resistant mutant can become

**FIGURE 8.5**    The effect of sampled virus population size on the dominance of an inhibitor-resistant mutant. The upper left population is composed of inhibitor-sensitive (blue spheres) and inhibitor-resistant (red spheres) viruses. If during passages P1 and P2, the amount of sampled virus is insufficient (upper three successive populations), the resistant virus mutant will not be enriched in the population. If during passages P1 and P2, the amount of sampled virus exceeds a critical value (bottom three successive populations), the population will be enriched in the resistant mutant. See text for numerical examples.

gradually dominant and can be isolated for further studies. To give another example, a single amino acid replacement that requires two mutations (the change from CAG to AUG to attain substitution Q151M in HIV-1 reverse transcriptase, associated with resistance to multiple antiretroviral nucleosides) will occur at a lower frequency than replacements that require a single mutation. If each of the two mutations reaches a frequency of $2 \times 10^{-4}$, the expected frequency of the drug-resistant genomes (ignoring fitness effects) will be $(2 \times 10^{-4}) \times (2 \times 10^{-4}) = 4 \times 10^{-8}$. Thus, at least $4 \times 10^8$ viral genomes must undergo one round of copying (or a lower number of genomes a proportionally higher number of rounds of copying) to approach a good probability to obtain a drug-resistant genome in that viral population. Population size limitation of a drug selection event is a specific example of how random events may intervene in the process of positive selection (compare with Section 3.2 and Fig. 3.2 in Chapter 3; in that figure, the random event that excludes the positively selected

population is conceptually equivalent to the insufficient population size depicted in the upper infection series of Fig. 8.5).

When two or more mutations are needed to confer the resistance phenotype, drug resistance will be less likely not only due to the lower probability of generating the two required mutations but also because of the increased chances of two mutations in the same genome entailing a fitness cost. A virus that requires three or more mutations to overcome a selective constraint may occur at a frequency in the range of $10^{-12}$ or lower which will often be insufficient for the mutant to be present in the mutant spectrum of the infected host (Fig. 8.6).

Failure to select for a drug-resistant mutant in cell culture does not necessarily mean that the resistant mutant is not present in the population. It may mean that due to a high genetic barrier, the selection experiment was designed to infect with an insufficient amount of virus. Similar and even more accentuated problems are encountered in selection experiments in vivo, since not only

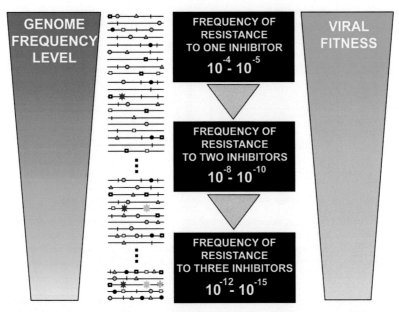

**FIGURE 8.6** Decreased frequency and fitness of mutant genomes resistant to one, two, or three inhibitors. The genome frequency level decreases by several orders of magnitude when resistance to one inhibitor (red asterisk in the upper mutant spectrum), or two inhibitors (red and green asterisks in the middle-mutant spectrum), or three inhibitors (red, green, and blue asterisk in the bottom mutant spectrum) must occur in the same genome (left part of the figure and numerical values at the center). Increased number of mutations generally implies fitness decrease (right part of the figure). See text for implications.

population size but also the source of the virus (blood or another organ) for the next infection may determine the inclusion or exclusion of the relevant mutation (compare Figs. 8.1 and 8.6).

### 8.4.4 Phenotypic barrier and selective strength

The phenotypic barrier or fitness cost inflicted by a drug-resistance mutation (Box 8.3) is often estimated empirically from the frequency of the relevant substitution in patients treated with the drug or in cell culture assays. An adequate procedure to quantify the phenotypic barrier to resistance is to determine the fitness of the virus expressing the protein with the wild-type amino acid (the one that confers drug sensitivity) relative to the virus expressing the protein with the substituted amino acid that confers resistance; fitness is measured in the absence

and presence of the drug (double assay). This is an extension of the determination of fitness vectors described in Section 5.1.1 of Chapter 5, as depicted in Fig. 8.7; the assays are best performed in cell culture, although the use of explants or in vivo assays is also feasible. Two parameters can be calculated: the fitness cost inflicted by the amino acid substitution associated with resistance, in the absence of the drug (reflected in a fitness value $f_{-DRUG} \leq 1$ relative to the wild type), and the selective advantage conferred by the substitution in the presence of the drug (reflected in a fitness value $f_{+DRUG} > 1$ relative to the wild type). The lower the value of $f_{-DRUG}$, the higher the fitness cost. We define the selective strength of the resistance mutation as $f_{+DRUG}/f_{-DRUG}$. For example, if we put arbitrary numbers (unrelated to values shown in ordinate) to the fitness values in the first graph of Fig. 8.7,

**FIGURE 8.7** Selective strength of an inhibitor-resistance mutant. A mutation may confer a different degree of resistance at a different fitness cost for the virus. The fitness vectors in the left panel denote limited resistance at a low fitness cost, whereas the vectors in the right panel denote high resistance despite a considerable fitness cost inflicted by the resistance mutation. The selective strength is defined as the ratio of fitness in the presence and absence of the drug, as explained in the text.

$f_{+DRUG} = 1.4$ and $f_{-DRUG} = 0.8$, we obtain a selective strength of 1.7. For the vectors in the second graph, if $f_{+DRUG} = 3.6$ and $f_{-DRUG} = 0.1$, the selective strength is 36. High selective strength means an important selective advantage conferred by the amino acid substitution for virus replication in the presence of the drug despite a high fitness cost inflicted by the substitution (compare with Section 4.2 in Chapter 4 for the trade-off and "no free lunch" concepts). If the substitution does not entail any fitness cost ($f_{-DRUG} = 1$), the fitness value in the presence of the drug equals the selective strength. Selective strength can be calculated for a mutation or group of mutations that confer resistance to a drug used at a given concentration in a defined environment [example in (de la Higuera et al., 2017)]. The limitations of fitness measurements (environment dependence, etc.) described in Section 5.1.2 of Chapter 5 apply here. Since viral genomic sequences may vary in the course of fitness assays, a limited number of passages and triplicate parallel assays are recommended. If a substitution entails a high fitness cost, direct reversion of the substitution or incorporation of compensatory mutations may occur. Nucleotide sequence

monitoring in the course of the assay should reinforce the conclusions.

## 8.4.5 Multiple pathways and evolutionary history in the acquisition of drug resistance

Most drug-resistance mutations inflict a fitness cost upon the virus and yet very rarely drug resistance represents an unsurmountable barrier to maintain viral infectivity. Several possibilities can account for the pertinacious occurrence and selection of drug-resistant, viable viral mutants. One possibility, supported by some experimental and clinical observations, is that a drug-resistant phenotype may be achieved through a number of alternative genetic modifications. Even if a specific amino acid substitution—that would serve as the most direct and effective determinant of drug resistance—were highly detrimental or lethal for the virus, alternative mutations can often be found that lead to a similar resistance phenotype, or at least a sufficient resistance to permit virus replication and exploration of sequence space to find compensatory mutations.

The connectivity among points of sequence space and the fact that several space positions map into the same (or similar) drug-resistance phenotype contribute to the extended occurrence of drug resistance. This phenotypic redundancy applies to both standard nonmutagenic inhibitors and to mutagenic inhibitors. The cascade of mutations that confer resistance of picornaviruses to the mutagenic purine analog ribavirin illustrates how alternative amino acid substitutions in the viral polymerase (some being genuine resistance mutations and others acting as compensatory substitutions to maintain polymerase function) can lead to the ribavirin-resistance phenotype (discussed in Chapter 9).

A speculative interpretation of the systematic occurrence of drug-resistant viral mutants is that the majority of the chemicals used in antiviral therapy (Figs. 8.2 and 8.3) have a structure which may be related to natural compounds that viruses and their ancestral replicative machineries encountered in their continuous struggle to survive. In this view, drug resistance would have been gradually built as a consequence of coevolution (Section 4.5 of Chapter 4) between virus replicative and gene expression machineries and the "space" of chemical compounds that interacted with them. Mechanisms of drug resistance might have had their roots in molecular events repeatedly experienced as viruses evolved in an interactive manner with protocellular and cellular metabolites in our biosphere. Discrimination in favor of small molecule substrates compatible with a flow of genome replication and gene expression and avoidance of perturbing intruders that could alter catalytic activities should have been positively selected. Unfortunately, this is a possibility we will never be able to test. Whatever the reasons behind, the unfortunate reality is that drug resistance is an extremely frequent event that complicates enormously the control of the viral disease.

## 8.5 Molecular mechanisms of antiviral resistance

Multiple drug evasion mechanisms have been identified or proposed for RNA and DNA viruses, and they can operate depending on basic replicative parameters in connection with drug levels and their variation with time. Delayed lysis of bacteriophage ΦX174 that reduced the replication rounds in the presence of 5-fluorouracil (FU) produced resistance to this analog (Pereira-Gómez and Sanjuán, 2014). In this line, synchronization of a virus life cycle so that replication occurs when drug levels are minimal has been proposed as a potential resistance mechanism (Neagu et al., 2018). Virus plasticity favors modification of life cycle parameters for virus survival in the presence of drugs, provided time for the relevant selection events is allowed once the treatment has been implemented. The great majority of inhibitor-resistance mechanisms involve amino acid substitutions in viral proteins that directly or indirectly diminish the binding of the drug to the viral target.

The following major mechanisms have been documented:

- Substitutions in the protein targeted by the drug that decreases the affinity of the protein for the drug. Mutations that modify nucleotide selectivity in viral polymerases belong to this group. Substitutions may affect the neighborhood of the polymerase catalytic site, other polymerase domains, or even nonstructural proteins that interact with the polymerase.
- If the inhibitor acts on a viral protein that itself has some other viral protein or genomic structure as a target, amino acid substitutions, or mutations that affect that target may also contribute to resistance. Correlated mutations

in the first and second target may also yield the resistance phenotype. This is the case of some protease inhibitor-resistance mutations in HIV-1.

- In the case of viral polymerases, some mutations permit the excision of a chain-terminating nucleotide at the 3′-end of the primer. It is achieved through phosphorolysis mediated by a pyrophosphate donor, probably ATP.

## 8.5.1　Some examples with HIV-1

Resistance to nucleoside/nucleotide reverse transcriptase (RT) inhibitors (NRTIs) is achieved by one of at least two mechanisms: (i) discrimination against the incorporation of the triphosphate form of the NRTI and (ii) excision of the chain-terminating nucleotide once incorporated at the 3′-end of the growing DNA chain. This occurs with thymidine analog resistance mutation (or TAMs) that are typically selected under treatment with AZT or d4T [formulae (10) and (13) in Fig. 8.3]. Groups of amino acid substitutions may yield a multidrug-resistance phenotype. A well-studied example in HIV-1 is the Q151M complex in the reverse transcriptase, which includes substitutions A62V, V75I, F77L, F116Y, and Q151M. The phenotype consists of the limitation of incorporation of several nucleotide analogs. These and other mutants are characterized by a decrease in the catalytic rate constant ($k_{pol}$) of incorporation of the analog or analogs relative to standard nucleotides (see Section 2.6 in Chapter 2 for the basic kinetic parameters for polymerase activity). The combination of enzymological and structural studies has provided a molecular interpretation of the mechanism of inhibition of HIV-1 by NRTIs (reviews in Menéndez-Arias, 2013; Menéndez-Arias and Alvarez, 2014; Clutter et al., 2016; Günthard et al., 2019); for a predictive model that includes nucleotide levels, see von Kleist et al. (2012).

Nonnucleoside RT inhibitors (NNRTIs) bind to an RT pocket 10Å away (a considerable distance) from the catalytic site composed of residues of the p66 and p51 subunits of the enzyme. The hydrophobic nature of NNRTIs is illustrated in Fig. 8.3 with the structures of nevirapine, efavirenz, and delavirdine [formulae (15), (16), and (17), respectively]. Several mechanisms have been proposed to explain their inhibitory activity, including alteration of the catalytic amino acids YMDD at the RT active site, distortion of the nucleotide-binding site, or modification of the position of the primer that receives the incoming nucleotides (Menéndez-Arias, 2013). Amino acid substitutions that confer resistance to NNRTIs block the access of the inhibitors to their binding sites or alter the conformation and volume of the binding pocket.

The essential viral proteases are a target for the development of specific antiviral inhibitors. Examples are saquinavir and ritonavir for the HIV-1 [formulae (18) and (19), respectively, in Fig. 8.3]. Multiple resistance mutations have been described for protease inhibitors. They can affect the substrate-binding site or neighbor positions, often accompanied of compensatory substitutions at distant positions including sites of the target viral protein [e.g., the Gag cleavage sites in HIV-1 (Fun et al., 2012; Flynn et al., 2015)]. Some HIV-1 protease inhibitor combinations display a high genetic barrier to resistance but, significantly for the capacity of viruses to explore sequence space, resistant mutants with 20—30 amino acid substitutions in the protease-coding region have been isolated (Rhee et al., 2010).

The reviews by Menéndez-Arias (2013) and Menéndez-Arias and Alvarez (2014) provide excellent and detailed accounts of mechanisms of resistance of HIV-1 and HIV-2 to inhibitors that include virus entry and integrase inhibitors, in addition to the RT and protease inhibitor briefly described here.

## 8.5.2 Mutation site and functional barrier

Because of the distance effects that can be exerted among amino acids on the structure of proteins, substitutions that confer resistance to inhibitors of viral enzymes may lie far from the catalytic site. The effect of drug-resistance mutations on the general catalytic efficiency of a viral enzyme is one of the determinants of the functional barrier to resistance, since it may diminish replicative fitness. When an enzyme activity assay in vitro is available, the effects of specific drug-resistance amino acid substitutions on enzyme activity can be tested, although the observed alteration may not be the only influence on the fitness modification of the corresponding mutant virus. The reason is that most viral proteins (including viral enzymes) are multifunctional, and an enzyme activity assay may not capture the range of influences exerted by the enzyme. Regarding direct-acting antiviral (DAA) agents for HCV, the inhibitors that target the HCV polymerase (NS5B) generally display a higher functional barrier to resistance than the protease inhibitors and manifest a broader genotype coverage. Fitness decreases entail reductions in viral load and, consequently, lower probability of viral breakthrough (treatment failure). Nucleotide analogs that bind to conserved residues at or near the active site of the viral polymerase tend to show subtype-independent antiviral activity. Because amino acid substitutions at or near the active site of viral enzymes are likely to inflict a fitness cost, such substitutions may not preexist in treatment-naive patients (Margeridon-Thermet and Shafer, 2010; Sarrazin and Zeuzem, 2010). Interestingly, in the case of HIV-1, mutations conferring resistance to RT inhibitors inflict a lower fitness cost than mutations that confer resistance to protease inhibitors (reviewed in Martinez-Picado and Martinez, 2008). Thus, enzymes that perform similar functions for different viruses may have evolved to display different tolerance to amino acid substitutions. It is not possible to generalize which types of resistance mutations will display high or low functional barriers.

## 8.5.3 Additional considerations on escape mutant frequencies

There is a broad range of frequencies of antibody- and drug-escape mutants in viral populations, although values of $10^{-4}$ to $10^{-6}$ mutants per infectious unit are frequent for many RNA and DNA viruses (see Table 7.2 in Chapter 7 for antibody-escape mutants). A few estimates for drug-escape mutants are listed in Table 8.1; several observations and characterization of escape mutants have not been accompanied by frequency measurements. A point worth emphasizing is that quantification of viruses harboring biologically relevant mutations has been possible because an adequate biological assay is available. An antiviral agent or a neutralizing antibody measures the proportion of infectious viral particles that differ from the majority of the population in the relevant resistance trait. There is no reason to suspect that the viral amino acid residues (that are the target of an inhibitor or an antibody) that are substituted to confer the resistance phenotype are more prone to accept variations than many other amino acids in viral proteins. If we had additional selective agents to probe other viral sites, we expect a similar range of variant amino acids than using inhibitors or antibodies. This quite straightforward prediction is another way to state that there is general agreement in the mutation rates and frequencies for viruses being in the range of $10^{-3}$ to $10^{-5}$ substitutions per nucleotide (s/nt), calculated using a variety of biochemical and genetic methods (Chapter 2).

A quite general observation is that in antibody neutralization experiments, a fraction of the virus population remains infectious despite the

**TABLE 8.1**  The frequency of some drug-escape mutants.

| Drug[a] | Virus[b] | Frequency | Comments | References |
|---|---|---|---|---|
| HBB[c] | CAV9 | $10^{-4}$ | Reversion from HBB dependence to HBB independence | Eggers and Tamm (1965) |
| Amantadine and rimantadine | IV | $1 \times 10^{-3}$ to $4 \times 10^{-4}$ | Measurements in cell culture | Appleyard (1977), Lubeck et al. (1978) |
| Rimantadine[a] | IV | 27% | Percentage of children treated with rimantadine that shed resistant IV | Belshe et al. (1988) |
| Disoxaril[a] | HRV14 | $1 \times 10^{-3}$ to $4 \times 10^{-4}$ | Low-level resistance in cell culture | Heinz et al. (1989) |
|  |  | $4 \times 10^{-5}$ | High-level resistance in cell culture |  |
| Guanidine[d] | PV | $1.8 \times 10^{-5}$ to $4 \times 10^{-8}$ | Measurements in cell culture | Pincus and Wimmer (1986) |

[a] *The formula of these drugs is included in Figure 8.2 with the following number in parenthesis: amantadine (6); rimantadine (7); disoxaril (2).*
[b] *The virus abbreviations are CAV9, coxsackievirus type A9; IV, influenza virus; HRV14, human rhinovirus type 14; PV, poliovirus.*
[c] *HBB is 2-(alpha-hydroxybenzyl)-benzimidazole.*
[d] *Guanidine is aminoformamidine.*

addition of high antibody concentrations. Although the resistance mechanism is unclear, a possibility is that the heterogeneous viral population includes a small proportion of antigenic variants with decreased affinity for antibodies. Incomplete neutralization with the nonsigmoidal slope in the neutralization curves has been characterized for broadly neutralizing antibodies directed to HIV-1 (McCoy et al., 2015). This general observation is an added complication to preventive designs that consider administration of neutralizing antibodies either as vaccine additives or in combination with antiviral inhibitors (Section 9.2 in Chapter 9).

The calculated frequencies of antibody- or drug-resistant mutants in viral populations may be lower than the rate at which they originate by mutation due to the fitness cost of the mutation (Section 8.4.4). The argument is parallel to that used to justify why mutation rates and frequencies differ due to the fitness effects of mutations (Chapter 2). Another prediction derived from the above considerations is that mutations conferring resistance to antiviral agents are expected to be detected in viral populations never exposed to the relevant drugs. All

forms of genetic variation of viruses can contribute to dominance and spread of drug-resistant mutants, including the combined effect of mutation, recombination, and genome segment reassortment (Richman, 1996; Neher and Leitner, 2010; Rogers et al., 2015). The basal level of mutational pressure may be sufficient to provide a detectable proportion of escape mutants without the need for selection by the selective agent. This is an important aspect of antiviral therapy that is addressed next.

## 8.6 Antiviral resistance without prior exposure to antiviral agents

The first demonstration that the baseline mutation level in viral quasispecies can include a detectable level of mutations that confer resistance to inhibitors in the absence of selection by the inhibitors, was obtained by D.D. Ho, I. Nájera, C. López-Galíndez, and their colleagues working with HIV-1 (Mohri et al., 1993; Nájera et al., 1994, 1995). One of the studies examined the *pol* gene of 60 HIV-1 genomes obtained directly from lymphocytes of infected patients. Mutation frequencies for independent viral isolates were in the range of $1.6.10^{-2}$ to $3.4.10^{-2}$ s/nt, while for mutant spectrum components of individual isolates the values were $3.6.10^{-3}$ to $1.1.10^{-2}$ s/nt. In the virus from these patients, mutation frequencies at the codons for amino acids involved in antiretroviral resistance were very similar to the average mutation frequency for the entire *pol* gene. Consistently with the mutation frequency values, several mutations that led to amino acid substitutions that conferred resistance to reverse transcriptase inhibitors were identified in patients not subjected to therapy. At the time of the study, the number of antiretroviral agents was still limited, and a considerable number of patients were not treated. The authors gave convincing epidemiological arguments that the background of mutations related to antiretroviral resistance was a consequence of high mutation rates and quasispecies dynamics, and not due to the transmission of resistant virus from individuals that had been subjected to therapy (primary resistance) (Nájera et al., 1995).

The presence of inhibitor-resistance mutations in viral populations never exposed to the corresponding inhibitor has been confirmed for HIV-1 and for several other viruses, including HCV, and it is supported by the calculated mutant frequencies in viral quasispecies (Havlir et al., 1996; Lech et al., 1996; Ribeiro et al., 1998; Ribeiro and Bonhoeffer, 2000; Cubero et al., 2008; Johnson et al., 2008; Toni et al., 2009; Tsibris et al., 2009; Peres-da-Silva et al., 2017). Ample support has also come from deep sequencing analyses of mutant spectra, opening a point of debate on the basal frequency of inhibitor-resistance mutations that constitutes an indication to avoid the use of the corresponding inhibitors in therapy. At least in the case of HCV (and probably applicable to other viruses), there is no basis to suggest that the presence of a drug resistance mutation below a certain level in a patient will not have relevance for treatment failure when the inhibitor is administered. The understanding of quasispecies dynamics cautions against such reductionist arguments

as evidenced by clinical cases (Perales et al., 2018). The data underline the relevance of mutant spectra as phenotypic reservoirs to confront selective constraints before constraints are in operation. For treatments, including a drug that has already been administered to a patient in the past, the influence of quasispecies memory should also be considered (Section 5.5.1 in Chapter 5). Mutant spectra can be viewed as an anticipatory reservoir of phenotypes.

In addition to the presence of drug resistance mutations in viral populations due to mutant frequency levels, the transmission of drug-resistant mutants from treated to naïve patients may contribute to epidemiological relevance of resistance mutations. Such primary resistance has been amply documented with HIV-1, and it appears to increase in the case of HCV (Franco et al., 2014; Echeverría et al., 2016; Huang et al., 2019). Higher levels of resistance mutations as a function of time in untreated patients is an indication that the mutations are not due to basal mutant frequencies but to the epidemiological expansion of virus mutants that originated in treated patients. From the clinical data available, the rate of expansion of virus harboring resistance mutations may vary depending on transmission and epidemiological features of each pathogen, but it seems unavoidable in the face of extended treatments for genetically variable pathogenic viruses. We confront a situation with parallels with antibiotic resistance in bacteria (Chapter 10).

## 8.7 Fitness or a fitness-associated trait as a multidrug-resistance mechanism

The major mechanism of drug resistance in viruses is based on amino acid substitutions that render the drug ineffective through the several molecular mechanisms summarized in Section 8.5. Application of the cell culture system of HCV replication in human hepatoma cells (Lindenbach et al., 2005; Wakita et al., 2005; Zhong et al., 2005) to examine the effects of long-term evolution has indicated that viral fitness can be an additional mechanism of drug resistance. The evidence was obtained when addressing the important issue of HCV resistance to interferon-alpha (IFN-α). IFN-α and ribavirin were the two components of the standard of care treatment against HCV infections until the advent of new therapies based on DAA agents in 2014. Natural HCV isolates differ in IFN-α sensitivity, and the molecular basis of the difference is largely unknown. The study in cell culture consisted in subjecting a clonal population of HCV (termed HCVp0, prepared by electroporation of hepatoma cells with RNA encoding the viral genome, transcribed from a plasmid) to 100 serial passages (of the type described in Section 6.1 of Chapter 6) in the absence or presence of increasing concentrations of IFN-α added to the culture medium. Several mutations scattered throughout the HCV genome were associated with IFN-α resistance (Perales et al., 2013). The selection of multiple alternative mutations is most likely influenced by the fact that IFN-α evokes a multicomponent antiviral response, which is not focused toward a single viral protein (Perales et al., 2014). Unexpectedly, even the control HCV populations (those passaged 100 times in the absence of IFN-α) displayed a partial (but statistically significant) resistance to IFN-α that could not be attributed to endogenous IFN production by the hepatoma cells (Perales et al., 2013).

In view of this intriguing result, the initial HCVp0 population and the HCV population passaged 45 and 100 times in the absence of IFN-α (termed HCVp45 and HCVp100, respectively) were tested for their resistance to other inhibitors of HCV replication: the protease inhibitor telaprevir [formula (20) in Fig. 8.3], the NS5A inhibitor daclatasvir, the cellular protein cyclophilin A inhibitor cyclosporin A, the mutagenic purine nucleoside ribavirin, and the high barrier inhibitor sofosbuvir [formula (21) in Fig. 8.3]. HCVp45 and HCVp100 displayed significantly increased resistance to all inhibitors tested, as compared with the parental population HCVp0 (Sheldon et al., 2014; Gallego et al., 2016) (Fig. 8.8). Passage of HCV entailed a 2- to 3-fold increase of viral fitness and a broadening of the mutant spectrum that might have increased the frequency of mutations associated with drug resistance, thus explaining the behavior of the multiply passaged HCV populations. The search for the resistant mutations was easier for telaprevir, daclatasvir, and cyclosporin A than for the other drugs because amino acid substitutions in the target protein had been previously identified as responsible for drug resistance. (In the case of cyclosporin A resistance, substitutions map in NS5A and NS5B, because the drug binds to cyclophilin A, which in turn interacts with NS5A). Analysis of the mutant spectra of HCVp45 and HCVp100 by molecular cloning and Sanger sequencing and by deep sequencing failed to identify specific drug-resistance mutations. Since it could not be excluded that the broadening of the mutant spectrum might have increased the frequency of resistance mutations still to be characterized, two additional tests were performed. One was to determine the kinetics of viral production over a 1000-fold range of MOI in the absence and presence of telaprevir. Both the unpassaged and multiply passaged HCV displayed parallel kinetics at the different MOIs, which excludes that drug resistance was due to the presence of resistance mutations in minority components of the mutant spectrum (Fig. 8.8). To further substantiate the findings, biological clones obtained by end-point dilution of the corresponding HCVp0 and HCVp100 populations were tested regarding drug resistance. A biological clone should have eliminated the minority genomes that harbored drug-resistance mutations since biological cloning is the most severe form of bottleneck event (Sections 6.2 and 6.5 in Chapter 6). The biological clones did not display any decrease in drug resistance as compared

**FIGURE 8.8** Multidrug resistance of hepatitis C virus passaged in human hepatoma cells. Top: Evolution of virus infectivity in the course of 100 serial passages. Middle: HCVp0, HCVp45, and HCVp100 are the initial hepatitis C virus, the HCV passaged 45 times and the virus passaged 100 times in the absence of any drug (samples from the experiment described at the top). The individual panels give the virus titer in the course of 10 passages either in the absence of any drug (No drug) or the presence of IFN-α, ribavirin (Rib), telaprevir (TPV), daclatasvir (DCV), or cyclosporin A (CsA). Note the resistance of HCVp45 and HCVp100 as compared to HCVp0 to several inhibitors. Bottom: Virus titer in the course of three serial passages of HCVp0 and HCVp100 (written in the filled boxes on the left) at the indicated multiplicity of infection (box). See text for the interpretation of these experiments and Sheldon et al. (2014) for details. *The figure has been modified from Sheldon, J., Beach, N.M., Moreno, E., Gallego, I., Pineiro, D., et al., 2014. Increased replicative fitness can lead to decreased drug sensitivity of hepatitis C virus. J. Virol. 88, 12098−12111, with permission from the American Society for Microbiology, Washington DC, USA.*

with their corresponding parental, uncloned populations (Sheldon et al., 2014).

The above observations have established viral fitness as a multidrug-resistance determinant in HCV that may also apply to other viruses. One possible molecular mechanism may consist of competition between replicative complexes and inhibitory molecules inside the infected cells. This model implies that fitness increase is reflected either in more replicating molecules per each replicative unit or in an increase in the number of replicative units per cell, without any influence on the number of inhibitor molecules that reach the replication sites. Exploration of this competition model and alternative models and the extension to other viral-host systems are important challenges in the field of antiviral research.

To sum up, mutant spectra and quasispecies dynamics can mediate antiviral resistance by at least two mechanisms: (i) by increases in the proportion of resistance mutations in the mutant spectra and (ii) by a fitness increase promoted by continued viral replication in the same environment. Both mechanisms may act conjointly during viral infections in vivo. Some studies with HCV have documented drug-resistance phenotypes in infected patients, in the absence of specific drug-resistance mutations (Sullivan et al., 2013; Svarovskaia et al., 2014; Sato et al., 2015; Stross et al., 2016; Di Maio et al., 2017; Dietz et al., 2018). In fact, prolonged chronic HCV infections represent an adequate scenario for fitness increase due to extended rounds of infections in the same host liver. As a consequence, chronic infections may be prone to display fitness-associated multidrug-resistance phenotypes in the absence of drug-resistance mutations. The multiple mechanisms of drug resistance related to quasispecies dynamics justify even further the need for new antiviral strategies, as presented in Chapter 9.

## 8.8 Viral load, fitness, and disease progression

High viral loads are predictors of disease progression. For HIV-1 and other lentiviruses, efficient early control of virus replication by the host immune response is generally associated with limited disease severity. The viral load that follows after the initial immune response to HIV-1 is referred to as the "viral set point." In the absence of early therapy, low set points in HIV-1 are generally attributed to a strong cellular immune response, likely influenced by additional host and viral factors. [This and other aspects of HIV-1 replication and pathogenesis have been reviewed in excellent monographs by Levy (2007, 2015)]. A low set point predicts an asymptomatic outcome, and this is generally the case for viruses that establish persistent rather than acute infections. High viral fitness during the early stages of viral replication can promote disease manifestations. This was suggested by the progression toward the disease of a cohort of individuals that were infected during blood transfusion with an HIV-1 containing a large deletion in Nef, an adaptor protein that mediates replication and pathogenesis (reviewed in Arien and Verhasselt, 2008). After more than 15 years, some of the infected individuals showed clinical signs, probably as a result of the accumulation of mutations in the HIV-1 genome that compensated for the lack of Nef. More generally, fitness-decreasing (but not lethal) genetic lesions in a viral genome may be compensated by additional genomic mutations that become increasingly dominant in the course of further viral replication. The kinetics of fitness gain will depend on the nature of the lesion and the functional implications of the altered protein or genomic regulatory region (Chapter 5).

Fitness, replicative capacity, and viral load are directly interconnected parameters, and they

**FIGURE 8.9** A schematic representation of interconnected parameters of viral replication that often relate to disease progression. An understanding of quasispecies dynamics has made it evident that the aim of antiviral therapy need be not only to directly diminish the viral load but also to affect other parameters that can then reduce the viral load. See text for justification and references and Chapter 9 for new antiviral strategies that follow the concept expressed in this figure.

affect disease progression (Domingo et al., 2012) (Fig. 8.9). Fitness gain will be more effective with a high load of actively replicating virus in the infected organism. Elevated replicative capacity and fitness sustain high viral loads. The reason for this basic feature of viral population dynamics is that given a basal mutation rate, a large number of replicating genomes entails a correspondingly higher probability that a required mutation for fitness gain can be produced. The events involved are a specific case of search for adaptive mutations in terms of exploration of sequence space, as discussed in Section 3.7 of Chapter 3. While active viral replication, high load, and high fitness favor progression of the infection and disease manifestations, the fourth parameter included in the large arrow of Fig. 8.9, mutant spectrum diversity, has an optimal range. Too low or too high intrapopulation diversity is detrimental to virus adaptability. Insufficient diversity limits adaptability to complex environments (Pfeiffer

and Kirkegaard, 2005; Vignuzzi et al., 2006), while excess diversity may lead the virus to cross an extinction threshold, and this is the basis of lethal mutagenesis as an antiviral therapy (Chapter 9). An additional implication of the parameters shown in Fig. 8.9 for antiviral interventions is that fitness decrease is recognized as an alternative to inhibition of viral replication to control viral infections[(Clementi, 2008; Clementi and Lazzarin, 2010); reviewed in (Domingo et al. 2012)]. Thus, key features of quasispecies dynamics have a direct implication on the management of viral infections.

## 8.9 Limitations of simplified reagents and small molecules as antiviral agents

External interventions that have been applied or have been envisaged to limit or suppress virus infection include not only vaccination and administration of antiviral agents as described in previous sections, but also passive immunotherapy, antisense RNAs, or oligonucleotides with various chemical modifications, interfering RNAs, ribozymes, or their combinations. Biotechnological developments have favored the design of chemically defined vaccines (consisting of expressed immunogenic proteins, synthetic peptides, or peptide arrays), without the need to handle or administer live virus. One of the most naïve manifestations of trust in biotechnology in the middle of the 20th century was the belief that a catalog of plasmids encoding the antigenic proteins of the circulating types of pathogenic viruses would suffice to prepare the required vaccine as needed. Concerning influenza vaccines, W.I. Beveridge wrote the following: "The first objective would be to capture the full range of influenza A subtypes. Their antigens would be studied by specialists at central laboratories and made available for the preparation of particular vaccines if and when required. It might be feasible to stockpile some vaccine against all the principal hemagglutinin antigens to be used in a fire brigade type of action as soon as an incipient pandemic is spotted" (Beveridge, 1977). Naivety is also perceived in current designs of universal vaccines based on conserved antigens, considering the different escape mechanisms that operate in viruses to elude neutralization by antibodies (Chai et al., 2016). It is remarkable how our present understanding of viral populations renders obsolete the views expressed in the W.I. Beveridge book, and in other writings at the boom of implementation of DNA recombinant techniques. Yet, attempts to produce vaccines against highly variable viruses, based on antigenic structures or some viral isolates, are ongoing (Christiansen et al., 2018). A critical issue is if the host immune response against a vaccine engineered with "universal" (conserved) antigenic motifs will be sufficient to prevent disease upon infection by other forms of the same pathogen (Freeman and Cox, 2016). From our current knowledge of viral population dynamics, it seems unlikely but time will tell, since efforts toward the manufacturing of universal antiviral vaccines are under way.

With a conceptual similarity to vaccines, in medical practice, monotherapy with an antiviral agent was traditionally preferred over drug mixtures. (The change of paradigm was largely a consequence of the AIDS epidemic, and it was publicly expressed by the pioneer hepatologist S. Sherlock in a summary address of an International Symposium on Viral Hepatitis held in Madrid in 1998). The change of perspective is clear.

Some antiviral strategies, such as antisense nucleic acids or virus-directed ribozymes, were intensely investigated decades ago. It is unlikely that when used in isolation, they can be converted into useful antiviral therapies because resistant mutants are likely to be selected. Yet, they could be part of combinations with other antiviral inhibitors to provide a larger antiviral barrier (Chapter 9). A similar fate is likely for interfering RNAs (Boden et al., 2003; Gitlin et al., 2005; Herrera-Carrillo and Berkhout, 2015;

McDonagh et al., 2015). To a large extent, the failures of defined chemical entities (oligonucleotides, ribozymes, small molecule inhibitors, etc.) to control virus replication and spread are a consequence of their targeting a very defined viral genomic sequence combined with the adaptability of viral populations. Combination of such multiple elements have been envisaged and tested, but off-target effects and the adaptive potential of viruses are likely to limit their efficacy. Unfortunately, biotechnological developments that have been so positive for many research areas and practical applications tend to simplify the types of agents to prevent disease or inhibit virus replication, ignoring the inherent complexity of the object to be controlled. Success is unlikely when "complexity" is combated with "simplicity." An increased understanding of viral population dynamics over the last decades has changed the picture dramatically by providing an interpretation of "virus escape" as a general and largely unavoidable phenomenon. Such awareness has pushed the development of new antiviral designs, which are fundamentally centered in two strategies: a combination of multiple, independently acting elements or fitness decrease through excess mutations (Chapter 9).

## 8.10 "Hit early, hit hard"

A pronouncement by P. Ehrlich at the International Congress of Medicine held in London in 1913, reflected old traditions on how to treat microbial infections ["Here, therefore, the old therapeutic motto is applicable: *Frapper fort et frapper vitte*" and "Therefore, it is in my opinion necessary to *allow the therapeutic treatment to come into action as early as possible, ......*"; sentences taken from (Ehrlich, 1913), with emphases as in the original text]. This pronouncement fits our current understanding of viral populations. Following Ehrlich, the full implications of quasispecies-mediated adaptation of viruses for antiviral therapy were expressed by D.D. Ho in an influential article entitled "Time to hit HIV, early and hard" (Ho, 1995). The article title captures what is needed to prevent adaptation of a virus in the infected host. Any opportunity to replicate is exploited by the virus to increase its fitness and to become less vulnerable to internal (intrahost) or external interventions such as antiviral therapy. Treatment interruptions during chronic infections, such as "drug holidays" that in the case of HIV-1-infected patients were justified to alleviate side effects associated with administration of antiretroviral agents, provided an opportunity for the virus to gain fitness. In principle, given our current understanding of viral quasispecies dynamics, the proposal of P. Ehrlich and D.D. Ho is applicable to other viral pathogens.

One argument that tones down the strength of the "hit early and hit hard" proposal is that some infected patients may not progress to disease, but maintain an asymptomatic lifelong persistent infection. This is the case with elite controllers in the case of HIV-1 infection, and individuals infected with HCV who will not progress toward liver disease. In cases in which such nonprogression can be anticipated by viral and host parameters, it may be justified to exclude some patients from aggressive interventions (Suthar and Harries, 2015; Casado et al., 2018). As a general rule, however, the potential benefits of early treatment are obvious not only to avoid disease on an individual basis, but also to diminish the chances of virus transmission (reviewed in Suthar and Harries, 2015; see also Sections 7.1 and 7.7 in Chapter 7 regarding the relevance of viral population numbers in transmission). Restricting the number of treated patients for economic reasons will result in more expensive public health interventions when infected individuals develop the disease.

Box 8.4 includes recommendations for the use of antiviral agents and recapitulates concepts explained in this and preceding sections.

## 8.11 Information and global action

Despite the emphasis on evolutionary aspects, prevention and treatment of viral disease have many other angles some of which were

---

### BOX 8.4

## SOME RECOMMENDATIONS FOR THE USE OF ANTIVIRAL AGENTS

- Avoid monotherapy. Ideally, use two or more antiviral agents which do not share a mechanism of action (Fig. 8.6).
- Treat as soon as possible after virus diagnosis, to avoid virus adaptability associated with high virus population size and to minimize transmission of inhibitor-resistant mutants.
- Individual patients should be treated only during the time at which the drug proves effective. When viral load rebounds, treatment should be discontinued.

- Use deep sequencing methodology to determine mutant spectrum composition for an adequate choice of inhibitors. The aim is to design personalized treatments that consider the probability of drug combination efficacy with minimal side effects.
- Consider temporary shelving of effective drugs when resistant mutants acquire epidemiological relevance.

considered in Chapter 7 in connection with factors of disease emergence (Smolinski et al., 2003). Three of them should be mentioned here because they are as important as the adequate treatment designs described in this chapter: (i) adequacy of public health measures, (ii) public information about virus sources and means of contagion and (iii) need of global political action. Information to the public should aim at limiting the spread of disease that is, undertaking personal and collective actions to reduce the $R_0$ value for a given virus (Chapter 7, Section 7.2). As an illustration of this key point, there was a quite extensive information campaign on HIV-1 and AIDS in developed countries during the early decades of HIV-1 spread, while the information about other potentially threatening viruses such as Ebola or the severe acute or the Middle East respiratory syndrome (SARS and MERS, respectively) coronaviruses was more limited.

The need of a global response to limit the extension of disease episodes at the sites where they are initiated has been recognized for a long time, but it became obvious with the 2014–15 West African Ebola epidemic (see Siedner et al., 2015). There is a need for international organizations and governments of developed countries to provide the health-care workforce to assist low- and middle-income countries to control viral episodes at an early stage. "Help early, help effectively" is the recognized need at a global scale, which is the parallel to "Hit early, hit hard" for the treatment of infected patients.

Global early action and adequate information can be as important as an adequate treatment design to control viral disease. It can restrict viral replication rounds and consequent adaptability. Information is thought to have been critical for the control of Ebola epidemic in Nigeria (Siedner et al., 2015). However, information must also be planned to reach the target population in a convincing manner, as learned from the poliovirus vaccination and eradication campaign (Renne, 2010). The uncertainties regarding whether an initial, limited episode of viral disease will expand or die out do not help in decision making. However, the best choice in the case of emerging and reemerging infections is to act assuming the worst scenario.

## 8.12 Overview and concluding remarks

Medical interventions represent a totally new set of selective constraints that viruses are facing only since decades ago, an infinitesimal time of their existence as biological entities. However, the evolutionary mechanisms available to viruses have successfully coped with many selective pressures, notably the effect of vaccination when a broad immune response is not evoked, or treatment with antiviral agents. A common way to proceed is to test a new vaccine with an animal host, be it the authentic host or an animal model, and obtain full protection when the animal is challenged with a virus that matches the antigenic composition of the vaccine. Following the initial excitement, very often the vaccine displays only partial protection when tested in the natural environment. Somewhat parallel arguments can be made about clinical trials for antiviral agents, usually performed initially with selected groups of patients. Clinicians have coined the term "real-life" treatment studies when patients are not selected for optimal results (often pushed by commercial and political interests).

This chapter has emphasized how the complexity of viral populations is a serious (often underestimated) difficulty to prevent and treat viral disease. Examination of the molecular mechanisms exploited by viruses to survive despite antiviral interventions suggests two major lines of action: first, more judicious use of existing tools that should consider the complexity of viral populations and their dynamics; complexity cannot be combated with simplicity. Second, the need to design new antiviral strategies, a topic addressed in Chapter 9.

Several interconnected parameters determine the probability of success of an antiviral intervention. Most of them follow from the general concepts of Darwinian evolution explained in preceding chapters. It is important, however, to quantify as much as possible the evolutionary events that determine therapy success or failure. For this reason, the importance of viral population size, basic probability calculations of developing resistance, and the selective strength of mutations, have been explained with numerical examples. Hopefully, these simple quantifications will permit a higher awareness of when and why treatment may succeed or fail.

We live in a very unequal society. The chapter closes with the recognition that there are many social economic issues that are as important as scientific planning to combat the pathogenic viruses around us (see Summary Box).

---

## SUMMARY BOX

- Medical interventions represent a new class of selective constraints acting on viral populations.
- Viral evolution affects antiviral preventive and treatment strategies in two different ways: (i) through the molecular mechanisms of short-term response in treated individuals that select escape mutants, and (ii) through the epidemiological impact of viruses that have acquired escape mutations.
- Ineffective vaccines can contribute to the selection of antigenic viral variants.
- Selection of viral mutants resistant to antiviral agents is a general phenomenon. Selection is favored by suboptimal treatments and is delayed by the combined administration of multiple inhibitors. Resistance may also occur in the absence of specific resistance mutations, and it is associated with viral fitness.
- The aim of therapy should be to increase the functional barrier to resistance and to give no opportunity to the virus to pursue replication that increases its replicative fitness.
- Replication rate, viral load, fitness, and mutant spectrum complexity are interconnected parameters that may tip the balance toward either control of the infection or disease progression. Each of these parameters can be targeted in an antiviral design.
- Health care resources, adequate public information, and firm political action are as important as antiviral designs to control virus infections at a global level.

---

# References

Ab Ghani, N.S., Ramlan, E.I., Firdaus-Raih, M., 2019. Drug ReposER: a web server for predicting similar amino acid arrangements to know drug binding interfaces for potential drug repositioning. Nucleic Acids Res. https://doi.org/10.1093/nar/gkz391 pii:gkz391.

Agudo, R., de la Higuera, I., Arias, A., Grande-Perez, A., Domingo, E., 2016. Involvement of a joker mutation in a polymerase-independent lethal mutagenesis escape mechanism. Virology 494, 257–266.

Appleyard, G., 1977. Amantadine-resistance as a genetic marker for influenza viruses. J. Gen. Virol. 36, 249–255.

Arien, K.K., Verhasselt, B., 2008. HIV Nef: role in pathogenesis and viral fitness. Curr. HIV Res. 6, 200–208.

Bailey, J., Blankson, J.N., Wind-Rotolo, M., Siliciano, R.F., 2004. Mechanisms of HIV-1 escape from immune responses and antiretroviral drugs. Curr. Opin. Immunol. 16, 470–476.

Beerenwinkel, N., Daumer, M., Sing, T., Rahnenfuhrer, J., Lengauer, T., et al., 2005. Estimating HIV evolutionary pathways and the genetic barrier to drug resistance. J. Infect. Dis. 191, 1953–1960.

Belshe, R.B., Smith, M.H., Hall, C.B., Betts, R., Hay, A.J., 1988. Genetic basis of resistance to rimantadine emerging during treatment of influenza virus infection. J. Virol. 62, 1508–1512.

Beveridge, W.I.B., 1977. Influenza: The Last Great Plague; and Unfinished Story of Discovery. Prodist, New York.

Bloom, B.R., Lambert, P.-H., 2003. The Vaccine Book. Academic Press, Elsevier, San Diego.

Boden, D., Pusch, O., Lee, F., Tucker, L., Ramratnam, B., 2003. Human immunodeficiency virus type 1 escape from RNA interference. J. Virol. 77, 11531–11535.

Bull, R.A., Leung, P., Gaudieri, S., Deshpande, P., Cameron, B., et al., 2015. Transmitted/founder viruses rapidly escape from CD8$^+$ T cell responses in acute hepatitis C virus infection. J. Virol. 89, 5478–5490.

Burnet, M., 1966. Natural History of Infections Disease. Cambridge University Press, London.

Cale, E.M., Hraber, P., Giorgi, E.E., Fischer, W., Bhattacharya, T., et al., 2011. Epitope-specific CD8$^+$ T lymphocytes cross-recognize mutant simian immunodeficiency virus (SIV) sequences but fail to contain very early evolution and eventual fixation of epitope escape mutations during SIV infection. J. Virol. 85, 3746–3757.

Casado, C., Marrero-Hernández, S., Marquez-Arce, D., Pernas, M., Marfil, S., et al., 2018. Viral characteristics associated with the clinical nonprogressor phenotype are inherited by viruses from a cluster of HIV-1 elite controllers. mBio 9 pii: e02389-17.

Chai, N., Swem, L.R., Reichelt, M., Chen-Harris, H., Luis, E., et al., 2016. Two escape mechanisms of influenza a virus to a broadly neutralizing stalk-binding antibody. PLoS Pathog. 12, e1005792.

Cheng, B.Y., Ortiz-Riano, E., Nogales, A., de la Torre, J.C., Martinez-Sobrido, L., 2015. Development of live-attenuated arenavirus vaccines based on codon deoptimization. J. Virol. 89, 3523–3533.

Christiansen, D., Earnest-Silveira, L., Chua, B., Meuleman, P., Boo, I., et al., 2018. Immunological responses following administration of a genotype 1a/1b/2/3a quadrivalent HCV VLP vaccine. Sci. Rep. 8, 6483.

Chumakov, K., Kew, O., 2010. The poliovirus eradication initiative. In: Ehrenfeld, E., Domingo, E., Roos, R.P. (Eds.), The Picornaviruses. ASM Press, Washington, DC, pp. 449–459.

Ciurea, A., Klenerman, P., Hunziker, L., Horvath, E., Senn, B.M., et al., 2000. Viral persistence in vivo through selection of neutralizing antibody- escape variants. Proc. Natl. Acad. Sci. U.S.A. 97, 2749–2754.

Ciurea, A., Hunziker, L., Martinic, M.M., Oxenius, A., Hengartner, H., et al., 2001. CD4+ T-cell-epitope escape mutant virus selected in vivo. Nat. Med. 7, 795–800.

Clementi, M., 2008. Perspectives and opportunities for novel antiviral treatments targeting virus fitness. Clin. Microbiol. Infect. 14, 629–631.

Clementi, M., Lazzarin, A., 2010. Human immunodeficiency virus type 1 fitness and tropism: concept, quantification, and clinical relevance. Clin. Microbiol. Infect. 16, 1532–1538.

Clutter, D.S., Jordan, M.R., Bertagnolio, S., Shafer, R.W., 2016. HIV-1 drug resistance and resistance testing. Infect. Genet. Evol. 46, 292–307.

Coleman, J.R., Papamichail, D., Skiena, S., Futcher, B., Wimmer, E., et al., 2008. Virus attenuation by genome-scale changes in codon pair bias. Science 320, 1784–1787.

Constans, M., Ssemadaali, M., Kolyvushko, O., Ramamoorthy, S., 2015. Antigenic determinants of possible vaccine escape by porcine circovirus subtype 2b viruses. Bioinf. Biol. Insights 9 (Suppl. 2), 1–12.

Crowder, S., Kirkegaard, K., 2005. Trans-dominant inhibition of RNA viral replication can slow growth of drug-resistant viruses. Nat. Genet. 37, 701–709.

Cubero, M., Esteban, J.I., Otero, T., Sauleda, S., Bes, M., et al., 2008. Naturally occurring NS3-protease-inhibitor resistant mutant A156T in the liver of an untreated chronic hepatitis C patient. Virology 370, 237–245.

Cunningham, A.L., Garçon, N., Leo, O., Friedland, L.R., Strugnell, R., et al., 2016. Vaccine development: from concept to early clinical testing. Vaccine 34, 6655–6664.

De la Higuera, I., Ferrer-Orta, C., de Avila, A.I., Perales, C., Sierra, M., et al., 2017. Molecular and functional bases of selection against a mutation bias in an RNA virus. Genome Biol. Evol. 9, 1212–1228.

Di Lello, F.A., Ridruejo, E., Martínez, A.P., Pérez, P.S., Campos, R.H., et al., 2019. Molecular epidemiology of hepatitis B virus mutants associated with vaccine escape, drug resistance and diagnosis failure. J. Viral Hepat. 26, 552–560.

Dietz, J., Susser, S., Vermehren, J., Peiffer, K.H., Grammatikos, G., et al., 2018. Patterns of resistance-associated substitutions in patients with chronic HCV infection following treatment with direct-acting antivirals. Gastroenterology 154, 976–988.

Di Maio, V.C., Cento, V., Lenci, I., Aragri, M., Rossi, P., et al., 2017. Multiclass HCV resistance to direct-acting antiviral failure in real-life patients advocates for tailored second-line therapies. Liver Int. 37, 514–528.

Domingo, E., 1989. RNA virus evolution and the control of viral disease. Prog. Drug Res. 33, 93–133.

Domingo, E., Holland, J.J., 1992. Complications of RNA heterogeneity for the engineering of virus vaccines and antiviral agents. Genet. Eng. 14, 13–31.

Domingo, E., Biebricher, C., Eigen, M., Holland, J.J., 2001a. Quasispecies and RNA Virus Evolution: Principles and Consequences. Landes Bioscience, Austin.

Domingo, E., Mas, A., Yuste, E., Pariente, N., Sierra, S., et al., 2001b. Virus population dynamics, fitness variations and the control of viral disease: an update. Prog. Drug Res. 57, 77–115.

Domingo, E., Sheldon, J., Perales, C., 2012. Viral quasispecies evolution. Microbiol. Mol. Biol. Rev. 76, 159–216.

Echeverría, N., Betancour, G., Gámbaro, F., Hernández, N., López, P., et al., 2016. Naturally occurring NS3 resistance-associated variants in hepatitis C virus genotype 1: their relevance for developing countries. Virus Res. 223, 140–146.

Eggers, H.J., Tamm, I., 1961. Spectrum and characteristics of the virus inhibitory action of 2-(alpha-hydroxybenzyl)-benzimidazole. J. Exp. Med. 113, 657–682.

Eggers, H.J., Tamm, I., 1965. Coxsackie A9 virus: mutation from drug dependence to drug independence. Science 148, 97–98.

Ehrlich, P., 1913. Address in Pathology on Chemotherapy. International Medical Congress, London, pp. 353–359.

Evans, A.S., Kaslow, R.A., 1997. Viral Infections of Humans. Epidemiology and Control. Plenum Medical Book Company, New York and London.

Fischer, W., Ganusov, V.V., Giorgi, E.E., Hraber, P.T., Keele, B.F., et al., 2010. Transmission of single HIV-1 genomes and dynamics of early immune escape revealed by ultra-deep sequencing. PLoS One 5, e12303.

Flynn, W.F., Chang, M.W., Tan, Z., Oliveira, G., Yuan, J., et al., 2015. Deep sequencing of protease inhibitor resistant HIV patient isolates reveals patterns of correlated mutations in gag and protease. PLoS Comput. Biol. 11, e1004249.

Franco, S., Tural, C., Nevot, M., Moltó, J., Rockstroh, J.K., et al., 2014. Detection of a sexually transmitted hepatitis C virus protease inhibitor-resistance variant in a human immunodeficiency virus-infected homosexual man. Gastroenterology 147, 599–601.

Freeman, Z.T., Cox, A.L., 2016. Lessons from nature: understanding immunity to HCV to guide vaccine design. PLoS Pathog. 12, e1005632.

Fun, A., Wensing, A.M., Verheyen, J., Nijhuis, M., 2012. Human immunodeficiency virus gag and protease: partners in resistance. Retrovirology 9, 63.

Gallego, I., Sheldon, J., Moreno, E., Gregori, J., Quer, J., et al., 2016. Barrier-independent, fitness-associated differences in sofosbuvir efficacy against hepatitis C virus. Antimicrob. Agents Chemother. 60, 3786–3793.

Gandon, S., Mackinnon, M., Nee, S., Read, A., 2003. Imperfect vaccination: some epidemiological and evolutionary consequences. Proc. Biol. Sci. 270, 1129–1136.

Gitlin, L., Stone, J.K., Andino, R., 2005. Poliovirus escape from RNA interference: short interfering RNA-target recognition and implications for therapeutic approaches. J. Virol. 79, 1027–1035.

Gnadig, N.F., Beaucourt, S., Campagnola, G., Borderia, A.V., Sanz-Ramos, M., et al., 2012. Coxsackievirus B3 mutator strains are attenuated in vivo. Proc. Natl. Acad. Sci. U.S.A. 109, E2294–E2303.

Golan, D.E., Tashjian, A.H. (Eds.), 2011. Principles of Pharmacology: The Pathophysiologic Basis of Drug Therapy. Lippincott Williams & Wilkins, Baltimore, Philadelphia.

Graham, R.L., Becker, M.M., Eckerle, L.D., Bolles, M., Denison, M.R., et al., 2012. A live, impaired-fidelity coronavirus vaccine protects in an aged, immunocompromised mouse model of lethal disease. Nat. Med. 18, 1820–1826.

Günthard, H.F., Calvez, V., Paredes, R., Pillay, D., Shaffer, R.W., et al., 2019. Human immunodeficiency virus drug rsistance: 2018 recommendations of the International Antiviral Society-USA panel. Clin. Infect. Dis. 68, 177–187.

Hagan, T., Nakaya, H.I., Subramaniam, S., Pulendran, B., 2015. Systems vaccinology: enabling rational vaccine design with systems biological approaches. Vaccine 33, 5294–5301.

Hamelaar, J., Elangovan, R., Yun, J., Dickson-Tetteh, L., Fleminger, I., et al., 2019. Global and regional molecular epidemiology of HIV-1, 1990-2015: a systematic review, global survey, and trend analysis. Lancet Infect. Dis. 19, 143–155.

Havlir, D.V., Eastman, S., Gamst, A., Richman, D.D., 1996. Nevirapine-resistant human immunodeficiency virus: kinetics of replication and estimated prevalence in untreated patients. J. Virol. 70, 7894–7899.

Hedskog, C., Dvory-Sobol, H., Gontcharova, V., Martin, R., Ouyang, W., et al., 2015. Evolution of the HCV viral population from a patient with S282T detected at relapse after sofosbuvir monotherapy. J. Viral Hepat. 22, 871–881.

Heinz, B.A., Rueckert, R.R., Shepard, D.A., Dutko, F.J., McKinlay, M.A., et al., 1989. Genetic and molecular analyses of spontaneous mutants of human rhinovirus 14 that are resistant to an antiviral compound. J. Virol. 63, 2476–2485.

Herrera-Carrillo, E., Berkhout, B., 2015. The impact of HIV-1 genetic diversity on the efficacy of a combinatorial RNAi-based gene therapy. Gene Ther. 22, 485–495.

Herrmann Jr., E.C., Herrmann, J.A., 1977. A working hypothesis—virus resistance development as an indicator of specific antiviral activity. Ann. N. Y. Acad. Sci. 284, 632–637.

Himman, A., 2017. The eradication of polio: have we succeeded? Vaccine 35, 5519–5521.

Ho, D.D., 1995. Time to hit HIV, early and hard. N. Engl. J. Med. 333, 450–451.

Hsu, H.Y., Chang, M.H., Liaw, S.H., Ni, Y.H., Chen, H.L., 1999. Changes of hepatitis B surface antigen variants in carrier children before and after universal vaccination in Taiwan. Hepatology 30, 1312–1317.

Huang, W., Wang, M., Gong, Q., Yu, D., Chen, P., et al., 2019. Comparison of naturally occurring resistance-associated substitutions between 2008 and 2016 in Chinese patients with chronic hepatitis C virus infection. Microb. Drug Resist. https://doi.org/10.1089/mdr.2018.0360.

Ji, W., Niu, D.D., Si, H.L., Ding, N.Z., He, C.Q., 2014. Vaccination influences the evolution of classical swine fever virus. Infect. Genet. Evol. 25, 69–77.

Johnson, J.A., Li, J.F., Wei, X., Lipscomb, J., Irlbeck, D., et al., 2008. Minority HIV-1 drug resistance mutations are present in antiretroviral treatment-naive populations and associate with reduced treatment efficacy. PLoS Med. 5, e158.

Kekarainen, T., Gonzalez, A., Llorens, A., Segales, J., 2014. Genetic variability of porcine circovirus 2 in vaccinating and non-vaccinating commercial farms. J. Gen. Virol. 95, 1734–1742.

Kew, O., Morris-Glasgow, V., Landaverde, M., Burns, C., Shaw, J., et al., 2002. Outbreak of poliomyelitis in Hispaniola associated with circulating type 1 vaccine-derived poliovirus. Science 296, 356–359.

Kilbourne, E.D., 1987. Influenza. Plenum Medical Book Company, New York.

Kirkegaard, K., van Buuren, N.J., Mateo, R., 2016. My cousin, my enemy: quasispecies suppression of drug resistance. Curr. Opin. Virol. 20, 106–111.

Kirkwood, C.D., 2010. Genetic and antigenic diversity of human rotaviruses: potential impact on vaccination programs. J. Infect. Dis. 202 (Suppl. l), S43–S48.

Korboukh, V.K., Lee, C.A., Acevedo, M., Xiao, Y., Arnold, J.J., et al., 2014. RNA virus population diversity, an optimum for maximal fitness and virulence. J. Biol. Chem. 289, 29531–29544.

Kortenhoeven, C., Joubert, F., Bastos, A., Abolnik, C., 2015. Virus genome dynamics under different propagation pressures: reconstruction of whole genome haplotypes of west nile viruses from NGS data. BMC Genomics 16, 118.

Lech, W.J., Wang, G., Yang, Y.L., Chee, Y., Dorman, K., et al., 1996. In vivo sequence diversity of the protease of human immunodeficiency virus type 1: presence of protease inhibitor-resistant variants in untreated subjects. J. Virol. 70, 2038–2043.

Levy, J.A., 2007. HIV and the Pathogenesis of AIDS. ASM Press, Washington, DC.

Levy, J.A., 2015. Dispelling myths and focusing on notable concepts in HIV pathogenesis. Trends Mol. Med. 21, 341–353.

Lindenbach, B.D., Evans, M.J., Syder, A.J., Wolk, B., Tellinghuisen, T.L., et al., 2005. Complete replication of hepatitis C virus in cell culture. Science 309, 623–626.

Loh, L., Petravic, J., Batten, C.J., Davenport, M.P., Kent, S.J., 2008. Vaccination and timing influence SIV immune escape viral dynamics in vivo. PLoS Pathog. 4, e12.

Lubeck, M.D., Schulman, J.L., Palese, P., 1978. Susceptibility of influenza A viruses to amantadine is influenced by the gene coding for M protein. J. Virol. 28, 710–716.

Margeridon-Thermet, S., Shafer, R.W., 2010. Comparison of the mechanisms of drug resistance among HIV, hepatitis B, and hepatitis C. Viruses 2, 2696–2739.

Martinez-Picado, J., Martinez, M.A., 2008. HIV-1 reverse transcriptase inhibitor resistance mutations and fitness: a view from the clinic and ex vivo. Virus Res. 134, 104–123.

McCoy, L.E., Falkowska, E., Doores, K.J., Le, K., Sok, D., et al., 2015. Incomplete neutralization and deviation from sigmoidal neutralization curves for HIV broadly neutralizing monoclonal antibodies. PLoS Pathog. 11, e1005110.

McDonagh, P., Sheehy, P.A., Norris, J.M., 2015. Combination siRNA therapy against feline coronavirus can delay the emergence of antiviral resistance in vitro. Vet. Microbiol. 176, 10–18.

Melnick, J.L., Crowther, D., Barrera-Oro, J., 1961. Rapid development of drug-resistant mutants of poliovirus. Science 134, 557.

Menéndez-Arias, L., 2013. Molecular basis of human immunodeficiency virus type 1 drug resistance: overview and recent developments. Antivir. Res. 98, 93–120.

Menendez-Arias, L., Alvarez, M., 2014. Antiretroviral therapy and drug resistance in human immunodeficiency virus type 2 infection. Antivir. Res. 102, 70–86.

Menendez-Arias, L., Richman, D.D., 2014. Editorial overview: antivirals and resistance: advances and challenges ahead. Curr. Opin. Virol. 8, iv–vii.

Mitchison, D.A., 1950. Development of streptomycin resistant strains of tubercle bacilli in pulmonary tuberculosis; results of simultaneous sensitivity tests in liquid and on solid media. Thorax 5, 144–161.

Mohri, H., Singh, M.K., Ching, W.T., Ho, D.D., 1993. Quantitation of zidovudine-resistant human immunodeficiency virus type 1 in the blood of treated and untreated patients. Proc. Natl. Acad. Sci. U.S.A. 90, 25–29.

Mokaya, J., McNaughton, A.L., Hadley, M.J., Beloukas, A., Geretti, A.M., et al., 2018. A systematic review of hepatitis B virus (HBV) drug and vaccine escape mutations in Africa: a call for urgent action. PLoS Neglected Trop. Dis. 12, e0006629.

Moreno, E., Ojosnegros, S., García—Arriaza, J., Escarmís, C., Domingo, E., et al., 2014. Exploration of sequence space as the basis of viral RNA genome segmentation. Proc. Natl. Acad. Sci. U.S.A. 111, 6678—6683.

Muylkens, B., Meurens, F., Schynts, F., Farnir, F., Pourchet, A., et al., 2006. Intraspecific bovine herpesvirus 1 recombinants carrying glycoprotein E deletion as a vaccine marker are virulent in cattle. J. Gen. Virol. 87, 2149—2154.

Nájera, I., Richman, D.D., Olivares, I., Rojas, J.M., Peinado, M.A., et al., 1994. Natural occurrence of drug resistance mutations in the reverse transcriptase of human immunodeficiency virus type 1 isolates. AIDS Res. Hum. Retrovir. 10, 1479—1488.

Nájera, I., Holguín, A., Quiñones-Mateu, M.E., Muñoz-Fernández, M.A., Nájera, R., et al., 1995. Pol gene quasispecies of human immunodeficiency virus: mutations associated with drug resistance in virus from patients undergoing no drug therapy. J. Virol. 69, 23—31.

Neagu, I.A., Olejarz, J., Freeman, M., Rosenbloom, D.I.S., Nowak, M.A., et al., 2018. Life cycle synchronization is a viral drug resistance mechanism. PLoS Comput. Biol. 14, e1005947.

Neher, R.A., Leitner, T., 2010. Recombination rate and selection strength in HIV intra-patient evolution. PLoS Comput. Biol. 6, e1000660.

Nitta, S., Asahina, Y., Kato, T., Tsuchiya, J., Inoue-Shinomiya, E., et al., 2019. Impact of novel NS5A resistance-associated substitutions of hepatitis C virus detected in treatment-experienced patients. Sci. Rep. 9, 5722.

Pawlotsky, J.M., 2006. Hepatitis C virus population dynamics during infection. Curr. Top. Microbiol. Immunol. 299, 261—284.

Pawlotsky, J.M., 2019. Retreatment of hepatitis C virus infected patients with direct-acting antiviral failures. Semin. Liver Dis. https://doi.org/10.1055/s-0039-1687823.

Perales, C., Agudo, R., Manrubia, S.C., Domingo, E., 2011. Influence of mutagenesis and viral load on the sustained low-level replication of an RNA virus. J. Mol. Biol. 407, 60—78.

Perales, C., Beach, N.M., Gallego, I., Soria, M.E., Quer, J., et al., 2013. Response of hepatitis C virus to long-term passage in the presence of alpha interferon: multiple mutations and a common phenotype. J. Virol. 87, 7593—7607.

Perales, C., Beach, N.M., Sheldon, J., Domingo, E., 2014. Molecular basis of interferon resistance in hepatitis C virus. Curr. Opin. Virol. 8, 38—44.

Perales, C., Chen, Q., Soria, M.E., Gregori, J., García-Cehic, D., et al., 2018. Baseline hepatitis C virus resistance-associated substitutions present at frequencies lower than 15% may be clinically significant. Infect. Drug Resist. 11, 2207—2210.

Pereira-Gómez, M., Sanjuán, R., 2014. Delayed lysis confers resistance to the nucleoside analogue 5-fluorouracil and alleviates mutation accumulation in the single-stranded DNA bacteriophage φX174. J. Virol. 88, 5042—5049.

Peres-da-Silva, A., Brandao-Mello, C.E., Lampe, E., 2017. Prevalence of sofosbuvir resistance-associated variants in Brazilian and worldwide NS5B sequences of genotype-1 HCV. Antivir. Ther. 22, 447—451.

Perez-Sautu, U., Costafreda, M.I., Cayla, J., Tortajada, C., Lite, J., et al., 2011. Hepatitis a virus vaccine escape variants and potential new serotype emergence. Emerg. Infect. Dis. 17, 734—737.

Pfeiffer, J.K., Kirkegaard, K., 2005. Increased fidelity reduces poliovirus fitness under selective pressure in mice. PLoS Pathog. 1, 102—110.

Pincus, S.E., Wimmer, E., 1986. Production of guanidine-resistant and -dependent poliovirus mutants from cloned cDNA: mutations in polypeptide 2C are directly responsible for altered guanidine sensitivity. J. Virol. 60, 793—796.

Pircher, H., Moskophidis, D., Rohrer, U., Burki, K., Hengartner, H., et al., 1990. Viral escape by selection of cytotoxic T cell-resistant virus variants in vivo. Nature 346, 629—633.

Read, A.F., Baigent, S.J., Powers, C., Kgosana, L.B., Blackwell, L., et al., 2015. Imperfect vaccination can enhance the transmission of highly virulent pathogens. PLoS Biol. 13, e1002198.

Renne, E.P., 2010. The Politics of Polio in Northern Nigeria. Indiana University Press, Bloomington.

Rhee, S.Y., Taylor, J., Fessel, W.J., Kaufman, D., Towner, W., et al., 2010. HIV-1 protease mutations and protease inhibitor cross-resistance. Antimicrob. Agents Chemother. 54, 4253—4261.

Ribeiro, R.M., Bonhoeffer, S., Nowak, M.A., 1998. The frequency of resistant mutant virus before antiviral therapy. AIDS 12, 461—465.

Ribeiro, R.M., Bonhoeffer, S., 2000. Production of resistant HIV mutants during antiretroviral therapy. Proc. Natl. Acad. Sci. U.S.A. 97, 7681—7686.

Richman, D.D., 1994. Resistance, drug failure, and disease progression. AIDS Res. Hum. Retrovir. 10, 901—905.

Richman, D.D. (Ed.), 1996. Antiviral Drug Resistance. John Wiley and Sons Inc., New York.

Richman, D.D., Wrin, T., Little, S.J., Petropoulos, C.J., 2003. Rapid evolution of the neutralizing antibody response to HIV type 1 infection. Proc. Natl. Acad. Sci. U.S.A. 100, 4144–4149.

Rodriguez-Calvo, T., Ojosnegros, S., Sanz-Ramos, M., García-Arriaza, J., Escarmís, C., et al., 2010. New vaccine design based on defective genomes that combines features of attenuated and inactivated vaccines. PLoS One 5, e10414.

Rogers, M.B., Song, T., Sebra, R., Greenbaum, B.D., Hamelin, M.E., et al., 2015. Intrahost dynamics of antiviral resistance in influenza A virus reflect complex patterns of segment linkage, reassortment and natural selection. mBio 6 pii: e02464-14.

Rolland, M., 2019. HIV-1 phylogenetics and vaccines. Curr. Opin. HIV AIDS 14, 227–232.

Rowlands, D.J., Minor, P.D., 2010. Vaccine strategies. In: Ehrenfeld, E., Domingo, E., Roos, R.P. (Eds.), The Picornaviruses. ASM Press, Washington, DC, pp. 431–448.

Rozen-Gagnon, K., Stapleford, K.A., Mongelli, V., Blanc, H., Failloux, A.B., et al., 2014. Alphavirus mutator variants present host-specific defects and attenuation in mammalian and insect models. PLoS Pathog. 10, e1003877.

Sarrazin, C., Zeuzem, S., 2010. Resistance to direct antiviral agents in patients with hepatitis C virus infection. Gastroenterology 138, 447–462.

Sato, M., Maekawa, S., Komatsu, N., Tatsumi, A., Miura, M., et al., 2015. Deep Sequencing and phylogenetic analysis of variants resistant to interferon-based protease inhibitor therapy in chronic hepatitis induced by genotype-1b hepatitis C virus. J. Virol. 89, 6105–6116.

Schat, K.A., Baranowski, E., 2007. Animal vaccination and the evolution of viral pathogens. Rev. Sci. Tech. 26, 327–338.

Sheldon, J., Beach, N.M., Moreno, E., Gallego, I., Pineiro, D., et al., 2014. Increased replicative fitness can lead to decreased drug sensitivity of hepatitis C virus. J. Virol. 88, 12098–12111.

Siedner, M.J., Gostin, L.O., Cranmer, H.H., Kraemer, J.D., 2015. Strengthening the detection of and early response to public health emergencies: lessons from the West African Ebola epidemic. PLoS Med. 12, e1001804.

Smith, E.C., Blanc, H., Vignuzzi, M., Denison, M.R., 2013. Coronaviruses lacking exoribonuclease activity are susceptible to lethal mutagenesis: evidence for proofreading and potential therapeutics. PLoS Pathog. 9, e1003565.

Smith, E.C., Case, J.B., Blanc, H., Isakov, O., Shomron, N., et al., 2015. Mutations in coronavirus nonstructural protein 10 decrease virus replication fidelity. J. Virol. 89, 6418–6426.

Smolinski, M.S., Hamburg, M.A., Lederberg, J. (Eds.), 2003. Microbial Threats to Health. Emergence, Detection and Response. The National Academies Press, Washington DC.

Stapleford, K.A., Rozen-Gagnon, K., Das, P.K., Saul, S., Poirier, E.Z., 2015. Viral polymerase-helicase complexes regulate replication fidelity to overcome intracellular nucleotide depletion. J. Virol. 89, 11233–11244.

Stross, C., Shimakami, T., Haselow, K., Ahmad, M.Q., Zeuzem, S., et al., 2016. Natural HCV variants with increased replicative fitness due to NS3 helicase mutations in the C-terminal helix α18. Sci. Rep. 6, 19526.

Sullivan, J.C., De Meyer, S., Bartels, D.J., Dierynck, I., Zhang, E.Z., et al., 2013. Evolution of treatment-emergent resistant variants in telaprevir phase 3 clinical trials. Clin. Infect. Dis. 57, 221–229.

Suthar, A.B., Harries, A.D., 2015. A public health approach to hepatitis C control in low- and middle-income countries. PLoS Med. 12, e1001795.

Svarovskaia, E.S., Dvory-Sobol, H., Parkin, N., Hebner, C., Gontcharova, V., et al., 2014. Infrequent development of resistance to genotype 1-6 hepatitis C virus-infected subjects treated with sofosbuvir in phase 2 and 3 clinical trials. Clin. Infect. Dis. 59, 1666–1674.

Swayne, D.E., Kapczynski, D., 2008. Strategies and challenges for eliciting immunity against avian influenza virus in birds. Immunol. Rev. 225, 314–331.

Taboga, O., Tami, C., Carrillo, E., Núñez, J.I., Rodríguez, A., et al., 1997. A large-scale evaluation of peptide vaccines against foot-and-mouth disease: lack of solid protection in cattle and isolation of escape mutants. J. Virol. 71, 2606–2614.

Tami, C., Taboga, O., Berinstein, A., Nuñez, J.I., Palma, E.L., et al., 2003. Evidence of the coevolution of antigenicity and host cell tropism of foot-and-mouth disease virus in vivo. J. Virol. 77, 1219–1226.

Thiry, E., Muylkens, B., Meurens, F., Gogev, S., Thiry, J., et al., 2006. Recombination in the alphaherpesvirus bovine herpesvirus 1. Vet. Microbiol. 113, 171–177.

Toni, T.A., Asahchop, E.L., Moisi, D., Ntemgwa, M., Oliveira, M., et al., 2009. Detection of human immunodeficiency virus (HIV) type 1 M184V and K103N minority variants in patients with primary HIV infection. Antimicrob. Agents Chemother. 53, 1670–1672.

Tsibris, A.M., Korber, B., Arnaout, R., Russ, C., Lo, C.C., et al., 2009. Quantitative deep sequencing reveals dynamic HIV-1 escape and large population shifts during CCR5 antagonist therapy in vivo. PLoS One 4, e5683.

Valarcher, J.F., Schelcher, F., Bourhy, H., 2000. Evolution of bovine respiratory syncytial virus. J. Virol. 74, 10714–10728.

Van Regenmortel, M.H.V., 2012. Basic research in HIV vaccinology is hampered by reductionist thinking. Front. Immunol. 3 article 194.

Van Slyke, G.A., Arnold, J.J., Lugo, A.J., Griesemer, S.B., Moustafa, I.M., et al., 2015. Sequence-specific fidelity alterations associated with West Nile virus attenuation in mosquitoes. PLoS Pathog. 11, e1005009.

Vignuzzi, M., Stone, J.K., Arnold, J.J., Cameron, C.E., Andino, R., 2006. Quasispecies diversity determines pathogenesis through cooperative interactions in a viral population. Nature 439, 344–348.

Vignuzzi, M., Wendt, E., Andino, R., 2008. Engineering attenuated virus vaccines by controlling replication fidelity. Nat. Med. 14, 154–161.

von Kleist, M., Metzner, P., Marquet, R., Schutte, C., 2012. HIV-1 polymerase inhibition by nucleoside analogs: cellular- and kinetic parameters of efficacy, susceptibility and resistance selection. PLoS Comput. Biol. 8, e1002359.

Wakita, T., Pietschmann, T., Kato, T., Date, T., Miyamoto, M., et al., 2005. Production of infectious hepatitis C virus in tissue culture from a cloned viral genome. Nat. Med. 11, 791–796.

Weidt, G., Deppert, W., Utermohlen, O., Heukeshoven, J., Lehmann-Grube, F., 1995. Emergence of virus escape mutants after immunization with epitope vaccine. J. Virol. 69, 7147–7151.

Yoo, S.J., Kwon, T., Kang, K., Kim, H., Kang, S.C., et al., 2018. Genetic evolution of classical swine fever virus under immune environments conditioned by genotype 1-based modified live virus vaccine. Transbound. Emerg. Dis. 65, 735–745.

Zhong, J., Gastaminza, P., Cheng, G., Kapadia, S., Kato, T., et al., 2005. Robust hepatitis C virus infection in vitro. Proc. Natl. Acad. Sci. U.S.A. 102, 9294–9299.

# Trends in antiviral strategies

## Abbreviations

**AIDS** acquired immunodeficiency syndrome
**APOBEC3G** apolipoprotein B mRNA editing complex 3G
**AZA-C** 5-azacytidine
**DAA** direct-acting antiviral
**DENV** Dengue virus
**EBV** Ebola virus
**FMDV** foot-and-mouth disease virus
**GBvB** GB virus B
**FU** 5-fluorouracil
**FUTP** 5-fluorouridine-triphosphate
**GTP** guanosine-5′-triphosphate
**5-OH-dC** 5-hydroxydeoxycytidine
**HAART** highly active antiretroviral therapy
**HCV** hepatitis C virus
**HIV-1** human immunodeficiency virus type 1
**IFN** interferon
**IMPDH** inosine monophosphate dehydrogenase
**ITP** inosine triphosphate
**IV** influenza virus
**LCMV** lymphocytic choriomeningitis virus
**MNV** murine norovirus
**NLS** nuclear localization signal
**PV** poliovirus
**Rib** ribavirin
**RMP** ribavirin-monophosphate
**RVFV** Rift Valley fever virus
**TMV** Tobacco mosaic virus
**USUV** Usutu virus
**VPA** valproic acid
**VSV** vesicular stomatitis virus
**WNV** West Nile virus
**ZIKV** Zika virus

## 9.1 The challenge

There is no general procedure that can effectively prevent or control viral infections, at least based on the strategies developed over the last century and a half: vaccination, immunotherapy, chemotherapy, or their combinations. One of the main reasons is that each pathogenic virus to be controlled has some unique features in its complex and multifaceted interaction with the host cell. Not only for viruses but also cellular pathogens (bacteria, fungi, and protozoa) it is amply recognized that current practices for prevention and treatment have clear limitations. One of the difficulties arises from the potential for variation and evolution of viral and cellular pathogens that can jeopardize control strategies. For many decades there has been limited awareness of the adaptive potential of pathogens, and preventive and therapeutic designs were implemented ignoring evolution. Different experts have expressed the need to search for new paradigms to approach infectious disease, with the incorporation of Darwinian principles together with concepts from evolutionary ecology, and considering also re-implementation of old methods, such as passive antibody therapy (Casadevall, 1996; Stearns, 1999; Williams, 2009; Casadevall and Pirofski, 2015). Only very recently, the adaptive potential of pathogens, in particular, viral quasispecies dynamics, with its conceptual extensions to other pathogenic entities (Chapter 10) have been considered as an aspect of the disease control problem.

The challenge to be confronted is symbolically portrayed in Fig. 9.1, using one of the ways to depict viral quasispecies throughout this book (lines to represent genomes and symbols on the lines to represent mutations). About 100 million sheets (!) similar to the one shown in the figure are necessary to represent the hepatitis C virus (HCV) genomes found in an acutely infected human liver at a given time point. The precise number, types, and distribution of mutations among genomes varies not only among patients but also as a function of time in each patient. A few minutes later, a slightly different image will be produced, with relative proportions of different genomes and the mutational pattern

modified. This statement is an inference from replicative and mutational parameters determined in experimental systems, but comparisons of genome composition at short-time intervals (seconds, minutes, hours) are needed to define the extent of short-term variations. The dynamic mutant cloud here drawn on the basis of estimated HCV population numbers (and probably with an underrepresentation of the number of mutations per genome!) is equally applicable to the majority of RNA viruses and error-prone DNA viruses. It should be quite obvious to the reader of previous chapters that there is a nonzero probability that mutations present in the mutant spectrum may contribute to drug resistance, immune evasion, or fitness enhancement. Even if they are hidden as a tiny minority, they are ready to be selected. The challenge is clear.

**1 to $10^{10}$-$10^{12}$**

FIGURE 9.1  A scheme of quasispecies intended to stress the challenge of controlling the replication of an immense and diverse entity. Each horizontal line represents a viral genome and symbols on the lines symbolize mutations, in a typical depiction of quasispecies used throughout the book. The discontinuous lines represent genomes with five or six mutations considered in this example a sufficient number to severely decrease fitness. Their fate is to attain very-low-frequency levels or be eliminated from the evolving population. About 100 million of displays similar to the one shown in this page are needed to depict the hepatitis C viral genomes present in an individual acutely infected with the virus.

## 9.1.1 Virus as moving targets

The potential deleteriousness of mutational load (excessive number of mutations in the same genome) is symbolized in Fig. 9.1 by the genomes drawn as discontinuous lines that contain five mutations, here, assumed to be a sufficiently large number to decrease fitness, and correspondingly, the frequency of their potential descendants in subsequent replication rounds when the five mutations are maintained. Quasispecies dynamics imposes, however, that those genomes that move from a high-frequency to a low-frequency level be replaced by newly arising mutants that display higher fitness. The mutant cloud is highly dynamic with newly arising genomes incessantly exposed to the scrutiny of selection and the lottery of random drift. The image portrayed here through mutations is even more complex if we add recombination and genome segment reassortment in the case of viruses with segmented or multipartite genomes (Chapters 2 and 3). From the point of view of antiviral interventions, viruses are true "moving targets," in the sense that the repertoire of variants that we should inhibit at a one-time point is not exactly the same to be inhibited at a subsequent time point. The difference may be irrelevant regarding the efficacy of antiviral treatment, or it may not. The impact of the dynamic change cannot be anticipated since mutations that compromise antiviral efficacy arise in an unpredictable way. Applied to clinical practice, it means that a specific inhibitor combination may be effective for many infected individuals for a long time, or only for a few individuals for a short time, with a range of possible intermediate outcomes. The new antiviral strategies discussed in this chapter take into consideration the moving target feature of the viruses to be controlled, and they have been proposed by an increasing number of experts aware of the quasispecies challenge.

## 9.2 Practiced and proposed strategies to confront the moving target challenge with antiviral inhibitors

The basic statistical considerations that justify the need to use multiepitopic vaccines to protect against variable viruses characterized by quasispecies behavior (keep in mind Fig. 9.1) were discussed in Section 8.3 of Chapter 8. In the present chapter, we focus on pharmacological approaches based on the administration antiviral inhibitors; immunotherapy is mentioned as a potential ingredient of combination therapies with antiviral inhibitors.

The systematic selection of drug (or multidrug)-resistant mutants in viral populations has encouraged the design of antiviral strategies intended to avoid viral breakthrough and treatment failure. The main options in clinical practice or under investigation using nonmutagenic and mutagenic antiviral inhibitors are summarized in Box 9.1. None of the listed strategies is totally free of the problem of selection of drug-resistant mutants, but the proposals are intended to avoid or delay such selection. Next, we discuss the major features of each of the six suggestions listed in Box 9.1.

### 9.2.1 Combination treatments

Studies on the advantage of the combined administration of two drugs were pioneered by H.J. Eggers and I. Tamm as part of their studies with picornaviruses (Eggers and Tamm, 1963; Eggers, 1976). Combination therapy applied to human immunodeficiency virus type 1 (HIV-1) has been the great success for the control of acquired immunodeficiency syndrome (AIDS) that has drastically reduced AIDS-related mortality. This type of treatment for HIV-1 is termed highly active antiretroviral therapy (HAART), and it is implemented with three (sometimes

---

**BOX 9.1**
<hr style="width:30%">

### Some antiviral strategies to control viral quasispecies

**Based on nonmutagenic inhibitors**

- Combination treatments. Use of two or more inhibitors directed to independent viral targets.
- Splitting the treatment into induction and different maintenance regimen.
- Targeting cellular functions.
- Use of drugs that stimulate the host innate immune system.

- Combined use of immunotherapy and chemotherapy.

**Based on mutagenic agents**

- Lethal mutagenesis in their two modes of sequential and combined administration of inhibitors and mutagens.

---

two) different antiretroviral agents. HAART efficacy has steadily improved due to the availability of new inhibitors directed to different HIV-1 targets, with application to rescue treatments (Gandhi et al., 2019). Success is only partial due to side effects (that may be derived from off-target activities of the drugs), incomplete patient adherence to the treatment, selection of multidrug-resistant viral variants, and above all, to the retroviral nature of HIV-1 that renders virus extinction from the organism a difficult endeavor. In retroviruses and hepadnaviruses, the viral DNA is not only hidden from antiviral drugs, but it can serve as an archive of genomes whose replication can be reactivated upon treatment discontinuation. If combination therapy could be coupled with an efficient elimination of the proviral reservoirs in host DNA (De Crignis and Mahmoudi, 2014; Sadowski and Hashemi, 2019), HIV-1 could probably be eliminated from the infected organism, as achieved in the case of HCV infections in many patients treated with direct-acting antiviral agents (DAAs).

The general advantage of combination therapy over monotherapy is largely a consequence of quasispecies dynamics, and it has been amply evidenced by clinical practice, and supported by straightforward statistical considerations and

theoretical models of virus dynamics (Domingo, 1989; Domingo and Holland, 1992; Ho, 1995; Bonhoeffer et al., 1997; Pol et al., 1999; Ribeiro and Bonhoeffer, 2000; Le Moing et al., 2002; Van Vaerenbergh et al., 2002; Domingo et al., 2008; Müller and Bonhoeffer, 2008; Nijhuis et al., 2009) (see also Section 8.4 in Chapter 8). Furthermore, use of combination therapies conforms to the "hit early, hit hard" dictum of P. Ehrlich and D.D. Ho, explained in Section 8.10 of Chapter 8. There is the misconception (among some experts, but mainly among politicians!) that only patients in advanced phases of a viral infection (i.e., only precirrhotic or cirrhotic patients infected with HCV) should be treated aggressively with drug combinations. This is not what evolutionary virology teaches us: patients should be treated as strongly as possible and as early as possible, provided side effects can be controlled. Viruses should be given no chance to walk in sequence space in search of adaptive pathways (Chapter 3). Adequate combination treatments can fulfill such purpose.

### 9.2.2 Split treatments

A second proposed strategy is to divide an antiviral treatment into two steps: first an induction, and then a maintenance step (von Kleist et al., 2011). It is based on a theoretical model developed to confront HIV-1 infections. The key argument is that when a treatment has to be changed due to virus rebound, mutants resistant to the ineffective drug have had the opportunity to replicate in the patient, increasing the viral load and supplying a proviral archive with inhibitor-resistant latent viruses, thus excluding the drug as a component of future combination treatments. The induction regimen should be followed by a maintenance regimen at a point in time in which the second drug finds a low viral load and limited numbers of mutants resistant to the initial treatment. Critical issues for the implementation of this proposal are the timing of treatment switch, the decrease of viral load as a result of the induction regimen, and the genetic and phenotypic barriers to the maintenance regimen. Clinical trials are needed to explore this interesting proposal, which is of potentially broad applicability.

## 9.2.3 Targeting cellular functions

Many cellular functions are needed to complete any step of a virus replication cycle. A good deal of research in virology has as its main objective to identify and characterize cellular functions that participate in virus entry into the cell, intracellular multiplication, release from cells, or in the immune response. Since cellular proteins cannot vary in response to the presence of an inhibitor (at least for common cell types within the time frame of a viral infection), an obvious thought is to administer inhibitors of those cellular functions (often proteins) that are needed to sustain viral replication; such inhibitors should suppress viral replication without selection of inhibitor-resistant mutants (Geller et al., 2007; Hopkins et al., 2010; Garbelli et al., 2011; Kumar et al., 2011; Vidalain et al., 2015; Hover et al., 2017; de Wilde et al., 2018; Ong et al., 2018). Despite potential benefits, two major problems may be encountered with antiviral agents directed to cellular proteins:

(i) toxic effects derived from suppression or alteration of cellular activities in which the target protein is involved, and (ii) selection of viral mutants that are insensitive to the presence of the inhibitor, despite the inhibitor not being directed to viral components. Insensitivity may come about by at least three different mechanisms. If the cellular protein which is the target of the inhibitor forms a complex with a viral protein in the course of viral replication, amino acid substitutions in the viral protein may permit progression of the infection in the presence of the inhibitor. This is the case of HCV resistance to the nonimmunosuppressive cyclophilin inhibitor SCY-635. NS5A interacts with cyclophilins and with NS5B; amino acid substitutions in NS5A (T17A, E295K, and V44A) and in NS5B (T77K and I432V) decrease the sensitivity of the virus to SCY-635, although the precise molecular mechanism of resistance has been debated (Chatterji et al., 2010; Sarrazin and Zeuzem, 2010; Delang et al., 2011; Kwong et al., 2011; Vermehren and Sarrazin, 2011).

Development of resistance may be host cell-dependent, as documented with poliovirus resistance to brefeldin A, a drug that targets a cellular guanine nucleotide exchange factor found in the replication complex. Resistant mutants were selected in HeLa cells but not in Vero cells; the degree of resistance of poliovirus mutants differed among cells of human origin, and resistance inflicted a fitness cost upon the virus (Viktorova et al., 2015).

When inhibitors are targeted to the cellular receptor for a virus, mutants may be selected that can enter cells through an alternative receptor. This has been documented with HIV-1 that can use coreceptor CCR5, CXCR4, or both (dual-tropic viruses). A modified RANTES [regulated on activation, normal T cell expressed, and secreted; also termed as chemokine (C—C motif) ligand 5 or CCL5] selected coreceptor switch variants in a SCID (severe combined immunodeficiency) mouse model (Mosier et al., 1999). The mutants with amino acid substitutions in loop

V3 of Env protein were selected to use CXCR4 rather than CCR5 as a coreceptor (see also Section 4.4 in Chapter 4). V3 loop mutants associated with maraviroc resistance may exhibit increase replicative capacity (Garcia-Perez et al., 2015).

An alternative mechanism to overcome the inhibition of a cellular protein is that viral mutants are selected that can utilize the cellular protein in complex with the inhibitor. This is the case of HIV-1 mutants resistant to small coreceptor CCR5-binding inhibitors; mutants with amino acid substitution in the V3 loop region or elsewhere in Env can use either free or inhibitor-bound CCR5 to enter cells, with an efficiency that depends on the CCR5 expression level and the host cell type (Pugach et al., 2007, 2009). These examples illustrate the multiple pathways that viruses can exploit to overcome inhibitors directed to cellular proteins. They constitute additional evidence that "abundant sources of genetic variation exist for viruses to learn new tricks, …" emphasized by J. Lederberg in connection with viral disease emergence and reemergence (also quoted in Section 7.7.1 of Chapter 7). Since many viruses can use alternative receptors for entry into cells (Section 4.4 in Chapter 4), inhibitors directed to viral receptors may promote the selection of virus subpopulations that can use a different receptor.

### 9.2.4 Use of drugs that stimulate the host innate immune system

Some inhibitors of enzymes of the de novo pyrimidine biosynthesis pathway (DD264, bequinar, and A3) stimulate the innate immune response and behave as broad-spectrum antiviral inhibitors (Lucas-Hourani et al., 2013; Munier-Lehmann et al., 2015; Vidalain et al., 2015). The observed effect is one among other connections that have been established between nucleotide and DNA metabolism and immune stimulation (Motani et al., 2015). For some of the pyrimidine biosynthesis inhibitors, additional mechanisms of

antiviral activity might be involved, as in the inhibition of lymphocytic choriomeningitis virus (LCMV) replication and transcription by A3 (Ortiz-Riano et al., 2014). Stimulation of the innate immune response may restrict the selection of escape mutants because the virus must mutate at several sites to overcome the different branches of the response, as is the case with interferon (IFN) resistance (Perales et al., 2014). Multifactorial antiviral responses increase the genetic and phenotypic barriers to resistance (Section 8.4.2 in Chapter 8).

### 9.2.5 Combined use of immunotherapy and chemotherapy

An extension of the advantage of combination therapy to decrease the selection of antiviral-resistant mutants consists in the combined use of immunotherapy (administration of neutralizing antibodies or other means of immune stimulation, such as vaccination) together with antiviral inhibitors. The concept was pioneered by R.G. Webster and colleagues and was proposed as a strategy for the control of influenza viruses (Webster et al., 1985). The authors showed that the simultaneous administration of inactivated H5N2 vaccine and the inhibitor amantadine conferred protection against H5N2 influenza virus A/Chick/Pennsylvania/83 in chickens (Webster et al., 1986). Related notions have been investigated with other viruses and different components of the immune response (Seiler et al., 2000; Li et al., 2005). However, additional model in vivo experiments with animals and clinical trials with patients are necessary to investigate the effectiveness of combined immunotherapeutic approaches, in line with related research in cancer therapy (Dalgleish, 2015).

As a general outlook on the strategies summarized in previous sections, the potential of combining two (or even more) of the proposals listed in Box 9.1 is encouraging, provided off-target effects of drugs or immune interventions can be controlled and side effects minimized. A trend toward "complex" treatment protocols is the expected response to the adaptive capacity of viral quasispecies. The introduction of mutagenic agents in antiviral designs is an important departure that exploits one of the corollaries of quasispecies behavior: the error threshold relationship (introduced in Section 3.6.3 of Chapter 3). It is the basis of lethal mutagenesis discussed next.

## 9.3 Lethal mutagenesis and the error threshold

Lethal mutagenesis is defined as the process of viral extinction due to an excess of mutations in the viral genome. J.J. Holland and colleagues pioneered studies on the adverse effects of several chemical mutagens on the yield of infectious poliovirus (PV) and vesicular stomatitis virus (VSV) in cell culture (Holland et al., 1990). The term lethal mutagenesis was first proposed by L. Loeb, J.I. Mullins, and colleagues in a study of the loss of replicative potential of HIV-1 upon multiplication in human CEM cells in the presence of the deoxynucleoside analog 5-hydroxydeoxycytidine (5-OH-dC) (Loeb et al., 1999). The experimental design of lethal mutagenesis was inspired in the error threshold relationship derived from the basic equation of quasispecies dynamics (Eigen and Schuster, 1979; Swetina and Schuster, 1982; Nowak and Schuster, 1989; Schuster, 2016). Initially elaborated on simple (single peak) fitness landscapes, quasispecies theory has as one of its main corollaries that the stability of genetic information during a replicative process is dependent on two parameters: the error rate during replication and the amount of genetic information to be maintained. The basic equations that describe the relationship between the maximum tolerable mutation rate ($\mu_{max}$) to ensure the transfer of genetic information to the next generations of genomes with length $\nu$, are included in Fig. 9.2. They are relevant to the application of the error threshold concept to virology (Eigen, 2002; Schuster, 2016). In the equations, an important variable is the superiority of the master sequence over its surrounding mutant spectrum, denoted by $\sigma_m$. Consistent with $\sigma_m$ being in the numerator of the equations, virologists will intuitively understand that a high superiority of the dominant sequence reinforces the stability position of the population (dominant or master sequence together with its surrounding mutant cloud) in that environment, and in the absence of external perturbations. A larger mutational input is necessary to destabilize the distribution and drive the mutant spectrum toward extinction.

There are additional factors that can modify the stability of a mutant distribution and the position of the error threshold; the mathematical justification and practical implications for virology have been detailed (Schuster, 2016). Of particular relevance to a virus population stability is the influence of the fitness landscape in the maintenance of genetic information. The first noteworthy result of the theoretical studies is that an error threshold is present in realistic, rugged fitness landscapes, as those proposed to best describe viruses on the basis of experimental evidence (see also Section 5.3 in Chapter 5). Second, the position of the error threshold moves toward lower mutation rates when the ruggedness of the fitness landscape (represented by a parameter that consists of a band of fitness values) is increased. How replication of viruses in variable fitness landscapes (e.g., under variable environmental conditions) versus a constant environment may affect the ease of extinction by increased mutagenesis is a largely unexplored question.

### 9.3.1 Reconciliation of theory and experiment: a proposal

Several theoretical models have been presented to explain the lethal mutagenesis of viruses (review in Tejero et al., 2016). The models are conceptually diverse, and at times with

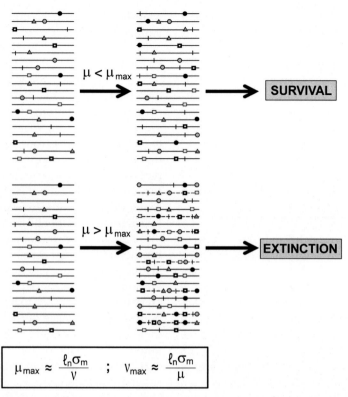

**FIGURE 9.2** A representation of virus entry into error catastrophe, and error threshold relationships of quasispecies theory. The mutant spectrum at the top replicates with a mutation rate ($\mu$) below the maximum compatible with the maintenance of genetic information ($\mu_{max}$), and the population remains stable (virus survival). The mutant spectrum below the first one replicates with a mutation rate ($\mu$) that exceeds the highest tolerable ($\mu_{max}$) for maintenance of the genetic information. As a consequence, there is a progressive deterioration of viral functions and genome replication, represented here by genomes drawn as discontinuous lines. Their number increases as replication proceeds under conditions of $\mu > \mu_{max}$ until the system collapses into a total loss of information (loss of virus identity and transition into replication-incompetent sequences: virus extinction). The box below the mutant distributions includes the basic mathematical formulae of the error threshold, in which $\sigma_m$ is the selectivity or superiority of the master sequence over its mutant spectrum, $\nu$ is the chain length of the replicating genome, and $\nu_{max}$ is the maximum length whose information can be maintained when replicating with mutation rate $\mu$. For the mathematical derivation of the error threshold relationships, see Schuster (2016). See text for the connection between the error threshold and lethal mutagenesis.

remarkably counterintuitive proposals. Some models deny a connection between the error threshold of quasispecies theory and the extinction of viruses by enhanced mutagenesis. In one of the models discussed by H. Tejero and colleagues, it was suggested that error catastrophe could not occur in the presence of lethal genotypes, a proposal that was considered peculiar by experimentalists, and that it was proven incorrect by Takeuchi and Hogeweg (2007) [see also references in that publication and in Tejero et al. (2016)]. Other models that have suggested that lethal mutagenesis is unrelated to the error threshold have defined an extinction threshold to mean the mutation rate at which a viral population goes extinct. A rather counterintuitive proposal is that the error threshold is caused by the "survival of the flattest" which means dominance of genomes with low replicative capacity and high tolerance to mutations (robustness) that would hinder virus extinction (Tejero et al., 2011) (see Section 5.7 in Chapter 5 on the advantage of the flattest in a fitness landscape).

C. Perales and I have carefully reviewed the main experimental results obtained in our and other laboratories on the molecular events that accompany the transition toward virus extinction and have tried to harmonize the experimental observations with the most realistic and significant theoretical models reviewed by Tejero et al. (2016). The mainstream of experimental results can be summarized as follows. The first studies that were an extension of those carried out in J.J. Holland's laboratory confirmed a connection between a mutagenic activity and decrease of viral infectivity and showed that low viral load and low viral fitness favored virus extinction (Loeb et al., 1999; Sierra et al., 2000). (That low population numbers increase the risk of extinction is accepted for any type of organism under any circumstance, and it is known as the Alee effect). Then, following the important discovery that the nucleoside analog ribavirin (1-β-D-ribofuranosyl-1-H-1,2,4-triazole-3-carboxamide)

is mutagenic for PV (Crotty et al., 2000) it was observed that ribavirin decreased PV-specific infectivity (the ratio of infectivity to the amount of viral RNA) (Crotty et al., 2001). A decrease of specific infectivity, together with an increase of mutation frequency, is now considered the standard way to distinguish lethal mutagenesis from mere inhibition (Box 9.2). [This is not to mean that decreases of specific infectivity are exclusive of lethal mutagenesis. They have also been documented as a result of codon deoptimization (Section 4.3 of Chapter 4) and in viral clones subjected to many plaque-to-plaque transfers (Section 6.5.2 in Chapter 6)].

In a subsequent study, A. Grande-Pérez and colleagues documented that treatment of cells persistently infected with LCMV with 5-fluorouracil [5-Fluoro-1H,3H-pyrimidine-2,4-dione (FU)] resulted in a decrease of infectivity that preceded the decrease of viral RNA (see Section 9.4 for the mutagenic mechanisms of ribavirin and FU). This result suggested that a class of replication-competent RNA that did not lead to infectious virus was generated during the

---

BOX 9.2

## Main observations on viral extinction by mutagenic agents

- Low viral load and low viral fitness favor extinction.
- During the transition toward extinction, the viral population:
    Decreases its specific infectivity.
    Increases its mutant spectrum complexity (movement toward usually unfavored regions of sequence space).
    Maintains an invariant consensus sequence.
- Reduction in viral load per se is not the mechanism of virus extinction. A load

decrease by a mutagenic agent may extinguish the virus while the same decrease with an inhibitor may not.
- Viruses displaying different replicative features respond to mutagenic agents in a very similar way: the strategy is of general applicability provided a virus-specific mutagenic agent is available.
- The use of mutagens as antiviral agents has been validated in vivo.

FU-mediated transition toward viral extinction. A theoretical model developed by S. Manrubia predicted that such defective genomes that were termed defectors were required to achieve LCMV extinction with a limited mutagenic intensity by FU (Grande-Pérez et al., 2005b). The proposal that loss of viral infectivity is due to the action of defective genomes produced by the mutagenic agent is termed the lethal defection model of virus extinction, and it is consistent with several observations on the extinction of other viruses (reviewed in Domingo et al., 2012). A diagnostic test is that loss of infectivity precedes loss of viral RNA (Fig. 9.3). Lethal defection can be regarded as an extreme outcome of interfering interactions that are exerted among components of the mutant spectrum when their mutational load increases (compare with Section 3.8 of Chapter 3).

The studies by A. Grande-Pérez, S. Manrubia, and colleagues distinguished two pathways that viruses can follow when subjected to mutagenesis: lethal defection at low mutagenic intensities, and overt lethality at high mutagenic intensities. Fig. 9.4 recapitulates our understanding of the steps involved in mutagenesis-driven virus extinction, with the important qualification that no sharp boundary exists between the lethal defection and the overt lethality phases. Studies with foot-and-mouth disease virus (FMDV) have provided further support to the lethal defection phase of extinction. Defective FMDV RNAs inhibited replication of standard FMDV RNA in a specific manner when co-electroporated into cells (Perales et al., 2007). Specificity means that the defective FMDV RNAs did not inhibit replication of the related encephalomyocarditis viral RNA. Specificity was also evidenced by the loss of interfering activity of a defector genome by the introduction of a mutation that prevented RNA replication. The requirement of replication of the defective RNAs suggests that a sufficient amount of expressed proteins is needed for interference (Perales et al., 2007). These results are consistent with enhanced and also specific interfering

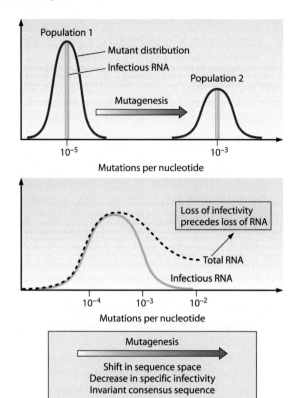

FIGURE 9.3 A summary of the main experimental observations upon mutagenesis of a viral population. Top: mutagenesis results in an expansion of mutant spectrum (*red* Gaussian distribution, that may be skewed in real populations), a 100-fold increase in the average number of mutations per nucleotide, and a decrease of infectious RNA (*blue thin column*) (specific values depend on each experimental system). Bottom: the increase in the average number of nucleotides leads to maintenance of viral RNA (*discontinuous curve*) despite the loss of infectivity (*blue curve*). The box at the bottom underlines the major changes underwent by the viral genomes as the result of mutagenesis; these points are also included in Box 9.2. *Figure reproduced from Domingo, E., Sheldon, J., Perales, C., 2012. Viral quasispecies evolution. Microbiol. Mol. Biol. Rev. 76, 159–216, with permission from the American Society for Microbiology, Washington DC, USA.*

activity exerted by mutagenized FMDV RNA (González-López et al., 2004). Thus, the available evidence suggests that mutant RNAs—whose frequency and mutational load increase as the mutagenesis proceeds—adversely affect the

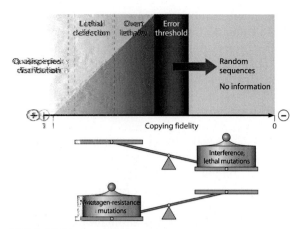

FIGURE 9.4   Scheme of the steps involved in virus extinction by enhanced mutagenesis. The graph includes a basic element of quasispecies fidelity: the horizontal black arrow that spans the entire possible range of copying fidelity values: from a value 1 (perfect copying fidelity, no mutations; incompatible with evolution of life) to 0 (complete infidelity, mutation as the norm; incompatible with maintenance of genetic information). On top of the horizontal axis, four successive recognized steps that are reached with decreasing copying fidelity, are represented: the region of maintenance of a dynamic quasispecies; the region of lethal defection, the region of overt lethality, and finally, the crossing of the error threshold (red arrow). The very last transition can be visualized as the viral genomic sequences having degenerated into random sequences; that is, a transition from information to no-information. The balances drawn at the bottom symbolize influences that may either favor or prevent virus extinction. Figure reproduced from Domingo, E., Sheldon, J., Perales, C., 2012. Viral quasispecies evolution. Microbiol. Mol. Biol. Rev. 76, 159–216, with permission from the American Society for Microbiology, Washington DC, USA.

replication of nonmutated RNAs that coexist in the same replicative ensemble. This is the major molecular event that has been recognized in the lethal defection phase of Fig. 9.4.

The infectious FMDV RNA that can be retrieved from a mutagenized population by dilution and plaque isolation display decreases in replicative fitness (up to 200-fold lower infectivity) relative to the parental, nonmutagenized virus (Arias et al., 2013) associated with an eight-fold increase in mutation frequency. This result reinforces the complexity of events during lethal mutagenesis, confirms that fitness may be impaired in clones that transiently survive the

mutagenic activity, and suggests an overlap between the lethal defection and overt lethality phases drawn in Fig. 9.4.

We can now consider theoretical models in light of the experimental results. First, the proposed interference by substituted, *trans*-acting proteins is in line with the early descriptions of fitness deterioration of cells due to the collapse of interdependent nucleic acid and protein *trans* networks in connection with the process of aging (Orgel, 1963). In the case of lethal defection, the effects of mutations have to be calibrated, keeping in mind the multifunctional nature of viral proteins (Section 3.8.1 in Chapter 3). When a protein is defective, it can jeopardize the activities of any other proteins that interact with it: a *trans* network can collapse by a domino effect. The possible influence of the topology of the network of interactions among genomes for maintenance of population stability is a largely unexplored possibility which is briefly addressed in the closing Chapter 10.

The notion that viral mutagenesis promotes drift in sequence space was shown by direct amplification of A, U-rich genomic sequences of FMDV subjected to ribavirin mutagenesis (Perales et al., 2011b). The main effect of ribavirin was to accelerate the occupation of A, U-rich regions of sequence space, presumably due to the tendency of this purine analog to produce an excess of $G \rightarrow A$ and $C \rightarrow U$ transitions (Section 9.4.). Analysis of the numbers and types of mutations suggests that the A, U-enriched portion of sequence space is detrimental to viral fitness. Movements toward unfavorable regions of sequence space are also suggested by mutant spectrum analyses of FMDV subjected to FU mutagenesis and other viruses subjected to other mutagenic agents (Grande-Pérez et al., 2002, 2005a; Agudo et al., 2008; Ortega-Prieto et al., 2013).

In view of the above evidence, any theoretical model of lethal mutagenesis that proposes a delocalization of the genome population in sequence space fits the experimental results of extinction. Specifically, models based on the advantage of the flattest that predict the absence of extinction (Tejero et al., 2016), in reality, predict the extinction of a real virus. This is because the mutagenesis-driven, astray walk in sequence space in the absence of a dominant master sequence should produce an increased number of defective genomes (lethal defection) in unfavorable regions of sequence space (such as the A, U-rich regions promoted by ribavirin). The net result should be not only lethal defection but also an increasingly frequent hitting of lethal portions of the space (overt lethality phase shown in Fig. 9.4). Thus, any theoretical models based on genome sequence delocalization fit the experimental observations (Perales and Domingo, 2016). How such delocalization can be turned into an antiviral strategy is discussed in the next sections.

## 9.4 Virus extinction by mutagenic agents

The pioneer experiments by J.J. Holland and colleagues demonstrated the adverse effects of mutagenic agents—including the base analog FU and the nucleoside analog 5-azacytidine [4-amino-1-$\beta$-D-ribofuranosyl-1,3,5-triazin-2(1$H$)-one (5-AZA-C)]—on the production of infectious PV and VSV progeny (Holland et al., 1990; Lee et al., 1997). These investigations were followed by several others that examined the effect of base and nucleoside analogs on virus survival. A few studies with animal viruses are summarized in Table 9.1 to illustrate that viruses displaying diverse replicative strategies (positive- and negative-strand RNA viruses and retroviruses) are vulnerable to increases in mutation rate. The investigations in cell culture or animal hosts employ mutagenic base or nucleoside analogs which are converted intracellularly into the nucleoside-triphosphate forms. The latter can be incorporated into RNA that in the next round of template copying will give rise to misincorporations (point mutations) and in some cases to

RNA chain termination. Mutagenesis is mainly due to ambiguous pairing between the mutagenic base and standard nucleotides, with the formation of Watson-Crick or wobble base pairs. Some base-pair structures of FU with A or G are drawn in Fig. 9.5 (Sowers et al., 1988; Yu et al., 1993), and of ribavirin and T-705 (favipiravir) (5-Fluoro-2-oxo-1$H$-pyrazine-3-carboxamide) in Fig. 9.6 (Crotty et al., 2000; Jin et al., 2013) (see Section 2.2 in Chapter 2 for the standard Watson-Crick and some wobble base pairs).

The possible point mutations generated as a consequence of FU or ribavirin incorporation during viral genome replication are depicted in Figs. 9.7 and 9.8, respectively. The figures represent the most frequent cause of the successive analog incorporation and subsequent events upon copying of positive- and negative-strand viral RNA, although the pathways are equally applicable to DNA in the case of deoxynucleotide analogs. The schemes can serve as a guide to anticipate the possible mutation types induced by other analogs provided their base-pairing behavior has been investigated by physicochemical procedures, considering that mutation preferences may be affected by environmental factors.

The mutant repertoire produced by a mutagenic nucleotide may be influenced by the structure and micro-environment of the viral polymerase in the neighborhood of the site that determines the base pairing of the template with the incoming nucleotide. Although polymerases share general structural features, nucleotide discrimination and incorporation is subtly dependent on the nature of surrounding amino acids, template-primer nucleotide sequences, or ionic composition (see Section 2.6 of Chapter 2 on polymerase structure and fidelity mutants, and Section 9.6 on interactions in viral polymerases associated with resistance to mutagenic nucleotides). For several viruses, the mutant spectra produced upon replication in the presence of FU include an excess of A $\rightarrow$ G and U $\rightarrow$ C, over G $\rightarrow$ A and C $\rightarrow$ U transitions

**TABLE 9.1** Some studies on lethal mutagenesis of viruses.

| Virus[a] | Base or nucleoside analog[b] | Main observations | Reference |
|---|---|---|---|
| DENV | T-705 | Several base and nucleoside analogss evaluated. Mutagenesis by T-1105 and T-1106 documented. Bias in favor of transitions | Qiu et al. (2018) |
| EBV | T-705 | Protection of macaques associated with mutagenic doses. Bias in favor of $G \rightarrow A$ and $C \rightarrow U$ | Guedj et al. (2018) |
| FMDV | FU 5-AZA-C | Extinction favored by low viral load and low viral fitness | Sierra et al. (2000); Pariente et al. (2001) |
| FMDV | Rib | Enhanced mutagenesis eliminated the virus from persistently infected cells | Airaksinen et al. (2003) |
| FMDV | FU | Virus extinction required mutagenesis and not only inhibition by guanidine hydrochloride and heparin | Pariente et al. (2003) |
| FMDV | FU | Extinction of a Rib-resistant mutant by the combination of FU and guanidine hydrochloride | Perales et al. (2009a) |
| FMDV | T-705 | New option for viral infections of veterinary impact. No obvious outstanding mutation type | De Avila et al. (2017) |
| GBvB | Rib | Evidence of error-prone replication induced by Rib in cell culture, but no significant reduction of viremia in vivo (tamarin model). RTP incorporated in viral RNA can induce misincorporations and an elongation block | Lanford et al. (2001); Maag et al. (2001) |
| HCV | Rib | Rib increased the mutation frequency in conserved regions of an HCV replicon genome | Contreras et al. (2002) |
| HCV | Rib | Rib-mediated extinction of HCV through an increase of mutant spectrum complexity and decrease of specific infectivity | Ortega-Prieto et al. (2013) |
| HCV | T-705 | New option for anti-HCV treatments. Bias in favor of $G \rightarrow A$ and $C \rightarrow U$ | De Avila et al. (2016) |
| Hantaan virus | Rib | Rib-induced high mutation frequency in S segment. A range of Rib concentrations revealed a nonlinear fit of mutation frequency and mutagen concentration | Severson et al. (2003); Chung et al. (2007) |
| HIV-1 | 5-OH-dC | Decrease of virus infectivity; 3-fold increase of mutation frequency, and 5.6-fold increase in $G \rightarrow A$ transitions | Loeb et al. (1999) |
| HIV-1 | 5-OH-dC | Efficient HIV-1 extinction by combinations of 5-OH-dC and 3'-azido-3'-deoxythymidine (AZT) | Tapia et al. (2005) |
| HIV-1 | 5-AZA-C | Lethal mutagenesis during reverse transcription is the main antiviral effect of 5-AZA-C, following reduction to the deoxycytidine form | Dapp et al. (2009) |
| HIV-1 | 5-AZA-C | Adverse effect of 5-AZA-C on viral infectivity, potentiated by the host factor APOBEC3G | Dapp et al. (2012) |
| IV | T-705 | First evidence that the broad-spectrum antiviral agent T-705 (favipiravir) can act as a lethal mutagen | Baranovich et al. (2013) |
| IV | FU 5-AZA-C Rib | Effective lethal mutagenesis of H3N2 and H1N1 IVs. Evidence of high barrier to resistance | Pauly and Lauring (2015) |

*(Continued)*

**TABLE 9.1**    Some studies on lethal mutagenesis of viruses.—cont'd

| Virus[a] | Base or nucleoside analog[b] | Main observations | Reference |
|---|---|---|---|
| LCMV | FU | Largest increases in mutation frequency did not predict virus extinction. Virus extinction occurred without modification of the consensus sequence | Grande-Pérez et al. (2002, 2005a) |
| LCMV | FU Rib | FU- and Rib-mediated extinction of LCMV in BHK-21 cells, associated with up to 10-fold increases of mutation frequency with FU, and inhibition of gene expression with RBV | Ruiz-Jarabo et al. (2003) |
| LCMV | Rib | Mutagenic activity documented in cell culture | Moreno et al. (2011) |
| MNV | T-705 | Efficient extinction in cell culture and in vivo. Bias in favor of A→G and U→C | Arias et al. (2014) |
| PV | Rib | Evidence of Rib-mediated error catastrophe of PV | Crotty et al. (2001) |
| PV | FU 5-AZA-C | Impairments of viral replication; 16- to 2.8-fold increases in mutation frequency | Holland et al. (1990) |
| RVFV | T-705 | Lethal mutagenesis in cell culture. Prospect of new treatments for this emerging disease. Significant increase in G→A and C→U transitions | Borrego et al. (2019) |
| TMV | FU | FU-mediated decrease of TMV infectivity, and increase of A→G, U→C transitions | Diaz-Martinez et al. (2018) |
| VSV | FU | Adverse effect of mutagenesis on virus viability and adaptability | Lee et al. (1997) |
| VSV | FU 5-AZA-C | Impairments of viral replication; 16- to 2.8-fold increases in mutation frequency | Holland et al. (1990) |
| WNV | Rib | Error-prone replication and transition to error catastrophe were dependent on the host cell line | Day et al. (2005) |
| WNV | T-705 | Adds to prospects of lethal mutagenesis to combat flaviviral infections. Bias in favor of G→A and A→G, to a lesser extent U→C | Escribano-Romero et al. (2017) |
| ZIKV | T-705 FU Rib | FU more effective against USUV than Zika, and more converse for Rib and T-705. Bias in favor of G→A and C→U | Bassi et al. (2018) |

[a] *The virus abbreviations are: DENV, Dengue virus; EBV, Ebola virus; FMDV, foot-and-mouth disease virus; GBvB, GB virus B; HCV, hepatitis C virus; HIV-1, human immunodeficiency virus type 1; LCMV, lymphocytic choriomeningitis virus; MNV, murine norovirus; PV, poliovirus; RVFV, Rift Valley fever virus; TMV, Tobacco mosaic virus; USUV, Usutu virus; VSV, vesicular stomatitis virus; WNV, West Nile virus; ZIKV, Zika virus.*
[b] *The abbreviations for drug names are: 5-AZA-C, 5-azacytidine; FU, 5-fluorouracil; 5-OH-dC, 5-hydroxydeoxycytidine; Rib, ribavirin; T-705, favipiravir.*

(Sierra et al., 2000; Grande-Pérez et al., 2002; Ruiz-Jarabo et al., 2003; Bassi et al., 2018; Diaz-Martínez et al., 2018; review on FU antiviral activity and mutagenesis in Agudo et al., 2009). The bias in favor of A → G and U → C mutations suggests that FU behaved as U rather than C when incorporated by the viral polymerases, and that, once FU was a template residue, it tended to behave as C as rather than U (Fig. 9.7).

Studies with several viruses have indicated that ribavirin mutagenesis yields mutant spectra with an excess of G → A and C → U over A → G and U → C (Fig. 9.8), although the incorporation bias depends on the virus [(Crotty et al., 2001; Airaksinen et al., 2003; Day et al., 2005; Chung et al., 2007; Agudo et al., 2010; Moreno et al., 2011; Dietz et al., 2013; Ortega-Prieto et al., 2013; Galli et al., 2018) among other

**Base pairing behavior
of nucleotide analogues**

**Base pairing behavior
of nucleotide analogues**

FIGURE 9.5 Some of the possible base pairs between 5-fluorouracil (FU) and the standard nucleotides A and G. Hydrogen bonds are indicated by discontinuous red lines; sugar and phosphate residues are not included.

FIGURE 9.6 Some of the possible base pairs between ribavirin (Rib) or favipiravir (T-705) and the standard nucleotides U and C. Hydrogen bonds are indicated by *discontinuous red lines*; sugar and phosphate residues are not included. The rotation about the carboxamide bond that contributes to alternative pairing with U or C is indicated by the small circular arrow.

studies; as reviews of the mutagenic activity of ribavirin see (Crotty et al., 2002; Beaucourt and Vignuzzi, 2014)].

The effect of a mutagenic nucleotide analog may be influenced by the position it occupies in the template. This was shown in the early site-directed mutagenesis experiments of bacteriophage Qβ RNA performed by R. Flavell, C. Weissmann, and colleagues (Chapter 3). When the pyrimidine analog $N^4$-hydroxy CMP was

present at the extracistronic position 15, it directed the incorporation of GMP slightly more efficiently than AMP, while at position 39 incorporation of AMP was three-fold higher than GMP (compare Flavell et al., 1974 and Domingo et al., 1976).

**FIGURE 9.7** Mutations produced by 5-fluorouracil (F) as a result of its incorporation into viral RNA. The RNA polarity is indicated on the left (+, positive-strand or genomic RNA in the case of positive-strand RNA viruses; −, negative-strand or complementary RNA; in the case of negative-strand RNA viruses the genomic RNA is of negative polarity). Downward arrows indicate standard base copying and F incorporation. The block on the left explains the consequences of incorporation of F instead of U either in the minus strand (second row) or in the plus strand (third row). The block on the right indicates the consequences of incorporation of F instead of C either into the minus row (second row) or into the plus strand (third row). The boxes at the bottom indicate the types of mutations expected in the mutant spectrum (see text for references).

**FIGURE 9.8** Mutations produced by ribavirin (R) as a result of its incorporation into viral RNA. The RNA polarity is indicated on the left (+, positive-strand or genomic RNA in the case of positive-strand RNA viruses; −, negative-strand or complementary RNA; in the case of negative-strand RNA viruses the genomic RNA is of negative polarity). Downward arrows indicate standard base copying and R incorporation. The block on the left explains the consequences of incorporation of R instead of A either into the minus strand (second row) or into the plus strand (third row). The block on the right indicates the consequences of incorporation of R instead of G either into the minus strand (second row) or into the plus strand (third row). The boxes at the bottom indicate the types of mutations expected in the mutant spectrum (see text for references).

## 9.4.1 The search for new mutagenic nucleotide analogs

New mutagenic nucleotides are currently being investigated as potential lethal mutagens for viruses (Harki et al., 2002, 2006, 2007; Graci and Cameron, 2004; Beach et al., 2014; Dapp et al., 2014; Vivet-Boudou et al., 2015; among other studies). There is active research to apply drugs (or their derivatives) used in antibacterial or anticancer therapy to lethal mutagenesis of viruses (drug repositioning or drug repurposing; see Section 8.4 in Chapter 8). L.M. Mansky and his associates have pioneered such efforts for the search of new antiretroviral agents, as well as the study of new combination therapies based

on lethal mutagenesis (Dapp et al., 2012, 2013; Bonnac et al., 2013; Rawson and Mansky, 2014). They studied the joint effect of 5-AZA-C and apolipoprotein B mRNA editing complex 3G (APOBEC3G) on the mutational spectrum of HIV-1. The results revealed unexpected changes in the mutational trend, particularly an increase in G → A transitions and a decrease of G → C transversions, a mutation type which is prominent with 5-AZA-C alone (Dapp et al., 2009). They have also defined a new class of antiviral agents termed "mild mutagens" that while exhibiting a limited mutagenic activity they may enhance the antiviral effect of other inhibitors (Rawson et al., 2013). It may be interesting to explore possible mild mutagenic activities for other analogs, such as shown with sofosbuvir for Zika virus (Sacramento et al., 2017).

The complexities of the interaction among mutagens and between inhibitors and mutagens in infected cells are still poorly understood. A few studies have provided evidence that some sequential inhibitor-mutagen treatments may have an advantage over the corresponding combinations, fundamentally because they avoid interferences derived from the simultaneous presence of a mutagen and inhibitor during viral replication (Section 9.8).

## 9.5 Lethal mutagenesis *in vivo*: complications derived from multiple mechanisms of drug action—the case of ribavirin

Most of the investigations summarized in the previous section were carried out in cell culture. Some animal experiments and a clinical trial have provided proof of principle of the feasibility of antiviral interventions based on lethal mutagenesis in vivo. J.C. de la Torre and colleagues documented that administration of FU to mice prevented the establishment of a persistent LCMV infection in the animals (Ruiz-Jarabo et al., 2003). J.I. Mullins and colleagues carried out the first phase II clinical trial using lethal mutagenesis by administering KP1461 (N4-heptyloxycarbonyl-5,6-dihydro-5-aza-2′-deoxycitidine), the prodrug of KP1212 (5,6- dihydro-5-aza-2′-deoxycitidine), to HIV-1 infected volunteers previously treated with antiretroviral agents (Mullins et al., 2011). No reduction in viral load or increase in average mutation frequencies was noted in the treated patients. However, mutations that likely occurred in HIV-1 in the course of the treatment were predominantly A → G and G → A transitions, as expected from the base-pairing behavior of the pyrimidine analog. The study validated lethal mutagenesis as an antiviral approach for human disease.

The purine analog favipiravir (T-705) (6-fluoro-3-hydroxy-2-pirazinecarboxamide; base pairing behavior shown in Fig. 9.6) acted as an effective antinorovirus agent in a mouse model, with features diagnostic of lethal mutagenesis: increases of mutation frequency and decreases of specific infectivity (Arias et al., 2014) (Box 9.2). Favipiravir and derivatives are interesting compounds because they share with ribavirin a broad antiviral spectrum of activity. Favipiravir has proven effective against several viruses in vivo, including West Nile infection of mice and hamsters (Morrey et al., 2008), lethal inhalation of Rift Valley fever virus in rats (Caroline et al., 2014), and highly pathogenic H5N1 IV in mice (Kiso et al., 2010). It is still an open question if it acts as a lethal mutagen in these in vivo systems as it does with IV in cell culture (Baranovich et al., 2013) and norovirus in cell culture and in vivo (Arias et al., 2014). (Favipiravir mutagenesis is discussed in the coming Section 9.5.1).

Ribavirin was used for many years in combination with pegylated IFN-α as the standard of care treatment for HCV infections (McHutchison et al., 1998; Cummings et al., 2001; Di Bisceglie et al., 2001). It is not clear whether lethal mutagenesis is part of the anti-HCV activity of ribavirin, with some studies favoring a mutagenic activity on the virus and others not (Gerotto et al., 1999; Querenghi et al., 2001; Sookoian et al., 2001; Dixit et al., 2004; Asahina et al., 2005; Chevaliez et al., 2007; Lutchman et al., 2007; Perelson and Layden, 2007; Cuevas et al., 2009; Dietz et al., 2013). The study by Dietz et al. (2013) involved deep sequencing analyses of HCV from patients subjected to ribavirin monotherapy, and the virus showed the mutational bias expected from ribavirin mutagenesis. It has been suggested that in cell culture experiments, the concentrations of ribavirin used are not attainable in vivo. Until measurements of the concentration of ribavirin nucleotides at the HCV replication complexes are available, it cannot be concluded that ribavirin concentrations in vivo are incompatible with mutagenesis; also, a broad range of ribavirin concentration in human serum have been reported during treatment (discussion and references in Ortega-Prieto et al., 2013). Furthermore, there are several reasons to miss a mutagenic activity in vivo with current analytical procedures. Ribavirin produced transient expansions followed by compression of mutant spectrum complexity in model studies with FMDV in cell culture (Ojosnegros et al., 2008; Perales et al., 2011b) (Fig. 9.9). A possible interpretation is that when the mutational load due to ribavirin mutagenesis surpasses some critical value, increasingly defective genomes cease to contribute to progeny, resulting in mutant spectrum compression. Additional variations in FMDV mutant spectrum complexity were observed upon other mutagenic treatments and

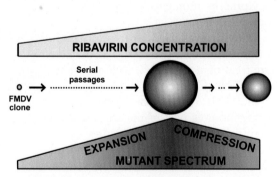

FIGURE 9.9 Expansion and compression of an FMDV mutant spectrum following ribavirin mutagenesis. A biological clone of FMDV (left) was subjected to serial passages in the presence of increasing concentrations of ribavirin (200 up to 5000 μM). The size of the three spheres quantifies in an approximate manner the complexity of the mutant spectrum, following the average distance parameter used to quantify genome subpopulations in the partition analysis of quasispecies (PAQ) clustering procedure of Baccam et al. (2001). The amplitude of the mutant spectrum increased first, but then it was compressed to a significant extent. The scheme is based on data reported in (Ojosnegros et al., 2008) and possible interpretations and implications are discussed in the text.

their interruption (discussed in Ojosnegros et al., 2008).

These model studies render unsurprising that no consistent expansions of HCV mutant spectra have been observed in patients subjected to ribavirin-based treatments or in chronically infected, untreated patients (compare, e.g., Farci et al., 2000; Duffy et al., 2002; Sullivan et al., 2007; Ramachandran et al., 2011; Palmer et al., 2014). Thus, even during ribavirin treatment alteration of mutant spectrum complexity may be missed depending on the time after treatment onset at which the viral sample is obtained. Also, it is necessary to quantify the amount of viral RNA to ensure that the sequence analyses either by standard molecular cloning-Sanger sequencing or deep sequencing provide a faithful representation of the biological sample (see Section 3.6.4 in Chapter 3). In our view, it is unlikely that a mutagenic activity readily observed in cell culture may be totally absent and bear no relationship with the activity of ribavirin in vivo, even though other mechanisms of ribavirin action may also play a role in its overall antiviral activity.

A mutagenic activity of ribavirin has been documented in cell culture with HCV and subgenomic replicons (Contreras et al., 2002; Zhou et al., 2003; Kanda et al., 2004; Ortega-Prieto et al., 2013), albeit with exceptions (Kato et al., 2005; Mori et al., 2011). The main problem to interpret the mechanism of antiHCV activity of ribavirin in vivo is that this antiviral agent can act through several nonexclusive mechanisms, including: (i) immunomodulatory activity with enhancement of the Th1 antiviral response (Hultgren et al., 1998; Ning et al., 1998); (ii) upregulation of expression of genes related to IFN signaling pathways (Zhang et al., 2003; Feld et al., 2007); (iii) depletion of intracellular guanosine-5'-triphosphate (GTP) levels associated with the inhibition of inosine-monophosphate dehydrogenase (IMPDH, the enzyme that converts inosine-monophosphate into xanthosine-monophosphate in the GTP biosynthesis pathway) by ribavirin-monophosphate (RMP) (Streeter et al., 1973), with a mutagenic effect of reduced inosine triphosphate (ITP) pyrophosphatase activity (which is associated with a common allelic variant in humans) (Nyström et al., 2018); (iv) inhibition of mRNA cap formation (Goswami et al., 1979); (v) inhibition of viral polymerases, independent of a mutagenic activity (Eriksson et al., 1977; Wray et al., 1985; Toltzis et al., 1988; Fernandez-Larsson et al., 1989; Maag et al., 2001; Bougie and Bisaillon, 2003); (vi) lethal mutagenesis, as evidenced with several RNA viruses (some of the studies are listed in Table 9.1), but not with other viruses (Leyssen et al., 2005, 2006; Kim and Lee, 2013). As general reviews of the antiviral properties of ribavirin, see (Snell, 2001; Graci and Cameron, 2002; Parker, 2005; Beaucourt and Vignuzzi, 2014).

It is not easy to separate an inhibitory from a mutagenic activity exerted by mutagenic base or nucleoside analogs and their intracellular derivatives. Inhibition may be a consequence of mutagenesis or inhibition and mutagenesis may have two totally separate modes of action. In the case of the antiviral activity of FU against FMDV, it was possible to show inhibition of the uridylylation of primer protein VPg (the initial step of picornavirus RNA synthesis) by 5-fluorouridine-triphosphate (FUTP), in addition to a mutagenic activity of FUTP during RNA elongation (Agudo et al., 2008). In the case of ribavirin, the situation is more complex, and participation of several of the six mechanisms listed in the previous paragraph in its anti-HCV activity, including human alleles that can potentiate mutagenesis (Nyström et al., 2018), should be considered. In particular, it would be interesting to reexamine the evidence for inhibition of viral polymerases to ascertain that the inhibition is independent of mutagenic activity. An interesting possibility is that inadvertently lethal mutagenesis might have been the mode of action of broad-spectrum antiviral agents used for decades, and traditionally considered standard nonmutagenic inhibitors.

## 9.5.1 Favipiravir as antiviral inhibitor and mutagen

Favipiravir was discovered by Y. Furuta and colleagues as an effective antiviral agent against IV (Furuta et al., 2002, 2017). It has proven effective against a broad range of RNA viruses, with documented inhibitory, RNA chain termination and mutagenic activities in cell culture and in vivo [several studies are included in Table 9.1, and have been reviewed (Perales et al., 2019). A significant result was reported by X. de Lamballerie and colleagues who showed that in experimental infections of macaques with Ebola virus, a protective activity of favipiravir was conferred by drug doses that evoked virus mutagenesis (Guedj et al., 2018).

The ongoing studies with favipiravir, and many others on ribavirin that can now be considered a classic in antiviral designs have elevated lethal mutagenesis to the rank of a realistic antiviral strategy. The concept offers promise for additional mutagenic nucleotide (perhaps new drugs designed to modify polymerase fidelity without themselves being nucleotide analogs) for emerging viral infections (that we know are going to happen despite our incapacity to anticipate which, how, and when) and established infections for which new treatment options may become a necessity. As with any type of antiviral agent, however, preclinical and clinical assays will precede their licensing for human use [discussed in (Perales et al., 2019)]. The challenge remains for the lethal mutagens-mediated selection of mutant viruses that display resistance either to the mutagenic activity itself or to the consequences of mutagenesis.

## 9.6 Virus resistance to mutagenic agents: multiple mechanisms and evidence of abortive escape pathways

It took more than 1 decade of use of ribavirin to obtain the first resistant mutants using Sindbis virus and mycophenolic acid (a nonmutagenic inhibitor of IMPDH); the selected mycophenolic acid-resistant mutants displayed cross-resistance to ribavirin (Scheidel et al., 1987; Scheidel and Stollar, 1991). A decade later the first ribavirin-resistant mutants of poliovirus selected in the laboratory (Pfeiffer and Kirkegaard, 2003), and of HCV from patients under ribavirin monotherapy were described (Young et al., 2003). Subsequent work has characterized viral mutants resistant to mutagenic agents, particularly ribavirin, FU, and 5-AZA-C (Pfeiffer and Kirkegaard, 2005b; Sierra et al., 2007; Arias et al., 2008; Agudo et al., 2010; Levi et al., 2010; Arribas et al., 2011; Feigelstock et al., 2011; Domingo-Calap et al., 2012; Sadeghipour et al., 2013; Zeng et al., 2013, 2014;

among other studies). The major overall conclusion of these investigations is that viruses can develop resistance to mutagenic nucleotide analogs as they do to nonmutagenic inhibitors.

Several mechanisms of resistance to mutagenic nucleotide analogs have been described, and the major ones are listed in Box 9.3. They can be broadly divided into two main categories: modifications of the viral polymerase and modification of other proteins. The first substitution to be described that conferred ribavirin resistance was G64S in the PV polymerase (3D) that resulted in an increase of polymerase template-copying fidelity. Studies with PV harboring this mutation have been instrumental in showing the relevance of mutant spectrum complexity in virus adaptability (Pfeiffer and Kirkegaard, 2005a; Vignuzzi et al., 2006) (see Section 2.6 in Chapter 2). These and other studies have established the important concept that to maintain adequate fitness and a good survival probability, a virus population must keep its

---

### BOX 9.3

#### Main molecular mechanisms of resistance to mutagenic nucleotide analogs

- Amino acid substitutions in the viral polymerase:
  Substitutions that increase the general copying fidelity of the enzyme.
  Substitutions that specifically limit the incorporation of the mutagenic nucleotide.
  Substitutions that modulate the relative incorporation of the standard nucleotides.
- Amino acid substitutions in other viral proteins:
  Substitutions in nonstructural viral proteins that participate in viral replication or modify polymerase fidelity or modulation of nucleotide incorporation.

- Combinations of some of these mechanisms.
- Fitness-enhancing mutations (unrelated to resistance per se) can contribute to the expression and stability of the resistance trait.

(Most virological studies have involved base or nucleoside analogs rather than nucleoside-triphosphates. It is assumed that at least for the mechanisms described to date, base and nucleosides are converted into the nucleoside-triphosphate derivatives which are responsible for the mutagenic activity. See text for references).

genome heterogeneity within a suitable range: too low a diversity impairs adaptability, and too high a diversity may approach the population to an extinction threshold (Smith et al., 2013; Smith and Denison, 2013; Zeng et al., 2014). Several studies have demonstrated an attenuation phenotype associated with low- or high-fidelity viral mutants (Borderia et al., 2016; Rai et al., 2017). In a clinical setting, maintaining mutant spectra of viruses within a suitable amplitude range is one of the predictors of viral survival and progression toward disease (Section 8.8 in Chapter 8).

## 9.6.1 Unpredictable effects of some polymerase substitutions

The same amino acid substitution in the polymerase of related viruses may have a different phenotypic effect. The ribavirin resistance, fidelity-enhancing PV substitution G64S in 3D was obtained independently in two different laboratories (Pfeiffer and Kirkegaard, 2003; Castro et al., 2005). This suggested that PV may have a very restricted number of mutations to attain ribavirin resistance, despite ample evidence that viruses generally display alternative mutational pathways toward resistance to nonmutagenic inhibitors (Section 8.4.5 in Chapter 8), and more generally, multiple pathways for fitness gain (Escarmís et al., 1999; Nguyen et al., 2012; Cabanillas et al., 2013; Arenas et al., 2016). When FMDV whose polymerase is closely related to that of PV (Ferrer-Orta et al., 2004) was passaged in the presence of ribavirin, replacement G62S (the one equivalent to G64S in PV) was not selected. Instead, 3D substitution M296I became dominant and conferred ribavirin resistance (Sierra et al., 2007). A mutant FMDV encoding G62S in 3D that was constructed by site-directed mutagenesis displayed a strong selective disadvantage relative to the standard virus, that was partially compensated by the presence of M296I. The mutant reverted upon

passage in cell culture while FMDV with substitution M296I in 3D was stable (Ferrer-Orta et al., 2010). The comparison of the enzymological properties of the FMDV polymerase (3D) with either G62S, M296I, or both indicated that G62S impairs RNA binding, RNA polymerization, and the incorporation of RMP into RNA. Therefore, despite G62S being of potential benefit for the replication of FMDV in the presence of ribavirin, its selection would be abortive. Despite the sites of substitution G62S and M296I being separated by 13.1 Å, a network of interactions allowed a cross-influence between the two sites, with an effect on the catalytic domain of the enzyme (Ferrer-Orta et al., 2010). Still, with FMDV, the same amino acid substitutions in 3D can have opposite effects on template copying fidelity depending on the viral serotype (sequence context) (Li et al., 2018). Some polymerase substitutions may evoke resistance to multiple mutagenic nucleotides, such as cross-resistance to ribavirin and the nucleoside derivative of T-1106 of St. Louis encephalitis virus mutants (Griesemer et al., 2017). Thus, due to distance effects and the subtle, sequence-dependent network connections among residues within the polymerase molecule, the same amino acid replacement may have disparate and unpredictable effects on the behavior of closely related polymerases.

## 9.6.2 Polymerase fidelity and modulation of nucleotide incorporation

Substitution M296I in 3D of FMDV offers an example of mutagen resistance that limited the incorporation of RMP in the viral RNA without significant alteration of the general template-copying fidelity of the enzyme (Sierra et al., 2007; Arias et al., 2008) (Box 9.3). Fitness of the mutant FMDV relative to the standard virus was 3.8 in the presence of 800 μM ribavirin, and 0.5 in the absence of ribavirin [these values represent a selective strength of 7.6 for substitution M296I (f $_{+\ \mathrm{Ribavirin}}$/f $_{-\ \mathrm{Ribavirin}}$), calculated as described in Section 8.4.4 of Chapter 8].

The selective strength conferred by substitution M296I was not sufficient to maintain this substitution as the only one in 3D for the virus to respond to higher ribavirin concentrations. Passage of the 3D M296I mutant virus in the presence of larger concentrations of ribavirin, resulted in the selection of two additional substitutions, P44S and P169S, to yield a virus with the triple substitution P44S, P169S, and M296I in 3D; this triple mutant was termed SSI (Agudo et al., 2010). The selective strength of the triple mutant measured in the presence of 800 μM ribavirin was 18.3, 2.4-fold higher than the selective strength conferred by M296I alone. Significantly, the most salient biochemical feature displayed by the mutant polymerase was that P44S restricted the incorporation of RMP more strongly opposite C than opposite U, as determined in several in vitro incorporation assays. As a consequence, during replication in the presence of ribavirin, mutant SSI could maintain a balance of the different transition types [measured as the ratio of (G → A) + (C → U) to (A → G) + (U → C) mutations] typical of the standard virus when replicating in the absence of mutagens (Agudo et al., 2010) (see Fig. 9.8 for the ribavirin-mediated mutational pathways). The modulation mechanism associated with 3D substitution P44S in FMDV is one of the mutagen resistance mechanisms listed in Box 9.3. Modulation of transition types allows the virus to maintain its typical mutant spectrum complexity with its corresponding adaptability. Interestingly, substitution P169S contributed a selective advantage only in the presence of very high (5000 μM) ribavirin concentrations, once the transition-modulating phenotype through P44S had already been acquired. Box 9.3 also lists as mutations contributing to mutagen escape those that increase fitness without being bona fide resistance mutations. Comparison of the fitness-enhancing mutations

described for antiviral resistance (Chapter 8) and the effect of P169S in 3D of FMDV suggests that such mutations can be divided into two classes: those that increase fitness generally and those that do so only in the presence of mutagen and even in the presence of some range of mutagen concentrations (Agudo et al., 2010).

The analyses of mutant spectra produced by FMDV with the standard and SSI 3D in the absence and presence of ribavirin suggest that a bias in favor of G → A and C → U may be detrimental to the virus presumably because it may modify RNA and protein composition, and probably also increase the frequency of defective genomes. The latter was suggested by the presence of a stop codon in one of the clones of the mutagenized standard virus population (Agudo et al., 2010).

The three-dimensional structures of the ribavirin-resistant polymerases revealed alterations in the N-terminal region of 3D (Agudo et al., 2010), affecting sites that belong to a nuclear localization signal (NLS) present in FMDV 3D (Sanchez-Aparicio et al., 2013). Amino acid substitutions within the NLS that diminished the transport to the nucleus of 3D and 3D3C (a functional precursor intermediate of 3D and the protease 3C) also modified the template binding and nucleotide recognition properties of 3D (Ferrer-Orta et al., 2015; de la Higuera et al., 2018). Interestingly, some replacements within the NLS increased the incorporation of RMP relative to standard substrates, suggesting that structural alterations in viral polymerases may enhance the vulnerability of viruses to nucleotide analogs. The results on the effect of amino acid substitutions within the NLS of FMDV 3D emphasize the multifunctional nature of this viral polymerase, in line with the recognized multifunctionality of many viral proteins, a feature that increases vulnerability to lethal mutagenesis. Despite uncertainties repeatedly exposed in this book (dependence of enzyme behavior on the protein and template sequence context, behavior modifications due to introduction of additional amino acid substitutions, etc.) it can be envisaged that through structure-based designs, new drugs could be found that are tailored to viral polymerases to enhance the incorporation of mutagenic nucleotides; they could be added to existing nucleotide analogs to formulate new antiviral combinations.

The comparison of the several amino acid and groups of amino acid substitutions that in 3D of FMDV can be involved in ribavirin resistance (Sierra et al., 2007; Agudo et al., 2010; Zeng et al., 2014; Ferrer-Orta et al., 2015) suggests that there are multiple, in some cases even independent, evolutionary pathways for a virus to achieve resistance to mutagenic agents, as there are multiple pathways toward resistance to standard, nonmutagenic inhibitors. In the case of PV, different ribavirin resistance substitutions would be expected if independent viral lineages were subjected to a range of ribavirin concentrations. Modulation of nucleotide incorporation is also a mechanism of FMDV resistance to FU through a single polymerase substitution that prevents the excess of A to G and U to C transitions induced by this analog (de la Higuera et al., 2017). An added advantage of the selection of mutants with altered mutational spectrum is that they may lead to virus attenuation in different hosts (Warmbrod et al., 2019). Thus, there is ample evidence with several viral systems that polymerase substitutions can modulate nucleotide incorporation, and that modulation per se is a mechanism of resistance to lethal mutagenesis.

The viral polymerases are not the only determinants of template-copying fidelity. The viral replicative machineries consist of a complex of several viral and host proteins, and it has been shown that proteins other than the polymerase can also contribute to polymerase fidelity and resistance to mutagenic agents (Box 9.3). When the triple SSI ribavirin-resistant mutant of FMDV (Agudo et al., 2010) was further passaged in the presence of high ribavirin concentrations,

a population that included replacement I248T in nonstructural protein 2C was selected (Agudo et al., 2016). The possibility that the read-through protein of bacteriophage Qβ may contribute to 5-AZA-C resistance was suggested by a study of E. Lázaro and colleagues (Arribas et al., 2011), although limited amino acid substitutions in the viral replicase can confer resistance (Cabanillas et al., 2014). The small, nonenzymatic coronavirus nonstructural protein 10 is involved in maintaining the replication fidelity of the virus (Smith et al., 2015). Two substitutions in protein NS5A of an HCV replicon conferred low-level resistance to ribavirin (Pfeiffer and Kirkegaard, 2005b). In the case of the DNA bacteriophage φX174 that does not encode its own polymerase, FU resistance was achieved through substitutions in the virus-coded lysis protein (Pereira-Gomez and Sanjuan, 2014). It was proposed that delayed cell lysis, increase in the amount of progeny virus per cell, and limitation of the number of infectious cycles reduced the chances of mutagenesis (also discussed in Section 8.5 of Chapter 8 with regard to general molecular mechanisms of antiviral resistance). Therefore, current evidence anticipates multiple adaptive mechanisms of virus resistance to mutagenic nucleotides, not necessarily confined to the polymerase as documented for standard inhibitors. There is, however, an interesting possibility of an initial advantage of mutagenic agents over standard inhibitors to impede selection of resistant mutants. Before justifying this suggestion, current therapeutic alternatives based on lethal mutagenesis are discussed.

## 9.7 Virus extinction as the outcome of replacement of virus subpopulations: tempo and mode of mutation acquisition

Some general considerations derived from the observations described in previous sections are worth commenting. The extinction of a virus is not only a consequence of the reduction of viral load (with insufficient $R_0$ value to ensure cell-to-cell transmission to sustain the infection; see Section 7.2 in Chapter 7 for the concept of $R_0$), or of a simple movement in sequence space, or even of an increase of mutational load in replicating genomes. It is the result of a combination of several influences under specific circumstances of the kinetics of accumulation of mutations in viral genomes. We consider these points in turn.

Studies by N. Pariente, C. Perales, and colleagues showed that a given reduction in viral load achieved through mutagenesis was sufficient to drive FMDV toward extinction, but that when the same reduction was achieved through inhibition, it did not lead to virus extinction (Pariente et al., 2001, 2003; 2005; Perales et al., 2011a). The main difference is that mutagenesis is a dynamic process of genome variation and reduction of viral load, while inhibition is basically a static process regarding the incorporation of mutations. Only when a decrease in viral load has already taken place, the virus becomes more vulnerable to mutagenesis (Sierra et al., 2000).

Movements in sequence space are the norm during virus replication and evolution, independently of being subjected to increased mutagenesis or not. An A, U-rich genome subpopulation that became dominant upon passage of FMDV in the presence of ribavirin also existed in the standard viral population, albeit at a lower level (Perales et al., 2011b). That is, the main effect of ribavirin was to shift the dominant part of the mutant spectrum into an A, U-rich region. This shift may increase the presence of defective genomes, including defectors that interfere with replication of the standard virus (Section 9.3.1). We encounter here again one of the main departures introduced by quasispecies in the understanding of virus evolution: replacement of some genome subpopulations by others. In the case of extinction, the movement consists of the replacement of standard subpopulations by others lying in unfavorable portions of sequence space. Expressed in another manner: mutagenesis

forces an uncontrolled, not a fitness gradient-mediated, relocalization in sequence space.

The mutational load, calculated as the number of mutations incorporated in an individual genome is not by itself a predictor of survival or extinction. It depends on the location of the mutations. Accumulation of mutations may reduce fitness and robustness to the effect of additional mutations (Section 6.7.1 in Chapter 6) but still allow replicative competence. An FMDV clone subjected to multiple plaque-to-plaque transfers accumulates mutations at a rate of 0.1−0.25 mutations per passage, which implies that viable FMDV genomes can accumulate a number of mutations equivalent to a mutation frequency of $10^{-2}$ substitutions per nucleotide (Escarmís et al., 2008). A 10-fold lower mutation frequency is sufficient to extinguish the same viral clone upon passage in the presence of a mutagenic agent. The tempo and mode of accumulation of mutations is the determinant factor. E. Lázaro and colleagues showed that 5-AZA-C enhanced Qβ replication during plaque development, but drove the virus to extinction during growth in the liquid culture medium unless a mutagen-resistant Qβ was selected (Cases-Gonzalez et al., 2008; Arribas et al., 2011). The experimental design in the plaque transfers is such that at each transfer a viral particle capable of developing a visible plaque on the cell monolayer is rescued no matter how many companions are left behind (extinguished). The applied selection is powerful. In contrast, during mutagenesis without intervening plaque isolations, no means to recover viable minority survivors are included in the design: if a rare potential survivor with a cluster of mutations arises transiently, it will be suppressed by the surrounding mutant spectrum or will perish due to the next rounds of mutagenesis. Thus, even if beneficial mutations are more likely to arise in low-fitness genomes, their contribution to survival is minimal under a continuing mutagenic environment.

A critical point for prediction of virus extinction is how close to an error (or extinction) threshold virus replication takes place. The evidence is that DNA genomes whose replication is catalyzed by a high-fidelity DNA polymerase may resist transient increases of mutation rate and even benefit from them (Cupples and Miller, 1989; Solé and Deisboeck, 2004; Springman et al., 2010). Thus, viral extinction by enhanced mutagenesis is conditioned by several factors related mainly to basal replicative parameters of the virus and the kinetics of mutagenesis.

## 9.8 The interplay between inhibitors and mutagenic agents in viral populations: sequential versus combination treatments

As a test of the paradigm of the general advantage of combination therapies over monotherapy, initial experiments were designed to study the efficacy of a combination of a mutagenic agent and an antiviral inhibitor. The results documented that extinction of high-fitness FMDV or HIV-1 could be achieved by a combination of a mutagenic agent and a nonmutagenic inhibitor, but not with the mutagenic agent alone (Pariente et al., 2001, 2005; Tapia et al., 2005). Despite being an expected result, additional experiments pursued by C. Perales with FMDV demonstrated that the response of the virus to the combined action of an inhibitor and a mutagenic agent is a bit more complex than initially thought. The production of infectivity and viral RNA was measured in the course of viral passages in the presence of the inhibitor guanidinium chloride alone, or ribavirin alone, or a combination of guanidinium chloride and ribavirin, or with an initial passage in the presence of guanidinium chloride, followed by passages in the presence of ribavirin alone (Fig. 9.10). The results of quantification of FMDV progeny and a sensitive RT-PCR test of virus extinction, documented that the design consisting of administering guanidine alone first, followed by additional passages in the presence of ribavirin alone was the most effective to reduce progeny production

and to drive the virus toward extinction (Fig. 9.10B−E) (Perales et al., 2009b). S. Manrubia developed a theoretical model whose main parameters were the concentration of standard and defective viruses sensitive and resistant to the inhibitor, the viral mutation rate, the rate of generation of inhibitor-resistant mutants, the number of standard and defective progeny genomes, and the number of infected cells and of infectious cycles per cell. The model (described in Perales et al., 2009b; Iranzo et al., 2011) explained the advantage of the sequential inhibitor-mutagen administration protocol over the corresponding combination in terms of replicative parameters and intensity of mutagenesis. When an inhibitor and a mutagen are present together during viral replication, their joint influence on the viral RNA may jeopardize virus extinction (possible mechanisms are summarized in Box 9.4). Notably, the presence of the mutagen may transiently increase the frequency of inhibitor-resistant mutants, and therefore, favoring virus escape and treatment failure. A second mechanism by which the simultaneous presence of an inhibitor and mutagen can impede virus extinction is that in the presence of the inhibitor, the RNA defectors that are generated by mutagenesis of the viral RNA cannot interfere with viral replication. The reason is that defector replication is needed for lethal defection (Section 9.3.1), presumably because replication increases the amount of RNA that encodes *trans*-acting substituted proteins that are detrimental to the virus life cycle, and responsible for lethal defection (Grande-Pérez et al., 2005b).

The advantage of sequential inhibitor-mutagen administration has also been evidenced with LCMV, taking advantage of the double inhibitor and mutagenic character of ribavirin, that depends on its concentration (Moreno et al., 2012). For each virus-host system, it will be necessary to carry out experiments to delimit the range of inhibitor and mutagen concentrations at which the sequential treatment is advantageous. The translation into clinical practice is of interest,

**FIGURE 9.10** Alternative antiviral designs using an inhibitor and a mutagenic agent. This model study was carried out with FMDV replicating in BHK-21 cells. The inhibitor used was guanidine hydrochloride (GU), and the mutagen was ribavirin (R). (A) The four types of serial passages (p indicates passage number) are, from top to bottom: passages in the presence of guanidine alone (*yellow diamond*); passages in the presence of ribavirin (R) alone (*red triangle*); passages in the presence of a mixture of GU and R (denoted as [+ GU + R]; red star; passage number is in quotes because the first passage with the mixture of GU and R was considered equivalent to the second passage in the presence of a single drug); finally, a first passage in the presence of GU, followed by four passages in the presence of R (*blue circle*). (B), (C), (D) Virus titer and viral RNA level in the course of six passages in the presence of 5 mM R and 16, 18, 20 mM GU in (B), (C), (D), respectively, with passage regimen code indicated in the box. Note that the highest reductions of FMDV progeny production are achieved using the sequential GU-R protocol (*blue circles*) and that infectivity and viral RNA are lost earlier at the highest GU concentration tested. (E) A highly sensitive RT-PCR amplification used as diagnostic of viral extinction confirmed the advantage of the sequential GU-ribavirin treatment. Symbols are as in (A). *Figure reproduced from Perales, C., Agudo, R., Tejero, H., Manrubia, S.C., Domingo, E., 2009b. Potential benefits of sequential inhibitor-mutagen treatments of RNA virus infections. PLoS Pathog. 5, e1000658 where additional experimental details can be found.*

---

BOX 9.4

Interactions between inhibitors and mutagenic nucleotides
that can affect the efficacy of combination treatments

- A mutagen can increase the frequency of inhibitor-resistant mutants.
- An inhibitor can prevent replication of interfering mutants that contribute to lethal defection.

- The mutant spectrum can suppress inhibitor-resistant mutants that affect a *trans*-complementable protein.
- A mutagenized mutant spectrum can suppress high-fitness genomes.

See text for justification and references.

---

but it will have the added complications derived from target compartmentalization and uneven inhibitor and mutagen concentrations in different compartments (Steinmeyer and Wilke, 2009). The model of S. Manrubia and J. Iranzo also predicts that when therapy is based on the use of either two mutagenic agents or two inhibitors, a combination treatment is always preferred over the sequential administration. Only a few studies have investigated the consequences of using two mutagens together, with or without inhibitors (Perales et al., 2009a; Dapp et al., 2012). The availability of two or more virus-specific mutagen agents of different mutational preferences has the advantage that they will target different regions of sequence space (related to network connections discussed in Chapter 10), although the resulting mutational patterns may be difficult to interpret (Dapp et al., 2012). Also, a mutagen may extinguish a virus that has acquired resistance to another mutagen as in the extinction of a ribavirin-resistant FMDV mutant by FU (Perales et al., 2009a). According to the model, the worst option is to administer first a mutagen and then an inhibitor. Unless the mutagen achieves extreme reductions of viral load, it may generate an expanded mutant spectrum from which inhibitor-resistant mutants may be selected (Iranzo et al., 2011; Perales et al., 2012).

## 9.9 Prospects for a clinical application of lethal mutagenesis

Lethal mutagenesis constitutes an example of how a fundamental theoretical concept initially unrelated to virology can unfold into a potential application in the form of antiviral designs. This transition is exemplified by M. Eigen, who established the basis of quasispecies theory (Eigen, 1971) and wrote a comment of a paper on virus extinction by mutagenic nucleotides, 30 years later (Eigen, 2002). Can you imagine a grant application to work on quasispecies with the aim of controlling viral infections? Some readers will probably object immediately: nothing has been achieved yet with lethal mutagenesis at the clinical level. True, unless ribavirin has been clearing (or helped the immune system to clear) HCV, with the participation of mutagenesis (see Section 9.5). To be able to talk about the prospect of an application is sufficient to make the point about the relevance of basic research, and it is the prospect that we address here.

Box 9.5 lists advantages and limitations of lethal mutagenesis treatments as compared with standard, nonmutagenic inhibitors. Quantifications of mutagen-escape mutant frequencies have not been performed. The derivation of

---

### BOX 9.5

## Advantages and limitations of antiviral treatments based on lethal mutagenesis

### Advantages

- A possible high barrier to resistance.
- Its mechanism of action favors the suppression of possible resistant variants by the mutagenized mutant spectrum within infected cells.
- Lethal mutagenesis may be included in sequential or combination designs with other classes of antiviral agents.
- Natural mechanisms of resistance to genetic parasites include lethal mutagenesis-like strategies: APOBEC, ADAR, RIP, etc.

### Limitations

- The mutagenic activity of the agents must be virus specific. It cannot mutagenize cellular nucleic acids.
- Possible off-target effects are still poorly understood.
- The number of available antiviral mutagenic agents is restricted.
- Additional experiments with animal models and preclinical and clinical trials are needed prior to possible therapy implementation.
- Resistance of expert panels to encourage the exploration of unconventional antiviral approaches both for scientific and commercial reasons.

---

some of the mutants that have been studied (Section 9.6) required a gradual increase of mutagen concentration in the course of selective passages. The difficulty in isolating mutants from a mutagen-treated viral population may stem from the suppressive environment created by the mutagenesis itself that may preclude or delay dominance of the resistant mutant.

The possibility of sequential inhibitor-mutagen treatments (Section 9.8) is clinically relevant because such designs may diminish the severity of side effects by avoiding the simultaneous presence of two drugs in treated patients. Shorter treatment duration of sequential treatments is also a possibility.

Some natural cellular mechanisms of defense against genetic parasites are based on producing an excess of mutations in the invader. In Chapter 2 (Section 2.7), APOBEC (*a*polipoprotein *B* mRNA *e*diting *c*omplex) and ADAR (*a*denosine *d*eaminase *a*cting on double-stranded *R*NA)

were discussed. They exert nucleic acid editing functions used by the cell that can be recruited as antiviral responses (Harris and Dudley, 2015; Venkatesan et al., 2018). Editing is part of the replication cycle of several viruses. It has been suggested that *Paramyxovirinae* might have evolved to possess a genome of polyhexameric length (known as the "rule of six") to avoid uncontrolled editing and error catastrophe of the virus (Kolakofsky et al., 2005). There are additional mutagenic-like activities that mimic lethal mutagenesis. One of them is termed RIP (*r*epeat-*i*nduced *p*oint mutations) that operates in some filamentous fungi to mutate genetic intruders, including transposable elements (Galagan and Selker, 2004; Clutterbuck, 2011; Braga et al., 2014; Amselem et al., 2015; Van de Wouw et al., 2019). Some experts regard as highly positive that an intended medical intervention resembles a natural process.

Box 9.5 also lists several limitations, some of which are obvious (need of specificity to mutagenize viral but not cellular nucleic acids, lack of information on off-target effects, and a necessity to explore treatment efficacy in vivo). It is likely that the number of virus-specific mutagenic agents will increase in the coming decades. If their efficacy in vivo can be properly documented, expert panels may moderate their attitude toward these and other new antiviral approaches [discussion of the licensing issue in (Perales et al., 2019)].

## 9.10 Some atypical proposals

Decisions on the suitability of new treatments are mainly based on the three basic parameters $CC_{50}$ (as a measure of toxicity), $IC_{50}$ (as a measure of inhibitory potential), and the therapeutic index that they yield ($TI = CC_{50}/IC_{50}$) (explained in Section 8.4.3 of Chapter 8). There is an additional parameter that should also be considered: the spectrum of antiviral activity, regarding the number of unrelated viral pathogens that are effectively inhibited by the treatment. Although evidence is still lacking, the possibility that broad-spectrum in the above sense may predict a capacity to inhibit broad repertoires of quasispecies swarms is appealing. The broadness of antiviral efficacy is dependent on the mechanism of antiviral activity. In this and the preceding chapters, we have described two classes of broad-spectrum antiviral inhibitors: those that stimulate the innate immune response (notably inhibitors of pyrimidine biosynthesis) and mutagenic nucleotide analogs (notably ribavirin and favipiravir as the most relevant examples, likely to be followed by additional ones). Thus, it would be interesting to investigate the joint use of these two classes of compounds (stimulators of immune responses and lethal mutagens) in sequential or combination protocols of the type described in Section 9.8.

Drug repositioning may offer additional possibilities. F. Sobrino and colleagues demonstrated that valproic acid (2-propylpentanoic acid, VPA) displays a broad antiviral activity against many enveloped viruses (Martin-Acebes et al., 2011; Vazquez-Calvo et al., 2013). VPA is used to treat epilepsy and bipolar mania, among other disorders. It is listed among the essential medicines by the World Health Organization. Despite its antiviral potency being modest, its inclusion as part of combined therapies should also be considered in view of its broad-spectrum of activity. Triple, broad-spectrum drug combinations appear as an attractive possibility provided antagonistic interactions are avoided, and the side effects are tolerable.

There is a long way to go before lethal mutagenesis or other new designs based on the concepts listed in Box 9.1 can be applied to the treatment of viral disease. However, the hope is that the experience gained mainly with HIV-1 but also with other error-prone viruses will favor explorations that regard the adaptive potential of viruses as the major challenge to be confronted.

## 9.11 Overview and concluding remarks

It is uncertain whether the possibilities presented here as new trends in antiviral strategies will satisfy the demands of new paradigms to approach infectious disease (emphasized in the opening paragraphs of this chapter), or it will be necessary to wait for more innovative advance. The available possibilities have as a common trend that they respond to the quasispecies challenge, and fulfill requirements expressed repeatedly in this book: viral populations to be controlled should be denied any opportunity to replicate, because replication entails the exploration of sequence space, which is a condition for adaptation. Viruses have to be hit hard and as soon as possible both at the level of individual infections and at the epidemiological level. To the extent that new treatment designs can fulfill this requirement, they are worth being pursued.

Probably, combinations of broad-spectrum antiviral agents, irrespective of their nature (lethal mutagens, stimulators of the immune response, mixtures of highly neutralizing antibodies, or their combinations) are among the best options presently available. Despite the general preference for combination treatments, once a mutagenic agent enters the antiviral formulations, the possibility of sequential inhibitor-mutagen treatments should be considered. They may be as effective as the corresponding combinations and may alleviate the burden of side effects, one of the problems of current treatments.

Given the challenge of existing viral diseases and the emergent viral diseases unavoidably to come, what should not be done is to minimize the importance of the challenge, and as a consequence, withdraw the focus from the essential issue. Antiviral designs must be planned by counting on quasispecies dynamics. Controversies about which theoretical models best describe extinction of viruses by lethal mutagenesis should be used to clarify antiviral mechanisms and find the adjustments for best efficacy. Controversies should not be presented as evidence against the real nature of dynamic quasispecies. With the application of deep sequencing to viral populations, the challenge has become dramatically evident. New antiviral targets, new drugs (mutagenic and nonmutagenic), and studies with animal models will hopefully contribute to reaching the goal (see Summary Box).

---

## Summary Box

- There is a need to develop new antiviral strategies to control viral pathogens characterized by quasispecies dynamics.
- Several new designs have been implemented or are under investigation. They aim at increasing the barrier to resistance or diminishing viral fitness.
- Lethal mutagenesis consists of virus extinction by excess mutations. Experimental evidence suggests that lethal mutagenesis is directly related to the error threshold concept of quasispecies theory.
- Mutagen-resistant viral mutants have been isolated, and their study has been instrumental in documenting the adaptive value conferred by a mutant spectrum of adequate complexity.

- There are several molecular mechanisms of virus resistance to mutagenic agents. One of them is modulation of nucleotide incorporation, selected to counteract the mutational bias evoked by a mutagen while maintaining the level of mutant spectrum diversity.
- Sequential inhibitor-mutagen treatments may have an advantage over the corresponding combination, depending on replicative parameters of the virus to be controlled and the inhibitory and mutagenic intensities.
- New treatments using broad-spectrum mutagenic and nonmutagenic drug combinations offer prospects for the control of error-prone viruses.

# References

Agudo, R., Arias, A., Pariente, N., Perales, C., Escarmis, C., et al., 2008. Molecular characterization of a dual inhibitory and mutagenic activity of 5-fluorouridine triphosphate on viral RNA synthesis. Implications for lethal mutagenesis. J. Mol. Biol. 382, 652−666.

Agudo, R., Arias, A., Domingo, E., 2009. 5-Fluorouracil in lethal mutagenesis of foot-and-mouth disease virus. Future Med. Chem. 1, 529−539.

Agudo, R., Ferrer-Orta, C., Arias, A., de la Higuera, I., Perales, C., et al., 2010. A multi-step process of viral adaptation to a mutagenic nucleoside analogue by modulation of transition types leads to extinction-escape. PLoS Pathog. 6, e1001072.

Agudo, R., de la Higuera, I., Arias, A., Grande-Perez, A., Domingo, E., 2016. Involvement of a joker mutation in a polymerase-independent lethal mutagenesis-escape mechanism. Virology 494, 257−266.

Airaksinen, A., Pariente, N., Menendez-Arias, L., Domingo, E., 2003. Curing of foot-and-mouth disease virus from persistently infected cells by ribavirin involves enhanced mutagenesis. Virology 311, 339−349.

Amselem, J., Lebrun, M.H., Quesneville, H., 2015. Whole genome comparative analysis of transposable elements provides new insight into mechanisms of their inactivation in fungal genomes. BMC Genomics 16, 141.

Arenas, M., Lorenzo-Redondo, R., Lopez-Galíndez, C., 2016. Influence of mutation and recombination on HIV-1 in vitro fitness recovery. Mol. Phylogent. Evol. 94, 264−270.

Arias, A., Arnold, J.J., Sierra, M., Smidansky, E.D., Domingo, E., et al., 2008. Determinants of RNA-dependent RNA polymerase (in)fidelity revealed by kinetic analysis of the polymerase encoded by a foot-and-mouth disease virus mutant with reduced sensitivity to ribavirin. J. Virol. 82, 12346−12355.

Arias, A., Isabel de Avila, A., Sanz-Ramos, M., Agudo, R., Escarmis, C., et al., 2013. Molecular dissection of a viral quasispecies under mutagenic treatment: positive correlation between fitness loss and mutational load. J. Gen. Virol. 94, 817−830.

Arias, A., Thorne, L., Goodfellow, I., 2014. Favipiravir elicits antiviral mutagenesis during virus replication in vivo. eLife 3, e03679.

Arribas, M., Cabanillas, L., Lazaro, E., 2011. Identification of mutations conferring 5-azacytidine resistance in bacteriophage Qbeta. Virology 417, 343−352.

Asahina, Y., Izumi, N., Enomoto, N., Uchihara, M., Kurosaki, M., et al., 2005. Mutagenic effects of ribavirin and response to interferon/ribavirin combination therapy in chronic hepatitis C. J. Hepatol. 43, 623−629.

Baccam, P., Thompson, R.J., Fedrigo, O., Carpenter, S., Cornette, J.L., 2001. PAQ: partition analysis of quasispecies. Bioinformatics 17, 16−22.

Baranovich, T., Wong, S.S., Armstrong, J., Marjuki, H., Webby, R.J., et al., 2013. T-705 (favipiravir) induces lethal mutagenesis in influenza A H1N1 viruses in vitro. J. Virol. 87, 3741−3751.

Bassi, M.R., Navarro Sempere, R., Meyn, P., Polacek, C., Arias, A., 2018. Extinction of Zika virus and Usutu virus by lethal mutagenesis revels different patterns of sensitivity to three mutagenic drugs. Antimicrob. Agents Chemother. 62, 00380−18.

Beach, L.B., Rawson, J.M., Kim, B., Patterson, S.E., Mansky, L.M., 2014. Novel inhibitors of human immunodeficiency virus type 2 infectivity. J. Gen. Virol. 95, 2778−2783.

Beaucourt, S., Vignuzzi, M., 2014. Ribavirin: a drug active against many viruses with multiple effects on virus replication and propagation. Molecular basis of ribavirin resistance. Curr. Opin. Virol. 8, 10−15.

Bonhoeffer, S., May, R.M., Shaw, G.M., Nowak, M.A., 1997. Virus dynamics and drug therapy. Proc. Natl. Acad. Sci. U.S.A. 94, 6971−6976.

Bonnac, L.F., Mansky, L.M., Patterson, S.E., 2013. Structure-activity relationships and design of viral mutagens and application to lethal mutagenesis. J. Med. Chem. 56, 9403−9414.

Borderia, A.V., Rozen-Gagnon, K., Vignuzzi, M., 2016. Fidelity variants and RNA quasispecies. Curr. Top. Microbiol. Immunol. 392, 303−322.

Borrego, B., de Avila, A.I., Domingo, E., Brun, A., 2019. Lethal mutagenesis of Rift Valley fever virus induced by favipiravir. Antimicrob. Agents Chemother. 63(8), e00669-19.

Bougie, I., Bisaillon, M., 2003. Initial binding of the broad spectrum antiviral nucleoside ribavirin to the hepatitis C virus RNA polymerase. J. Biol. Chem. 278, 52471−52478.

Braga, R.M., Santana, M.F., Veras da Costa, R., Brommonschenkel, S.H., de Araujo, E.F., et al., 2014. Transposable elements belonging to the Tc1-Mariner superfamily are heavily mutated in Colletotrichum graminicola. Mycologia 106, 629−641.

Cabanillas, L., Arribas, M., Lázaro, E., 2013. Evolution at increased error rates leads to coexistence of multiple adaptive pathways in an RNA virus. BMC Evol. Biol. 13, 11.

Cabanillas, L., Sanjuan, R., Lazaro, E., 2014. Changes in protein domains outside the catalytic site of the bacteriophage Qbeta replicase reduce the mutagenic effect of 5-azacytidine. J. Virol. 88, 10480−10487.

Caroline, A.L., Powell, D.S., Bethel, L.M., Oury, T.D., Reed, D.S., et al., 2014. Broad spectrum antiviral activity of favipiravir (T-705): protection from highly lethal inhalational Rift Valley Fever. PLoS Negl. Trop. Dis. 8, e2790.

Casadevall, A., 1996. Crisis in infectious diseases: time for a new paradigm? Clin. Infect. Dis. 23, 790−794.

Casadevall, A., Pirofski, L.A., 2015. The ebola epidemic crystallizes the potential of passive antibody therapy for infectious diseases. PLoS Pathog. 11, e1004717.

Cases-Gonzalez, C., Arribas, M., Domingo, E., Lazaro, E., 2008. Beneficial effects of population bottlenecks in an RNA virus evolving at increased error rate. J. Mol. Biol. 384, 1120–1129.

Castro, C., Arnold, J.J., Cameron, C.E., 2005. Incorporation fidelity of the viral RNA-dependent RNA polymerase: a kinetic, thermodynamic and structural perspective. Virus Res. 107, 141–149.

Chatterji, U., Lim, P., Bobardt, M.D., Wieland, S., Cordek, D.G., et al., 2010. HCV resistance to cyclosporin A does not correlate with a resistance of the NS5A-cyclophilin A interaction to cyclophilin inhibitors. J. Hepatol. 53, 50–56.

Chevaliez, S., Brillet, R., Lazaro, E., Hezode, C., Pawlotsky, J.M., 2007. Analysis of ribavirin mutagenicity in human hepatitis C virus infection. J. Virol. 81, 7732–7741.

Chung, D.H., Sun, Y., Parker, W.B., Arterburn, J.B., Bartolucci, A., et al., 2007. Ribavirin reveals a lethal threshold of allowable mutation frequency for Hantaan virus. J. Virol. 81, 11722–11729.

Clutterbuck, A.J., 2011. Genomic evidence of repeat-induced point mutation (RIP) in filamentous ascomycetes. Fungal Genet. Biol. 48, 306–326.

Contreras, A.M., Hiasa, Y., He, W., Terella, A., Schmidt, E.V., et al., 2002. Viral RNA mutations are region specific and increased by ribavirin in a full-length hepatitis C virus replication system. J. Virol. 76, 8505–8517.

Crotty, S., Maag, D., Arnold, J.J., Zhong, W., Lau, J.Y., et al., 2000. The broad-spectrum antiviral ribonucleoside ribavirin is an RNA virus mutagen. Nat. Med. 6, 1375–1379.

Crotty, S., Cameron, C.E., Andino, R., 2001. RNA virus error catastrophe: direct molecular test by using ribavirin. Proc. Natl. Acad. Sci. U.S.A. 98, 6895–6900.

Crotty, S., Cameron, C., Andino, R., 2002. Ribavirin's antiviral mechanism of action: lethal mutagenesis? J. Mol. Med. 80, 86–95.

Cuevas, J.M., González-Candelas, F., Moya, A., Sanjuán, R., 2009. Effect of ribavirin on the mutation rate and spectrum of hepatitis C virus in vivo. J. Virol. 83, 5760–5764.

Cummings, K.J., Lee, S.M., West, E.S., Cid-Ruzafa, J., Fein, S.G., et al., 2001. Interferon and ribavirin vs interferon alone in the re-treatment of chronic hepatitis C previously nonresponsive to interferon: a meta-analysis of randomized trials. J. Am. Med. Assoc. 285, 193–199.

Cupples, C.G., Miller, J.H., 1989. A set of lacZ mutations in *Escherichia coli* that allow rapid detection of each of the six base substitutions. Proc. Natl. Acad. Sci. U.S.A. 86, 5345–5349.

Dalgleish, A.G., 2015. Rationale for combining immunotherapy with chemotherapy. Immunotherapy 7, 309–316.

Dapp, M.J., Clouser, C.L., Patterson, S., Mansky, L.M., 2009. 5-Azacytidine can induce lethal mutagenesis in human immunodeficiency virus type 1. J. Virol. 83, 11950–11958.

Dapp, M.J., Holtz, C.M., Mansky, L.M., 2012. Concomitant lethal mutagenesis of human immunodeficiency virus type 1. J. Mol. Biol. 419, 158–170.

Dapp, M.J., Patterson, S.E., Mansky, L.M., 2013. Back to the future: revisiting HIV-1 lethal mutagenesis. Trends Microbiol. 21, 56–62.

Dapp, M.J., Bonnac, L., Patterson, S.E., Mansky, L.M., 2014. Discovery of novel ribonucleoside analogs with activity against human immunodeficiency virus type 1. J. Virol. 88, 354–363.

Day, C.W., Smee, D.F., Julander, J.G., Yamshchikov, V.F., Sidwell, R.W., et al., 2005. Error-prone replication of West Nile virus caused by ribavirin. Antiviral Res 67, 38–45.

De Avila, A.I., Gallego, I., Soria, M.E., Gregori, J., Quer, J., Esteban, J.I., Rice, C.M., Domingo, E., Perales, C., 2016. Lethal mutagenesis of hepatitis C virus induced by favipiravir. PLoS One 11, e0164691.

De Avila, A.I., Moreno, E., Perales, C., Domingo, E., 2017. Favipiravir can evoke lethal mutagenesis and extinction of foot-and-mouth disease virus. Virus Res. 233, 105–112.

De Crignis, E., Mahmoudi, T., 2014. HIV eradication: combinatorial approaches to activate latent viruses. Viruses 6, 4581–4608.

De la Higuera, I., Ferrer-Orta, C., de Avila, A.I., Perales, C., Sierra, M., et al., 2017. Molecular and functional bases of selection against a mutation bias in an RNA virus. Genome Biol. Evol. 9, 1212–1228.

De la Higuera, I., Ferrer-Orta, C., Moreno, E., de Avila, A.I., Soria, M.E., et al., 2018. Contribution of a multifunctional polymerase region of foot-and-mouth disease virus to lethal mutagenesis. J. Virol. 92, e01119–18.

de Wilde, A.H., Pham, U., Posthuma, C.C., Snijder, E.J., 2018. Cyclophilins and cyclophilin inhibitors in nidovirus replication. Virology 522, 46–55.

Delang, L., Vliegen, I., Froeyen, M., Neyts, J., 2011. Comparative study of the genetic barriers and pathways towards resistance of selective inhibitors of hepatitis C virus replication. Antimicrob. Agents Chemother. 55, 4103–4113.

Di Bisceglie, A.M., Thompson, J., Smith-Wilkaitis, N., Brunt, E.M., Bacon, B.R., 2001. Combination of interferon and ribavirin in chronic hepatitis C: re-treatment of nonresponders to interferon. Hepatology 33, 704–707.

Diaz-Martinez, L., Brichette-Mieg, I., Pineno-Ramos, A., Dominguez-Huerta, G., Grande-Perez, A., 2018. Lethal mutagenesis of an RNA plant virus via lethal defection. Sci. Rep. 8, 1444.

Dietz, J., Schelhorn, S.E., Fitting, D., Mihm, U., Susser, S., et al., 2013. Deep sequencing reveals mutagenic effects of ribavirin during monotherapy of hepatitis C virus genotype 1-infected patients. J. Virol. 87, 6172–6181.

Dixit, N.M., Layden-Almer, J.E., Layden, T.J., Perelson, A.S., 2004. Modelling how ribavirin improves interferon response rates in hepatitis C virus infection. Nature 432, 922–924.

Domingo, E., 1989. RNA virus evolution and the control of viral disease. Prog. Drug Res. 33, 93–133.

Domingo, E., Holland, J.J., 1992. Complications of RNA heterogeneity for the engineering of virus vaccines and antiviral agents. Genet. Eng. 14, 13–31.

Domingo, E., Flavell, R.A., Weissmann, C., 1976. In vitro site-directed mutagenesis: generation and properties of an infectious extracistronic mutant of bacteriophage Qβ. Gene 1, 3–25.

Domingo, E., Grande-Perez, A., Martin, V., 2008. Future prospects for the treatment of rapidly evolving viral pathogens: insights from evolutionary biology. Expert Opin. Biol. Ther. 8, 1455–1460.

Domingo, E., Sheldon, J., Perales, C., 2012. Viral quasispecies evolution. Microbiol. Mol. Biol. Rev. 76, 159–216.

Domingo-Calap, P., Pereira-Gomez, M., Sanjuan, R., 2012. Nucleoside analogue mutagenesis of a single-stranded DNA virus: evolution and resistance. J. Virol. 86, 9640–9646.

Duffy, M., Salemi, M., Sheehy, N., Vandamme, A.M., Hegarty, J., et al., 2002. Comparative rates of nucleotide sequence variation in the hypervariable region of E1/E2 and the NS5b region of hepatitis C virus in patients with a spectrum of liver disease resulting from a common source of infection. Virology 301, 354–364.

Eggers, H.J., 1976. Successful treatment of enterovirus-infected mice by 2-(alpha-hydroxybenzyl)-benzimidazole and guanidine. J. Exp. Med. 143, 1367–1381.

Eggers, H.J., Tamm, I., 1963. Synergistic effect of 2-(alpha-hydroxybenzyl)-benzimidazole and guanidine on picornavirus reproduction. Nature 199, 513–514.

Eigen, M., 1971. Self-organization of matter and the evolution of biological macromolecules. Naturwissenschaften 58, 465–523.

Eigen, M., 2002. Error catastrophe and antiviral strategy. Proc. Natl. Acad. Sci. U.S.A. 99, 13374–13376.

Eigen, M., Schuster, P., 1979. The Hypercycle. A Principle of Natural Self-Organization. Springer, Berlin.

Eriksson, B., Helgstrand, E., Johansson, N.G., Larsson, A., Misiorny, A., et al., 1977. Inhibition of influenza virus ribonucleic acid polymerase by ribavirin triphosphate. Antimicrob. Agents Chemother. 11, 946–951.

Escarmís, C., Dávila, M., Domingo, E., 1999. Multiple molecular pathways for fitness recovery of an RNA virus debilitated by operation of Muller's ratchet. J. Mol. Biol. 285, 495–505.

Escarmís, C., Lazaro, E., Arias, A., Domingo, E., 2008. Repeated bottleneck transfers can lead to non-cytocidal forms of a cytopathic virus: implications for viral extinction. J. Mol. Biol. 376, 367–379.

Escribano-Romero, E., Jimenez de Oya, N., Domingo, E., Saiz, J.C., 2017. Extinction of West Nile virus by favipiravir through lethal mutagenesis. Antimicrob. Agents Chemother. 61, e01400–e01417.

Farci, P., Shimoda, A., Coiana, A., Diaz, G., Peddis, G., et al., 2000. The outcome of acute hepatitis C predicted by the evolution of the viral quasispecies. Science 288, 339–344.

Feigelstock, D.A., Mihalik, K.B., Feinstone, S.M., 2011. Selection of hepatitis C virus resistant to ribavirin. Virol. J. 8, 402.

Feld, J.J., Nanda, S., Huang, Y., Chen, W., Cam, M., et al., 2007. Hepatic gene expression during treatment with peginterferon and ribavirin: identifying molecular pathways for treatment response. Hepatology 46, 1548–1563.

Fernandez-Larsson, R., O'Connell, K., Koumans, E., Patterson, J.L., 1989. Molecular analysis of the inhibitory effect of phosphorylated ribavirin on the vesicular stomatitis virus in vitro polymerase reaction. Antimicrob. Agents Chemother. 33, 1668–1673.

Ferrer-Orta, C., Arias, A., Perez-Luque, R., Escarmis, C., Domingo, E., et al., 2004. Structure of foot-and-mouth disease virus RNA-dependent RNA polymerase and its complex with a template-primer RNA. J. Biol. Chem. 279, 47212–47221.

Ferrer-Orta, C., Sierra, M., Agudo, R., de la Higuera, I., Arias, A., et al., 2010. Structure of foot-and-mouth disease virus mutant polymerases with reduced sensitivity to ribavirin. J. Virol. 84, 6188–6199.

Ferrer-Orta, C., de la Higuera, I., Caridi, F., Sanchez-Aparicio, M.T., Moreno, E., et al., 2015. Multifunctionality of a picornavirus polymerase domain: nuclear localization signal and nucleotide recognition. J. Virol. 89, 6848–6859.

Flavell, R.A., Sabo, D.L., Bandle, E.F., Weissmann, C., 1974. Site-directed mutagenesis: generation of an extracistronic mutation in bacteriophage Q beta RNA. J. Mol. Biol. 89, 255–272.

Furuta, Y., Takahashi, K., Fukuda, Y., Kuno, M., Kamiyama, T., et al., 2002. In vitro and in vivo activities of anti-influenza virus compound T-705. Antimicrob. Agents Chemother. 46, 977–981.

Furuta, Y., Komeno, T., Nakamura, T., 2017. Favipiravir (T-705), a broad spectrum inhibitor of viral polymerase. Proc. Jpn. Acad. Ser. B Phys. Biol. Sci. 93, 449–463.

Gandhi, R.T., Tashima, K.T., Smeaton, L.M., Vu, V., Ritz, J., et al., 2019. Long-term outcomes in a large randomized trial of HIV-1 salvage therapy: 96-week results of AIDS clinical trials group A5241 (OPTIONS). J. Infect. Dis. https://doi.org/10.1093/infdis/jiz281 pii: jiz.281.

Galagan, J.E., Selker, E.U., 2004. RIP: the evolutionary cost of genome defense. Trends Genet. 20, 417–423.

Galli, A., Mens, H., Gottwein, J.M., Gerstoft, J., Bukh, J., 2018. Antiviral effect of ribavirin against HCV associated with increased frequency of G-to-A and C-to-U transitions in infectious cell culture model. Sci. Rep. 8, 4619.

Garbelli, A., Radi, M., Falchi, F., Beermann, S., Zanoli, S., et al., 2011. Targeting the human DEAD-box polypeptide 3 (DDX3) RNA helicase as a novel strategy to inhibit viral replication. Curr. Med. Chem. 18, 3015–3027.

Garcia-Perez, J., Staropoli, I., Azoulay, S., Heinrich, J.T., Cascajero, A., et al., 2015. A single-residue change in the HIV-1 V3 loop associated with maraviroc resistance impairs CCR5 binding affinity while increasing replicative capacity. Retrovirology 12, 50.

Geller, R., Vignuzzi, M., Andino, R., Frydman, J., 2007. Evolutionary constraints on chaperone-mediated folding provide an antiviral approach refractory to development of drug resistance. Genes Dev. 21, 195–205.

Gerotto, M., Sullivan, D.G., Polyak, S.J., Chemello, L., Cavalletto, L., et al., 1999. Effect of retreatment with interferon alone or interferon plus ribavirin on hepatitis C virus quasispecies diversification in nonresponder patients with chronic hepatitis C. J. Virol. 73, 7241–7247.

González-López, C., Arias, A., Pariente, N., Gómez-Mariano, G., Domingo, E., 2004. Preextinction viral RNA can interfere with infectivity. J. Virol. 78, 3319–3324.

Goswami, B.B., Borek, E., Sharma, O.K., Fujitaki, J., Smith, R.A., 1979. The broad spectrum antiviral agent ribavirin inhibits capping of mRNA. Biochem. Biophys. Res. Commun. 89, 830–836.

Graci, J.D., Cameron, C.E., 2002. Quasispecies, error catastrophe, and the antiviral activity of ribavirin. Virology 298, 175–180.

Graci, J.D., Cameron, C.E., 2004. Challenges for the development of ribonucleoside analogues as inducers of error catastrophe. Antivir. Chem. Chemother. 15, 1–13.

Grande-Pérez, A., Sierra, S., Castro, M.G., Domingo, E., Lowenstein, P.R., 2002. Molecular indetermination in the transition to error catastrophe: systematic elimination of lymphocytic choriomeningitis virus through mutagenesis does not correlate linearly with large increases in mutant spectrum complexity. Proc. Natl. Acad. Sci. U.S.A. 99, 12938–12943.

Grande-Pérez, A., Gómez-Mariano, G., Lowenstein, P.R., Domingo, E., 2005a. Mutagenesis-induced, large fitness variations with an invariant arenavirus consensus genomic nucleotide sequence. J. Virol. 79, 10451–10459.

Grande-Pérez, A., Lázaro, E., Lowenstein, P., Domingo, E., Manrubia, S.C., 2005b. Suppression of viral infectivity through lethal defection. Proc. Natl. Acad. Sci. U.S.A. 102, 4448–4452.

Griesemer, S.B., Kramer, L.D., Van Slyke, G.A., Pata, J.D., Gohara, D.W., et al., 2017. Mutagen resistance and mutation restriction of St. Louis encephalitis virus. J. Gen. Virol. 98, 201–211.

Guedj, J., Piorkowski, G., Jacquot, F., Madelain, V., Nguyen, T.H.T., 2018. Antiviral efficacy of favipiravir against Ebola virus: a translational study in cynomolgus macaques. PLoS Med. 15, e1002535.

Harki, D.A., Graci, J.D., Korneeva, V.S., Ghosh, S.K., Hong, Z., et al., 2002. Synthesis and antiviral evaluation of a mutagenic and non-hydrogen bonding ribonucleoside analogue: 1-beta-D-Ribofuranosyl-3-nitropyrrole. Biochemistry 41, 9026–9033.

Harki, D.A., Graci, J.D., Galarraga, J.E., Chain, W.J., Cameron, C.E., et al., 2006. Synthesis and antiviral activity of 5-substituted cytidine analogues: identification of a potent inhibitor of viral RNA-dependent RNA polymerases. J. Med. Chem. 49, 6166–6169.

Harki, D.A., Graci, J.D., Edathil, J.P., Castro, C., Cameron, C.E., et al., 2007. Synthesis of a universal 5-nitroindole ribonucleotide and incorporation into RNA by a viral RNA-dependent RNA polymerase. Chembiochem 8, 1359–1362.

Harris, R.S., Dudley, J.P., 2015. APOBECs and virus restriction. Virology 479–480C, 131–145.

Ho, D.D., 1995. Time to hit HIV, early and hard. N. Engl. J. Med. 333, 450–451.

Holland, J.J., Domingo, E., de la Torre, J.C., Steinhauer, D.A., 1990. Mutation frequencies at defined single codon sites in vesicular stomatitis virus and poliovirus can be increased only slightly by chemical mutagenesis. J. Virol. 64, 3960–3962.

Hopkins, S., Scorneaux, B., Huang, Z., Murray, M.G., Wring, S., et al., 2010. SCY-635, a novel nonimmunosuppressive analog of cyclosporine that exhibits potent inhibition of hepatitis C virus RNA replication in vitro. Antimicrob. Agents Chemother. 54, 660–672.

Hover, S., Foster, B., Barr, J.N., Mankouri, j., 2017. Viral dependence on cellular ion channels-an emergent antiviral target? J. Gen. Virol. 98, 345–351.

Hultgren, C., Milich, D.R., Weiland, O., Sallberg, M., 1998. The antiviral compound ribavirin modulates the T helper (Th) 1/Th2 subset balance in hepatitis B and C virus-specific immune responses. J. Gen. Virol. 79(Pt 10), 2381–2391.

Iranzo, J., Perales, C., Domingo, E., Manrubia, S.C., 2011. Tempo and mode of inhibitor-mutagen antiviral therapies: a multidisciplinary approach. Proc. Natl. Acad. Sci. U.S.A. 108, 16008–16013.

Jin, Z., Smith, L.K., Rajwanshi, V.K., Kim, B., Deval, J., 2013. The ambiguous base-pairing and high substrate efficiency of T-705 (Favipiravir) Ribofuranosyl 5'-triphosphate towards influenza A virus polymerase. PLoS One 8, e68347.

Kanda, T., Yokosuka, O., Imazeki, F., Tanaka, M., Shino, Y., et al., 2004. Inhibition of subgenomic hepatitis C virus RNA in Huh-7 cells: ribavirin induces mutagenesis in HCV RNA. J. Viral Hepat. 11, 479–487.

Kato, T., Date, T., Miyamoto, M., Sugiyama, M., Tanaka, Y., et al., 2005. Detection of anti-hepatitis C virus effects of interferon and ribavirin by a sensitive replicon system. J. Clin. Microbiol. 43, 5679–5684.

Kim, Y., Lee, C., 2013. Ribavirin efficiently suppresses porcine nidovirus replication. Virus Res. 171, 44–53.

Kiso, M., Takahashi, K., Sakai-Tagawa, Y., Shinya, K., Sakabe, S., et al., 2010. T-705 (favipiravir) activity against lethal H5N1 influenza A viruses. Proc. Natl. Acad. Sci. U.S.A. 107, 882–887.

Kolakofsky, D., Roux, L., Garcin, D., Ruigrok, R.W., 2005. Paramyxovirus mRNA editing, the "rule of six" and error catastrophe: a hypothesis. J. Gen. Virol. 86, 1869–1877.

Kumar, N., Liang, Y., Parslow, T.G., Liang, Y., 2011. Receptor tyrosine kinase inhibitors block multiple steps of influenza a virus replication. J. Virol. 85, 2818–2827.

Kwong, A.D., Najera, I., Bechtel, J., Bowden, S., Fitzgibbon, J., et al., 2011. Sequence and phenotypic analysis for resistance monitoring in hepatitis C virus drug development: recommendations from the HCV DRAG. Gastroenterology 140, 755–760.

Lanford, R.E., Chavez, D., Guerra, B., Lau, J.Y., Hong, Z., et al., 2001. Ribavirin induces error-prone replication of GB virus B in primary tamarin hepatocytes. J. Virol. 75, 8074–8081.

Le Moing, V., Chene, G., Carrieri, M.P., Alioum, A., Brun-Vezinet, F., et al., 2002. Predictors of virological rebound in HIV-1-infected patients initiating a protease inhibitor-containing regimen. AIDS 16, 21–29.

Lee, C.H., Gilbertson, D.L., Novella, I.S., Huerta, R., Domingo, E., et al., 1997. Negative effects of chemical mutagenesis on the adaptive behavior of vesicular stomatitis virus. J. Virol. 71, 3636–3640.

Levi, L.I., Gnadig, N.F., Beaucourt, S., McPherson, M.J., Baron, B., et al., 2010. Fidelity variants of RNA dependent RNA polymerases uncover an indirect, mutagenic activity of amiloride compounds. PLoS Pathog. 6, e1001163.

Leyssen, P., Balzarini, J., De Clercq, E., Neyts, J., 2005. The predominant mechanism by which ribavirin exerts its antiviral activity in vitro against flaviviruses and paramyxoviruses is mediated by inhibition of IMP dehydrogenase. J. Virol. 79, 1943–1947.

Leyssen, P., De Clercq, E., Neyts, J., 2006. The anti-yellow fever virus activity of ribavirin is independent of error-prone replication. Mol. Pharmacol. 69, 1461–1467.

Li, C., Wang, H., Yuan, T., Woodman, A., Yang, D., et al., 2018. Foot-and-mouth disease virus type O specific mutations determine RNA-dependent RNA polymerase fidelity and virus attenuation. Virology 518, 87–94.

Li, M.J., Kim, J., Li, S., Zaia, J., Yee, J.K., et al., 2005. Long-term inhibition of HIV-1 infection in primary hematopoietic cells by lentiviral vector delivery of a triple combination of anti-HIV shRNA, anti-CCR5 ribozyme, and a nucleolar-localizing TAR decoy. Mol. Ther. 12, 900–909.

Loeb, L.A., Essigmann, J.M., Kazazi, F., Zhang, J., Rose, K.D., et al., 1999. Lethal mutagenesis of HIV with mutagenic nucleoside analogs. Proc. Natl. Acad. Sci. U.S.A. 96, 1492–1497.

Lucas-Hourani, M., Dauzonne, D., Jorda, P., Cousin, G., Lupan, A., et al., 2013. Inhibition of pyrimidine biosynthesis pathway suppresses viral growth through innate immunity. PLoS Pathog. 9, e1003678.

Lutchman, G., Danehower, S., Song, B.C., Liang, T.J., Hoofnagle, J.H., et al., 2007. Mutation rate of the hepatitis C virus NS5B in patients undergoing treatment with ribavirin monotherapy. Gastroenterology 132, 1757–1766.

Maag, D., Castro, C., Hong, Z., Cameron, C.E., 2001. Hepatitis C virus RNA-dependent RNA polymerase (NS5B) as a mediator of the antiviral activity of ribavirin. J. Biol. Chem. 276, 46094–46098.

Martin-Acebes, M.A., Vazquez-Calvo, A., Rincon, V., Mateu, M.G., Sobrino, F., 2011. A single amino acid substitution in the capsid of foot-and-mouth disease virus can increase acid resistance. J. Virol. 85, 2733–2740.

McHutchison, J.G., Gordon, S.C., Schiff, E.R., Shiffman, M.L., Lee, W.M., et al., 1998. Interferon alfa-2b alone or in combination with ribavirin as initial treatment for chronic hepatitis C. Hepatitis Interventional Therapy Group. N. Engl. J. Med. 339, 1485–1492.

Moreno, H., Gallego, I., Sevilla, N., de la Torre, J.C., Domingo, E., et al., 2011. Ribavirin can be mutagenic for arenaviruses. J. Virol. 85, 7246–7255.

Moreno, H., Grande-Pérez, A., Domingo, E., Martín, V., 2012. Arenaviruses and lethal mutagenesis. Prospects for new ribavirin-based interventions. Viruses 4, 2786–2805.

Mori, K., Ikeda, M., Ariumi, Y., Dansako, H., Wakita, T., et al., 2011. Mechanism of action of ribavirin in a novel hepatitis C virus replication cell system. Virus Res. 157, 61–70.

Morrey, J.D., Taro, B.S., Siddharthan, V., Wang, H., Smee, D.F., et al., 2008. Efficacy of orally administered T-705 pyrazine analog on lethal West Nile virus infection in rodents. Antiviral Res 80, 377–379.

Mosier, D.E., Picchio, G.R., Gulizia, R.J., Sabbe, R., Poignard, P., et al., 1999. Highly potent RANTES analogues either prevent CCR5-using human immunodeficiency virus type 1 infection in vivo or rapidly select for CXCR4-using variants. J. Virol. 73, 3544–3550.

Motani, K., Ito, S., Nagata, S., 2015. DNA-mediated cyclic GMP-AMP synthase-dependent and -independent regulation of innate immune responses. J. Immunol. 194, 4914–4923.

Müller, V., Bonhoeffer, S., 2008. Intra-host dynamics and evolution of HIV infections. In: Domingo, E., Parrish, C.R., Holland, J.J. (Eds.), Origin and Evolution of Viruses, second ed. Elsevier, London, pp. 279–302.

Mullins, J.I., Heath, L., Hughes, J.P., Kicha, J., Styrchak, S., et al., 2011. Mutation of HIV-1 genomes in a clinical population treated with the mutagenic nucleoside KP1461. PLoS One 6, e15135.

Munier-Lehmann, H., Lucas-Hourani, M., Guillou, S., Helynck, O., Zanghi, G., et al., 2015. Original 2-(3-alkoxy-1H-pyrazol-1-yl)pyrimidine derivatives as inhibitors of human dihydroorotate dehydrogenase (DHODH). J. Med. Chem. 58, 860–877.

Nguyen, A.H., Molineux, I.J., Springman, R., Bull, J.J., 2012. Multiple genetic pathways to similar fitness limits during viral adaptation to a new host. Evolution 66, 363–374.

Nijhuis, M., van Maarseveen, N.M., Boucher, C.A., 2009. Antiviral resistance and impact on viral replication capacity: evolution of viruses under antiviral pressure occurs in three phases. Handb. Exp. Pharmacol. 189, 299–320.

Ning, Q., Brown, D., Parodo, J., Cattral, M., Gorczynski, R., et al., 1998. Ribavirin inhibits viral-induced macrophage production of TNF, IL-1, the procoagulant fgl2 prothrombinase and preserves Th1 cytokine production but inhibits Th2 cytokine response. J. Immunol. 160, 3487–3493.

Nowak, M., Schuster, P., 1989. Error thresholds of replication in finite populations mutation frequencies and the onset of Muller's ratchet. J. Theor. Biol. 137, 375–395.

Nyström, K., Wanrooij, P.H., Waldenström, J., Adamek, L., Brunet, S., et al., 2018. Inosine triphosphate pyrophosphatase dephosphorylates ribavirin triphosphate and reduced enzymatic activity potentiates mutagenesis in hepatitis C virus. J. Virol. 92, e01087–18.

Ojosnegros, S., Agudo, R., Sierra, M., Briones, C., Sierra, S., et al., 2008. Topology of evolving, mutagenized viral populations: quasispecies expansion, compression, and operation of negative selection. BMC Evol. Biol. 8, 207.

Ong, H.K., Soo, B.P.C., Chu, K.L., Chao, S.H., 2018. XBP-1, a cellular target for the development of novel anti-viral strategies. Curr. Prot. Pept. Sci. 19, 145–154.

Orgel, L.E., 1963. The maintenance of the accuracy of protein synthesis and its relevance to ageing. Proc. Natl. Acad. Sci. U.S.A. 49, 517–521.

Ortega-Prieto, A.M., Sheldon, J., Grande-Pérez, A., Tejero, H., Gregori, J., et al., 2013. Extinction of hepatitis C virus by ribavirin in hepatoma cells involves lethal mutagenesis. PLoS One 8, e71039.

Ortiz-Riano, E., Ngo, N., Devito, S., Eggink, D., Munger, J., et al., 2014. Inhibition of arenavirus by A3, a pyrimidine biosynthesis inhibitor. J. Virol. 88, 878–889.

Palmer, B.A., Dimitrova, Z., Skums, P., Crosbie, O., Kenny-Walsh, E., et al., 2014. Analysis of the evolution and structure of a complex intrahost viral population in chronic hepatitis C virus mapped by ultradeep pyrosequencing. J. Virol. 88, 13709–13721.

Pariente, N., Sierra, S., Lowenstein, P.R., Domingo, E., 2001. Efficient virus extinction by combinations of a mutagen and antiviral inhibitors. J. Virol. 75, 9723–9730.

Pariente, N., Airaksinen, A., Domingo, E., 2003. Mutagenesis versus inhibition in the efficiency of extinction of foot-and-mouth disease virus. J. Virol. 77, 7131–7138.

Pariente, N., Sierra, S., Airaksinen, A., 2005. Action of mutagenic agents and antiviral inhibitors on foot-and-mouth disease virus. Virus Res. 107, 183–193.

Parker, W.B., 2005. Metabolism and antiviral activity of ribavirin. Virus Res. 107, 165–171.

Pauly, M.D., Lauring, A.S., 2015. Effective lethal mutagenesis of influenza virus by three nucleoside analogs. J. Virol. 89, 3584–3597.

Perales, C., Domingo, E., 2016. Antiviral strategies based on lethal mutagenesis and error threshold. Curr. Top. Microbiol. Immunol. 392, 323–339.

Perales, C., Mateo, R., Mateu, M.G., Domingo, E., 2007. Insights into RNA virus mutant spectrum and lethal mutagenesis events: replicative interference and complementation by multiple point mutants. J. Mol. Biol. 369, 985–1000.

Perales, C., Agudo, R., Domingo, E., 2009a. Counteracting quasispecies adaptability: extinction of a ribavirin-resistant virus mutant by an alternative mutagenic treatment. PLoS One 4, e5554.

Perales, C., Agudo, R., Tejero, H., Manrubia, S.C., Domingo, E., 2009b. Potential benefits of sequential inhibitor-mutagen treatments of RNA virus infections. PLoS Pathog. 5, e1000658.

Perales, C., Agudo, R., Manrubia, S.C., Domingo, E., 2011a. Influence of mutagenesis and viral load on the sustained low-level replication of an RNA virus. J. Mol. Biol. 407, 60–78.

Perales, C., Henry, M., Domingo, E., Wain-Hobson, S., Vartanian, J.P., 2011b. Lethal mutagenesis of foot-and-mouth disease virus involves shifts in sequence space. J. Virol. 85, 12227–12240.

Perales, C., Iranzo, J., Manrubia, S.C., Domingo, E., 2012. The impact of quasispecies dynamics on the use of therapeutics. Trends Microbiol. 20, 595–603.

Perales, C., Beach, N.M., Sheldon, J., Domingo, E., 2014. Molecular basis of interferon resistance in hepatitis C virus. Curr. Opin. Virol. 8, 38–44.

Perales, C., Gallego, I., De Avila, A.I., Soria, M.E., Gregori, J., et al., 2019. The increasing impact of lethal mutagenesis of viruses. Future Med. Chem. in press.

Pereira-Gomez, M., Sanjuan, R., 2014. Delayed lysis confers resistance to the nucleoside analogue 5-fluorouracil and alleviates mutation accumulation in the single-stranded DNA bacteriophage varphiX174. J. Virol. 88, 5042–5049.

Perelson, A.S., Layden, T.J., 2007. Ribavirin: is it a mutagen for hepatitis C virus? Gastroenterology 132, 2050–2052.

Pfeiffer, J.K., Kirkegaard, K., 2003. A single mutation in poliovirus RNA-dependent RNA polymerase confers resistance to mutagenic nucleotide analogs via increased fidelity. Proc. Natl. Acad. Sci. U.S.A. 100, 7289–7294.

Pfeiffer, J.K., Kirkegaard, K., 2005a. Increased fidelity reduces poliovirus fitness under selective pressure in mice. PLoS Pathog. 1, 102–110.

Pfeiffer, J.K., Kirkegaard, K., 2005b. Ribavirin resistance in hepatitis C virus replicon-containing cell lines conferred by changes in the cell line or mutations in the replicon RNA. J. Virol. 79, 2346–2355.

Pol, S., Couzigou, P., Bourliere, M., Abergel, A., Combis, J.M., et al., 1999. A randomized trial of ribavirin and interferon-alpha vs. interferon-alpha alone in patients with chronic hepatitis C who were non-responders to a previous treatment. Multicenter Study Group under the coordination of the Necker Hospital, Paris, France. J. Hepatol. 31, 1–7.

Pugach, P., Marozsan, A.J., Ketas, T.J., Landes, E.L., Moore, J.P., et al., 2007. HIV-1 clones resistant to a small molecule CCR5 inhibitor use the inhibitor-bound form of CCR5 for entry. Virology 361, 212–228.

Pugach, P., Ray, N., Klasse, P.J., Ketas, T.J., Michael, E., et al., 2009. Inefficient entry of vicriviroc-resistant HIV-1 via the inhibitor-CCR5 complex at low cell surface CCR5 densities. Virology 387, 296–302.

Qiu, L., Patterson, S.E., Bonnac, L.F., Geraghty, R.J., 2018. Nucleobases and corresponding nucleosides display potent antiviral activities against dengue virus possibly through viral lethal mutagenesis. PLoS Negl. Trop. Dis. 12, e0006421.

Querenghi, F., Yu, Q., Billaud, G., Maertens, G., Trepo, C., et al., 2001. Evolution of hepatitis C virus genome in chronically infected patients receiving ribavirin monotherapy. J. Viral Hepat. 8, 120–131.

Rai, D.K., Diaz-San Segundo, F., Campagnola, G., Keith, A., Schafer, E.A., et al., 2017. Attenuation of foot-and-mouth disease virus by engineering viral polymerase fidelity. J. Virol. 91 pii: e00081-17.

Ramachandran, S., Campo, D.S., Dimitrova, Z.E., Xia, G.L., Purdy, M.A., et al., 2011. Temporal variations in the hepatitis C virus intrahost population during chronic infection. J. Virol. 85, 6369–6380.

Rawson, J.M., Heineman, R.H., Beach, L.B., Martin, J.L., Schettler, E.K., et al., 2013. 5,6-dihydro-5-aza-2'-deoxycytidine potentiates the anti-HIV-1 activity of ribonucleotide reductase inhibitors. Bioorg. Med. Chem. 21, 7222–7228.

Rawson, J.M., Mansky, L.M., 2014. Retroviral vectors for analysis of viral mutagenesis and recombination. Viruses 6, 3612–3642.

Ribeiro, R.M., Bonhoeffer, S., 2000. Production of resistant HIV mutants during antiretroviral therapy. Proc. Natl. Acad. Sci. U.S.A. 97, 7681–7686.

Ruiz-Jarabo, C.M., Ly, C., Domingo, E., de la Torre, J.C., 2003. Lethal mutagenesis of the prototypic arenavirus lymphocytic choriomeningitis virus (LCMV). Virology 308, 37–47.

Sacramento, C.Q., De Melo, G.R., De Freitas, C.S., Rocha, N., Hoelz, L.V.B., et al., 2017. The clinically approved antiviral drug sofosbuvir inhibits Zika virus replication. Sci. Rep. 7, 40920.

Sadeghipour, S., Bek, E.J., McMinn, P.C., 2013. Ribavirin-resistant mutants of human enterovirus 71 express a high replication fidelity phenotype during growth in cell culture. J. Virol. 87, 1759–1769.

Sadowski, J., Hashemi, F.B., 2019. Strategies to eradicate HIV from infected patients: elimination of latent provirus reservoirs. Cell. Mol. Life Sci. https://doi.org/10.1007/s00018-019-03156-8.

Sanchez-Aparicio, M.T., Rosas, M.F., Sobrino, F., 2013. Characterization of a nuclear localization signal in the foot-and-mouth disease virus polymerase. Virology 444, 203–210.

Sarrazin, C., Zeuzem, S., 2010. Resistance to direct antiviral agents in patients with hepatitis C virus infection. Gastroenterology 138, 447–462.

Scheidel, L.M., Stollar, V., 1991. Mutations that confer resistance to mycophenolic acid and ribavirin on Sindbis virus map to the nonstructural protein nsP1. Virology 181, 490–499.

Scheidel, L.M., Durbin, R.K., Stollar, V., 1987. Sindbis virus mutants resistant to mycophenolic acid and ribavirin. Virology 158, 1–7.

Schuster, P., 2016. Quasispecies on fitness landscapes. Curr. Top. Microbiol. Immunol. 392, 61–120.

Seiler, P., Senn, B.M., Klenerman, P., Kalinke, U., Hengartner, H., et al., 2000. Additive effect of neutralizing antibody and antiviral drug treatment in preventing virus escape and persistence. J. Virol. 74, 5896–5901.

Severson, W.E., Schmaljohn, C.S., Javadian, A., Jonsson, C.B., 2003. Ribavirin causes error catastrophe during Hantaan virus replication. J. Virol. 77, 481–488.

Sierra, S., Davila, M., Lowenstein, P.R., Domingo, E., 2000. Response of foot-and-mouth disease virus to increased mutagenesis: influence of viral load and fitness in loss of infectivity. J. Virol. 74, 8316–8323.

Sierra, M., Airaksinen, A., González-López, C., Agudo, R., Arias, A., et al., 2007. Foot-and-mouth disease virus mutant with decreased sensitivity to ribavirin: implications for error catastrophe. J. Virol. 81, 2012−2024.

Smith, E.C., Denison, M.R., 2013. Coronaviruses as DNA wannabes: a new model for the regulation of RNA virus replication fidelity. PLoS Pathog. 9, e1003760.

Smith, E.C., Blanc, H., Vignuzzi, M., Denison, M.R., 2013. Coronaviruses lacking exoribonuclease activity are susceptible to lethal mutagenesis: evidence for proofreading and potential therapeutics. PLoS Pathog. 9, e1003565.

Smith, E.C., Case, J.B., Blanc, H., Isakov, O., Shomron, N., et al., 2015. Mutations in coronavirus nonstructural protein 10 decrease virus replication fidelity. J. Virol. 89, 6418−6426.

Snell, N.J., 2001. Ribavirin-current status of a broad spectrum antiviral agent. Expert Opin. Pharmacother. 2, 1317−1324.

Solé, R.V., Deisboeck, T.S., 2004. An error catastrophe in cancer? J. Theor. Biol. 228, 47−54.

Sookoian, S., Castano, G., Frider, B., Cello, J., Campos, R., et al., 2001. Combined therapy with interferon and ribavirin in chronic hepatitis C does not affect serum quasispecies diversity. Dig. Dis. Sci. 46, 1067−1071.

Sowers, L.C., Eritja, R., Kaplan, B., Goodman, M.F., Fazakerly, G.V., 1988. Equilibrium between a wobble and ionized base pair formed between fluorouracil and guanine in DNA as studied by proton and fluorine NMR. J. Biol. Chem. 263, 14794−14801.

Springman, R., Keller, T., Molineux, I.J., Bull, J.J., 2010. Evolution at a high imposed mutation rate: adaptation obscures the load in phage T7. Genetics 184, 221−232.

Stearns, S.C., 1999. Evolution in Health and Disease. Oxford University Press, Oxford.

Steinmeyer, S.H., Wilke, C.O., 2009. Lethal mutagenesis in a structured environment. J. Theor. Biol. 261, 67−73.

Streeter, D.G., Witkowski, J.T., Khare, G.P., Sidwell, R.W., Bauer, R.J., et al., 1973. Mechanism of action of 1- -D-ribofuranosyl-1,2,4-triazole-3-carboxamide (Virazole), a new broad-spectrum antiviral agent. Proc. Natl. Acad. Sci. U.S.A. 70, 1174−1178.

Sullivan, D.G., Bruden, D., Deubner, H., McArdle, S., Chung, M., et al., 2007. Hepatitis C virus dynamics during natural infection are associated with long-term histological outcome of chronic hepatitis C disease. J. Infect. Dis. 196, 239−248.

Swetina, J., Schuster, P., 1982. Self-replication with errors. A model for polynucleotide replication. Biophys. Chem. 16, 329−345.

Takeuchi, N., Hogeweg, P., 2007. Error-threshold exists in fitness landscapes with lethal mutants. BMC Evol. Biol. 7, 15.

Tapia, N., Fernandez, G., Parera, M., Gomez-Mariano, G., Clotet, B., et al., 2005. Combination of a mutagenic agent with a reverse transcriptase inhibitor results in systematic inhibition of HIV-1 infection. Virology 338, 1−8.

Tejero, H., Marin, A., Montero, F., 2011. The relationship between the error catastrophe, survival of the flattest, and natural selection. BMC Evol. Biol. 11, 2.

Tejero, H., Montero, F., Nuño, J.C., 2016. Theories of lethal mutagenesis: from error catastrophe to lethal defection. Curr. Top. Microb. Immunol. 392, 161−179.

Toltzis, P., O'Connell, K., Patterson, J.L., 1988. Effect of phosphorylated ribavirin on vesicular stomatitis virus transcription. Antimicrob. Agents Chemother. 32, 492−497.

Van de Wouw, A.P., Elliot, C.E., Popa, K.M., Idnurm, A., 2019. Analysis of repeat induced point (RIP) mutations in *Leptosphaeria maculans* indicates variability in the RIP process between fungal species. Genetics 211, 89−104.

Van Vaerenbergh, K., Harrer, T., Schmit, J.C., Carbonez, A., Fontaine, E., et al., 2002. Initiation of HAART in drug-naive HIV type 1 patients prevents viral breakthrough for a median period of 35.5 months in 60% of the patients. AIDS Res. Hum. Retroviruses 18, 419−426.

Vazquez-Calvo, A., Martin-Acebes, M.A., Saiz, J.C., Ngo, N., Sobrino, F., et al., 2013. Inhibition of multiplication of the prototypic arenavirus LCMV by valproic acid. Antiviral Res 99, 172−179.

Venkatesan, S., Rosenthal, R., Kanu, N., McGranahan, N., Bartek, J., et al., 2018. Perspective: APOBEC mutagenesis in drug resistance and immune escape in HIV and cancer evolution. Ann. Oncol. 29, 563−572.

Vermehren, J., Sarrazin, C., 2011. New HCV therapies on the horizon. Clin. Microbiol. Infect. 17, 122−134.

Vidalain, P.O., Lucas-Hourani, M., Helynck, O., Tangy, F., Munier-Lehmann, H., 2015. Stimulation of the antiviral innate immune response by pyrimidine biosynthesis inhibitors: a surprise of phenotypic screening. Med. Sci. (Paris) 31, 98−104.

Vignuzzi, M., Stone, J.K., Arnold, J.J., Cameron, C.E., Andino, R., 2006. Quasispecies diversity determines pathogenesis through cooperative interactions in a viral population. Nature 439, 344−348.

Viktorova, E.G., Nchoutmboube, J., Ford-Siltz, L.A., Belov, G.A., 2015. Cell-specific establishment of poliovirus resistance to an inhibitor targeting a cellular protein. J. Virol. 89, 4372−4386.

Vivet-Boudou, V., Isel, C., El Safadi, Y., Smyth, R.P., Laumond, G., et al., 2015. Evaluation of anti-HIV-1 mutagenic nucleoside analogues. J. Biol. Chem. 290, 371−383.

von Kleist, M., Menz, S., Stocker, H., Arasteh, K., Schutte, C., et al., 2011. HIV quasispecies dynamics during pro-active treatment switching: impact on multi-drug resistance and resistance archiving in latent reservoirs. PLoS One 6, e18204.

Warmbrod, K.L., Patterson, E.I., Kautz, T.F., Stanton, A., Rockx-Brouwer, D., et al., 2019. Viral RNA-dependent RNA polymerase mutants display an altered mutation spectrum resulting in attenuation in both mosquito and vertebrate hosts. PLoS Pathog. 15, e1007610.

Webster, R.G., Kawaoka, Y., Bean, W.J., Beard, C.W., Brugh, M., 1985. Chemotherapy and vaccination: a possible strategy for the control of highly virulent influenza virus. J. Virol. 55, 173–176.

Webster, R.G., Kawaoka, Y., Bean, W.J., 1986. Vaccination as a strategy to reduce the emergence of amantadine- and rimantadine-resistant strains of A/Chick/Pennsylvania/83 (H5N2) influenza virus. J. Antimicrob. Chemother. 18, 157–164.

Williams, P.D., 2009. Darwinian interventions: taming pathogens through evolutionary ecology. Trends Parasitol. 26, 83–92.

Wray, S.K., Gilbert, B.E., Noall, M.W., Knight, V., 1985. Mode of action of ribavirin: effect of nucleotide pool alterations on influenza virus ribonucleoprotein synthesis. Antiviral Res 5, 29–37.

Young, K.C., Lindsay, K.L., Lee, K.J., Liu, W.C., He, J.W., et al., 2003. Identification of a ribavirin-resistant NS5B mutation of hepatitis C virus during ribavirin monotherapy. Hepatology 38, 869–878.

Yu, H., Eritja, R., Bloom, L.B., Goodman, M.F., 1993. Ionization of bromouracil and fluorouracil stimulates base mispairing frequencies with guanine. J. Biol. Chem. 268, 15935–15943.

Zeng, J., Wang, H., Xie, X., Yang, D., Zhou, G., et al., 2013. An increased replication fidelity mutant of foot-and-mouth disease virus retains fitness in vitro and virulence in vivo. Antiviral Res 100, 1–7.

Zeng, J., Wang, H., Xie, X., Li, C., Zhou, G., et al., 2014. Ribavirin-resistant variants of foot-and-mouth disease virus: the effect of restricted quasispecies diversity on viral virulence. J. Virol. 88, 4008–4020.

Zhang, Y., Jamaluddin, M., Wang, S., Tian, B., Garofalo, R.P., et al., 2003. Ribavirin treatment up-regulates antiviral gene expression via the interferon-stimulated response element in respiratory syncytial virus-infected epithelial cells. J. Virol. 77, 5933–5947.

Zhou, S., Liu, R., Baroudy, B.M., Malcolm, B.A., Reyes, G.R., 2003. The effect of ribavirin and IMPDH inhibitors on hepatitis C virus subgenomic replicon RNA. Virology 310, 333–342.

# Collective population effects in nonviral systems

## Abbreviations

**FMDV** foot-and-mouth disease virus
**LTR** long terminal repeat
**MOI** multiplicity of infection
**www** World Wide Web

## 10.1 Concept generalization

The major concepts that support the population structure and dynamics of replicating viruses are of a general nature and can be applied to any biological system. They include extended Darwinian principles of genetic variation, competition, selection, and random events, error-prone replication, and information theory and network connectivity. Specifically, current quasispecies research spans topics as diverse as the origin of life, replicator networks, RNA-RNA interactions, chemical mutagenesis and proofreading activities, evolution of tumor cells, chromosomal instability, bacterial and stem cell populations, and the error threshold for maintenance of genetic information in variable fitness landscapes, among others (Domingo and Schuster, 2016; Domingo and Perales, 2019). A requirement for conceptual extensions from one domain of research to another is to adjust parameters to realistic values for each particular system.

Examination of these major concepts leads in quite a straightforward manner to the identification of several interesting implications for nonviral systems.

The scope of Darwinian principles was documented by prebiotic and biological activities summarized in Chapters 1 and 6. As a relevant modern example, the external (human-made) intervention in the administration of antiviral agents to infected patients (Chapters 8 and 9) has a parallel in the use of antibiotics to control pathogenic bacteria. The differences are obviously not in the general principle of natural selection that pushes bacteria to evolve antibiotic resistance but in the molecular mechanisms of acquisition and spread of the resistance phenotype. Similarities and differences between antiviral and antibiotic resistance are described in Section 10.2.1.

The second major notion is error-prone replication which is the basis of the potential rapid evolution of RNA genetic elements, or any genetic system whose replication is catalyzed by a low fidelity nucleic acid polymerase (Chapters 2 and 3). It is based on imperfect template copying, unchecked by repair activities. Limited copying fidelity is ultimately due to atomic fluctuations (the consequence of quantum mechanical uncertainties associated with the electronic distributions among interacting atoms; see Chapter 2) that restrict recognition accuracies. Inexact recognition and errors during replication have been the basis of the origin of life and its subsequent evolution and diversification into multiple subcellular, precellular, cellular, and multicellular forms (Chapter 1). Therefore, quantum mechanical uncertainties constitute a second basic concept that can be generalized to all biological systems, even nongenetic systems (Section 10.6).

Moving to the third major concept, the theory of information was one of the key ingredients in the formulation of quasispecies theory by M. Eigen and P. Schuster (Chapter 3). The connectivity among related genomes in the form of walks in sequence space is recognized as one of the bases of virus adaptability. A collection of related viral genomes, such as those that constitute a mutant spectrum are connected through mutation and recombination. They are a specific case of links among a set of individual entities, as also recognized with other classes of connected elements, such as proteins within a cell, biological species in an ecosystem, or sites (nodes or hubs) in the World Wide Web (www or, simply, the web or Internet). Because of its connectivity and self-organized nature, the Internet can be studied as a complex evolving system. Self-organized systems are those in which some order emerges from interactions among their individual components, without the participation of an external actor. They have been described in many disciplines ranging from physics to robotics, and they are relevant to several areas of biology (Witzany, 2014; Wedlich-Söldner and Betz, 2018). Identification of critical thresholds can define parameters of stability versus vulnerability of self-organized systems. While the web topology is known to affect connectivity, accessibility, and survival of its components, little is known of alternative topologies of connected genomes within viral quasispecies. The error threshold notion derived from quasispecies theory and relevant for the

development of new antiviral strategies (Chapters 3, 8 and 9) is of broad applicability to systems consisting of connected elements. Prospects for future research in this area fueled by the application of deep sequencing methods are examined in Section 10.8. Thus, network connectivity stands as a generalizable feature that permits comparing intrapopulation virus interactions with those of other genetic and nongenetic systems. The relationship among major notions and some of their practical implications are represented schematically in Fig. 10.1. We examine some of them beginning with the most immediate: a comparison of viruses and cells, and of antiviral resistance in viruses with antibiotic resistance in bacteria, both events occurring as a consequence of positive selection in response to medical intervention.

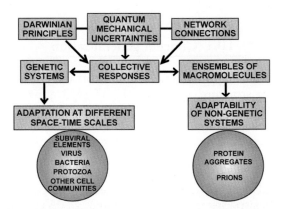

FIGURE 10.1   Some general concepts and their implications. The three top boxes indicate three generalizable notions that permit extending features of viral populations to nonviral systems. The major consequence of the interaction among the three notions is the collective responses observed in genetic systems and in macromolecular assemblages. Genetic systems display adaptability at different space and time scales. Collective responses in populations of macromolecules (nongenetic systems) display Darwinian behavior, such as in the case of prions. The behavior of cellular and macromolecular systems is analyzed and compared with virus behavior in several sections of the present chapter.

## 10.2 Viruses and cells: the genome size-mutation-time coordinates revisited

High mutation rates and the dynamic mutant cloud nature of viral quasispecies—that renders the use of consensus sequences insufficient to describe viruses at the population and behavioral levels—have as one of its consequence the efficient exploration of sequence space whose speed can increase depending on the fitness gradient acting on the population. The relative exploration efficiency of other biological systems is largely dictated by genome size, mutation rate, and the time frame in which events develop. Section 3.7.2 of Chapter 3 compares the orders of magnitude difference that exists in the capacity of sequence space occupation and exploration between cells and viruses.

For the entire population of human genomes to be perceived as a dynamic mutant cloud, the time frame between the primitive humans (about 2.8 million years ago with the first members of *Homo habilis*) and us should be compressed by a factor of $10^3$. Then your admired Egyptian pharaoh Ramses II and yourself would be separated by about 3 years, and those humans between Ramses II and you and surrounding relatives would appear as a dynamic mutant cloud with mutations and other genetic lesions coming and going in minutes. If human genome sequencing could have been applied retrospectively and interpreted with the knowledge that we will likely acquire during the present century, you could even trace the origin of the set of mutations responsible for your uncle's propensity to suffer cancer or your in-law's friend's schizophrenia. The accelerated motion would, however, show an important difference with the viral clouds that populate infected hosts: in the course of human history, a great majority of points in sequence space could not be reached. The total number of human beings that have been born on Earth has been estimated in $1.08 \times 10^{11}$, less than the

number of infectious HCV or HIV-1 particles in many infected patients. At most given times of our compressed human history, the number of mutants would be inferior to the total number of possible single mutants in the human genome (parameters in Box 3.3 of Chapter 3).

The discovery of the extreme heterogeneity of RNA virus populations last century —when nucleotide sequencing procedures were barely available— preceded the uncovering of remarkable degrees of heterogeneity within cell populations (in culture but also within tissues and organs). They are often due to epigenetic modifications, and they have extended the notion of complexity (understood as differences among individual components of a collectivity) to the cellular world, despite the differences in genome size and replication time frame underlined in the previous paragraph. Single-cell variations at the genotypic and phenotypic levels should soon permeate studies on virus-cell interactions with the perspective that both virus and cell populations offer heterogeneities that may modify their interactions [see (Satija et al., 2014; Konry et al., 2016; Liu et al., 2018; Min and Spencer, 2019) as examples of many studies that have addressed cell population complexity]. The origin and biological impact of cell heterogeneity is an active field of research.

Bacterial genome size lies somewhere in between that of viruses and multicellular differentiated organisms (Chapter 1), and so is the capacity of bacteria to explore sequence space through point mutations. Some acceleration in exploration rate is permitted in mutator bacteria (Mehta et al., 2019), a phenotype that will probably be encountered in other microbial communities, as procedures to calculate error rates and frequencies are applied to unicellular eukaryotes. However, a large genome size limits the real number of mutants visited by the cellular clouds versus the total number possible. Not surprisingly, bacterial adaptive mechanisms have been enriched with many possibilities of gene transfers, and some of this enrichment is manifested in the mechanisms of selection of antibiotic-resistant bacteria, the phenomenon parallel to antiviral resistance described in Chapter 8.

## 10.2.1 A comparison of antiviral and antibiotic resistance

In addition to mutation and several forms of recombination (homologous, nonhomologous, mobile elements such as transposons, and insertion elements with generation of insertions and deletions), bacteria can engage in conjugation, fusion, assimilation, infection, transduction, and transformation (review in Miller and Day, 2004; Read and Massey, 2014). These multiple mechanisms provide adaptive potential that can generate and spread antibiotic-resistant forms among different bacterial species. Antibiotic resistance represents a growing health problem, as evidenced by the 25,000 people that die each year in the European Union as a consequence of sepsis (pathology derived from the presence of microorganisms in the blood and immune system reactions) due to antibiotic-resistant bacteria [November 2014 report of the European Academies Science Advisory Council (www.easac.eu)].

Antibiotic resistance has similarities and differences with antiviral resistance (Box 10.1). Some of the common features justify the extension to antibiotic use of the recommendations proposed to minimize the effects of antiviral resistance in clinical practice (Section 8.10 and Box 8.4 in Chapter 8). Specific suggestions are the occasional shelving of antibiotics to repopulate natural bacterial communities with sensitive forms, avoidance of the indiscriminate use of antibiotics, and the need to fund research for the discovery of new antibiotics to expand the available repertoire to control bacterial forms resistant to the classical antibiotics. Mutation, which plays a crucial role in the development of

antiviral resistance, also participates in antibiotic resistance, including mutational modulations of the type related to those associated with virus resistance to lethal mutagenesis (Section 9.6.2 in Chapter 9). Specifically, the generally observed bias of transitions over transversions has been reported to facilitate resistance to multiple antibiotics in *Mycobacterium tuberculosis* (Payne et al., 2019). It is not surprising that adaptive events take advantage of the variant types that the replicative machinery can offer. Resistance to lethal mutagenesis is due to amino acid substitutions (in the viral polymerase or other nonstructural proteins) that modify nucleotide preferences for incorporation into viral RNA. One driver of antibiotic resistance is the bias toward transition mutations observed in metagenomic analyses, and experimental evolution studies with microorganisms.

Shelving of antibiotics to replenish natural populations of bacteria with antibiotic-sensitive forms is unlikely to be effective. Antibiotic-resistant bacteria generally do not outcompete sensitive forms in the field, and both forms coexist. Strain ecology-based models that take into account fitness effects of antibiotic resistance explain such a coexistence, and the preferential acquisition of new resistances by those bacteria that display a multidrug resistance phenotype (Lehtinen et al., 2019).

Several features that distinguish antiviral resistance from antibiotic resistance can be attributed to the variety of molecular mechanisms that bacteria have for the acquisition and transfer of genetic material. While antiviral resistance tends to be confined to the virus that develops it, antibiotic resistance can spread to cognate and to phylogenetically distant bacteria even beyond the clinical setting of the treated patients. Antiviral resistance is generally transmitted with the virus that incorporated the trait. Antibiotic resistance can be acquired and transmitted by bacteria other than the one that was selected upon exposure to antibiotic treatment. Multidrug-resistant viruses can be

---

BOX 10.1

## Some similarities and differences between antibiotic (AB) resistance in bacteria and antiviral (AV) resistance in viruses

### Common features

- Resistant forms gradually replace sensitive forms upon the use of ABs or AVs.
- Low effective concentrations of AB or AV either in an organism or at specific sites within the organism favor selection of resistant pathogens.
- Resistant pathogens often display lower fitness than their sensitive counterparts, when fitness is measured in the absence of AB or AV. In consequence, parental sensitive pathogens often replace resistant forms in the absence of AB or AV.
- Fitness decrease associated with resistance can be compensated by an additional genetic change in the pathogen.
- The degree of resistance may increase with the continued presence of the AB or AV.
- Multidrug-resistant forms can be selected.

### Distinctive features

| Antibiotics | Antiviral agents |
|---|---|
| Resistance maps in chromosomal genes, plasmids, or mobile elements | Resistance maps in the viral genome, often the gene encoding the target of the AV |
| Resistance is transmitted among bacterial species | Resistance is often restricted to the virus isolate that acquired the resistance and its progeny |
| AB use selects for resistance in nonpathogenic bacteria | AV resistant forms generally remain in the treated patients and their contacts. No selection of resistance in nonpathogenic viruses has been described yet |
| Resistance can be transferred from nonpathogenic to pathogenic bacteria | Transference of AV resistance from one virus to another is unlikely. Recombination-mediated acquisition of resistance necessitates the involved viruses to be closely related |

*Expanded from Domingo, E., Biebricher, C., Eigen, M., Holland, J.J., 2001. Quasispecies and RNA Virus Evolution: Principles and Consequences. Landes Bioscience, Austin. and other studies quoted in the text.*

---

selected by recombination and then spread to new hosts. However, recombination is relatively restricted in the field (Section 10.7), and recombination-mediated dissemination of antiviral resistance is less efficient than the spread of antibiotic resistance through genetic exchanges. Moreover, while the use of antiviral agents is restricted to treat infected patients, antibiotics are used for agricultural purposes (treatment of farm animals and plants) so that resistance can evolve in nonpathogenic bacteria and then be transferred into bacterial pathogens. No parallel event in terms of environmental penetration has yet been described for antiviral agents.

The comparison between antiviral and antibiotic resistance mechanisms (Box 10.1) constitutes a very clear example of how the molecular mechanisms of genome variation and mobility available to a biological system influence the way selection events unfold. High mutation rates set the scene for the evolution of error-prone replicating viruses, and the general escape strategy or acquisition of mutation-dependent phenotypes is widespread among them. In contrast, for more complex genomes that must refrain from generalized high mutation rates to avoid lethality, mechanisms based on genome rearrangements and transfers are the molecular frame for evolutionary events. Complex DNA viruses exploit genome expansions, rearrangements, and exchanges as adaptive mechanisms, close to the strategies employed by bacteria.

## 10.3 Darwinian principles and intrapopulation interactions acting on bacterial cell populations

The emphasis on mutation as being inherent to quasispecies theory, and less relevant for other mathematical formulations of evolutionary dynamics (Section 3.1 in Chapter 3) justifies quasispecies as a descriptor of bacteria in at least three different but related aspects:

- The recognized genetic heterogeneity of bacterial populations which is accentuated in mutator bacteria, those that display mutation rates which are $10^2$- to $10^3$-fold higher than the standard values for bacteria. Virtually, any individual bacterium is likely to vary genetically from any other individual bacterium of the same species.
- The dynamics of genetic variation, competition, and selection that render bacterial populations highly dynamic and adaptable.

- Intrapopulation interactions by cellular signals that modify the behavior of bacteria when they are organized as a collectivity, as opposed to the same bacteria acting in isolation.

The first point listed above is a straightforward extension of mutant generation in bacteria which occurs with average rates that are $10^2$- to $10^3$-fold lower than those of RNA viruses, leading to quasispecies-like populations (Covacci and Rappouli, 1998; Farzi et al., 2015; Bertels et al., 2017). Metagenomic surveys of bacterial populations have revealed what the authors called a "long tail" of rare representatives of phyla (Sharon et al., 2015). Some of the minority components are enriched in some enzymes, suggesting that they may confer an adaptive role to the entire community. The reader may have established a connection between the term "long tail" used by F. Banfield and colleagues and "comet tail" used by M. Eigen to refer to the error copies of primitive replicons in quasispecies theory (Section 3.6.1 in Chapter 3).

The second point is a statement of general adaptive potential, with the genome size-mutation-time frame implications for the limitation of sequence space exploration (Section 10.1). The third point is an extension to bacterial populations of the interactions established among components of a mutant spectrum in viruses extensively discussed in previous chapters, but with bacteria using different mechanisms. In the case of viruses, the interactions are frequently mediated by *trans*-acting gene products that confer a viral population the status of a unit of selection. In the case of bacteria, the intrapopulation interactions are based on elaborate systems of intercellular communication which are increasingly recognized as relevant to the behavior of the microbial world generally (Foster, 2011; Gibbs and Federle, 2015).

The regulated, cell density-dependent response of bacterial collectivities is known as quorum sensing (Nealson and Hastings, 1979;

Fuqua et al., 1994; Miller and Bassler, 2001; Ueda et al., 2019). Quorum sensing is based on signal transduction schemes that allow the bacterial population to produce a coordinated response to external stimuli. The result is the possibility to modulate bacterial virulence, which has conceptual parallelism in the case of viruses when their populations differentiate as a result of serial infections (Rumbaugh et al., 2009; Ojosnegros et al., 2010a, b; Vignuzzi and Andino, 2010). Bacterial biofilms are communities of microorganisms supported by some types of extracellular polymeric matrices in which the participating bacteria are modified physiologically (Costerton et al., 1978, 1987, 1999). Biofilms may display mechanisms of antibiotic resistance additional to those that operate in isolated bacteria (Section 10.2), including a physical impairment to the diffusion of the antibiotic to reach the target bacteria (Stewart and Costerton, 2001; Hall and Mah, 2017; Rojo-Molinero et al., 2019). The presence or absence of bacteriophage in a biofilm may also affect bacterial virulence (Rice et al., 2009). Biofilms, being a class of bacterial consortium, represent a manifestation of social behavior of bacteria, rendering a collectivity the target of selection (Caldwell and Costerton, 1996), a feature shared with viral populations.

## 10.4 The dynamics of unicellular parasites in the control of parasitic disease

The selection of drug resistance in viruses and bacteria summarized in Section 10.2.1 is also a common occurrence among pathogenic unicellular eukaryotes, including fungi, protozoans, such as amebas, the malaria parasite *Plasmodium falciparum*, or different *Leishmania* species, the causative agents of several forms of leishmaniasis and kala-azar. The molecular mechanisms of resistance range from point mutations in the drug target genes to activation of the efflux pumps that prevent the transport of drug into

the cells, or changes in the "transportome" that affect one or multiple drugs (multidrug resistance) (Wahlgren et al., 1999; Koenderink et al., 2010; Naidoo and Roper, 2010; Sanchez et al., 2010; Munday et al., 2015; Dhingra et al., 2019). Drug resistance has been associated with a defective mismatch repair system in *Plasmodium falciparum* (Ahmad and Tuteja, 2014). Natural populations of unicellular parasites are highly polymorphic (Escalante et al., 1998; Rottschaefer et al., 2011; Nkhoma et al., 2012). Extensive intrastrain genomic heterogeneity in *Leishmania*, termed mosaic aneuploidy, is essential for parasite adaptability and selection of drug resistance mutations (Mannaert et al., 2012; Sterkers et al., 2012; Lachaud et al., 2014). Alterations of growth conditions may perturb the population equilibrium reflected in the replacement of some parasite subpopulations by others (Andrews et al., 1992; Valadares et al., 2012; Duncan et al., 2013). As in the case of viruses, cellular parasites can escape antibodies or CTLs (De La Cruz et al., 1989; Baltz et al., 1991; al-Khedery et al., 1999), in particular, immune evasion through recombination-mediated switching of components of the variant surface glycoprotein coat (Horn and McCulloch, 2010; da Silva et al., 2018; Zhang et al., 2019). Subpopulations of *Plasmodium falciparum* with decreased sensitivity to monoclonal antibodies have been selected in vitro (Iqbal et al., 1997). The mechanism involved may be the selection of preexisting parasite subpopulations with reduced constitutive expression of the relevant antigens, or antibody-induced downregulation of antigen expression. Induction of complement resistance in vivo and in vitro has been described for some amebas (Hamelmann et al., 1993). Although later than for viruses, the need to consider parasite genetic heterogeneity in vaccine design has been emphasized (Gleichsner et al., 2015).

The trend in the last decades has been to unveil an increasing level of genetic heterogeneity in protozoan populations, with similarities with the conceptual changes that occurred with viruses a few decades earlier [i.e., genomic differences among subtypes (Gentekaki et al., 2017)]. Except for the differences in genome size-population size relationship between cells and viruses (emphasized in general terms in Section 10.2), intrapopulation heterogeneity, the coexistence of different subpopulations, and replacement of some subpopulations by others described in protozoa are clearly parallel features and events to those detailed for viruses in preceding chapters.

Application of whole-genome sequencing and new bioinformatic procedures to *Plasmodium falciparum* is currently used to obtain accurate genomic data sets, and to identify positive selection, in particular, *loci* involved in immune evasion and drug resistance (Samad et al., 2015). As an outcome of the implications of population complexity, the advantage of double- and triple-drug therapies is also increasingly recognized (Shanks et al., 2015). Therefore, endowed with mutation and recombination as major molecular mechanisms of genome variation, cellular parasites display features of quasispecies dynamics, in support of the broad relevance of quasispecies in biological systems (Domingo et al., 2001; Domingo and Schuster, 2016) (see Section 10.7 for a discussion of clonality in the microbial world).

## 10.5 Cancer dynamics: heterogeneity and group behavior

Cancer cells are extremely heterogeneous entities that can be regarded as cellular parasites of differentiated organisms that have acquired autonomous behavior. Already several decades ago, tumor heterogeneity was considered an important adaptive trait relevant to cancer metastasis (secondary growth in a part of the organism different from the original tumor site) (Hansemann, 1890; Boveri, 1914; Nowell, 1976;

Poste et al., 1982; Nicolson, 1987; Chin et al., 2011; Hanahan and Weinberg, 2011; Raynaud et al., 2018; Sardanyés and Alarcón, 2018; Gerlee and Johansson, 2019). Cell heterogeneity and modified behavior of cancer cells are mediated by point mutations, chromosomal rearrangements, and epigenetic modifications (trait modifications not due to changes in DNA sequence; examples are DNA methylation and histone modifications, such as phosphorylation). Most tumors are characterized by high mutation rates, although the necessity of high mutation rates for tumorigenesis is a debated issue. In a study by W.F. Bodmer and colleagues, the dual participation of mutation and selection in tumor progression was considered. It was argued that the standard cellular mutation rates (in the order of $10^{-8}$ per *locus* per cell per generation) may be sufficient to explain the key mutations associated with the transformed phenotype, although in the course of tumor progression, mutation rates can increase up to $10^4$-fold (Tomlinson et al., 1996). These authors consider cell selection more important than an initial high mutation rate to trigger tumor initiation. The rate of tumor progression may be favored by increases in the mutation rate as tumor cells replicate and diversify. Such enhancement of mutation rate presents a situation of genome instability, which has not yet been described during virus infections: an increase in mutation rate as the virus replicates in an infected host or in cell culture. This is a possibility that cannot be excluded given the evidence that multiple viral proteins, not only the viral polymerases, can be involved in template-copying fidelity (Section 9.6 in Chapter 9).

A difficulty, which is encountered in viruses as well as in tumors or other cellular collectivities, is the distinction between mutation rate (occurrence of mutations) and mutation frequency (the combined outcome of occurrence and selection; Section 2.5 in Chapter 2). Two main possibilities can be envisaged: an increase in mutation rate is necessary either to reach a phenotypic trait required for tumor progression or not to reach it but to increase the rate of tumor progression. In the latter scenario, the initial relevant trait would be equally accessible with the standard cellular mutation rate. A mutator phenotype would act as an evolutionary motor to increase the chances of finding the most advantageous mutational pathway for tumor growth and dispersal (Nicolson, 1987; Loeb, 2001; Bielas et al., 2006). The mutator phenotype in some types of cancer may be linked to APOBEC-mediated deamination (Vartanian et al., 2008; Caval et al., 2014; Venkatesan et al., 2018).

R.A. Gatenby and B.R. Frieden have emphasized the loss of information during carcinogenesis, reflected in a progressive deprivation of cellular differentiation, expressed as differentiation "information catastrophe." This is a Darwinian process toward maximum proliferation together with decreased cellular specialization (Gatenby and Frieden, 2002, 2004). The priority for proliferation at the cost of the specialized function is conceptually parallel to the selection of rapidly replicating bacteriophage Qβ RNA in vitro at the cost of RNA infectivity, in the classic S. Spiegelman-C. Weissmann experiments (Section 2.8 in Chapter 2).

The contribution of multiple mechanisms of DNA modification and damage for the cell to reach a level of transformation is subject to population size constraints similar to those that may limit the selection of multidrug resistance in viruses (Chapter 8). Alternative phenotypes may confer tumor progression and metastatic potential to different degrees. If approximation to the most effective phenotype is considered akin to fitness increase, attaining the required constellation of genetic alterations will depend on the tumor size (number of cells) on which genetic variation acts. Variation, in this case, includes point mutations, recombination and rearrangements, genetic modifications mediated by cellular factors, and epigenetic changes.

Mutations affecting repair genes, and proteins that bind to DNA, transcriptional regulators, or proteins involved in signaling pathways, among others, have been implicated in the gain of an oncogene or loss of a tumor suppressor (Futreal et al., 2004; Greenman et al., 2007; Chin et al., 2011; Shlien et al., 2015). These observations suggest that evolving tumor cells display features of complex adaptive systems sustained by cellular heterogeneity and clonal evolution (Nowell, 1976; Deisboeck and Couzin, 2009) (see also Section 10.8). This dynamic cellular system evolves, fueled by genetic and epigenetic change, following the general Darwinian principles of competition and selection (Maley and Forrest, 2000; Maley et al., 2006; Merlo et al., 2006). Cell populations may also be subjected to complementation or suppression interactions that may modulate tumor characteristics.

The similarity between cancer and RNA virus evolution has long been recognized (Gonzalez-Garcia et al., 2002; Brumer et al., 2006; Tannenbaum et al., 2006; Fox and Loeb, 2010); reviewed in (Ojosnegros et al., 2011; Domingo and Schuster, 2016). The parallels between the two systems extend to the proposal of an error threshold in cancer (Solé and Deisboeck, 2004; Fox and Loeb, 2010; Sardanyés et al., 2017). Cells, like viruses, have a maximum tolerable error rate above which maintenance of information for cell viability is no longer possible (Fig. 10.2). Genomic mutations (in general, genomic lesions) in cancer cells are several orders of magnitude higher than in normal cells ($10^3$-fold according to the data of Fox and Loeb, 2010 used in Fig. 10.2). In parallel with viruses, relatively modest increases in mutation rate may drive cancer cells into error catastrophe (Fox and Loeb, 2010; Loeb, 2011). This important conceptual parallelism opens anticancer treatment options based on the administration of nucleotide analogs, or synergistic combinations between mutagenic agents and inhibitors directed to critical functions that mediate cellular transformation.

FIGURE 10.2    Diagram depicting error threshold and lethal mutagenesis in cancer. The average number of somatic mutations per genome is taken as 50 for normal cells (distribution on the left). This number increases to 50,000 for cancer cells (second distribution). If the number of mutations is increased twofold, an error threshold is crossed that results in loss of information and cancer extinction. *Data copied and figure adapted from Fox, E.J., Loeb, L.A., 2010. Lethal mutagenesis: targeting the mutator phenotype in cancer. Semin. Cancer Biol. 20, 353—359, with permission from the authors.*

## 10.5.1 The two-component theory of cancer: similarities with other biological systems, and therapeutic implications

Some models distinguish two components regarding the genetic information in cancer cells: a robust component that ensures stability, and a variable component that permits phenotypic explorations that confer adaptability (Solé et al., 2008). The distinction is again parallel to one found in viruses: the core information that preserves virus identity versus flexible information that generates diversity to cope with environmental demands. This two-component theory is reminiscent of the "pan-genome" concept in bacteria. It proposes that individuals of a given bacterial species share a core component, while different representatives of the same species differ in part of their genetic information (Whitaker and Banfield, 2006; Mira et al., 2010). The genome complement of the bacterial species as a whole is larger than the genome portion

present in individual strains within the species. Events, such as genetic exchanges among bacteria, contribute to enrich the flexible component while preserving bacterial identity as well as the minimal ingredients for survival.

It is expected that the phenotypic pluripotency of cancer cells, which is largely derived from their flexible cellular moiety, will rapidly select for cell mutants resistant to chemotherapeutic agents, particularly during monotherapy (Luo et al., 2009). Suppression of tumor growth to minimize selection of resistant cancer cell subpopulations (as was the case with a selection of viral subpopulations from mutant spectra) is more likely using combination therapies [(Smidt et al., 2018; Chao et al., 2019) among many other chemotherapeutic combination protocols]. The objective is to achieve highly active anticancer treatments similar to the highly active treatments used for virus infections. Designs to suppress chemoresistant populations have been proposed (Gatenby et al., 2009). Cancer therapies require addressing different targets in a personalized manner. Tumor cell heterogeneity also represents a limitation to the efficacy of oncolytic viruses (Forbes et al., 2013). The requirement of combination treatments to control viruses emphasized in Chapters 8 and 9 is recognized as a valid approach also for cellular pathogens (Avner et al., 2012).

We reach the unifying notion for viruses and cellular collectivities that, as proposed in the quasispecies theory, the unit of selection can be an ensemble of individuals. In addition to the acquisition of new phenotypes through interactions among individual components (emerging features), minorities can provide the raw material for survival when the majority is hit by external aggression. Interestingly, a collective behavior may also be manifested in protein aggregates, such as prions, with Darwinian behavior despite not being a genetic system.

## 10.6 Collective behavior of prions

The extensions of quasispecies theory outlined in previous sections have been applied to systems in which inheritance in the form of template copying is the key ingredient, common to viral and cellular lifestyles. Yet, interestingly, C. Weissmann and colleagues have provided direct evidence that prions (protein-only pathogenic entities) display features of quasispecies. Prions are the causative agents of transmissible spongiform encephalopathies (Alper et al., 1967; Griffith, 1967; Bolton et al., 1982; Prusiner, 1982, 1991; Walker and Jucker, 2015). The prion-associated diseases include Creutzfeldt-Jakob disease, fatal familial insomnia, and Gerstmann-Sträussler-Scheinker disease in humans, as well as scrapie in sheep, bovine spongiform encephalopathy in cattle, and chronic wasting disease in cervids. The basis of prion disease is the propagation of $PrP^{Sc}$ protein aggregates. $PrP^{Sc}$ differs in conformation from a normal host glycoprotein termed $PrP^{C}$ (C stands for *common* or *cellular*, Sc for *scrapie*). Protein aggregation is increasingly viewed as a normal physiological process so that beneficial aggregation should be distinguished from pathologic aggregation often associated with neurodegeneration (Tyedmers et al., 2008; Jarosz et al., 2010; Tuite and Serio, 2010; Falsone and Falsone, 2015).

Protein conformation defines the strain characteristics and host range of prions. Different conformations can be regarded as "prion mutants." In a strikingly parallel behavior of RNA viruses, prions cloned by end-point dilution in cell culture can become heterogeneous by increasing the repertoire of conformers in the prion population (Li et al., 2010). Expressed in a virus-centric fashion, the passage of prions increases their mutation frequency (frequency of different conformations). Selective pressures result in the selection of variants, including drug-resistant mutants (Ghaemmaghami et al., 2009; Li et al., 2010; Mahal et al., 2010; Shorter, 2010; Weissmann et al., 2011). Prion behavior can be modified depending on the population size (large populations vs. bottleneck events produced by limiting dilution) in the protein misfolding cyclic amplification assay (the parallel to serial viral passages described in Section 6.2 of Chapter 6). Conformational variants are identified only after large population passages, and a decline in variant amplification rates is observed upon bottleneck passages (Vanni et al., 2014). Thus, Darwinian behavior and a fitness effect of passage regime are properties of prions despite being a nongenetic macromolecular population system.

The events that underlie the quasispecies behavior of prions, reflected in the dynamics of conformation heterogeneity, competition, and selection, are largely unknown. They may involve differences in glycosylation or transitions among conformations triggered by environmental factors, and perhaps with participation of truncated forms of the protein (Helmus et al., 2008; Notari et al., 2008; Zanusso et al., 2014; Igel-Egalon et al., 2019). We have proposed that irrespective of likely effects of the molecular composition of the prion proteins and environmental factors in modifying conformational preferences, some basic mechanisms acting on molecular fluctuations may also play a role (Ojosnegros et al., 2011). Protein conformation is the result of multiple amino acid-amino acid interactions and atomic modifications of the type described to explain the origin of mutations through nonstandard base pairs among nucleotides (Chapters 2 and 9).

In the case of protein aggregates, the adoption of one versus alternative conformations may depend on atom ionization, and ionic and hydrophobic contacts, all of which depend in turn on bond torsion angles subjected to thermal fluctuations. The different influences at play may also determine the presence of minority conformations at different frequencies. Structural heterogeneities and transitions among conformations are widespread in protein populations (Smith et al., 2015). Such transitions may play physiological roles, and only occasionally they may be involved in pathological conditions. The relevance of alternative conformations in prions became apparent because they can give rise to disease, but in other cases, they may remain unnoticed. L.C. Walker and M. Jucker have proposed to define prions as "proteinaceous nucleating particles" and in fact "nucleation" may be one of the events responsible of their quasispecies behavior (Walker and Juker, 2015; Huang et al., 2019). A possibility is that a specific conformation variant can nucleate the conversion of additional molecules into the same conformation to yield aggregates enriched in the new variants [see (Bernacki and Murphy, 2009; Kastelic et al., 2015) and references therein for some theoretical aspects of protein aggregation]. It would be extremely interesting to develop a theory of quasispecies for macromolecular populations (nongenetic systems) that would serve as a framework to penetrate into prion-related disease. Such a theory may help understanding pathologies related to altered aggregation behavior of cellular proteins, as the molecular quasispecies of replicative systems helped to interpret virus behavior (Mas et al., 2010; Ojosnegros et al., 2011).

## 10.7 Molecular mechanisms of variation and clonality in evolution

In this and previous chapters, it has been emphasized that while point mutations are a universal means of genetic variation, changes based on loss, gain, or rearrangements of genomic portions or segments are used unevenly by different biological systems. For bacteria, protozoa, and cancer cells, large genome modifications are widely exploited as adaptive mechanisms. This poses the question if their evolution is predominantly clonal; in this context, clonality means that individual lineages evolve separately and that either sex or recombination plays no role or only a minimal role in their mode of life and survival strategies. The debate bears on the historical or present advantage of sex as a reproductive strategy (Barton and Charlesworth, 1998; Agrawal, 2006), discussed in Section 6.5.1 of Chapter 6 as a mechanism to counteract the fitness-reducing effects of Muller's ratchet.

The controversy on clonality has been quite vivid for protozoa, with interesting studies and controversial proposals by many authors. The clonality model has been championed by M. Tibayrenc and F.J. Ayala and is suggested to be applicable not only to protozoa but also to viruses and bacteria (Tibayrenc and Ayala, 2012, 2014; 2017). The concept of predominant clonal evolution proposed by these authors is based on significant linkage disequilibrium (nonrandom association of genotypes at different loci), and maintenance of the identity of different lineages or genetic subdivisions in the form of recognized clades (or near-clades) that are stable in space and time, among other arguments. The absence of recombination could be either because of lack of molecular mechanisms to perform it or because of the scarcity of double infections during which two genetically distinct protozoans from the same species could exchange genetic material (the "starving sex" hypothesis). Other authors consider that there are data in favor of nonclonality and of an active role of recombination during the evolution of protozoa and other cellular entities (Heitman, 2010; Ramirez and Llewellyn, 2014; Rogers et al., 2014; Berry et al., 2019). In other cases, such as in some yeasts, there is evidence of both clonality and recombination in the field (Lott et al., 2010). Rather than going deeper in the controversy with cellular organisms, here we will briefly examine clonality in viral systems, since they may shed light on clonality as a main mechanism of evolution despite the presence of active recombination machinery.

The term clonality applied to virus populations, and evolution has at least three different meanings. The one more distant from the concept under discussion is that it may refer to biological clones of viruses, to describe populations that are the progeny of a single genome. We sometimes speak of a clonal population of the virus to signify a population derived from a single genome (obtained from a single viral plaque or end-point dilution series). Biological cloning is the most severe form of a genetic bottleneck, and the effects of bottlenecks are extensively discussed in other chapters of this book. A second meaning is the clonal expansion of viral genomes integrated with cellular DNA, accompanying the multiplication of the cells that harbor them. Some retroviruses may integrate their genomes for considerable time periods so that they follow the pattern of variation typical of the host cell rather than the error-prone replication of the virus in the process of progeny production. This is the case of the human T-cell lymphotropic virus types 1 and 2 (HTLV-1 and HTLV-2). They have two routes for expanding their genetic material: production of viral particles by activating viral expression from the latent reservoir, or duplication of their genetic material as part of cellular DNA [see (Melamed et al., 2014; Kulkami and Bangham, 2018) and references therein]. The latter has been termed the mitotic stage of the replication cycle of these viruses. As evidence that cell multiplication confers a selective advantage to the virus, there are viral functions that drive cellular proliferation. The latter should be distinguished from virus-driven malignancy (cellular transformation) that involves other mechanisms. In this second meaning of clonality, the viruses undergo clonal expansion through the proliferation of their carrying cells.

The third meaning of clonality in viruses that do not integrate DNA copies of their genome in the cellular DNA means the evolution of separate lineages, with little (or no) role of genetic exchanges among lineages; it corresponds closely to the meaning of clonal evolution of cellular parasites. Evidence of clonality in viruses is provided by the existence of near-clades and consistent linkage disequilibrium. However, there is extensive evidence of recombination in most viruses, including RNA viruses (Section 2.9 in Chapter 2). The question is as to how can clonality be maintained and considered the predominant mode of evolution in the face of extensive recombination. The proposal to solve this issue has been to distinguish unproductive or inconsequential recombination from biologically meaningful recombination (Perales et al., 2015). Current evidence is that only a minority of all the recombination events undergone by viruses have biological consequences. Despite the recombination machinery (which is closely ingrained into the replication machinery) remaining intact and active, most recombination events go unnoticed due to lack of appropriate markers to distinguish parental from recombinant genomes, and, more relevant biologically, absence of opportunity of fitness increase through recombination. Recombination in HIV-1 illustrates these points. Recombination is probably a very frequent occurrence in HIV-1 (Shriner et al., 2004) yet its biological effects regarding fitness gains of mosaic genomes were only noticed when HIV-1 lineages had previously diversified through point mutations (Thomson et al., 2002; Song et al., 2018).

Identification of recombination intermediates in enterovirus-infected cells using deep sequencing illustrates an additional level of unproductive recombination (Lowry et al., 2014). The study by D.J. Evans and colleagues unveiled multiple imperfect recombination events during infection, in what was termed the step of generation of recombinant genomes. It was followed by the resolution step in which a few successful recombinants gave rise to viral progeny. This is yet another example of negative selection acting

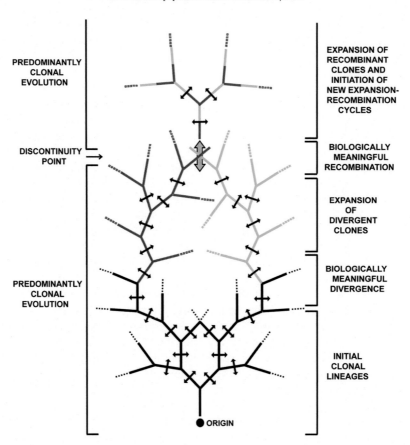

PREDOMINANTLY
CLONAL
EVOLUTION

EXPANSION OF
RECOMBINANT
CLONES AND
INITIATION OF
NEW EXPANSION-
RECOMBINATION
CYCLES

DISCONTINUITY
POINT

BIOLOGICALLY
MEANINGFUL
RECOMBINATION

EXPANSION
OF
DIVERGENT
CLONES

BIOLOGICALLY
MEANINGFUL
DIVERGENCE

PREDOMINANTLY
CLONAL
EVOLUTION

INITIAL
CLONAL
LINEAGES

● ORIGIN

**FIGURE 10.3** A schematic representation of the clonal evolution of a virus despite continuous, active recombination. At an initial point (Origin), virus expands through replication (*straight black lines*), with continuous occurrence of inconsequential recombination (*double arrows perpendicular to the replication lines*). Discontinuous lines indicate that multiple divergent lineages are formed and pursue expansion. At a certain point, two sublineages diversify by mutation yielding viruses with potentially relevant differences (*the red and blue lineages*). In turn, each of the two new lineages proceeds with their diversification. At a critical or discontinuity point (*thick vertical double arrow*), a biologically meaningful recombination event occurs that gives rise to a mosaic virus (*red-blue chimera*) that initiates its clonal propagation. The major events in the process are indicated at the left and right margins of the diagram. See text for further clarification and implications. *Figure reproduced from Perales, C., Moreno, E., Domingo, E., 2015. Clonality and intracellular polyploidy in virus evolution and pathogenesis. Proc. Natl. Acad. Sci. U.S.A. 112, 8887–8892, with permission from the National Academy of Sciences USA.*

on nascent variant molecules to allow survival of only a few out of many genetic variants (Section 3.4 in Chapter 3). This example emphasizes the distinction between "occurrence" and "biological consequences" of recombination in viruses. There are many examples of biologically relevant recombination in viruses, and some of them were reviewed in Chapter 2, including the establishment of vaccine-derived poliovirus and HIV-1 recombinants at the epidemiological level, as well as a salient transition toward genome segmentation of FMDV upon large MOI passages in cell culture. Since recombination requires coinfection of the same cell with two (or more) virus types that can recombine, the "starving sex hypothesis" may be fulfilled due to a low probability of cell coinfections in differentiated organisms. In addition, coinfection may be prevented by mechanisms of superinfection exclusion (related to cross-protection) by which active replication of a virus blocks a second infection by a related virus [(Webster et al., 2013; Zhang et al., 2018) and references therein]. In favor of the possibility of recombination is the observation that in host organisms, some cell subsets may be more prone than average to capture viral particles, therefore, increasing the probability of coinfection to higher values than expected by chance (Cicin-Sain et al., 2005; Chohan et al., 2005).

In view of the above considerations, we have suggested a model for virus clonality whose main ingredients are summarized in Box 10.2, and it is diagrammatically represented in Fig. 10.3 (Perales et al., 2015). The critical point is that despite recombination offering the virus a means to explore distant regions of sequence space (essential for the occupation of new niches and long-term survival), it has not been established as a necessity to complete the replication cycle of viruses. Strand transfers during reverse transcription, or discontinuous RNA synthesis mechanisms during RNA replication or transcription, can be regarded as mechanistically akin to recombination, but they do not affect the genetic composition of progeny except in the case of coinfections. In contrast, recombination is an intrinsic necessity of all organisms in which sexual mechanisms have been established. The scarcity of biologically meaningful recombination maintains stable near-clades in viruses, as documented with many phylogenetic analyses. As indicated in Fig. 10.3, occasional discontinuity points (highlighted in the figure by the selection of mosaic genomes) may be biologically very relevant, but they are not part of the way of life of viruses. The scheme with discontinuity points does not imply any space or temporal scale. It may refer either to intrahost or interhost events or to a single infection or multiple infections (host-to-host transmissions at an epidemiological stage). Thus, clonality can coexist with recombination in the evolution of cells and viruses. What the model we propose for virus stresses is that recombination can play an occasional evolutionary role (Fig. 10.3) but, contrary to sexual reproduction, recombination is not a norm imposed upon viruses to complete their replication cycles.

---

## BOX 10.2

### Maintenance of clonality in virus evolution

- Genetic recombination is frequent but biologically meaningful recombination is limited.
- Recombination events are probably frequent during intracellular virus replication, with many recombination intermediates subjected to negative selection.
- Biologically meaningful recombination is limited by:

    — The need of prior diversification of parental viruses by mutation.
    — The requirement of coinfection of the same cell by the two distinguishable viruses.
    — The fitness requirements of the recombinants that must coexist with or outcompete the parental viruses.

## 10.8 Genomes, clones, consortia, and networks

There are several levels of organization and interactions in biological and nonbiological systems that affect their behavior and stability. Classical ones at the level of cells and organisms are symbiosis (including endosymbiosis), parasitism, commensalism, and mutualism. In the flow of increasing complexity in viral genomes that are subjected to error-prone replication, the following order can be established: individual genome → multiple genomes → quasispecies as a unit of selection → multiple interacting quasispecies. While the consideration of viral populations as interacting sets of individuals is now well established (Domingo et al., 2012; Villarreal and Witzany, 2013; Domingo and Schuster, 2016), the interactions among different viral quasispecies and their possible significance are still largely unexplored.

Viral quasispecies can behave as complex adaptive systems as evidenced by the collective properties discussed in this book, notably the presence of a molecular memory of past evolutionary events, akin to memory in the immune system (Section 5.5 in Chapter 5). The critical ingredient of quasispecies memory or a mutant ensemble directing behavior is a network of interactions established among members of a mutant spectrum. Two categories of interactions can be distinguished: those due to the mutations per se, and those due to gene expression products, in particular, altered proteins that act in *trans* on other proteins or genomes of the same replicative unit. Mutation per se allows connections among component of the mutant spectrum, including mutational transitions between the master or dominant genome and its surrounding genomes located at 1, 2, etc. mutations distance in sequence space. We have described that an apparently disorganized mutant spectrum may, in reality, be composed of multiple subclasses of genomes, as revealed, for example, by partition methods that should be increasingly applied to analyze mutant spectra dissected by new sequencing methodologies (Section 3.6.4 in Chapter 3). A still largely unexplored subject is the extent of quasispecies substructuring in viral samples in vivo (a viral population from a cell, or a group of cells, an organ, etc.) and the network of interactions among them. Before addressing this point and its implications, we should briefly comment on networks.

### 10.8.1 Interacting networks and power laws

Many natural and artificial systems have as one of their defining features being self-organized in a network manner, in a way which is often not random. They are widespread in the sense that they participate in disparate categories of interactions, some familiar to biologists and others not. Interacting networks are established among metabolites and metabolic reactions, or proteins within a cell, or organisms within an ecological system. There are also social networks, for example, the interactions among your relatives or your friends and their acquaintances, or the nodes and hubs that conform the www [for an overview that includes specific references, see (Barabási, 2014)]. Regarding biological implications, network structure can predict conditions of network stability (Yan et al., 2017; Delmas et al., 2018; Santolini and Barabasi, 2018) or choice of drug combinations to treat multiple diseases (Cheng et al., 2019).

The second point is that the connections in these types of self-organized networks follow some patterns. This can be intuitively understood: the proteins within a cell do not interact at random; some proteins have many partners while the majority has only a few partners. Nodes in the web have unequal weight in the interactions: some important hubs have many interactions while the majority has few interactions. This type of network topology is termed scale-free, and it is mathematically described as a power law, meaning that when two quantities are compared (number of proteins vs. the number of their interactions in the cell, or number of nodes vs. the number of their interactions in the web), one of the quantities varies as a power of the other quantity. Events described by a power law are frequent in nature (Bak, 1994, 1996; Newman, 2005). They describe apparently disparate events, such as the intensity of earthquakes as a function of their numbers, or the number of inhabitants in a city versus the number of cities, among many other examples. A scale-free (nonrandom) organization described by a power law generally reflects an underlying dynamic process in its construction: how cellular evolution has adjusted the expression level and interaction among proteins or how dynamic sociological aspects have generated megacities in our planet. A power law is not the only statistical distribution that echoes an underlying mechanism. In Section 6.6 of Chapter 6, we encountered the Weibull distribution that characterized viral titers in successive plaque transfers, and that probably reflected the multiple cell-virus interactions that participate in the burst of viral progeny (Lázaro et al., 2003).

A few studies have addressed power laws in relation to viral dynamics. A. Moya and his colleagues described network dynamics of eukaryotic long terminal repeat (LTR) retroelements of eukaryotes. The connectivity among phenotypic markers was distributed according to a power law, that manifested a mode of evolution guided by an intrinsic capacity of LTR phenotypes to self-organize following a pattern characteristic of complex systems (Llorens et al., 2009). In a deep sequencing analysis to examine the mutational load of HCV in patients

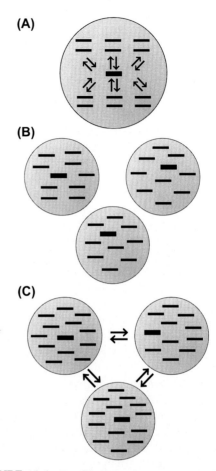

FIGURE 10.4 Possible interactions among quasispecies. Different mutant distributions (*circles*) are depicted in a simplified manner with genomes as horizontal lines (symbols that represent mutations have been omitted). The master or dominant genome is represented as a *thicker line*. (A) A single quasispecies in which interactions (mutational transitions) between the master sequence and surrounding members of the mutant spectrum are indicated by *arrows*. (B) Three quasispecies in which the interactions between the master sequence and mutant spectrum components have been omitted for clarity. In this case, the three mutant distributions are isolated by a mutational barrier. (C) Same as (B) except that in this case, interactions (mutational transitions) are established among the three mutant distributions. See text for implications.

subjected to peginterferon-alpha 2a-ribavirin therapy, A.M. Di Bisceglie and colleagues found that 36,818 mutations detected in 56 patients displayed a power-law distribution of mutation frequencies (Wang et al., 2014). A power-law behavior was also identified for a late phase of an Ebola virus epidemics (Gustafson and Proctor, 2017). Also, clusters of similar sequences in virus phylogeny follow a power-law behavior, with small numbers associated with large clusters (Dearlove et al., 2017). These analyses have implications for antiviral interventions. Of relevance for marine ecology, virus abundances in the ocean are a power-law function of microbial cell abundances (Wigington et al., 2017).

Self-organized systems display critical thresholds for spreading efficacy. For example, to be implemented, an innovation (scientific or technological) must propagate through the network of potential recipients above a critical threshold number of contacts (Abrahamson and Rosenkopf, 1997). We discussed the error threshold concept in Chapter 9 as a limit for maintenance of genetic information. In complex systems, there are critical thresholds needed to achieve a result, such as the dissemination of an epidemics or the successful implementation of an innovation. Networks have additional features that may be relevant to virus population stability.

## 10.9 An additional level of virus vulnerability?

Self-organized networks have two additional features that may offer unexplored possibilities to drive viral populations toward an externally-mediated extinction: substructuring and target-dependent vulnerability. An apparently homogeneous network of connections, such as the www is in reality divided into subcompartments [also termed continents; see (Barabási, 2014) for review and references]. The stability of the ensemble depends not only on maintaining

sufficient numbers of connections but also on maintaining some key connections; some minor nodes can be eliminated with no effect on the ensemble while highly connected hubs must be preserved to ensure stability. Mutational links, translated to viral populations, are established between the master (or dominant) genomic sequences and their neighbors in sequence space, and also among the minority neighbors. Not all points in sequence space have the same weight regarding quasispecies stability. If we could specifically hit the master sequence (by some "finely directed" mutagenesis procedure), we could destabilize the ensemble in a much more effective way than just hitting at random any genome from the population.

The issue of sub-structuring within viral populations is a largely unexplored one. The increasing application of powerful sequencing platforms should allow determination of multitudes of whole-genome sequences and may clarify the possible presence of substructures within viral populations as they replicate at different size-time scales (within single cells, sets of cells, cell cultures under various environmental influences, etc.). Some possibilities of absence or presence of different types of substructures are schematically presented in Fig. 10.4. The assumption is often made that a viral quasispecies is dominated by a master sequence which is surrounded by a mutant cloud that influences the performance of the ensemble (Fig. 10.4A). We know, however, that a viral population can be dominated by two or more master sequences. In these cases, there might be several mutant clouds isolated one from the other, with borders determined by mutational distance (Fig. 10.4B). Alternatively, the different quasispecies may be connected among them by accessible mutational distances (Fig. 10.4C). We may envisage different topologies with various degrees of complexity that should be revealed by deep sequencing analyses combined with bioinformatics procedures with the aim of defining genomic subgroups within viral populations.

Two general outcomes can be anticipated: the absence of substructuring or presence of multiple clusters ("nodes" or "hubs") of genomes with different numbers of connections and, thus, a different contribution to the stability of the system as a whole. A specific antiviral activity directed to debilitate the genomes with the maximum number of connections would be much more effective than a general antiviral activity directed to any genome in an indiscriminate manner. Again, debilitating the masters is more effective than debilitating minorities. The obvious difficulty is how to achieve such specific targeting. One possibility is to pursue designs of the type defined as treatment splitting into an induction and a maintenance regimen (Box 9.1 in Chapter 9) proposed by M. von Kleist and colleagues (von Kleist et al., 2011) and adapted to a second mutagenesis step. The induction phase should enrich the population in the most frequent and connected components of the mutant distributions, enhancing their exposure to mutagenic agents in the second treatment phase. Another largely unexplored possibility which is complementary to the previous one is to choose the adequate mutagenic agents to produce the highest frequency of deleterious mutations. It is known that different mutagenic agents evoke different repertoires of mutations, shifting sequence space into different directions (Section 9.4 in Chapter 9). The base composition and codon usage in the genomes that conform the mutant distributions may suggest a mutagen-dependent efficacy of viral load reduction. At the risk of failing in scientific predictions (as many colleagues have), I end this section by suggesting that an increased number of viral-specific mutagenic agents, together with a deeper knowledge of the network organization of viral quasispecies (Fig. 10.4), may reveal new vulnerabilities and help in the design of more effective antiviral strategies.

## 10.10 Overview and concluding remarks

This chapter, as it also happens with Chapter 1, deviates somehow from the mainstream of the book. Several concepts relevant to the understanding of viruses as populations do not apply obviously solely to viruses. The reason is that the relevant concepts are of a very general nature. They can be translated to other biological systems, provided the key parameters are adjusted. The comparative analysis has served to emphasize that different molecular mechanisms of genetic variation (with genomic transfers and rearrangements becoming more prominent relative to mutation as genome complexity increases) are used to achieve similar features and goals. Features are collective behavior (with several cell-to-cell communication mechanisms) and selection of cellular subpopulations in the face of external constraints. This has been illustrated with striking similar mechanisms of resistance to antibiotics and drugs displayed by bacteria and unicellular eukaryotes. If anything, the observations reinforce the conclusions reached for viruses in preceding chapters, and indicate that control of viruses and pathogenic cells share the need for new paradigms.

A well-accepted parallelism is the one between RNA virus evolution and cancer development, both being a Darwinian process of accelerated genetic variation, competition, and selection of the most suitable individuals in each environment. Cancer metastasis is the typical manifestation of tumor cell adaptability, or capacity of cells to grow in tissues and organs different from the one where the tumor originated. The process may be regarded as a restatement of the difficulties to control tumor progression. However, there is another side to it: the existence of an error threshold for maintenance of the information needed for cellular viability. Therefore, a new possible therapeutic design for cancer is on the horizon: lethal mutagenesis, following the strategy for RNA viruses.

A remarkable finding is that prions and possibly other proteins that function as macromolecular aggregates display features of quasispecies despite not being genetic systems. It means that Darwinian behavior does not require genetic modification to operate, as we already encountered in Chapter 1 with selective events in a prebiotic scenario. Quasispecies dynamics in prions necessitates a theoretical development because it will help in approaching only-protein-based pathologies, as quasispecies theory helped to explain adaptability of pathogenic viruses and open new ways to combat them.

We have also addressed an existing controversy on clonality in microbial evolution. Centered on viruses, the proposal made in this chapter is that clonality (absence of recombination as a way of life) can be the norm despite the continuous operation of recombination. The main point to solve this paradox is a distinction between unproductive or inconsequential recombination and biologically meaningful recombination. A deeper penetration into the replicating structures of viruses and cellular parasites should tell us soon if the model proposed for viruses is valid and if it is applicable also to cellular microbes.

Finally, this chapter (and the book) ends by acknowledging the importance of network interactions at different levels. It is worth noting that possible substructuring of viral quasispecies should be amenable to analysis during the coming years mainly as a result of massive whole-genome sequencing. The feeling is that a new window of exciting information about virus population complexity will soon open. At the risk of falling victim to the inability of scientists to predict the future, I have adventured possible ways by which detailed knowledge of quasispecies subpopulations (if such subpopulations exist and can be defined) might provide new approaches to combat viral disease. It looks still far, but unless some new paradigm takes the scene, it seems like a worth exploration, in line with what this book has aimed at transmitting (See Summary Box).

---

### Summary box

## Extensions to Genetic and Nongenetic systems

- Darwinian principles, quantum mechanic uncertainties, and information theory are the pillars on which viral population dynamics rests. These three concepts, being of a fundamental nature, can be applied to a variety of genetic and nongenetic systems.
- Bacteria and cancer cells can display collective features parallel to those described for viral populations: population heterogeneity, intrapopulation interactions, and collective responses to external stimuli. Single-cell analyses provide increasing evidence of genetic and phenotypic heterogeneity among cells of the same population.

- Cancer cells have a high mutational load and multiply close to an error threshold for maintenance of genetic information. Lethal mutagenesis by mutagenic nucleotides is a therapeutic option to treat cancer.
- Viruses evolve in a predominantly clonal way. Despite active recombination, biologically meaningful recombination is not the norm in the replication cycle of viruses.
- Scale-free, interacting networks describe many natural and artificial systems. Possible network connections within viral populations may unveil new points of virus vulnerability.

# References

Abrahamson, E., Rosenkopf, L., 1997. Social network effects on the extent of innovation diffusion: a computer simulation. Organ. Sci. 8, 289–309.

Agrawal, A.F., 2006. Evolution of sex: why do organisms shuffle their genotypes? Curr. Biol. 16, R696–R704.

Ahmad, M., Tuteja, R., 2014. Emerging importance of mismatch repair components including UvrD helicase and their cross-talk with the development of drug resistance in malaria parasite. Mutat. Res. 770, 54–60.

al-Khedery, B., Barnwell, J.W., Galinski, M.R., 1999. Antigenic variation in malaria: a 3′ genomic alteration associated with the expression of a P. knowlesi variant antigen. Mol. Cell 3, 131–141.

Alper, T., Cramp, W.A., Haig, D.A., Clarke, M.C., 1967. Does the agent of scrapie replicate without nucleic acid? Nature 214, 764–766.

Andrews, R.H., Chilton, N.B., Mayrhofer, G., 1992. Selection of specific genotypes of Giardia intestinalis by growth in vitro and in vivo. Parasitology 105(Pt 103), 375–386.

Avner, B.S., Fialho, A.M., Chakrabarty, A.M., 2012. Overcoming drug resistance in multi-drug resistant cancers and microorganisms: a conceptual framework. Bioengineered 3, 262–270.

Bak, P., 1994. Self-organized criticality: a holistic view of nature. In: Cowan, G.A., Pines, D., Meltzer, D. (Eds.), Complexity. Metaphors, Models and Reality. Addison-Wesley Publishing Co., Reading, MA, pp. 477–496.

Bak, P., 1996. How Nature Works. Springer-Verlag New York, Inc., New York.

Baltz, T., Giroud, C., Bringaud, F., Eisen, H., Jacquemot, C., et al., 1991. Exposed epitopes on a Trypanosoma equiperdum variant surface glycoprotein altered by point mutations. EMBO J. 10, 1653–1659.

Barabási, A.-L., 2014. Linked: How Everything Is Connected to Everything Else and What it Means for Business, Science, and Everyday Life. Basic Books, New York.

Barton, N.H., Charlesworth, B., 1998. Why sex and recombination? Science 281, 1986–1990.

Bernacki, J.P., Murphy, R.M., 2009. Model discrimination and mechanistic interpretation of kinetic data in protein aggregation studies. Biophys. J. 96, 2871–2887.

Berry, A.S.F., Salazar-Sanchez, R., Castillo-Neyra, R., Borrini-Mayori, K., Chipana-Ramos, C., et al., 2019. Sexual reproduction in a natural Trypanosoma cruzi population. PLoS Negl. Trop. Dis. 13, e0007392.

Bertels, F., Gokhale, C.S., Traulsen, A., 2017. Discovering complete quasispecies in bacterial genomes. Genetics 206, 2149–2157.

Bielas, J.H., Loeb, K.R., Rubin, B.P., True, L.D., Loeb, L.A., 2006. Human cancers express a mutator phenotype. Proc. Natl. Acad. Sci. U.S.A. 103, 18238–18242.

Bolton, D.C., McKinley, M.P., Prusiner, S.B., 1982. Identification of a protein that purifies with the scrapie prion. Science 218, 1309–1311.

Boveri, T., 1914. Zur Frage der Entstehung Maligner Tumoren. Verlag, Jena.

Brumer, Y., Michor, F., Shakhnovich, E.I., 2006. Genetic instability and the quasispecies model. J. Theor. Biol. 241, 216–222.

Caldwell, D.E., Costerton, J.W., 1996. Are bacterial biofilms constrained to Darwin's concept of evolution through natural selection? Microbiologia 12, 347–358.

Caval, V., Suspene, R., Shapira, M., Vartanian, J.P., Wain-Hobson, S., 2014. A prevalent cancer susceptibility APOBEC3A hybrid allele bearing APOBEC3B 3′UTR enhances chromosomal DNA damage. Nat. Commun. 5, 5129.

Chao, M.-W., Chang, L.-H., Tu, H.-J., Chang, C.-D., Lai, M.-J., et al., 2019. Combination treatment strategy for pancreatic cancer involving the novel HDAC inhibitor MPT0E028 with MEK inhibitors beyond K-Ras status. Clin. Epigenet. 11, 35.

Cheng, F., Kovács, I.A., Barabasi, A.-L., 2019. Network-based prediction of drug combinations. Nat. Commun. 10, 1197.

Chin, L., Hahn, W.C., Getz, G., Meyerson, M., 2011. Making sense of cancer genomic data. Genes Dev. 25, 534–555.

Chohan, B., Lavreys, L., Rainwater, S.M., Overbaugh, J., 2005. Evidence for frequent reinfection with human immunodeficiency virus type 1 of a different subtype. J. Virol. 79, 10701–10708.

Cicin-Sain, L., Podlech, J., Messerle, M., Reddehase, M.J., Koszinowski, U.H., 2005. Frequent coinfection of cells explains functional in vivo complementation between cytomegalovirus variants in the multiply infected host. J. Virol. 79, 9492–9502.

Costerton, J.W., Geesey, G.G., Cheng, K.J., 1978. How bacteria stick. Sci. Am. 238, 86–95.

Costerton, J.W., Cheng, K.J., Geesey, G.G., Ladd, T.I., Nickel, J.C., et al., 1987. Bacterial biofilms in nature and disease. Annu. Rev. Microbiol. 41, 435–464.

Costerton, J.W., Stewart, P.S., Greenberg, E.P., 1999. Bacterial biofilms: a common cause of persistent infections. Science 284, 1318–1322.

Covacci, A., Rappuoli, R., 1998. Helicobacter pylory: molecular evolution of a bacterial quasi-species. Curr. Opin. Microbiol. 1, 96–102.

da Silva, M.S., Hovel-Miner, G.A., Briggs, E.M., Elias, M.C., McCulloch, R., 2018. Evaluation of mechanisms that may generate DNA lesions triggering antigenic variation in African trypanosomes. PLoS Pathog. 14, e1007321.

De La Cruz, V.F., Maloy, W.L., Miller, L.H., Good, M.F., McCutchan, T.F., 1989. The immunologic significance of variation within malaria circumsporozoite protein sequences. J. Immunol. 142, 3568–3575.

Dearlove, B.L., Xiang, F., Frost, S.D.W., 2017. Biased phylodynamic inferences from analyzing clusters of viral sequences. Virus Evol. 3, vex020.

Deisboeck, T.S., Couzin, I.D., 2009. Collective behavior in cancer cell populations. Bioessays 31, 190–197.

Delmas, E., Besson, M., Brice, M.H., Burkle, R.A., Dalla Riva, G.V., et al., 2018. Analysing ecological networks of species interactions. Biol. Rev. Camb. Philos. Soc. https://doi.org/10.1111/brv.12433.

Dhingra, S.K., Gabryszewski, S.J., Small-Saunders, J.L., Yeo, T., Henrich, P.P., et al., 2019. Global spread of mutant PfCRT and its pleiotropic impact on *Plasmodium falciparum* multidrug resistance and fitness. MBio 10 pii: e02731-18.

Domingo, E., Perales, C., 2019. Viral quasispecies. PLoS Genet. In press.

Domingo, E., Schuster, P., 2016. Quasispecies: from theory to experimental systems. Curr. Top. Microbiol. Immunol. 392.

Domingo, E., Biebricher, C., Eigen, M., Holland, J.J., 2001. Quasispecies and RNA Virus Evolution: Principles and Consequences. Landes Bioscience, Austin.

Domingo, E., Sheldon, J., Perales, C., 2012. Viral quasispecies evolution. Microbiol. Mol. Biol. Rev. 76, 159–216.

Duncan, A.B., Gonzalez, A., Kaltz, O., 2013. Stochastic environmental fluctuations drive epidemiology in experimental host-parasite metapopulations. Proc. Biol. Sci. 280, 20131747.

Escalante, A.A., Lal, A.A., Ayala, F.J., 1998. Genetic polymorphism and natural selection in the malaria parasite *Plasmodium falciparum*. Genetics 149, 189–202.

Falsone, A., Falsone, S.F., 2015. Legal but lethal: functional protein aggregation at the verge of toxicity. Front. Cell. Neurosci. 9, 45.

Farzi, N., Malekian, T., Alebouyeh, M., Vaziri, F., Zali, M.R., 2015. Genotype diversity and quasispecies development of *Helicobacter pylori* in a single host. Jpn. J. Infect. Dis. 68, 176–180.

Forbes, N.E., Abdelbary, H., Lupien, M., Bell, J.C., Diallo, J.S., 2013. Exploiting tumor epigenetics to improve oncolytic virotherapy. Front. Genet. 4, 184.

Foster, K.R., 2011. The sociobiology of molecular systems. Nat. Rev. 12, 193–203.

Fox, E.J., Loeb, L.A., 2010. Lethal mutagenesis: targeting the mutator phenotype in cancer. Semin. Cancer Biol. 20, 353–359.

Fuqua, W.C., Winans, S.C., Greenberg, E.P., 1994. Quorum sensing in bacteria: the LuxR-LuxI family of cell density-responsive transcriptional regulators. J. Bacteriol. 176, 269–275.

Futreal, P.A., Coin, L., Marshall, M., Down, T., Hubbard, T., et al., 2004. A census of human cancer genes. Nat. Rev. Cancer 4, 177–183.

Gatenby, R.A., Frieden, B.R., 2002. Application of information theory and extreme physical information to carcinogenesis. Cancer Res. 62, 3674–3684.

Gatenby, R.A., Frieden, B.R., 2004. Information dynamics in carcinogenesis and tumor growth. Mutat. Res. 568, 259–273.

Gatenby, R.A., Silva, A.S., Gillies, R.J., Frieden, B.R., 2009. Adaptive therapy. Cancer Res. 69, 4894–4903.

Gentekaki, E., Curtis, B.A., Stairs, C.W., Klimes, V., Elias, M., et al., 2017. Extreme genome diversity in the hyperprevalent parasitic eukaryote *Blastocystis*. PLoS Biol. 15, e2003769.

Gerlee, P., Johansson, M., 2019. Inferring rates of metastatic dissemination using stochastic network models. PLoS Comput. Biol. 15, e1006868.

Ghaemmaghami, S., Ahn, M., Lessard, P., Giles, K., Legname, G., et al., 2009. Continuous quinacrine treatment results in the formation of drug-resistant prions. PLoS Pathog. 5, e1000673.

Gibbs, K.A., Federle, M.J., 2015. A social medium: ASM's 5th cell-cell communication in bacteria meeting in review. J. Bacteriol. 197(13), 2084–2093.

Gleichsner, A.M., Thiele, E.A., Minchella, D.J., 2015. It's all about those bases: the need for incorporating parasite genetic heterogeneity into the development of Schistosome vaccines. PLoS Negl. Trop. Dis. 9, e0003805.

Gonzalez-Garcia, I., Sole, R.V., Costa, J., 2002. Metapopulation dynamics and spatial heterogeneity in cancer. Proc. Natl. Acad. Sci. U.S.A. 99, 13085–13089.

Greenman, C., Stephens, P., Smith, R., Dalgliesh, G.L., Hunter, C., et al., 2007. Patterns of somatic mutation in human cancer genomes. Nature 446, 153–158.

Griffith, J.S., 1967. Self-replication and scrapie. Nature 215, 1043–1044.

Gustafson, K.B., Proctor, J.L., 2017. Identifying spatiotemporal dynamics of Ebola in Sierra Leone using virus genomes. J. R. Soc. Interface 14 pii: 20170583.

Hall, C.W., Mah, T.F., 2017. Molecular mechanisms of biofilm-based antibiotic resistance and tolerance in pathogenic bacteria. FEMS Microbiol. Rev. 41, 276–301.

Hamelmann, C., Foerster, B., Burchard, G.D., Shetty, N., Horstmann, R.D., 1993. Induction of complement resistance in cloned pathogenic Entamoeba histolytica. Parasite Immunol. 15, 223–228.

Hanahan, D., Weinberg, R.A., 2011. Hallmarks of cancer: the next generation. Cell 144, 646–674.

Hansemann, D.V., 1890. Ueber asymmetrische Zelltheilung in epithel Krebsen und deren biologische Bedeutung. Virchow's Arch. Path. Anat. 119, 299–326.

Heitman, J., 2010. Evolution of eukaryotic microbial pathogens via covert sexual reproduction. Cell Host Microbe 8, 86–99.

Helmus, J.J., Surewicz, K., Nadaud, P.S., Surewicz, W.K., Jaroniec, C.P., 2008. Molecular conformation and dynamics of the Y145Stop variant of human prion protein in amyloid fibrils. Proc. Natl. Acad. Sci. U.S.A. 105, 6284–6289.

Horn, D., McCulloch, R., 2010. Molecular mechanisms underlying the control of antigenic variation in African trypanosomes. Curr. Opin. Microbiol. 13, 700–705.

Huang, Q., Wang, H., Gao, H., Cheng, P., Zhu, L., et al., 2019. In situ observation of amyloid nucleation and fibrillation by FastScan atomic force microscopy. J. Phys. Chem. Lett. 10, 214–222.

Igel-Egalon, A., Bohl, J., Moudjou, M., Herzog, L., Reine, F., et al., 2019. Heterogeneity and architecture of pathological prion protein assemblies: time to revisit the molecular basis of the prion replication process? Viruses 11, 429.

Iqbal, J., Siripoon, N., Snounou, V.G., Perlmann, P., Berzins, K., 1997. *Plasmodium falciparum*: selection of parasite subpopulations with decreased sensitivity for antibody-mediated growth inhibition in vitro. Parasitology 114(Pt 4), 317–324.

Jarosz, D.F., Taipale, M., Lindquist, S., 2010. Protein homeostasis and the phenotypic manifestation of genetic diversity: principles and mechanisms. Annu. Rev. Genet. 44, 189–216.

Kastelic, M., Kalyuzhnyi, Y.V., Hribar-Lee, B., Dill, K.A., Vlachy, V., 2015. Protein aggregation in salt solutions. Proc. Natl. Acad. Sci. U.S.A. 112, 6766–6770.

Koenderink, J.B., Kavishe, R.A., Rijpma, S.R., Russel, F.G., 2010. The ABCs of multidrug resistance in malaria. Trends Parasitol. 26, 440–446.

Konry, T., Sarkar, S., Sabhachandani, P., Cohen, N., 2016. Innovative tools and technology for analysis of single cells and cell-cell interaction. Annu. Rev. Biomed. Eng. 18, 259–284.

Kulkami, A., Bangham, C.R.M., 2018. HTLV-1: regulating the balance between proviral latency and reactivation. Front. Microbiol. 9, 449.

Lachaud, L., Bourgeois, N., Kuk, N., Morelle, C., Crobu, L., et al., 2014. Constitutive mosaic aneuploidy is a unique genetic feature widespread in the Leishmania genus. Microbes Infect 16, 61–66.

Lázaro, E., Escarmis, C., Perez-Mercader, J., Manrubia, S.C., Domingo, E., 2003. Resistance of virus to extinction on bottleneck passages: study of a decaying and fluctuating pattern of fitness loss. Proc. Natl. Acad. Sci. U.S.A. 100, 10830–10835.

Lehtinen, S., Blanquart, F., Lipsitch, M., Fraser, C., 2019. On the evolutionary ecology of multidrug resistance in bacteria. PLoS Pathog. 15, e1007763.

Li, J., Browning, S., Mahal, S.P., Oelschlegel, A.M., Weissmann, C., 2010. Darwinian evolution of prions in cell culture. Science 327, 869–872.

Liu, Q., Herring, C.A., Sheng, Q., Ping, J., Simmons, A.J., et al., 2018. Quantitative assessment of cell population diversity in single-cell landscapes. PLoS Biol. 16, e2006687.

Llorens, C., Munoz-Pomer, A., Bernad, L., Botella, H., Moya, A., 2009. Network dynamics of eukaryotic LTR retroelements beyond phylogenetic trees. Biol. Direct 4, 41.

Loeb, L.A., 2001. A mutator phenotype in cancer. Cancer Res. 61, 3230–3239.

Loeb, L.A., 2011. Human cancers express mutator phenotypes: origin, consequences and targeting. Nat. Rev. Cancer 11, 450–457.

Lott, T.J., Frade, J.P., Lockhart, S.R., 2010. Multilocus sequence type analysis reveals both clonality and recombination in populations of Candida glabrata bloodstream isolates from U.S. surveillance studies. Eukaryot. Cell 9, 619–625.

Lowry, K., Woodman, A., Cook, J., Evans, D.J., 2014. Recombination in enteroviruses is a biphasic replicative process involving the generation of greater-than genome length "imprecise" intermediates. PLoS Pathog. 10, e1004191.

Luo, J., Solimini, N.L., Elledge, S.J., 2009. Principles of cancer therapy: oncogene and non-oncogene addiction. Cell 136, 823–837.

Mahal, S.P., Browning, S., Li, J., Suponitsky-Kroyter, I., Weissmann, C., 2010. Transfer of a prion strain to different hosts leads to emergence of strain variants. Proc. Natl. Acad. Sci. U.S.A. 107, 22653–22658.

Maley, C.C., Forrest, S., 2000. Exploring the relationship between neutral and selective mutations in cancer. Artif. Life 6, 325–345.

Maley, C.C., Galipeau, P.C., Finley, J.C., Wongsurawat, V.J., Li, X., et al., 2006. Genetic clonal diversity predicts progression to esophageal adenocarcinoma. Nat. Genet. 38, 468–473.

Mannaert, A., Downing, T., Imamura, H., Dujardin, J.C., 2012. Adaptive mechanisms in pathogens: universal aneuploidy in Leishmania. Trends Parasitol. 28, 370–376.

Mas, A., Lopez-Galindez, C., Cacho, I., Gomez, J., Martinez, M.A., 2010. Unfinished stories on viral quasispecies and Darwinian views of evolution. J. Mol. Biol. 397, 865–877.

Mehta, H.H., Prater, A.G., Beabout, K., Elworth, R.A.L., Karavis, M., et al., 2019. The essential role of hypermutation in rapid adaptation to antibiotic stress. Antimicrob. Agents Chemother. https://doi.org/10.1128/AAC.00744-19.

Melamed, A., Witkover, A.D., Laydon, D.J., Brown, R., Ladell, K., et al., 2014. Clonality of HTLV-2 in natural infection. PLoS Pathog. 10, e1004006.

Merlo, L.M., Pepper, J.W., Reid, B.J., Maley, C.C., 2006. Cancer as an evolutionary and ecological process. Nat. Rev. Cancer 6, 924–935.

Miller, M.B., Bassler, B.L., 2001. Quorum sensing in bacteria. Annu. Rev. Microbiol. 55, 165–199.

Miller, R.V., Day, M.J., 2004. Microbial Evolution. Gene Establishment, Survival and Exchange. ASM Press, Washington, DC.

Min, M., Spencer, S.L., 2019. Spontaneous slow-cycling subpopulatioins of human cells originate from activation of stress-response pathways. PLoS Biol. 17, e3000178.

Mira, A., Martin-Cuadrado, A.B., D'Auria, G., Rodriguez-Valera, F., 2010. The bacterial pan-genome: a new paradigm in microbiology. Int. Microbiol. 13, 45–57.

Munday, J.C., Settimo, L., de Koning, H.P., 2015. Transport proteins determine drug sensitivity and resistance in a protozoan parasite, *Trypanosoma brucei*. Front. Pharmacol. 6, 32.

Naidoo, I., Roper, C., 2010. Following the path of most resistance: dhps K540E dispersal in African *Plasmodium falciparum*. Trends Parasitol. 26, 447–456.

Nealson, K.H., Hastings, J.W., 1979. Bacterial bioluminescence: its control and ecological significance. Microbiol. Rev. 43, 496–518.

Newman, M.E.J., 2005. Power laws. Pareto distributions and Zipf's law. Contemp. Phys. 46, 323–351.

Nicolson, G.L., 1987. Differential growth properties of metastatic large-cell lymphoma cells in target organ-conditioned medium. Exp. Cell Res. 168, 572–577.

Nkhoma, S.C., Nair, S., Cheeseman, I.H., Rohr-Allegrini, C., Singlam, S., et al., 2012. Close kinship within multiple-genotype malaria parasite infections. Proc. Biol. Sci. 279, 2589–2598.

Notari, S., Strammiello, R., Capellari, S., Giese, A., Cescatti, M., et al., 2008. Characterization of truncated forms of abnormal prion protein in Creutzfeldt-Jakob disease. J. Biol. Chem. 283, 30557–30565.

Nowell, P.C., 1976. The clonal evolution of tumor cell populations. Science 194, 23–28.

Ojosnegros, S., Beerenwinkel, N., Antal, T., Nowak, M.A., Escarmis, C., et al., 2010a. Competition-colonization dynamics in an RNA virus. Proc. Natl. Acad. Sci. U.S.A. 107, 2108–2112.

Ojosnegros, S., Beerenwinkel, N., Domingo, E., 2010b. Competition-colonization dynamics: an ecology approach to quasispecies dynamics and virulence evolution in RNA viruses. Commun. Integr. Biol. 107, 2108–2112.

Ojosnegros, S., Perales, C., Mas, A., Domingo, E., 2011. Quasispecies as a matter of fact: viruses and beyond. Virus Res. 162, 203–215.

Payne, J.L., Menardo, F., Trauner, A., Borrell, S., Gyghi, S.M., et al., 2019. Transition bias influences the evolution of antibiotic resistance in *Mycobacterium tuberculosis*. PLoS Biol. 17, e3000265.

Perales, C., Moreno, E., Domingo, E., 2015. Clonality and intracellular polyploidy in virus evolution and pathogenesis. Proc. Natl. Acad. Sci. U.S.A. 112, 8887–8892.

Poste, G., Tzeng, J., Doll, J., Greig, R., Rieman, D., et al., 1982. Evolution of tumor cell heterogeneity during progressive growth of individual lung metastases. Proc. Natl. Acad. Sci. U.S.A. 79, 6574–6578.

Prusiner, S.B., 1982. Novel proteinaceous infectious particles cause scrapie. Science 216, 136–144.

Prusiner, S.B., 1991. Molecular biology of prion diseases. Science 252, 1515–1522.

Ramirez, J.D., Llewellyn, M.S., 2014. Reproductive clonality in protozoan pathogens—truth or artefact? Mol. Ecol. 23, 4195–4202.

Raynaud, F., Mina, M., Tavernari, D., Ciriello, G., 2018. Pan-cancer inference of intra-tumor heterogeneity reveals associations with different forms of genomic instability. PLoS Genet. 14, e1007669.

Read, T.D., Massey, R.C., 2014. Characterizing the genetic basis of bacterial phenotypes using genome-wide association studies: a new direction for bacteriology. Genome Med. 6, 109.

Rice, S.A., Tan, C.H., Mikkelsen, P.J., Kung, V., Woo, J., et al., 2009. The biofilm life cycle and virulence of *Pseudomonas aeruginosa* are dependent on a filamentous prophage. ISME J. 3, 271–282.

Rogers, M.B., Downing, T., Smith, B.A., Imamura, H., Sanders, M., et al., 2014. Genomic confirmation of hybridisation and recent inbreeding in a vector-isolated Leishmania population. PLoS Genet. 10, e1004092.

Rojo-Molinero, E., Macià, M.D., Oliver, A., 2019. Social behavior of antibiotic resistant mutants within *Pseudomonas aeruginosa* biofilm communities. Front. Microbiol. 10, 570.

Rottschaefer, S.M., Riehle, M.M., Coulibaly, B., Sacko, M., Niare, O., et al., 2011. Exceptional diversity, maintenance of polymorphism, and recent directional selection on the APL1 malaria resistance genes of *Anopheles gambiae*. PLoS Biol. 9, e1000600.

Rumbaugh, K.P., Diggle, S.P., Watters, C.M., Ross-Gillespie, A., Griffin, A.S., et al., 2009. Quorum sensing and the social evolution of bacterial virulence. Curr. Biol. 19, 341–345.

Samad, H., Coll, F., Preston, M.D., Ocholla, H., Fairhurst, R.M., et al., 2015. Imputation-based population genetics analysis of *Plasmodium falciparum* malaria parasites. PLoS Genet. 11, e1005131.

Sanchez, C.P., Dave, A., Stein, W.D., Lanzer, M., 2010. Transporters as mediators of drug resistance in *Plasmodium falciparum*. Int. J. Parasitol. 40, 1109–1118.

Santolini, M., Barabasi, A.-L., 2018. Predicting perturbation patterns from the topology of biological networks. Proc. Natl. Acad. Sci. U.S.A. 115, E6375–E6383.

Sardanyés, J., Alarcón, T., 2018. Noise-induced bistability in the fate of cancer phenotypic quasispecies: a bit-strings approach. Sci. Rep. 8, 1027.

Sardanyés, J., Martínez, R., Simó, C., Solé, R., 2017. Abrupt transitions to tumor extinction: a phenotypic quasispecies model. J. Math. Biol. 74, 1589—1609.

Satija, R., Shalek, A.R., 2014. Heterogeneity in immune responses: from populations to single cells. Trends Immunol. 35, 219—229.

Schmidt, E.M., Lamprecht, S., Blaj, C., Schaaf, C., Krebs, S., et al., 2018. Targeting tumor cell plasticity by combined inhibition of NOTCH and MAPK signaling in colon cancer. J. Exp. Med. 215, 1693—1708.

Shanks, G.D., Edstein, M.D., Jacobus, D., 2015. Evolution from double to triple-antimalarial drug combinations. Trans. R. Soc. Trop. Med. Hyg. 109, 182—188.

Sharon, I., Kertesz, M., Hug, L.A., Pushkarev, D., Blauwkamp, T.A., et al., 2015. Accurate, multi-Kb reads resolve complex populations and detect rare microorganisms. Genome Res. 25, 534—543.

Shlien, A., Campbell, B.B., de Borja, R., Alexandrov, L.B., Merico, D., et al., 2015. Combined hereditary and somatic mutations of replication error repair genes result in rapid onset of ultra-hypermutated cancers. Nat. Genet. 47, 257—262.

Shorter, J., 2010. Emergence and natural selection of drug-resistant prions. Mol. Biosyst. 6, 1115—1130.

Shriner, D., Rodrigo, A.G., Nickle, D.C., Mullins, J.I., 2004. Pervasive genomic recombination of HIV-1 in vivo. Genetics 167, 1573—1583.

Smith, C.A., Ban, D., Pratihar, S., Giller, K., Schwiegk, C., et al., 2015. Population shuffling of protein conformations. Angew. Chem. Int. Ed. Engl. 54, 207—210.

Solé, R.V., Deisboeck, T.S., 2004. An error catastrophe in cancer? J. Theor. Biol. 228, 47—54.

Solé, R.V., Rodriguez-Caso, C., Deisboeck, T.S., Saldaña, J., 2008. Cancer stem cells as the engine of unstable tumor progression. J. Theor. Biol. 253, 629—637.

Song, H., Giorgi, E.E., Ganusov, V.V., Cai, F., Athreya, G., et al., 2018. Tracking HIV-1 recombination to resolve its contribution to HIV-1 evolution in natural infection. Nat. Commun. 9, 1928.

Sterkers, Y., Lachaud, L., Bourgeois, N., Crobu, L., Bastien, P., et al., 2012. Novel insights into genome plasticity in Eukaryotes: mosaic aneuploidy in Leishmania. Mol. Microbiol. 86, 15—23.

Stewart, P.S., Costerton, J.W., 2001. Antibiotic resistance of bacteria in biofilms. Lancet 358, 135—138.

Tannenbaum, E., Sherley, J.L., Shakhnovich, E.I., 2006. Semiconservative quasispecies equations for polysomic genomes: the haploid case. J. Theor. Biol. 241, 791—805.

Thomson, M.M., Perez-Alvarez, L., Najera, R., 2002. Molecular epidemiology of HIV-1 genetic forms and its significance for vaccine development and therapy. Lancet Infect. Dis. 2, 461—471.

Tibayrenc, M., Ayala, F.J., 2012. Reproductive clonality of pathogens: a perspective on pathogenic viruses, bacteria, fungi, and parasitic protozoa. Proc. Natl. Acad. Sci. U.S.A. 109, E3305—E3313.

Tibayrenc, M., Ayala, F.J., 2014. Cryptosporidium, Giardia, Cryptococcus, Pneumocystis genetic variability: cryptic biological species or clonal near-clades? PLoS Pathog. 10, e1003908.

Tibayrenc, M., Ayala, F.J., 2017. Is predominant clonal evolution a common evolutionary adaptation to parasitism in pathogenic parasitic protozoa, fungi, bacteria, and viruses? Adv. Parasitol. 97, 243—325.

Tomlinson, I.P., Novelli, M.R., Bodmer, W.F., 1996. The mutation rate and cancer. Proc. Natl. Acad. Sci. U.S.A. 93, 14800—14803.

Tuite, M.F., Serio, T.R., 2010. The prion hypothesis: from biological anomaly to basic regulatory mechanism. Nat. Rev. Mol. Cell Biol. 11, 823—833.

Tyedmers, J., Madariaga, M.L., Lindquist, S., 2008. Prion switching in response to environmental stress. PLoS Biol. 6, e294.

Ueda, H., Stephens, K., Trivisa, K., Bentley, W.E., 2019. Bacteria floc, but do they flock? Insights from population interaction models of quorum sensing. MBio 10 pii: e00972-19.

Valadares, H.M., Pimenta, J.R., Segatto, M., Veloso, V.M., Gomes, M.L., et al., 2012. Unequivocal identification of subpopulations in putative multiclonal *Trypanosoma cruzi* strains by FACs single cell sorting and genotyping. PLoS Negl. Trop. Dis. 6, e1722.

Vanni, I., Di Bari, M.A., Pirisinu, L., D'Agostino, C., Agrimi, U., et al., 2014. In vitro replication highlights the mutability of prions. Prion 8, 154—160.

Vartanian, J.P., Guetard, D., Henry, M., Wain-Hobson, S., 2008. Evidence for editing of human papillomavirus DNA by APOBEC3 in benign and precancerous lesions. Science 320, 230—233.

Venkatesan, S., Rosenthal, R., Kanu, N., McGranahan, N., Bartek, J., et al., 2018. Perspective: APOBEC mutagenesis in drug resistance and immune escape in HIV and cancer evolution. Ann. Oncol. 29, 563—572.

Vignuzzi, M., Andino, R., 2010. Biological implications of picornavirus fidelity mutants. In: Ehrenfeld, E., Domingo, E., Roos, R.P. (Eds.), The Picornaviruses. ASM Press, Washington DC, pp. 213—228.

Villarreal, L.P., Witzany, G., 2013. Rethinking quasispecies theory: from fittest type to cooperative consortia. World J. Biol. Chem. 4, 79—90.

von Kleist, M., Menz, S., Stocker, H., Arasteh, K., Schutte, C., et al., 2011. HIV quasispecies dynamics during pro-active treatment switching: impact on multi-drug resistance and resistance archiving in latent reservoirs. PLoS One 6, e18204.

Wahlgren, M., Fernandez, V., Chen, Q., Svard, S., Hagblom, P., 1999. Waves of malarial variations. Cell 96, 603–606.

Walker, L.C., Jucker, M., 2015. Neurodegenerative diseases: expanding the prion concept. Annu. Rev. Neurosci. 38, 87–103.

Wang, W., Zhang, X., Xu, Y., Weinstock, G.M., Di Bisceglie, A.M., et al., 2014. High-resolution quantification of hepatitis C virus genome-wide mutation load and its correlation with the outcome of peginterferon-alpha2a and ribavirin combination therapy. PLoS One 9, e100131.

Webster, B., Ott, M., Greene, W.C., 2013. Evasion of superinfection exclusion and elimination of primary viral RNA by an adapted strain of hepatitis C virus. J. Virol. 87, 13354–13369.

Wedlich-Söldner, R., Betz, T., 2018. Self-organization: the fundament of cell biology. Philos. Trans. R. Soc. Lond. B Biol. Sci. 373 pii: 20170103.

Weissmann, C., Li, J., Mahal, S.P., Browning, S., 2011. Prions on the move. EMBO Rep. 12, 1109–1117.

Whitaker, R.J., Banfield, J.F., 2006. Population genomics in natural microbial communities. Trends Ecol. Evol. 21, 508–516.

Wigington, C.H., Sonderegger, D., Brussaard, C.P., Buchan, A., Finke, J.F., et al., 2017. Re-examination of the relationship between marine virus and microbial cell abundances. Nat. Microbial. 1, 15024.

Witzany, G., 2014. Biological self-organization. Int. J. Signs Semiot. Syst. 3, 1–11.

Yan, G., Martinez, N.D., Liu, Y.Y., 2017. Degree heterogeneity and stability of ecological networks. J. R. Soc. Interface 14 pii:20170189.

Zanusso, G., Florini, M., Ferrari, S., Meade-White, K., Barbieri, I., et al., 2014. Gerstmann-Sträusler-Scheinker disease and "anchorless prion protein" mice share prion conformation properties diverging from sporadic Creutzfeldt-Jakob disease. J. Biol. Chem. 289, 4870–4881.

Zhang, X.F., Zhang, S., Guo, Q., Sun, R., Wei, T., et al., 2018. A new mechanistic model for viral cross protection and superinfection exclusion. Front. Plant Sci. 9, 40.

Zhang, X., Alexander, N., Leonardi, I., Mason, C., Kirkman, L.A., et al., 2019. Rapid antigenic diversification through mitotic recombination in the human malaria parasite Plasmodium falciparum. PLoS Biol. 17, e3000271.

# Subject Index

*Note:* 'Page numbers followed by "f" indicate figures and "t" indicate tables'.

# Author Index

*Note*: 'Page numbers followed by *"b"* indicate boxes, *"f"* indicate figures and *"t"* indicate tables'.

Printed in the United States
By Bookmasters